CONTEMPORARY
MATHEMATICS

478

Representation Theory

Fourth International Conference
on Representation Theory
July 16–20, 2007
Lhasa, China

Zongzhu Lin
Jianpan Wang
Editors

American Mathematical Society
Providence, Rhode Island

Editorial Board

Dennis DeTurck, managing editor

George Andrews   Abel Klein   Martin J. Strauss

2000 *Mathematics Subject Classification.* Primary 16Gxx, 17Bxx, 20Cxx, 20Gxx; Secondary 17B10, 17B20, 17B37, 17B45, 17B56, 20G05, 20G10, 20G42, 20C05, 20C08, 20C30.

Library of Congress Cataloging-in-Publication Data

Representation theory / Zongzhu Lin, Jianpan Wang, editors.
    p. cm. — (Contemporary mathematics ; v. 478)
    Includes bibliographical references.
    ISBN 978-0-8218-4555-4 (alk. paper)
    1. Representations of algebras—Congresses. 2. Representations of groups—Congresses.
I. Lin, Zongzhu. II. Wang, Jianpan, 1949– III. Title.

QA150.R46 2008
515′.7223—dc22

         2008034291

The 4th International Conference on Representation Theory
第四届国际表示论会议

July 16-20, 2007 Lhasa, Tibet, China

Contents

Preface

The fourth **International Conference on Representation Theory** (ICRT-IV) was held in Lhasa, China, during July 16–20, 2007. The first three conferences were held in Shanghai (1998), Kunming (2001), and Chengdu (2004). The main goal of the ICRT is to bring together representation theorists of various subjects such as algebraic groups, quantum groups, finite groups, Lie algebras, vertex operator algebras, Hecke algebras and complex reflection groups, quivers, finite dimensional algebras, Hall algebras, and other related topics. The representation theory has evolved to connect many different fields of mathematics and the ICRT intends to communicate the common ideas and approaches to common questions. During the five day programs, participants have endured both mathematical and physical challenges. There was also a two day sub-conference on Mathematical History which covered many interesting historical topics such as early mathematics text books in Tibet and the first Chinese national who ever obtained formal Ph.D. in mathematics.

This volume contains eighteen papers that more or less reflect the topics of the conference. All papers were carefully refereed. Several referees read several versions of the manuscripts and made critical comments to ensure the high quality of the papers. The editors want to express their sincere appreciation to all referees for their anonymous contribution. Due to page limitation, the editors regret that not all submitted papers are included and thank all authors for their contribution to this proceedings.

It was a challenge from many respects to bring the conference to this "high level". The conference would not be possible without many people's hardwork. Jianpan Wang and Hebing Rui had traveled to Lhasa several times to arrange hotel and lecture facilities. Both of them have been working for more than a year raising fund for the conference and making travel arrangements for international participants. Many participants took trains through the newly built railroad at 5 km altitude and the train ride lasts for more than two days. Making such arrangement was impossible without the help of many local mathematicians at various cities. The mathematicians in Tibet University, Da Luosang Langjie and Yutian Fei, provided invaluable help making local arrangement including the post-conference tours. The volunteer staff member, Xin Wang (a graduate student at University of Virginia), provided an impeccable professional-type services during conference.

The financial support from the following organizations, projects, and individuals are greatly appreciated:

— Tianyuan Mathematical Fund, National Natural Science Foundation of China

— "985" fund, East China Normal University

— "111 Project", Department of Mathematics, East China Normal University

— Changjiang Scholars and Innovative Research Team, Department of Mathematics, East China Normal University

— Jun Hu, "New Century Excellent Talents in University", Beijing Institute of Technology

— Fang Li, "New Century Excellent Talents in University", Zhejiang University

— Hebing Rui, "New Century Excellent Talents in University", East China Normal University

— Yucai Su, "One-Hundred Talent Program", University of Science and Technology of China

— Jianpan Wang, Fund of Science and Technology Commission of Shanghai Municipality, No. 06JC 14024

— Jianpan Wang, fund of National Natural Science Foundation of China, No. 10631010

The ICRT-IV was co-hosted by East China Normal University and the Tibet University at Lhasa. The organizing committee consists of Jianpan Wang (Chair, East China Normal University), Da Luosang Langjie (Co-chair, Tibet University at Lhasa), Bangming Deng (Beijing Normal University), Yutian Fei (Tibet University), Jun Hu (Beijing Institute of Technology), Ruyun Ma (Northwest Normal University), Liangang Peng (Sichuan University), Hebing Rui (East China Normal University).

The program committee consists of Jianpan Wang (Chair, East China Normal University), Jiping Zhang (Co-chair, Peking University), Chongying Dong (UC Santa Cruz), Jie Du (University of New South Wales), Zongzhu Lin (Kansas State University), Yucai Su (The University of Science and Technology of Chinese), Weiqiang Wang (University of Virginia), Nanhua Xi (Academy Sinica), Jie Xiao (Tsinghua University).

Last but not least, the editors want to thank Christine M. Thivierge, Associate Editor for Proceedings at American Mathematical Society who worked so closely and patiently with the editors to ensure the timely publication of this proceedings.

Zongzhu Lin
Jianpan Wang

June 2008

List of Talks

Principal Speakers

Henning H. Andersen
   Sum formulas and Ext-groups
Susumu Ariki
   Crystal theory and hecke algebras of type B_n
Michel Broué
   Families of unipotent characters and cyclotomic algebras
Jonathan Brundan
   Cyclotomic Hecke algebras and parabolic category $\mathcal{O}$
Jon Carlson
   Modules of constant Jordan type
Shun-Jen Cheng
   Kostants homology formula for infinite-dimensional lie superalgebras
Chongying Dong
   Representation theory for vertex operator algebras
Yun Gao
   Irreducible Wakimoto-like modules for lie algebras of type A
Shrawan Kumar
   Special isogenies and tensor product multiplicities
George Lusztig
   Unipotent elements in small characteristic
Brian Parshall
   Some results on quantum and algebraic group cohomology and applications
Liangang Peng
   Conical extensions of derived categories
Claus M. Ringel
   The relevance and the ubiquity of prüfer modules
Olivier Schiffmann
   Geometric construction of macdonald polynomials
Leonard Scott
   Some $\mathbb{Z}/2$-graded representation theory
Toshiaki Shoji,
   Lusztigs conjecture for finite classical groups
Yucai Su
   Quasifinite representations of some lie algebras containing the Virasoro algebra

Toshiyuki Tanisaki
Differential operators on quantized flag manifolds at roots of 1

Nanhau Xi
Some maximal elements in a baby Verma module

Contributed 30 Minute Talks

Thomas Brüstle
Gentle algebras given by surface triangulations

Vlastimil Dlab
Standardly stratified approximations

Jie Du
Quantum $\mathfrak{gl}_\infty$, infinite q-Schur algebras and their representations

Ming Fang
Dominant dimensions and double centralizer properties

Andrew Francis
Symmetric polynomials of Jucys-Murphy elements and the centre of the Iwahori-Hecke algebra of type A

Xuhua He
Minimal length elements and G-stable pieces

Jun Hu
Morita equivalences of cyclotomic hecke algebras of type $G(r, p, n)$

Lizhen Ji
Duality of arithmetic groups, mapping class groups and outer automorphisms groups of free groups

Otto Kerner
Cluster tilted algebras of rank three

Yanan Lin,
Generic sheaves over elliptic curves

Zongzhu Lin
Canonical bases for algebras arising from quivers with loops

Gongxiang Liu
Classification of finite dimensional basic hopf algebras according to their representation type

Alistair Savage
A geometric construction of a crystal graph commutor

Bin Shu
Representations and cohomology of Jacobson-Witt algebras

Mei Si,
Discriminants of Brauer algebras

Alexander Stolin
Classification of quantum groups

Liping Wang
Leading coefficients of the Kazhdan-Lusztig polynomials of the affine Weyl group of type $\tilde{B}_2$

Adrian Leonard Williams
Minimal dimensions of some irreducible representations of the symmetric group

Jiping Zhang
The block union of finite groups

Mathematical History Talks

Tom Archibald
Chinese mathematicians and the international research community: the case of Hu Mingfu and integral equations

Elena Ausejo
Commercial arithmetic in the Spanish renaissance

Jose A. Cervera
The Chou Suan by Giacomo Rho: An example of mathematical adaptation in China

Joseph W. Dauben
Zhu Shijie and the Jade Mirror of the Four Unknowns

Qi Han
Antoine Thomas (1644C1709) and the first introduction of western algebra into china

Da Luosang Langjie
Duchung Zurtsi : A mathematics textbook for secular and official schools of Tibet

Karen H. Parshall
4000 Years of algebra: an historical tour from BM 13901 to Moderne Algebra

Yongdong Peng
The early communication of cybernetics in China (1929C1966)

Yibao Xu
Chinese gougu theory versus Euclidean geometry: views of a seventeenth-century Chinese mathematician

David E. Zitarelli
Miss Mullikin and the internationalization of topology

List of Participants

Henning H. Andersen
Aarhus University, DENMARK

Tom Archibald
Simon Fraser University, CANADA

Susumu Ariki
Kyoto University, JAPAN

Elena Ausejo
University of Zaragoza, SPAIN

Xiaotang Bai
Nankai University, Tianjing

Stephen Berman
University of Saskatchewan, CANADA

Michel Broué
Institut Henri-Poincaré Paris, FRANCE

Jonathan Brundan
University of Oregon, USA

Thomas Brüstle
Université de Sherbrooke, CANADA

Jon F. Carlson
University of Georgia, USA

Bintao Cao
Institute of Mathematics, Academia Sinica, Beijing

Jose A. Cervera
ITESM, Campus Monterrey, MEXICO

Shun-Jen Cheng
Academia Sinica (Taipei), Taipei

Joseph W. Dauben,
The City University of New York, USA

Bangming Deng
Beijing Normal University, Beijing

Vlastimil Dlab
Carleton University, CANADA

Chongying Dong
University of California at Santa Cruz, USA

Jie Du
University of New South Wales, AUSTRALIA

Zhoutian Fan
Beijing Normal University, Beijing

Ming Fang
Beijing University of Technology, Beijing

Yutian Fei
Tibet University, Lhasa

Andrew Francis
University of West Sydney, AUSTRALIA

Qiang Fu
Tongji University, Shanghai

Yun Gao
York University, CANADA

Jingyun Guo
Xiangtan University, Xiangtan

Qi Han
Institute for History of Natural Science, Academia Sinica, Beijing

Yang Han
Institute of Systems Science, Academia Sinica, Beijing

Xuhua He
Stony Brook University, USA

Terrell L. Hodge
Western Michigan University, USA

Masaharu Kaneda
Osaka City University, JAPAN

Shrawan Kumar
University of North Carolina, USA

Jiuzu Hong
Institute of Mathematics, Academia Sinica, Beijing

Jun Hu
Beijing Institute of Technology, Beijing

Lizhen Ji
University of Michigan, USA

Otto T. Kerner
Heinrich-Heine-Universität, GERMANY

Jonathan Kujawa
University of Georgia, USA

Da Luosang Langjie
Tibet University, Lhasa

Fang Li
Zhejiang University, Hangzhou

Lei Lin
East China Normal University, Shanghai

Yanan Lin
Xiamen University, Xiamen

Zongzhu Lin
Kansas State University, USA

Gongxiang Liu
Institute of Mathematics, Academia Sinica, Beijing

Li Luo
Institute of Mathematics, Academia Sinica, Beijing

George Lusztig
MIT, USA

František Marko
Penn State University, Hazleton, USA

Qingnian Pan
Huizhou University, Huizhou

Brian J. Parshall
University of Virginia, USA

Karen H. Parshall
University of Virginia, USA

Lianggang Peng
Sichuan University, Chengdu

Yongdong Peng
Hubei Education Press, Wuhan

Claus M. Ringel
University of Bielefeld, GERMANY

Hebing Rui
East China Normal University, Shanghai

Alistair Savage
University of Ottawa, CANADA

Olivier Schiffmann
École Normale Supérieure, FRANCE

Leonard Scott
University of Virginia, USA

Jianyi Shi
East China Normal University, Shanghai

Toshiaki Shoji
Nagoya University, JAPAN

Bin Shu
East China Normal University, Shanghai

Mei Si
East China Normal University, Shanghai

Alexander Stolin
University of Gothenburg, SWEDEN

Yucai Su
University of Science and Technology of China, Hefei

Toshiyuki Tanisaki
Osaka City University, JAPAN

Jianpan Wang
East China Normal University,
Shanghai

Li Wang
Shanghai Normal University, Shanghai

Liping Wang
Institute of Mathematics, Academia
Sinica, Beijing

Weiqiang Wang
University of Virginia, USA

Xiaoming Wang
East China Normal University,
Shanghai

Xin Wang
University of Virginia, USA

Adrian Williams
Impare College, University of London,
UK

Nanhua Xi
Academia Sinica, Beijing

Yibao Xu
The City University of New York, USA

Rong Yan
Institute of Mathematics, Academia
Sinica, Beijing

Ziting Zeng
Beijing Normal University, Beijing

Jiping Zhang
Peking University, Beijing

Pu Zhang
Shanghai Jiaotong University, Shanghai

Qinhai Zhang
Shanxi Normal University, Xi'an

Shizhuo Zhang
Beijing University of Technology,
Beijing

David E. Zitarelli,
Temple University, USA

Contemporary Mathematics
Volume **478**, 2009

Sum formulas and Ext-groups

Henning Haahr Andersen

1. Introduction

Let G be a reductive algebraic group over a field k of characteristic $p > 0$. In [**4**] we gave a unified proof for the sum formulas for the terms in the Jantzen filtrations of Weyl modules on the one hand side and on the terms of the corresponding filtrations involving tilting modules on the other hand. At the same time our approach in loc. cit. applies to the quantum group U_q associated with G when q is a root of unity. One advantage of these sum formulas is that they do not require any restrictions on the prime p or on the order of the root of unity in question. In contrast, many other methods and results on Weyl modules and tilting modules are only valid in the range $p \geq h$, the Coxeter number of G. Sometimes we even require $p \gg 0$ with no specified bound, see e.g. [**6**].

The Jantzen filtration for Weyl modules as well as their counterparts involving tilting modules are constructed by passing to the Chevalley groups $G_{\mathbb{Z}}$ for G, respectively the Lusztig form $U_{k[v,v^{-1}]}$ of the quantum group U_q. Our approach in [**4**] is based on an Euler type formula for Ext-groups between certain $G_{\mathbb{Z}}$-modules, respectively $U_{k[v,v^{-1}]}$-modules. In this paper we shall briefly recall the key ingredients of this method. In particular, we shall focus on the Ext-groups between (dual) Weyl modules for $G_{\mathbb{Z}}$. We shall illustrate how these groups contain a lot of information relevant for the representation theory of G. At the same time we will discover that the individual groups are very hard to compute. As an explicit example we make a detailed study of the case where $G = GL_n$ and where the highest weights of the Weyl modules involved differ by a single root. In this situation we manage to calculate all Ext-groups (they all vanish except the first). This case was handled by Kulkarni in [**11**] by rather long direct computations. Our approach avoids most of these and have also the advantage that it works without modifications in the corresponding quantum case.

Math Subject Classification numbers: 20G05, 17B37.

As a further illustration of how to obtain information about individual such Ext-groups we show how the translation arguments in [**2**] can be applied to give similar calculations for adjacent Weyl modules for $G_{\mathbb{Z}_p}$ in the general case. This however requires $p \geq h$.

Just like the approach in [**4**] our calculations here are carried out in such a way that they immediately generalize to quantum groups at roots of 1. So if U_q is the quantum group corresponding to a root system R of type A then we determine $\text{Ext}^i_{U_{k[v,v^{-1}]}}(\nabla_{k[v,v^{-1}]}(\lambda), \nabla_{k[v,v^{-1}]}(\lambda - \beta))$ for all $i \in \mathbb{N}, \lambda \in X^+$ and $\beta \in R^+$. Likewise for a general quantum group we handle adjacent dual Weyl modules at a root of unity $q \in k$ of order at least the Coxeter number.

The results in this note go somewhat further in the direction of explicit determination of Ext-groups than what I presented in my lecture at ICRT-IV in Lhasa. I would like again to thank the organizers of this extremely nice and stimulating conference.

2. Euler type formulas

2.1. Notation. As in the introduction we let G denote a reductive algebraic group over a field k. We set $p = \text{char}(k)$ and assume throughout that $p > 0$. We choose a maximal torus T in G and a Borel subgroup B containing T. Then R will be the root system for (G, T). We fix a set of simple roots S in R by requiring that the roots of B are the corresponding negative roots $-R^+$. The number of positive roots is called N. This is also the dimension of the flag variety G/B.

The character group for T is denoted X. We let X^+ be the set of dominant characters, i.e., $X^+ = \{\lambda \in X \mid \langle \lambda, \alpha^\vee \rangle \geq 0 \text{ for all } \alpha \in R^+\}$. Note that characters of T extend to B. So for $\lambda \in X$ we often denote by the same symbol the 1-dimensional B-module obtained from this character.

The Weyl group $W = N_G(T)/Z_G(T)$ for G acts naturally on X. If $\alpha \in R$ then the reflection $s_\alpha \in W$ corresponding to α is given by $s_\alpha(\lambda) = \lambda - \langle \lambda, \alpha^\vee \rangle \alpha$ for all $\lambda \in X$. We shall also use the 'dot-action' defined by $w \cdot \lambda = w(\lambda + \rho) - \rho, w \in W, \lambda \in X$. Here ρ is the half sum of the positive roots.

Each element $w \in W$ is a product of simple reflections (reflections for simple roots) and we have the corresponding length function l on W taking w into the minimal number of such simple reflections needed to express w. The unique longest element in W is denoted w_0. It has length $l(w_0) = N$.

If M is a finite dimensional T-module and $\lambda \in X$ then the weight space M_λ is defined by $M_\lambda = \{m \in M \mid tm = \lambda(t)m \text{ for all } t \in T\}$. We say that λ is a weight of M if $M_\lambda \neq 0$. The character ch M is ch $M = \sum_{\lambda \in X} (\dim M_\lambda) e^\lambda \in \mathbb{Z}[X]$.

For each $\lambda \in X^+$ we have a Weyl module $\Delta(\lambda)$ for G with highest weight λ. Its contragredient dual $\Delta(\lambda)^*$ is denoted $\nabla(-w_0\lambda)$. Note that then the dual Weyl module $\nabla(\mu)$ attached to $\mu \in X^+$ has highest weight μ (because $w_0(\lambda)$ is the smallest weight of $\Delta(\lambda)$).

Each $\nabla(\mu)$ contains a unique simple submodule which we denote $L(\mu)$. Then $L(\mu)$ is also the unique simple quotient of $\Delta(\mu)$. The collection $\{L(\mu)\}_{\mu \in X^+}$ is up to isomorphisms a complete list of finite dimensional simple G-modules.

2.2. Cohomology modules.

2.2. Cohomology modules. Let M be a finite dimensional B-module. Then we will write $H^0(M)$ for the G-module $\mathrm{Ind}_B^G M$ induced by M. This is also the 0-th cohomology (i.e., the set of global sections) for the vector bundle on G/B associated with M. More generally, we denote by $H^i(M)$ the i-th cohomology of this bundle, or alternatively the value of the i-th right derived functor $R^i \mathrm{Ind}_B^G$ on M. It is well known (as G/B is a projective variety) that the cohomology $H^\bullet(M)$ is finite dimensional, and that $H^i(M) = 0$ for $i > N$.

The Euler character of a B-module M is given by

$$\chi(M) = \sum_{\mu \in X} (-1)^i \mathrm{ch}(H^i(M)).$$

Note that χ is additive, i.e., if $0 \to M_1 \to M \to M_2 \to 0$ is a short exact sequence of finite dimensional B-modules then $\chi(M) = \chi(M_1) + \chi(M_2)$.

Recall that the Weyl modules in Section 2.1 are special instances of cohomology modules. To be precise, we have $\Delta(\lambda) \simeq H^N(w_0 \cdot \lambda)$ and $\nabla(\lambda) \simeq H^0(\lambda)$ for all $\lambda \in X^+$. Kempf's vanishing theorem [**10**] implies that $\chi(\lambda) = \mathrm{ch}\,\Delta(\lambda) = \mathrm{ch}\,\nabla(\lambda)$.

2.3. Chevalley groups.

2.3. Chevalley groups. Let $G_{\mathbb{Z}}$ be a split and connected reductive algebraic group scheme over $\mathbb{Z}$ corresponding to G. In other words $G_{\mathbb{Z}}$ is the associated Chevalley group. Then G is obtained from $G_{\mathbb{Z}}$ by extending scalars to k. More generally, we write G_A for the group scheme over an arbitrary commutative ring A obtained via the base change $\mathbb{Z} \to A$. (The case $A = \mathbb{Z}_p$, the ring of p-adic integers, will be needed in Chapter 5.) We use similar notation relative to the subgroups T and B. In particular, $T_{\mathbb{Z}}$ is a split maximal torus in $G_{\mathbb{Z}}$ with $T_k = T$.

Note that for a G_A-module V that is free of finite rank as an A-module, $\mathrm{ch}(V)$ makes sense by considering ranks of weight spaces. If our field k is an A-algebra then we have for such a module $\mathrm{ch}(V) = \mathrm{ch}(V \otimes_A k)$.

For any commutative ring A and any B_A-module M we write $H_A^i(M)$ for the G_A-module $R^i \mathrm{Ind}_{B_A}^{G_A} M$. See [**9**], I.5 for the general properties of these modules. In particular, we recall that if A is noetherian and M is finitely generated over A, then $H_A^i(M)$ is also finitely generated over A, see [**9**], Proposition I.5.12 c).

Given any commutative ring A, for each $\lambda \in X^+$ we have the following two G_A-modules: the Weyl module $\Delta_A(\lambda)$ and the dual Weyl module $\nabla_A(\lambda)$. These modules are characteristic-free, i.e., as A-modules both are free of rank equal to $\dim \Delta(\lambda)$ and we have G_A-module isomorphisms $\Delta_A(\lambda) \simeq \Delta_{\mathbb{Z}}(\lambda) \otimes A$ and $\nabla_A(\lambda) \simeq \nabla_{\mathbb{Z}}(\lambda) \otimes A$. Just as for G, we have $\nabla_A(\lambda) = H_A^0(\lambda)$ and $\Delta_A(\lambda) \simeq H_A^N(w_0 \cdot \lambda)$.

2.4. Ext groups. Consider finitely generated $G_{\mathbb{Z}}$-modules M and N. By [9], II.B, the groups $\mathrm{Ext}^i_{G_{\mathbb{Z}}}(M, N)$ are finitely generated and vanish for large enough i. We will also need the following special cases of some vanishing results which can also be found in loc. cit.

PROPOSITION 2.1. *For $\lambda, \mu \in X^+$,*

    a) $\mathrm{Ext}^i_{G_{\mathbb{Z}}}(\Delta_{\mathbb{Z}}(\mu), \nabla_{\mathbb{Z}}(\lambda)) = 0$ *unless $\mu = \lambda$ and $i = 0$.*
       $\mathrm{Hom}_{G_{\mathbb{Z}}}(\Delta_{\mathbb{Z}}(\lambda), \nabla_{\mathbb{Z}}(\lambda)) = \mathbb{Z}$.
    b) $\mathrm{Ext}^i_{G_{\mathbb{Z}}}(\nabla_{\mathbb{Z}}(\mu), \nabla_{\mathbb{Z}}(\lambda)) = 0$ *unless $\mu \geq \lambda$ and $\mathrm{ht}(\mu - \lambda) \geq i$.*
       $\mathrm{Hom}_{G_{\mathbb{Z}}}(\nabla_{\mathbb{Z}}(\lambda), \nabla_{\mathbb{Z}}(\lambda)) = \mathbb{Z}$.

The universal coefficient theorem [9], Proposition I.4.18 a) gives analogous results over G_A for other commutative rings A. In particular the proposition stays valid after replacing each $\mathbb{Z}$ by the ring of p-adic integers $\mathbb{Z}_p$.

2.5. Euler coefficients. If $n = \pm p_1^{a_1} p_2^{a_2} \cdots p_r^{a_r}$ is the decomposition of $n \in \mathbb{Z}$ into different prime powers then we set

$$\mathrm{div}(n) = \sum_{i=1}^{r} a_i [p_i] \in \mathcal{D}(\mathbb{Z})$$

where $\mathcal{D}(\mathbb{Z})$ is the divisor group for $\mathbb{Z}$. When M is a finite $\mathbb{Z}$-module of order $|M|$ then we set $\mathrm{div}(M) = \mathrm{div}(|M|)$. We use this to define the following Euler coefficients of $G_{\mathbb{Z}}$ and $B_{\mathbb{Z}}$-modules.

Let V and V' be $G_{\mathbb{Z}}$-modules, both finitely generated over $\mathbb{Z}$. Then $\mathrm{Ext}^i_{G_{\mathbb{Z}}}(V, V')$ is finite for all $i > 0$. This follows from Section 2.4 and the universal coefficient theorem [9], Proposition I.4.18 a), because $\mathrm{Ext}^i_{G_{\mathbb{C}}}(A, B) = 0$ for all $i > 0$ and for any two rational $G_{\mathbb{C}}$-modules A and B ($G_{\mathbb{C}}$ being reductive). If the $G_{\mathbb{C}}$-modules $V_{\mathbb{C}} = V \otimes_{\mathbb{Z}} \mathbb{C}$ and $V' \otimes_{\mathbb{Z}} \mathbb{C}$ do not have an isomorphic simple summand, then $\mathrm{Hom}_{G_{\mathbb{Z}}}(V, V')$ is finite. This happens in particular when V or V' is finite. By Section 2.4 we have in any case $\mathrm{Ext}^i_{G_{\mathbb{Z}}}(V, V') = 0$ when $i \gg 0$. So whenever $\lambda \in X^+$ and V is a $G_{\mathbb{Z}}$-module such that $V_{\mathbb{C}}$ does not contain the irreducible module with highest weight λ (this is always the case when V is a finite $G_{\mathbb{Z}}$-module because then $V_{\mathbb{C}} = 0$), then the following expression gives a well defined element in $\mathcal{D}(\mathbb{Z})$

$$e_\lambda^G(V) = \sum_{i \geq 0} (-1)^i \mathrm{div}(\mathrm{Ext}^i_{G_{\mathbb{Z}}}(\Delta_{\mathbb{Z}}(\lambda), V)).$$

Clearly, e_λ^G is additive on exact sequences of such $G_{\mathbb{Z}}$-modules (in particular finite $G_{\mathbb{Z}}$-modules). For more details see [4], Section 4.1.

2.6. The alternating formulas. In [4] we computed the Euler coefficients $e_\lambda^G(V)$ from Section 2.5 in the case where V is the cokernel $Q(\mu)$ of the natural homomorphism $\Delta_{\mathbb{Z}}(\mu) \to \nabla_{\mathbb{Z}}(\mu)$ (These formulas were the key that made us able to deduce the sum formulas). Note that because of Proposition 2.1 a) if $\lambda \neq \mu$ then we have $e_\lambda^G(Q(\mu)) = -e_\lambda^G(\Delta_{\mathbb{Z}}(\mu))$. This means that the formulas in loc .cit. give an expression for the alternating sum of $\mathrm{div}(\mathrm{Ext}^i_{G_{\mathbb{Z}}}(\Delta_{\mathbb{Z}}(\lambda), \Delta_{\mathbb{Z}}(\mu)))$. It is convenient for our purposes here to have the corresponding statements for dual Weyl modules. This goes as follows:

Let $\lambda, \mu \in X^+$ with $\lambda \neq \mu$. For $\beta, \gamma \in R^+$ we set

$$V_\beta(\lambda, \mu) = \{(x, m) \mid x \in W, 0 < m < \langle \mu + \rho, \beta^\vee \rangle \text{ with } x \cdot \lambda = \mu - m\beta\}.$$

and

$$U_\gamma(\lambda, \mu) = \{(w, n) \mid w \in W, n < 0 \text{ or } n > \langle \lambda + \rho, \gamma^\vee \rangle, w \cdot \mu = \lambda - n\gamma\}.$$

Replacing Weyl modules by dual Weyl modules in [4] we can now state the following two formulas for the alternating sum of the Ext-groups between dual Weyl modules

THEOREM 2.2. *If λ and μ are two different dominant weights then*

$$\sum_{i \geq 0} (-1)^i \, \mathrm{div}(\mathrm{Ext}^i_{G_{\mathbb{Z}}}(\nabla_{\mathbb{Z}}(\mu), \nabla_{\mathbb{Z}}(\lambda))) = \sum_{\beta \in R^+} \sum_{(x,m) \in V_\beta(\lambda,\mu)} (-1)^{l(x)} \, \mathrm{div}(m)$$

and

$$\sum_{i \geq 0} (-1)^i \, \mathrm{div}(\mathrm{Ext}^i_{G_{\mathbb{Z}}}(\nabla_{\mathbb{Z}}(\mu), \nabla_{\mathbb{Z}}(\lambda))) = \sum_{\gamma \in R^+} \sum_{(w,n) \in U_\gamma(\mu,\lambda)} (-1)^{l(w)} \, \mathrm{div}(n).$$

REMARK 2.3. In [4] we computed the size of $V_\beta(\lambda, \mu)$ (and equivalently of $U_\gamma(\lambda, \mu)$) for most root systems. They turn out to be quite small. In particular if R is of type A then there is for a given pair (λ, μ) at most one $\beta \in R^+$ for which $V_\beta(\lambda, \mu)$ is non-empty and if non-empty $V_\beta(\lambda, \mu)$ has just two elements. This makes the right hand sides of the formulas in this theorem easy to compute. We shall take advantage of this in the next section.

3. The general linear group

3.1. The natural module. In this section we consider the case $G = GL(V)$ where V is an $(n+1)$-dimensional vector space over k. We identify G with $GL_{n+1}(k)$ by choosing a basis $\{e_1, e_2, \cdots, e_{n+1}\}$ for V. Then we let T, respectively B denote the subgroup consisting of all diagonal, respectively lower triangular matrices in G. We let $\epsilon_i \in X$ denote the character of T (and of B) which projects an element onto its i-th diagonal entry. Then $\epsilon_1, \epsilon_2, \cdots, \epsilon_{n+1}$ are the weights of V and we choose the enumeration of the basis above such that e_i has weight ϵ_i. The root system for (G, T) consists of the roots $\epsilon_i - \epsilon_j$ where $1 \leq i \neq j \leq n + 1$ and the simple roots are $\epsilon_i - \epsilon_{i+1}$, $i = 1, \cdots, n$.

Whenever $\lambda \in X$ we write $\lambda = \sum_{i=1}^{n+1} \lambda_i \epsilon_i$. Then $\lambda \in X^+$ if and only if $\lambda_1 \geq \lambda_2 \geq \cdots \geq \lambda_{n+1}$.

We choose a $\mathbb{Z}$-form $V_{\mathbb{Z}}$ of V by setting $V_{\mathbb{Z}} = \mathbb{Z}e_1 \oplus \cdots \oplus \mathbb{Z}e_{n+1}$. Then $G_{\mathbb{Z}} = GL(V_{\mathbb{Z}})$ is the $\mathbb{Z}$-group functor whose value at a ring A is $G_{\mathbb{Z}}(A) = GL(V_A) = GL_{n+1}(A)$ where $V_A = V_{\mathbb{Z}} \otimes_{\mathbb{Z}} A$. Also $T_{\mathbb{Z}}$ and $B_{\mathbb{Z}}$ are the corresponding subgroup functors.

3.2. The GL_2 case. Let $n = 1$, i.e. $G = GL_2(k)$.

LEMMA 3.1. *If $\lambda, \mu \in X^+$ satisfy $\mu \le \lambda + \epsilon_1$ then*

a) $\operatorname{Ext}^i_{G_{\mathbb{Z}}}(\nabla_{\mathbb{Z}}(\mu), \nabla_{\mathbb{Z}}(\lambda) \otimes V_{\mathbb{Z}}) = 0$ *for all $i > 0$,*

b) $\operatorname{Ext}^i_{G_{\mathbb{Z}}}(\nabla_{\mathbb{Z}}(\lambda + \epsilon_1), \nabla_{\mathbb{Z}}(\lambda + \epsilon_2)) \simeq \begin{cases} \mathbb{Z}/(\lambda_1 - \lambda_2 + 1)\mathbb{Z} & \text{for } i = 1, \\ 0 & \text{otherwise.} \end{cases}$

Proof: The tensor identity tells us that $\nabla_{\mathbb{Z}}(\lambda) \otimes_{\mathbb{Z}} V_{\mathbb{Z}} \simeq H^0_{\mathbb{Z}}(\lambda \otimes_{\mathbb{Z}} V_{\mathbb{Z}})$. The $B_{\mathbb{Z}}$-sequence

$$0 \to \epsilon_2 \to V_{\mathbb{Z}} \to \epsilon_1 \to 0$$

gives a short exact sequence of $G_{\mathbb{Z}}$-modules

$$0 \to \nabla_{\mathbb{Z}}(\lambda + \epsilon_2) \to \nabla_{\mathbb{Z}}(\lambda) \otimes_{\mathbb{Z}} V_{\mathbb{Z}} \to \nabla_{\mathbb{Z}}(\lambda + \epsilon_1) \to 0$$

because $H^1_{\mathbb{Z}}(\lambda + \epsilon_2) = 0$ (note that $\langle \lambda + \epsilon_2, \alpha_1^{\vee} \rangle = \lambda_1 - \lambda_2 - 1 \ge -1$).

Now Proposition 2.1 ensures the vanishing of both $\operatorname{Ext}^i_{G_{\mathbb{Z}}}(\nabla_{\mathbb{Z}}(\mu), \nabla_{\mathbb{Z}}(\lambda + \epsilon_2))$ and $\operatorname{Ext}^i_{G_{\mathbb{Z}}}(\nabla_{\mathbb{Z}}(\mu), \nabla_{\mathbb{Z}}(\lambda + \epsilon_1))$ for all $i \ge 1$ when $\mu < \lambda + \epsilon_1$. Hence we have proved a) except when $\mu = \lambda + \epsilon_1$. In that case our arguments above still gives the vanishing for $i > 1$ and by observing that $\operatorname{Hom}_{G_{\mathbb{Z}}}(\nabla_{\mathbb{Z}}(\lambda + \epsilon_1), \nabla_{\mathbb{Z}}(\lambda) \otimes V_{\mathbb{Z}}) \simeq \mathbb{Z} \simeq \operatorname{Hom}_{G_{\mathbb{Z}}}(\nabla_{\mathbb{Z}}(\lambda + \epsilon_1), \nabla_{\mathbb{Z}}(\lambda + \epsilon_1))$ we also get the following exact sequence

$$0 \to \mathbb{Z} \xrightarrow{\phi} \mathbb{Z} \to \operatorname{Ext}^1_{G_{\mathbb{Z}}}(\nabla_{\mathbb{Z}}(\lambda+\epsilon_1), \nabla_{\mathbb{Z}}(\lambda+\epsilon_2)) \to \operatorname{Ext}^1_{G_{\mathbb{Z}}}(\nabla_{\mathbb{Z}}(\lambda_1+\epsilon_1), \nabla_{\mathbb{Z}}(\lambda)\otimes_{\mathbb{Z}} V_{\mathbb{Z}}) \to 0.$$

Since $\lambda + \epsilon_1 - (\lambda + \epsilon_2) = \alpha_1$ has height 1 we get from Proposition 2.1 b) the vanishing of $\operatorname{Ext}^i_{G_{\mathbb{Z}}}(\nabla_{\mathbb{Z}}(\lambda + \epsilon_1), \nabla_{\mathbb{Z}}(\lambda + \epsilon_2))$ for $i > 1$. Therefore the left hand side in the alternating formula (Theorem 2.2) consists in our case just of 1 term and hence tells us that $|\operatorname{Ext}^1_{G_{\mathbb{Z}}}(\nabla_{\mathbb{Z}}(\lambda + \epsilon_1), \nabla_{\mathbb{Z}}(\lambda + \epsilon_2))| = \lambda_1 - \lambda_2 + 1$. In view of the above exact sequence we may finish the proof of both a) and b) by checking that the map ϕ is multiplication by $\lambda_1 - \lambda_2 + 1$.

Set $r = \lambda_1 - \lambda_2$. Recall that $\nabla_{\mathbb{Z}}(\lambda)$ may be identified with the r-th symmetric power $S^r(V_{\mathbb{Z}})$. We set $v_i = e_1^{r-i}e_2^i$, $i = 0, 1, \cdots, r$. These elements form then a $\mathbb{Z}$-basis for $\nabla_{\mathbb{Z}}(\lambda)$. We denote by v'_i, $i = 0, 1, \cdots, r+1$ the analogously defined basis for $\nabla_{\mathbb{Z}}(\lambda + \epsilon_1)$. It is then an easy exercise to check that $\operatorname{Hom}_{G_{\mathbb{Z}}}(\nabla_{\mathbb{Z}}(\lambda + \epsilon_1), \nabla_{\mathbb{Z}}(\lambda) \otimes V_{\mathbb{Z}})$ is generated by the following map

$$v'_i \mapsto (r + 1 - i)v_i \otimes e_1 + iv_{i-1} \otimes e_2, \quad i = 0, 1, \cdots, r+1.$$

Here we have set $v_{-1} = v_{r+1} = 0$. On the other hand, the natural map $v_i \otimes e_1 \mapsto v_i$, $v_i \otimes e_2 \mapsto v'_{i+1}$ generates $\operatorname{Hom}_{G_{\mathbb{Z}}}(\nabla_{\mathbb{Z}}(\lambda) \otimes V_{\mathbb{Z}}, \nabla_{\mathbb{Z}}(\lambda + \epsilon_1))$. It follows that ϕ is indeed multiplication by $r + 1$.

REMARK 3.2. Note that the second part of this lemma could maybe more appropriately be stated as follows

$$\mathrm{Ext}^i_{G_{\mathbb{Z}}}(\nabla_{\mathbb{Z}}(\lambda), \nabla_{\mathbb{Z}}(\lambda - \alpha_1)) \simeq \begin{cases} \mathbb{Z}/\langle\lambda, \alpha_1^\vee\rangle\mathbb{Z} & \text{for } i = 1, \\ 0 & \text{otherwise.} \end{cases}$$

for all λ with $\lambda - \alpha_1 \in X^+$. This is the $n = 1$ case of our main result in this section.

3.3. Restrictions to parabolic subgroups.
We return to the case of a general n. Here we set P_1 equal to the standard maximal parabolic subgroup corresponding to $\{\alpha_2, \alpha_3, \cdots, \alpha_n\}$. Then ϵ_1 is a character of P_1 and we have $V \simeq H^0(G/P_1, \epsilon_1)$. Moreover, we have a short exact sequence of P_1-modules

$$(1) \qquad\qquad 0 \to H^0(P_1/B, \epsilon_2) \to V \to \epsilon_1 \to 0.$$

Analogously we have the standard maximal parabolic subgroup P_2 corresponding to the subset $\{\alpha_1, \alpha_2, \cdots, \alpha_{n-1}\}$ of simple roots. This time ϵ_{n+1} is a character of P_2 and we have an exact P_2-sequence

$$(2) \qquad\qquad 0 \to \epsilon_{n+1} \to V \to H^0(P_2/B, \epsilon_1) \to 0.$$

Let $L_1 \leq P_1$ be the standard Levi subgroup of P_1. Then $L_1 \simeq T_1 \times G_1$ where T_1 is the 1-dimensional torus (embedded in G as the first diagonal entry) and G_1 is the subgroup of G consisting of those matrices which have 0 in all entries of the first row and column except for the first entry which is 1. Clearly, $G_1 \simeq GL_n(k)$.

We let V_1 denote the natural module for G_1. This extends to a module for P_1 and in fact $V_1 = H^0(P_1/B, \epsilon_2)_{|G_1}$. More generally, if for $\lambda \in X^+$ we set $\nabla_{L_1}(\lambda) = H^0(L_1/L_1 \cap B, \lambda)$ then this L_1-module extends to P_1 and we have $H^0(P_1/B, \lambda)_{|L_1} \simeq \nabla_{L_1}(\lambda)$.

This implies that if also $\mu \in X^+$ then

$$(3) \qquad \mathrm{Ext}^i_{P_1}(H^0(P_1/B, \mu), H^0(P_1/B, \lambda)) \simeq \mathrm{Ext}^i_{L_1}(\nabla_{L_1}(\mu), \nabla_{L_1}(\lambda))$$

for all i. On the other hand, the decomposition $L_1 = T_1 \times G_1$ gives $\nabla_{L_1}(\lambda) \simeq \lambda_1\epsilon_1 \otimes \nabla_1(\lambda')$ where $\lambda' = \sum_{i \geq 2} \lambda_i\epsilon_i$ and $\nabla_1(\lambda')$ is the dual Weyl module for G_1 with highest weight λ'. This implies

$$(4) \qquad \mathrm{Ext}^i_{L_1}(\nabla_{L_1}(\mu), \nabla_{L_1}(\lambda)) \simeq \begin{cases} \mathrm{Ext}^i_{G_1}(\nabla_1(\mu'), \nabla_1(\lambda')) & \text{if } \mu_1 = \lambda_1, \\ 0 & \text{otherwise.} \end{cases}$$

Analogously, we have for the Levi subgroup $L_2 \leq P_2$ that $L_2 \simeq G_2 \times T_1$ where $G_2 \simeq GL_n(k)$ and T_1 this time is embedded in G as the last diagonal entry. The natural module V_2 for G_2 extends to a P_2-module and we have $V_2 = H^0(P_2/B, \epsilon_1)_{|G_2}$. Also if for $\lambda \in X^+$ we write $\lambda = \lambda'' + \lambda_{n+1}\epsilon_{n+1}$ then in notation analogous to the above we have $H^0(P_2/B, \lambda)_{|L_2} \simeq \nabla_{L_2}(\lambda)$ and we get just as above

$$(5) \qquad \mathrm{Ext}^i_{L_2}(\nabla_{L_2}(\mu), \nabla_{L_2}(\lambda)) \simeq \begin{cases} \mathrm{Ext}^i_{G_2}(\nabla_2(\mu''), \nabla_2(\lambda'')) & \text{if } \mu_{n+1} = \lambda_{n+1}, \\ 0 & \text{otherwise.} \end{cases}$$

More generally, associated to any subset $I \subset \{\alpha_1, \alpha_2, \cdots, \alpha_n\}$ we have a standard maximal parabolic subgroup P_I with Levi subgroup L_I. If $\lambda \in X$ we set $H_I^0(\lambda) = H^0(P_I/B, \lambda)$. Then $H_I^0(\lambda) \neq 0$ if and only if $\langle \lambda, \alpha_i^\vee \rangle \geq 0$ for all $i \in I$. We set $\nabla_I(\lambda) = H^0(L_I/B \cap L_I, \lambda) = H_I^0(\lambda)_{|L_I}$. When $\lambda \in X^+$ the natural P_I-homomorphism $\nabla(\lambda) \to \nabla_I(\lambda)$ is surjective, i.e. we have a short exact sequence of P_I-modules

$$0 \to N_I(\lambda) \to \nabla(\lambda) \to \nabla_I(\lambda) \to 0.$$

Here all weights μ of $N_I(\lambda)$ satisfy $\mu = \lambda - \sum_i a_i \alpha_i$ with $a_i \in \mathbb{N}$ and $a_i \neq 0$ for some $i \notin I$.

For $\lambda \in X$ we set $\lambda^I = \sum \lambda_j \epsilon_j$ with the sum extending over those j for which α_j or α_{j-1} belong to I. In analogy with (4) and (5) we get then (in the obvious notation)

$$(6) \quad \mathrm{Ext}_{L_I}^i(\nabla_I(\mu), \nabla_I(\lambda)) \simeq \begin{cases} \mathrm{Ext}_{G_I}^i(\nabla_I(\mu^I), \nabla_I(\lambda^I)) & \text{if } \mu - \mu^I = \lambda - \lambda^I, \\ 0 & \text{otherwise;} \end{cases}$$

for all $\mu, \lambda \in X$ for which $\langle \mu, \alpha_i^\vee \rangle, \langle \lambda, \alpha_i^\vee \rangle \geq 0$ for all $\alpha_i \in I$.

For all this see [9] Chapter II.4.

When restricted to P_I the G-module $\nabla(\lambda)$ has a good filtration [8]. By the above $\nabla_I(\lambda)$ occurs exactly once as a quotient in such a filtration and all other factors $\nabla_I(\mu)$ have μ of the above form.

REMARK 3.3. All the results in this subsection have $\mathbb{Z}$-versions. This will be important in the following computations. It is left to the reader to place the relevant subscripts $\mathbb{Z}$ in the statements to get their formulations over $\mathbb{Z}$.

3.4. A key vanishing result. We remain in the case $G = GL(V)$ with V having dimension $n + 1$. Now we extend the result in Lemma 3.1 a) to the case of general n.

PROPOSITION 3.4. *Let $\mu, \lambda \in X^+$ with $\mu \leq \lambda + \epsilon_1$. Then*

$$\mathrm{Ext}_{G_\mathbb{Z}}^i(\nabla_\mathbb{Z}(\mu), \nabla_\mathbb{Z}(\lambda) \otimes_\mathbb{Z} V_\mathbb{Z}) = 0$$

for all $i > 0$.

Proof: To simplify notation we omit all subscripts $\mathbb{Z}$ in this proof. We use the notation from Section 3.3 above.

By the tensor identity we have $V \otimes \nabla(\lambda) \simeq H^0(V \otimes \lambda)$. Hence by Kempf's vanishing theorem we get

$$\mathrm{Ext}_G^i(\nabla(\mu), V \otimes \nabla(\lambda)) \simeq \mathrm{Ext}_B^i(\nabla(\mu), V \otimes \lambda)$$

for all i. The short exact P_1-sequence (1) therefore leads to the long exact sequence of $\mathbb{Z}$-modules

$$\cdots \to \mathrm{Ext}_B^i(\nabla(\mu), H^0(P_1/B, \epsilon_2) \otimes \lambda) \to \mathrm{Ext}_G^i(\nabla(\mu), V \otimes \nabla(\lambda)) \to \mathrm{Ext}_B^i(\nabla(\mu), \lambda + \epsilon_1) \to \cdots.$$

Since $\mu \not\geq \lambda + \epsilon_1$ we have $\mathrm{Ext}_B^i(\nabla(\mu), \lambda + \epsilon_1) = 0$ for all $i > 0$. Moreover, $\mathrm{Ext}_B^i(\nabla(\mu), H^0(P_1/B, \epsilon_2) \otimes \lambda) \simeq \mathrm{Ext}_{P_1}^i(\nabla(\mu), H^0(P_1/B, \epsilon_2) \otimes H^0(P_1/B, \lambda)) \simeq \mathrm{Ext}_{L_1}^i(\nabla(\mu)_{|L_1}, \nabla_{L_1}(\epsilon_2) \otimes \nabla_{L_1}(\lambda))$. Here the first isomorphism again relies on Kempf's vanishing theorem whereas the last one comes from (3).

As recalled in Section 3.3 $\nabla(\mu)$ has a good filtration as a P_1-module. The factors occurring have the form $\nabla_{L_1}(\nu)$ with $\nu \leq \mu$. Hence we are done if we prove
$$\mathrm{Ext}_{L_1}^i(\nabla_{L_1}(\nu), \nabla_{L_1}(\epsilon_2) \otimes \nabla_{L_1}(\lambda)) = 0 \text{ for all } i > 0$$
whenever $\nu \leq \lambda + \epsilon_1$. But by (4) we have
$$\mathrm{Ext}_{L_1}^i((\nabla_{L_1}(\nu), \nabla_{L_1}(\epsilon_2) \otimes \nabla_{L_1}(\lambda)) \simeq \begin{cases} \mathrm{Ext}_{G_1}^i(\nabla_1(\nu'), \nabla_1(\epsilon_2) \otimes \nabla_1(\lambda')) & \text{if } \nu_1 = \lambda_1, \\ 0 & \text{otherwise.} \end{cases}$$
Recall also from the previous section that $\nabla_1(\epsilon_2) \simeq V_1$. Hence the desired vanishing follows via our induction on n (the induction start is provided by Lemma 3.1) by noting that if $\nu_1 = \lambda_1$ then $\nu' \leq \lambda' + \epsilon_2$ for all $\nu \leq \lambda + \epsilon_1$. In fact, if $\nu = \lambda + \epsilon_1 - \sum_{i=1}^n a_i \alpha_i$ with $a_i \in \mathbb{N}$ then $\nu_1 = \lambda_1$ means $\lambda_1 = \lambda_1 + 1 - a_1$, i.e. $a_1 = 1$. But then $\nu' = \lambda' + a_1 \epsilon_2 - \sum_{i=2}^n a_i \alpha_i \leq \lambda' + \epsilon_2$.

3.5. The main theorem.

We can now state the main result of this section.

THEOREM 3.5. *Let β be an arbitrary positive root. Then for any $\lambda \in X^+$ for which $\lambda - \beta$ is also in X^+ we have*
$$\mathrm{Ext}_{G_\mathbb{Z}}^i(\nabla_\mathbb{Z}(\lambda), \nabla_\mathbb{Z}(\lambda - \beta)) \simeq \begin{cases} \mathbb{Z}/(\langle \lambda + \rho, \beta^\vee \rangle - 1)\mathbb{Z} & \text{if } i = 1, \\ 0 & \text{otherwise.} \end{cases}$$

Proof: Again in this proof we omit all $\mathbb{Z}$ subscripts.

There exist $1 \leq r \leq s \leq n$ such that $\beta = \alpha_r + \alpha_{r+1} + \cdots + \alpha_s = \epsilon_r - \epsilon_{s+1}$. We first reduce to the case $r = 1, s = n$:

Set $I = \{\alpha_r, \alpha_{r+1}, \cdots, \alpha_s\}$. As in Section 3.3 we let P_I, respectively L_I denote the corresponding standard parabolic subgroup, respectively its Levi part. Note that we have $L_I \simeq T_{r-1} \times GL_{s+2-r} \times T_{n-s}$ where the tori T_{r-1} and T_{n-s} have dimensions given by their index. Then we have via the results in Section 3.3
$$\mathrm{Ext}_G^i(\nabla(\lambda), \nabla(\lambda - \beta)) \simeq \mathrm{Ext}_{P_I}^i(\nabla_I(\lambda), \nabla_I(\lambda - \beta)) \simeq$$
(7) $\qquad \mathrm{Ext}_{L_I}^i(\nabla_I(\lambda), \nabla_I(\lambda - \beta)) \simeq \mathrm{Ext}_{G_I}^i(\nabla_I(\lambda^I), \nabla_I((\lambda - \beta)^I)).$

Here $G_I = GL_{s+2-r}$ and the last isomorphism follows from (6) by noting that $\lambda_j = (\lambda - \beta)_j$ for all j with $j < r$ or $j > s + 1$.

This means that we may replace G by G_I. Hence we may indeed assume $r = 1$ and $s = n$.

We proceed now by induction on $\mathrm{ht}(\beta) = n$. If $n = 1$ then we are in the GL_2-case and we may appeal to Lemma 3.1 for the result. So assume $n > 1$ and observe that since ϵ_{n+1} is the lowest weight of V the module $V \otimes (\lambda - \epsilon_1)$ contains $\lambda - \epsilon_1 + \epsilon_{n+1} = \lambda - \beta$ as a B-submodule. We therefore have an exact sequence of G-modules

$$(8) \qquad\qquad 0 \to \nabla(\lambda - \beta) \to V \otimes \nabla(\lambda - \epsilon_1) \to Q \to 0$$

where Q has a good G-filtration with factors $\nabla(\lambda - \sum_{i=1}^{n-1} \alpha_i), \nabla(\lambda - \sum_{i=1}^{n-2} \alpha_i), \cdots, \nabla(\lambda)$. Hence $\mathrm{Hom}_G(\nabla(\lambda), Q) \simeq \mathbb{Z}$ whereas $\mathrm{Ext}_G^i(\nabla(\lambda), Q) = 0$ for $i > 1$ by our induction on $\mathrm{ht}(\beta)$.

We claim that also $\mathrm{Ext}_G^1(\nabla(\lambda), Q)$ is zero. To see this we use the sequence (2) tensored by $\lambda - \epsilon_1$. When we induce the resulting sequence to G we get the exact sequence $0 \to \nabla(\lambda - \beta) \to V \otimes \nabla(\lambda - \epsilon_1) \to H^0(G/B, H^0(P_2/B, \epsilon_1) \otimes (\lambda - \epsilon_1)) \to 0$. This means that $Q \simeq H^0(G/B, H^0(P_2/B, \epsilon_1) \otimes (\lambda - \epsilon_1))$ and we get (using the same line of arguments as above)

$$\mathrm{Ext}_G^1(\nabla(\lambda), Q) \simeq \mathrm{Ext}_B^1(\nabla(\lambda), H^0(P_2/B, \epsilon_1) \otimes (\lambda - \epsilon_1)) \simeq$$

$$\mathrm{Ext}_{P_2}^1(\nabla(\lambda), H^0(P_2/B, \epsilon_1) \otimes H^0(P_2/B, \lambda - \epsilon_1)) \simeq$$

$$\mathrm{Ext}_{L_2}^1(\nabla_{L_2}(\lambda), \nabla(\epsilon_1) \otimes \nabla_{L_2}(\lambda - \epsilon_1)) \simeq \mathrm{Ext}_{G_2}^1(\nabla_2(\lambda''), V_2 \otimes \nabla_2(\lambda'' - \epsilon_1)).$$

The last term vanish by Proposition 3.4 applied to $G_2 \simeq GL_n$.

This means in view of (8) and Proposition 3.4 that $\mathrm{Ext}_G^i(\nabla(\lambda), \nabla(\lambda - \beta)) \simeq \mathrm{Ext}_G^{i-1}(\nabla(\lambda), Q) = 0$ for $i > 1$ and that we have an exact sequence

$$0 \to \mathbb{Z} \to \mathbb{Z} \to \mathrm{Ext}_G^1(\nabla(\lambda), \nabla(\lambda - \beta)) \to 0.$$

It follows that $\mathrm{Ext}_G^1(\nabla(\lambda), \nabla(\lambda - \beta))$ is cyclic. Now we apply the alternating formula Theorem 2.2 to conclude that its order is $\langle \lambda + \rho, \beta^\vee \rangle - 1$ as desired.

REMARK 3.6. The result in (7) holds more generally. The same arguments as we used above show that $\mathrm{Ext}_G^i(\nabla(\lambda), \nabla(\mu)) \simeq \mathrm{Ext}_{G_I}^i(\nabla_I(\lambda^I), \nabla_I(\mu^I))$ whenever $\mu \in \lambda - \mathbb{N}I$. Stated in 'partition language' this says that when computing Ext-groups between two (dual) Weyl modules given by the two partitions λ and μ we may strip off the first rows in λ and μ whenever their lengths coincide. The same goes for the last rows.

Similarly we can also strip off the first columns if they are identical. In fact, the character $\delta = \epsilon_1 + \epsilon_2 + \cdots + \epsilon_{n+1}$ is the determinant on T. It extends to G and hence for any $\lambda \in X^+$ we have $\nabla(\delta + \lambda) = \delta \otimes \nabla(\lambda)$. It follows that $\mathrm{Ext}_G^i(\nabla(\delta + \mu), \nabla(\delta + \lambda)) \simeq \mathrm{Ext}_G^i(\nabla(\mu), \nabla(\lambda))$ for all i.

4. Adjacent Weyl modules

4.1. In this section we discuss a result from [3] concerning extensions between adjacent Weyl modules. We consider here the case of an arbitrary connected reductive algebraic group G and we assume that the defining field k has $\mathrm{char}(k) = p \geq h$ where h is the Coxeter number for the root system of G.

If β is a positive root and $n \in \mathbb{Z}$ then we denote by $H_{\beta,n} = \{\lambda \in X \mid \langle \lambda + \rho, \beta^\vee \rangle = np\}$ the affine hyperplane associated to (β, n). We say that two alcoves C

and C' in X are adjacent if there is a unique pair (β, n) such that $H_{\beta,n}$ separates C and C'. If $\lambda \in C$ has mirror image $\lambda' \in C'$, i.e. $\lambda' = s_{\beta,n} \cdot \lambda$ then we also say that (λ, λ') are adjacent p-regular weights. Equivalently, λ' is maximal among all weights strongly linked to the p-regular weight λ, cf. [1].

Denote by $\mathbb{Z}_p$ the ring of p-adic integers. Then we can write any $m \in \mathbb{Z}_p$ in the form $m = up^{\nu_p(m)}$ with $u \in \mathbb{Z}_p$ a unit and $\nu_p : \mathbb{Z}_p \to \mathbb{N}$ the p-adic valuation. In this notation we have

THEOREM 4.1. *Let (λ, λ') be a pair of adjacent p-regular dominant weights separated by the hyperplane $H_{\beta,n}$. If $\lambda > \lambda'$ then*

$$\mathrm{Ext}^i_{G_{\mathbb{Z}_p}}(\nabla_{\mathbb{Z}_p}(\lambda), \nabla_{\mathbb{Z}_p}(\lambda')) \simeq \begin{cases} \mathbb{Z}/p^{\nu_p(np)}\mathbb{Z} & \text{if } i = 1, \\ 0 & \text{otherwise.} \end{cases}$$

In [3] we proved this result by exploiting properties of translation functors over $\mathbb{Z}_p$. Here we shall give an alternative proof which deduces it from our alternating formula Theorem 2.2.

4.2. The field case. We first treat the corresponding problem for Ext-groups for G. The arguments in this case were first given in [2]

PROPOSITION 4.2. *If (λ, λ') is a pair of adjacent p-regular dominant weights with $\lambda > \lambda'$ then*

$$\mathrm{Ext}^i_G(\nabla(\lambda), \nabla(\lambda')) \simeq \begin{cases} k & \text{if } i = 0, 1, \\ 0 & \text{otherwise.} \end{cases}$$

Proof: Let C and C' denote the alcoves containing λ and λ' respectively. Choose $\mu \in \bar{C} \cap \bar{C}'$ in such a way that $H_{\beta,n}$ is the only hyperplane containing μ. Denote by T_λ^μ, respectively T_μ^λ the translation functor from the λ-block to the μ-block, respectively vice versa (see [9] for the definition as well as the following properties of translation functors). Then we have

$$(9) \qquad\qquad T_\lambda^\mu \nabla(\lambda) \simeq \nabla(\mu) \simeq T_\lambda^\mu \nabla(\lambda'),$$

and there is an exact sequence

$$(10) \qquad\qquad 0 \to \nabla(\lambda') \to T_\mu^\lambda \nabla(\mu) \to \nabla(\lambda) \to 0.$$

This gives $\mathrm{Hom}_G(\nabla(\lambda), \nabla(\lambda')) \subset \mathrm{Hom}_G(\nabla(\lambda), T_\mu^\lambda \nabla(\mu)) \simeq \mathrm{Hom}_G(T_\lambda^\mu \nabla(\lambda), \nabla(\mu)) \simeq \mathrm{End}_G(\nabla(\mu)) \simeq k$. If $\mathrm{Hom}_G(\nabla(\lambda), \nabla(\lambda')) = 0$ then the sequence (10) splits and the simple module $L(\lambda)$ would be a submodule of $T_\mu^\lambda \nabla(\mu)$. But $T_\lambda^\mu L(\lambda) = 0$ and hence $\mathrm{Hom}_G(L(\lambda), T_\mu^\lambda \nabla(\mu)) \simeq \mathrm{Hom}_G(T_\lambda^\mu L(\lambda), \nabla(\mu)) = 0$.

These considerations show that the long exact sequence obtained from (10) by applying $\mathrm{Hom}_G(\nabla(\lambda), -)$ starts out as follows

$$(11) \qquad 0 \to k \to k \to k \to \mathrm{Ext}^1_G(\nabla(\lambda), \nabla(\lambda')) \to \mathrm{Ext}^1_G(\nabla(\lambda), T_\mu^\lambda \nabla(\lambda')) \to \cdots$$

As $\text{Ext}^i_G(\nabla(\lambda), T^\lambda_\mu \nabla(\lambda')) \simeq \text{Ext}^i_G(T^\mu_\lambda \nabla(\lambda), \nabla(\mu)) \simeq \text{Ext}^i_G(\nabla(\mu), \nabla(\mu)) = 0$ for all $i > 0$ we get from (11) $\text{Ext}^i_G(\nabla(\lambda), \nabla(\lambda')) \simeq \text{Ext}^{i-1}_G(\nabla(\lambda), \nabla(\lambda)) = 0$ for $i > 1$. The proposition follows.

4.3. Proof of Theorem 4.1. Recall that the universal coefficient theorem gives for arbitrary $\eta, \nu \in X^+$ and $i \in \mathbb{N}$

$$0 \to \text{Ext}^i_{G_{\mathbb{Z}_p}}(\nabla_{\mathbb{Z}_p}(\eta), \nabla_{\mathbb{Z}_p}(\nu)) \otimes_{\mathbb{Z}_p} k \to \text{Ext}^i_G(\nabla(\eta), \nabla(\nu)) \to$$

$$\text{Tor}^{\mathbb{Z}_p}_1(\text{Ext}^{i+1}_{G_{\mathbb{Z}_p}}(\nabla_{\mathbb{Z}_p}(\eta), \nabla_{Z_p}(\nu)), k) \to 0.$$

Taking $\eta = \lambda$ and $\nu = \lambda'$ we get from this and Proposition 4.2

$$\text{Ext}^i_{G_{\mathbb{Z}_p}}(\nabla_{\mathbb{Z}_p}(\lambda), \nabla_{\mathbb{Z}_p}(\lambda')) = 0 \text{ for all } i \geq 2.$$

Observe that also $\text{Hom}_{G_{\mathbb{Z}_p}}(\nabla_{\mathbb{Z}_p}(\lambda), \nabla_{\mathbb{Z}_p}(\lambda')) = 0$. So Theorem 2.2 gives that the order of $\text{Ext}^1_{G_{\mathbb{Z}_p}}(\nabla_{\mathbb{Z}_p}(\lambda), \nabla_{\mathbb{Z}_p}(\lambda'))$ is $p^{\nu_p(np)}$. In fact, since λ and λ' are adjacent we have $V_\alpha(\lambda', \lambda) = 0$ for $\alpha \neq \beta$ and $V_\beta(\lambda', \lambda) = \{d, m\}$ where $d = \langle \lambda + \rho, \beta^\vee \rangle - np < p$ and $m = np$. Hence we are done once we check that $\text{Ext}^1_{G_{\mathbb{Z}_p}}(\nabla_{\mathbb{Z}_p}(\lambda), \nabla_{\mathbb{Z}_p}(\lambda'))$ is cyclic. However, if not then $\text{Ext}^1_G(\nabla(\lambda), \nabla(\lambda'))$ (as well as $\text{Hom}_G(\nabla(\lambda), \nabla(\lambda'))$) would have dimension higher than 1. Proposition 4.2 says that this is not so.

REMARK 4.3. The general question of determining $\text{Ext}^i_G(\nabla(\mu), \nabla(\lambda))$ for $i \in \mathbb{N}, \lambda, \mu \in X^+$ is wide open. Even the case $i = 0$ is very far from being solved. Our results in this section says that $\text{Hom}_G(\nabla(\lambda), \nabla(\lambda'))$ is 1-dimensional whenever the two weights λ and λ' are adjacent. It follows from our results in Section 3 that when $G = GL(V)$ and β is a positive root then $\text{Hom}_G(\nabla(\lambda), \nabla(\lambda - \beta)) \simeq k$ if p divides $\langle \lambda + \rho, \beta^\vee \rangle - 1$ and 0 in all other cases. Some further results for $GL(V)$ can be found in [7].

QUESTION 4.4. Does there exist $\lambda, \mu \in X^+$ such that $\text{Hom}_G(\nabla(\lambda), \nabla(\mu))$ has dimension bigger than 1? Equivalently, are there examples where the finite abelian group $\text{Ext}^1_{G_{\mathbb{Z}}}(\nabla_{\mathbb{Z}}(\mu), \nabla_{\mathbb{Z}}(\lambda))$ is non-cyclic?

5. The quantum case

In this section k will still denote a field but we allow its characteristic p to be 0. On the other hand we exclude $p = 2$ and if the root system R contains a copy of G_2 we also exclude $p = 3$. We fix a root of unity $q \in k$ of order l or $2l$ with l odd. Let v be an indeterminate and denote by U_v the quantum group over $\mathbb{Q}(v)$ associated to the root system R. Set $A = k[v, v^{-1}]$ and let U_A be the A-form of U_v (via the Lusztig divided power construction, see [5]). We make k into an A-algebra by sending v to q and set then $U_q = U_A \otimes_A k$. More generally, for any A-algebra A' we set $U_{A'} = U_A \otimes_A A'$.

Without going into details (for those we refer to [5]) we just mention the categories $\mathcal{C}_A$ and $\mathcal{C}_q$ consisting of integrable U_A-, respectively U_q-modules. If $V \in \mathcal{C}_A$

is a finitely generated torsion A-module then we set in analogy with the definition of e_λ^G in Section 2.5

$$e_\lambda^U(V) = \sum_{i \geq 0} (-1)^i \operatorname{div}(\operatorname{Ext}_{\mathcal{C}_A}^i(\Delta_A(\lambda), V)).$$

Here $\lambda \in X^+$, $\Delta_A(\lambda)$ denotes the Weyl module for U_A with highest weight λ and $\operatorname{div} : A \to \mathcal{D}(A)$ is defined as in [**4**].

As in Section 2 this leads to the following alternating formula

THEOREM 5.1. *Let $\lambda, \mu \in X^+$ be two different dominant weights. Then*

a) $\sum_{i \geq 0} (-1)^i \operatorname{div}(\operatorname{Ext}_{\mathcal{C}_A}^i(\nabla_A(\mu), \nabla_A(\lambda))) = \sum_{\beta \in R^+} \sum_{(x,m) \in V_\beta(\lambda,\mu)} (-1)^{l(x)} \operatorname{div}([m]_{d_\beta})$.

b) $\sum_{i \geq 0} (-1)^i \operatorname{div}(\operatorname{Ext}_{\mathcal{C}_A}^i(\nabla_A(\mu), \nabla_A(\lambda))) = \sum_{\gamma \in R^+} \sum_{(w,n) \in U_\gamma(\mu,\lambda)} (-1)^{l(w)} \operatorname{div}([n]_{d_\gamma})$.

5.1. Type A. In this subsection we assume that R is a root system of type A. In particular all roots have the same length and all d_β are 1. Arguing as is Section 3 we then obtain

THEOREM 5.2. *Let β be an arbitrary positive root. Then for any $\lambda \in X^+$ for which $\lambda - \beta$ is also in X^+ we have*

$$\operatorname{Ext}_{\mathcal{C}_A}^i(\nabla_A(\lambda), \nabla_A(\lambda - \beta)) \simeq \begin{cases} A/([\langle \lambda + \rho, \beta^\vee \rangle - 1])A & \text{if } i = 1, \\ 0 & \text{otherwise.} \end{cases}$$

5.2. Adjacent Weyl modules. Let R again be arbitrary. Denote by $A_{(v-q)}$ the localization of A at the maximal ideal generated by $v - q$. We then have the following direct analogue of Theorem 4.1.

THEOREM 5.3. *Suppose $l \geq h$. Let (λ, λ') be a pair of adjacent l-regular dominant weights separated by the hyperplane $H_{\beta,n}$ and with $\lambda > \lambda'$.*

a) *If char $k = 0$ then*
$$\operatorname{Ext}_{\mathcal{C}_{A_{(v-q)}}}^i(\nabla_{A_{(v-q)}}(\lambda), \nabla_{A_{(v-q)}}(\lambda')) \simeq \begin{cases} k & \text{if } i = 1, \\ 0 & \text{otherwise.} \end{cases}$$

b) *If char $k = p > 0$ then*
$$\operatorname{Ext}_{\mathcal{C}_{A_{(v-q)}}}^i(\nabla_{A_{(v-q)}}(\lambda), \nabla_{A_{(v-q)}}(\lambda')) \simeq \begin{cases} A/(v-q)^{p^{\nu_p(n)}}A & \text{if } i = 1, \\ 0 & \text{otherwise.} \end{cases}$$

References

[1] Henning Haahr Andersen, *The strong linkage principle*, J. Reine Ang. Math. **315** (1980), 53–59.

[2] Henning Haahr Andersen, *On the structure of the cohomology of line bundles on G/B*, J. Algebra **71** (1981), 245–258.

[3] Henning Haahr Andersen, *Filtrations and tilting modules*, Ann. scient. Èc. Norm. Sup. **30** (1997), 353–366.

[4] Henning Haahr Andersen and Upendra Kulkarni, *Sum formulas for reductive algebraic groups,* Advances in Math. **217** (2008), 419–447.

[5] Henning Haahr Andersen, Patrick Polo and Wen Kexin, *Representations of quantum algebras,* Invent. math. **104** (1991), 1–59.

[6] Henning Haahr Andersen, Jens Carsten Jantzen and Wolfgang Soergel, *Representations of quantum groups at a $p-$th root of unity and of semisimple groups in characteristic p: Independence of p,* Asterisque **220** (1994), pp. 1–321.

[7] Roger Carter and M. T. J. Payne, *On homomorphisms between Weyl modules and Specht modules,* Math. Proc. Camb. Phil. Soc. **87** (1980), 419–425.

[8] Stephen Donkin, *Rational representations of algebraic groups,* Lecture Notes in Mathematics **1140** (Springer 1985).

[9] Jens Carsten Jantzen, *Representations of algebraic groups.* Second edition. Mathematical Surveys and Monographs, **107**. American Mathematical Society, Providence, RI, 2003.

[10] George Kempf, *Linear systems on homogeneous spaces,* Ann. Math. (2) **103** (1976), 557-591.

[11] Upendra Kulkarni, *On the Ext groups between Weyl modules for* GL_n. J. Algebra **304** (2006), 510–542.

DEPARTMENT OF MATHEMATICS, UNIVERSITY OF AARHUS, BUILDING 530, NY MUNKEGADE,8000 AARHUS C, DENMARK

E-mail address: mathha@imf.au.dk

Contemporary Mathematics
Volume **478**, 2009

Schur–Weyl duality in positive characteristic

Stephen Doty

ABSTRACT. Complete proofs of Schur–Weyl duality in positive characteristic
are scarce in the literature. The purpose of this survey is to write out the details
of such a proof, deriving the result in positive characteristic from the classical
result in characteristic zero, using only known facts from representation theory.

1. Introduction

Given a set A write $\mathfrak{S}_A$ for the symmetric group on A, i.e., the group of
bijections of A. For $\sigma \in \mathfrak{S}_A$ and $a \in A$ we always write $a\sigma$ for the image of a under
σ. In other words, we choose to write maps in $\mathfrak{S}_A$ on the right of their argument.
This means that $\sigma\tau$ (for $\sigma, \tau \in \mathfrak{S}_A$) is defined by $a(\sigma\tau) = (a\sigma)\tau$.

We will write $\mathfrak{S}_r$ as a shorthand for $\mathfrak{S}_{\{1,\dots,r\}}$.

Consider the group $\Gamma = \mathsf{GL}(V)$ of linear automorphisms on an n-dimensional
vector space V over a field K. We write elements $g \in \Gamma$ on the left of their argument.
(Indeed, maps are generally written on the left in this article, except when they
belong to a symmetric group.) The given action $(g, v) \to g(v)$ of Γ on V induces a
corresponding action on a tensor power $V^{\otimes r}$, with Γ acting the same in each tensor
position: $g(u_1 \otimes \cdots \otimes u_r) = (g(u_1)) \otimes \cdots \otimes (g(u_r))$, for $g \in \Gamma$, $u_i \in V$. Evidently the
action of Γ commutes with the "place permutation" action of $\mathfrak{S}_r$, acting on $V^{\otimes r}$
on the right via the rule $(u_1 \otimes \cdots \otimes u_r)\sigma = u_{1\sigma^{-1}} \otimes \cdots \otimes u_{r\sigma^{-1}}$. In this action,
a vector that started in tensor position $i\sigma^{-1}$ ends up in tensor position i, thus a
vector that started in tensor position i ends up in tensor position $i\sigma$.

We write KG for the group algebra of a group G. The fact that the two actions
commute means that the corresponding representations

$$(1.1) \qquad \Psi : K\Gamma \to \operatorname{End}_K(V^{\otimes r}); \qquad \Phi : K\mathfrak{S}_r \to \operatorname{End}_K(V^{\otimes r})$$

induce inclusions

$$(1.2) \qquad \Psi(K\Gamma) \subseteq \operatorname{End}_{\mathfrak{S}_r}(V^{\otimes r}); \qquad \Phi(K\mathfrak{S}_r) \subseteq \operatorname{End}_\Gamma(V^{\otimes r})$$

where $\operatorname{End}_{\mathfrak{S}_r}(V^{\otimes r})$ (respectively, $\operatorname{End}_\Gamma(V^{\otimes r})$) is defined to be the algebra of linear
operators on $V^{\otimes r}$ commuting with all operators in $\Phi(\mathfrak{S}_r)$ (respectively, $\Psi(\Gamma)$).

The author is grateful to Jun Hu for bringing reference [**12**] to his attention, and to the
referee for useful suggestions.

Equivalently, the commutativity of the two actions says that the representations in (1.1) induce algebra homomorphisms

$$(1.3) \qquad \overline{\Psi} : K\Gamma \to \mathrm{End}_{\mathfrak{S}_r}(V^{\otimes r}); \qquad \overline{\Phi} : K\mathfrak{S}_r \to \mathrm{End}_\Gamma(V^{\otimes r}).$$

The statement that has come to be known as "Schur–Weyl duality" is the following.

THEOREM 1 (Schur–Weyl duality). *For any infinite field K, the inclusions in (1.2) are actually equalities. Equivalently, the induced maps in (1.3) are surjective.*

In case $K = \mathbb{C}$ this goes back to a classic paper of Schur [21].[1] The main purpose of this survey is to write out a complete proof of the theorem for an arbitrary infinite field, *assuming* the truth of the result in case $K = \mathbb{C}$. The strategy, suggested by S. Koenig, is to argue that the dimension of each of the four algebras in the inclusions (1.2) is independent of the characteristic of the infinite field K. The claim for a general infinite field K then follows immediately from the classical result over $\mathbb{C}$, by dimension comparison.

We make no claim that this strategy is "best" in any sense; it is merely one possible approach. For a completely different recent approach, see [16].

2. Surjectivity of $\overline{\Psi}$

Let us first establish half of Theorem 1, namely the surjectivity of the induced map $\overline{\Psi} : K\Gamma \to \mathrm{End}_{\mathfrak{S}_r}(V^{\otimes r})$ in (1.3). For a very direct (and shorter) approach to this result, see the argument on page 210 of [1]. As already stated, the strategy followed here is to argue that the algebras $\Psi(K\Gamma)$, $\mathrm{End}_{\mathfrak{S}_r}(V^{\otimes r})$ have dimension (as vector spaces over K) which is independent of the characteristic of the infinite field K.

We first establish that $\dim_K \Psi(K\Gamma)$ is independent of K (so long as K is infinite). For this we need a general principal, which states that the "envelope" and "coefficient space" of a representation are dual to one another. To formulate the principle, let Γ be any semigroup and K any field (not necessarily infinite). Denote by K^Γ the K-algebra of K-valued functions on Γ, with the usual product and sum of elements f, f' of K^Γ given by $(ff')(g) = f(g)f'(g)$, $(f + f')(g) = f(g) + f'(g)$, for $g \in \Gamma$.

Given a representation $\tau : \Gamma \to \mathrm{End}_K(M)$ in a K-vector space M, the *coefficient space* of the representation is by definition the subspace $\mathrm{cf}_\Gamma M$ of K^Γ spanned by the coefficients $\{r_{ab}\}$ of the representation. The coefficients $r_{ab} \in K^\Gamma$ are determined relative to a choice of basis v_a $(a \in I)$ for M by the equations

$$(2.1) \qquad \tau(g)\, v_b = \sum_{a \in I} r_{ab}(g)\, v_a$$

for $g \in \Gamma$, $b \in I$.

Let $K\Gamma$ be the semigroup algebra of Γ. Elements of $K\Gamma$ are sums of the form $\sum_{g \in \Gamma} a_g g$ $(a_g \in K)$ with finitely many $a_g \neq 0$. The group multiplication extends by linearity to $K\Gamma$. The given representation $\tau : \Gamma \to \mathrm{End}_K(M)$ extends by linearity to an algebra homomorphism $K\Gamma \to \mathrm{End}_K(M)$; by abuse of notation we denote this

---

[1] A proof of Schur–Weyl duality over $\mathbb{C}$ can be extracted from Weyl's book [25]. A detailed and accessible proof is written out in [11, Theorem 3.3.8].

extended map also by τ. The *envelope*[2] of the representation τ is by definition the subalgebra $\tau(K\Gamma)$ of $\mathrm{End}_K(M)$. The representation τ factors through its envelope; that is, we have a commutative diagram

(2.2)

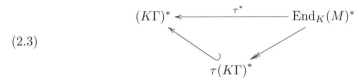

in which the leftmost and rightmost diagonal arrows are a surjection and injection, respectively. Taking linear duals, the above commutative diagram induces another one

(2.3)

$$(K\Gamma)^* \xleftarrow{\quad \tau^* \quad} \mathrm{End}_K(M)^*$$
$$\tau(K\Gamma)^*$$

in which the leftmost and rightmost diagonal arrows are now an injection and surjection, respectively. There is a natural isomorphism of vector spaces $(K\Gamma)^* \simeq K^\Gamma$, given by restricting a linear K-valued map on $K\Gamma$ to Γ; its inverse is given by the process of linearly extending a K-valued map on Γ to $K\Gamma$.

LEMMA 2 ([**4**, Lemma 1.2]). *The coefficient space* $\mathrm{cf}_\Gamma(M)$ *may be identified with the image of* τ^*, *so there is an isomorphism of vector spaces* $(\tau(K\Gamma))^* \simeq \mathrm{cf}_\Gamma M$.

PROOF. Relative to the basis v_a $(a \in I)$ the algebra $\mathrm{End}_K(M)$ has basis e_{ab} $(a, b \in I)$, where e_{ab} is the linear endomorphism of M taking v_b to v_a and taking all other v_c, for $c \neq b$, to 0. In terms of this notation, equation (2.1) is equivalent with the equality

(2.4)
$$\tau(g) = \sum_{a,b\in I} r_{ab}(g)\, e_{ab}.$$

Let e'_{ab} be the basis of $\mathrm{End}_K(M)^*$ dual to the basis e_{ab}, so that e'_{ab} is the linear functional on $\mathrm{End}_K(M)$ taking the value 1 on e_{ab} and taking the value 0 on all other e_{cd}. Then one checks that τ^* carries e'_{ab} onto r_{ab}. This proves that $\mathrm{cf}_\Gamma(M)$ may be identified with the image of τ^*, as desired.   □

We apply the preceding lemma to the representation $M = V^{\otimes r}$ of $\Gamma = \mathsf{GL}(V)$, to conclude that $\dim_K \Psi(K\Gamma)$ is equal to $\dim_K \mathrm{cf}_\Gamma(V^{\otimes r})$. Now the reader may easily check that coefficient spaces are *multiplicative*, i.e., $\mathrm{cf}_\Gamma(M \otimes N) = \mathrm{cf}_\Gamma(M) \cdot \mathrm{cf}_\Gamma(N)$. Here the multiplication takes place in K^Γ. We will apply this fact to compute the dimension of $\mathrm{cf}_\Gamma(V^{\otimes r}) = (\mathrm{cf}_\Gamma(V))^r$.

From now on we choose (and fix) a basis $\{v_1, \ldots, v_n\}$ of V and identify V with K^n and Γ with $\mathsf{GL}_n(K)$, by means of the chosen basis. Then the action of Γ on V is by matrix multiplication.

---

[2]This terminology is adapted from [**25**], where Weyl writes about the "enveloping algebra" of a group representation as the algebra generated by the endomorphisms on the representing space coming from the action of all group elements. In modern terminology, this is just the image of the representation's linear extension to the group algebra.

LEMMA 3. *For* $\Gamma = \mathsf{GL}_n(K)$ *and* K *any infinite field,* $\mathrm{cf}_\Gamma(V^{\otimes r})$ *is the vector space* $A_K(n,r)$ *consisting of all homogeneous polynomial functions on* Γ *of degree* r. *We have* $\dim_K A_K(n,r) = \binom{n^2+r-1}{r} = \dim_K \Psi(K\Gamma)$.

PROOF. Let $c_{ij} \in K^\Gamma$ be the function which maps a matrix $g \in \Gamma$ onto its (i,j)th matrix entry. By definition, a function $f \in K^\Gamma$ is polynomial[3] if it belongs to the polynomial algebra $K[c_{ij} : 1 \leqslant i,j \leqslant n]$. The c_{ij} are algebraically independent since K is infinite. Note that the c_{ij} are the coefficients of Γ on V, i.e., $\mathrm{cf}_\Gamma V = \sum_{1 \leqslant i,j \leqslant n} K c_{ij}$.

An element $f \in K[c_{ij} : 1 \leqslant i,j \leqslant n]$ is homogeneous of degree r if $f(ag) = a^r f(g)$ for all $a \in K$ and all $g \in \Gamma$. Here we define ag to be the matrix obtained from g by multiplying each entry by the scalar a.

Now from the equality $\mathrm{cf}_\Gamma V = \sum_{1 \leqslant i,j \leqslant n} K c_{ij}$ and the multiplicativity of coefficient spaces, it follows that $\mathrm{cf}_\Gamma(V^{\otimes r})$ is the vector space $A_K(n,r)$ consisting of all homogeneous polynomial functions on Γ of degree r. The equality $\dim_K A_K(n,r) = \binom{n^2+r-1}{r}$, now follows by an easy dimension count (or one can look at [**10**, §2.1]), and this is the same as $\dim_K \Psi(K\Gamma)$ by Lemma 2.   □

The preceding lemma establishes the fact that $\dim_K \mathrm{cf}_\Gamma(V^{\otimes r})$ is independent of the characteristic of K (so long as K is infinite). So we turn now to the task of establishing a similar independence statement for $\dim_K \mathrm{End}_{\mathfrak{S}_r}(V^{\otimes r})$.

Let us restrict the action of Γ to the "maximal torus" $T \subset \Gamma$ given by all diagonal matrices in $\Gamma = \mathsf{GL}_n(K)$. The abelian group T is isomorphic to the direct product $(K^\times) \times \cdots \times (K^\times)$ of n copies of the multiplicative group $K^\times$ of the field K, so its irreducible representations are one-dimensional, given on a basis element z by the rule $\mathrm{diag}(a_1,\ldots,a_n)(z) = a_1^{\lambda_1} \cdots a_n^{\lambda_n} z$, for various $\lambda_i \in \mathbb{N}$. For convenience of notation, write $t = \mathrm{diag}(a_1,\ldots,a_n)$, $\lambda = (\lambda_1,\ldots,\lambda_n)$, and $t^\lambda = a_1^{\lambda_1} \cdots a_n^{\lambda_n}$. Now T acts semisimply on $V^{\otimes r}$, and we have a "weight space decomposition"

$$(2.5) \qquad\qquad V^{\otimes r} = \bigoplus_{\lambda \in \mathbb{N}^n} (V^{\otimes r})_\lambda$$

where $(V^{\otimes r})_\lambda = \{m \in V^{\otimes r} : tm = t^\lambda m, \text{ for all } t \in T\}$.

Since the action of T on $V^{\otimes r}$ commutes with the place permutation action of $\mathfrak{S}_r$, it follows that each weight space $(V^{\otimes r})_\lambda$ is a $K\mathfrak{S}_r$-module. It is easy to write out a basis for $(V^{\otimes r})_\lambda$ in terms of the given basis $\{v_1,\ldots,v_n\}$ of V. Clearly $V^{\otimes r}$ has a basis consisting of simple tensors of the form $v_{i_1} \otimes \cdots \otimes v_{i_r}$ for various multi-indices $(i_1,\ldots,i_r)$ satisfying the condition $i_j \in \{1,\ldots,n\}$ for each $1 \leqslant j \leqslant r$. Each simple tensor $v_{i_1} \otimes \cdots \otimes v_{i_r}$ has weight $\lambda = (\lambda_1,\ldots,\lambda_n)$ where λ_i counts the number of indices j such that $i_j = i$. Thus it follows that $\sum_i \lambda_i = r$. Let us write $\Lambda(n,r)$ for the set of all $\lambda \in \mathbb{N}^n$ such that $\sum_i \lambda_i = r$. Then each summand $(V^{\otimes r})_\lambda$ is zero unless $\lambda \in \Lambda(n,r)$, so we may replace $\mathbb{N}^n$ by $\Lambda(n,r)$ in the decomposition (2.5).

From the above it follows that a basis of $(V^{\otimes r})_\lambda$, for any $\lambda \in \Lambda(n,r)$, is given by the set of all $v_{i_1} \otimes \cdots \otimes v_{i_r}$ of weight λ.

As a $K\mathfrak{S}_r$-module, the weight space $(V^{\otimes r})_\lambda$ may be identified with a "permutation" module M^λ. Typically, M^λ is defined as the induced module $\mathbf{1} \otimes_{(K\mathfrak{S}_\lambda)} (K\mathfrak{S}_r)$,

---

[3]The notion of "polynomial" functions on general linear groups goes back (at least) to Schur's 1901 dissertation.

where by $\mathbf{1}$ we mean the one dimensional module K with trivial action, and where $\mathfrak{S}_\lambda$ is the Young subgroup

$$\mathfrak{S}_{\{1,\ldots,\lambda_1\}} \times \mathfrak{S}_{\{\lambda_1+1,\ldots,\lambda_1+\lambda_2\}} \times \cdots \times \mathfrak{S}_{\{\lambda_{n-1}+1,\ldots,\lambda_{n-1}+\lambda_n\}}$$

of $\mathfrak{S}_r$ determined by $\lambda = (\lambda_1, \ldots, \lambda_n)$. By [**2**, §12D] this has a basis (over K) indexed by any set of right[4] coset representatives of $\mathfrak{S}_\lambda$ in $\mathfrak{S}_r$.

LEMMA 4. *For any field K, $\dim_K \operatorname{End}_{\mathfrak{S}_r}(V^{\otimes r})$ is independent of K.*

PROOF. From the decomposition (2.5) it follows that we have a direct sum decomposition of $\operatorname{End}_{\mathfrak{S}_r}(V^{\otimes r}) = \operatorname{Hom}_{\mathfrak{S}_r}(V^{\otimes r}, V^{\otimes r})$ of the form

$$\operatorname{End}_{\mathfrak{S}_r}(V^{\otimes r}) = \bigoplus_{\lambda,\mu\in\Lambda(n,r)} \operatorname{Hom}_{\mathfrak{S}_r}((V^{\otimes r})_\lambda, (V^{\otimes r})_\mu).$$

By Lemma 7(b) in the next section, we may identify

$$\operatorname{Hom}_{\mathfrak{S}_r}((V^{\otimes r})_\lambda, (V^{\otimes r})_\mu) \simeq \operatorname{Hom}_{\mathfrak{S}_r}(M^\lambda, M^\mu)$$

for any $\lambda, \mu \in \Lambda(n,r)$. By Mackey's theorem (see [**2**, §44] or combine [**22**, Proposition 22] with Frobenius reciprocity), it follows that $\dim_K \operatorname{Hom}_{\mathfrak{S}_r}(M^\lambda, M^\mu)$ is equal to the number of $(\mathfrak{S}_\lambda, \mathfrak{S}_\mu)$-double cosets in $\mathfrak{S}_r$, which is independent of K. This proves the claim. Alternatively, one can avoid the Mackey theorem by applying James [**13**, Theorem 13.19] directly (see also [**7**, Proposition 3.5]). □

Now we can obtain the main result of this section, which proves half of Schur–Weyl duality in positive characteristic. We remind the reader that the validity of Theorem 1 for $K = \mathbb{C}$ is assumed, so in particular $\Psi(\mathbb{C}\Gamma) = \operatorname{End}_{\mathfrak{S}_r}((\mathbb{C}^n)^{\otimes r})$.

PROPOSITION 5. *For any infinite field K, the image $\Psi(K\Gamma)$ of the representation Ψ is equal to the centralizer algebra $\operatorname{End}_{\mathfrak{S}_r}(V^{\otimes r})$, so the map $\overline{\Psi}$ in (1.3) is surjective.*

PROOF. By Lemmas 3 and 4 we have equalities

$$\dim_K \Psi(K\Gamma) = \dim_{\mathbb{C}} \Psi(\mathbb{C}\Gamma),$$

$$\dim_K \operatorname{End}_{\mathfrak{S}_r}((K^n)^{\otimes r}) = \dim_{\mathbb{C}} \operatorname{End}_{\mathfrak{S}_r}((\mathbb{C}^n)^{\otimes r})$$

for any infinite field K. Since $\Psi(\mathbb{C}\Gamma) = \operatorname{End}_{\mathfrak{S}_r}((\mathbb{C}^n)^{\otimes r})$ it follows that $\dim_K \Psi(K\Gamma) = \dim_K \operatorname{End}_{\mathfrak{S}_r}((K^n)^{\otimes r})$ for any infinite field K, and thus by comparison of dimensions the first inclusion in (1.2) must be an equality. Equivalently, the map $\overline{\Psi}$ in (1.3) is surjective. □

3. Surjectivity of $\overline{\Phi}$

It remains to establish the surjectivity of the induced map $\overline{\Phi}$ in (1.3). This surjectivity was first established in positive characteristic in [**3**, Theorem 4.1].[5] We will outline an alternative proof here, following our avowed strategy of showing that the dimensions of $\Phi(K\mathfrak{S}_r)$, $\operatorname{End}_\Gamma(V^{\otimes r})$ are independent of the characteristic of the infinite field K.

In order to establish the independence statement for $\Phi(K\mathfrak{S}_r)$ we apply results of Murphy and Härterich in order to compute the annihilator of the action of $\mathfrak{S}_r$

---

[4]Reference [**2**] works with left modules instead of right ones, so for our purposes left and right need to be interchanged there.

[5]The statement of Theorem 4.1 in [**3**] is actually much more general.

on $V^{\otimes r}$. Note that Murphy and Härterich worked with the Iwahori–Hecke algebra (with parameter q) in type A, so one needs to take $q = 1$ in their formulas in order to get corresponding results for the group algebra $K\mathfrak{S}_r$. The results of Murphy and Härterich hold over an arbitrary commutative integral domain, so K does not need to be an infinite field in this part. So we assume from now on, until the paragraph after Corollary 12, that K is a commutative integral domain.

Let λ be a composition of r. We regard λ as an infinite sequence $(\lambda_1, \lambda_2, \dots)$ of nonnegative integers such that $\sum \lambda_i = r$. The individual λ_i are the parts of λ, and the largest index ℓ such that $\lambda_\ell = 0$ and $\lambda_j = 0$ for all $j > \ell$ is the length, or number of parts, of λ. Any composition λ may be sorted into a partition λ^+, in which the parts are non-strictly decreasing. When writing compositions or partitions, trailing zero parts are usually omitted. If λ is a partition, we generally write λ' for the transposed (or conjugate) partition, corresponding to writing the rows of the Young diagram as columns.

Given a composition $\lambda = (\lambda_1, \dots, \lambda_\ell)$ of r, a Young diagram of shape λ is an arrangement of boxes into rows with λ_i boxes in the ith row. A λ-tableau T is a numbering of the boxes in the Young diagram of shape λ by the numbers $1, \dots, r$ so that each number appears just once. In other words, it is a bijection between the boxes in the Young diagram and the set $\{1, \dots, r\}$. Such a T is row standard if the numbers in each row are increasing when read from left to right, and standard if row standard and the numbers in each column are increasing when read from top to bottom.

The group $\mathfrak{S}_r$ acts naturally on tableaux, on the right, by permuting the entries. Given a tableau T, we define the row stabilizer of T to be the subgroup $R(T)$ of $\mathfrak{S}_r$ consisting of those permutations that permute entries in each row of T amongst themselves, similarly the column stabilizer is the subgroup $C(T)$ consisting of those permutations that permute entries in each column of T amongst themselves.

Let λ be a composition of r. Let T^λ be the λ-tableau in which the numbers $1, \dots, r$ have been inserted in the boxes in order from left to right along rows, read from top to bottom. Set $\mathfrak{S}_\lambda = R(T^\lambda)$. This is the same as the Young subgroup

$$\mathfrak{S}_{\{1,\dots,\lambda_1\}} \times \mathfrak{S}_{\{\lambda_1+1,\dots,\lambda_1+\lambda_2\}} \times \cdots$$

of $\mathfrak{S}_r$ defined by the composition λ. Given a row standard λ-tableau T, we define $d(T)$ to be the unique element of $\mathfrak{S}_r$ such that $T = T^\lambda d(T)$. Given any pair S, T of row standard λ-tableaux, following Murphy [19] we set

(3.1) $x_{ST} = d(S)^{-1} x_\lambda d(T); \quad y_{ST} = d(S)^{-1} y_\lambda d(T).$

where $x_\lambda = \sum_{w \in \mathfrak{S}_\lambda} w$ and $y_\lambda = \sum_{w \in \mathfrak{S}_\lambda} (\operatorname{sgn} w)\, w$.

THEOREM 6 (Murphy). *Let K be a commutative integral domain. Each of the sets $\{x_{ST}\}$ and $\{y_{ST}\}$, as (S, T) ranges over the set of all ordered pairs of standard λ-tableaux for all partitions λ of r, is a K-basis of the group algebra $A = K\mathfrak{S}_r$.*

Note that x_{ST} and y_{ST} are interchanged by the K-linear ring involution of $K\mathfrak{S}_r$ which sends w to $(\operatorname{sgn} w)w$, for $w \in \mathfrak{S}_r$. This gives a trivial way of converting results about one basis into results about the other.

We will need several equivalent descriptions of the permutation modules M^λ, which we now formulate. Let λ be a composition of r. Recall that $M^\lambda = \mathbf{1} \otimes_{(K\mathfrak{S}_\lambda)} (K\mathfrak{S}_r)$, where $\mathbf{1}$ is the one dimensional module K with trivial action. In [13,

Definition 4.1], an alternative combinatorial description of M^λ is given in terms of "tabloids" (certain equivalence classes of tableaux), and in [**5**, (1.3)] the authors write out an explicit isomorphism between these two descriptions. The following gives two additional descriptions of M^λ, the second of which was used already in the previous section.

LEMMA 7. *For any composition λ of r, the permutation module M^λ is isomorphic (as a right $K\mathfrak{S}_r$-module) with either of*

(a) the right ideal $x_\lambda(K\mathfrak{S}_r)$ of $K\mathfrak{S}_r$;

(b) the weight space $(V^{\otimes r})_\lambda$ in $V^{\otimes r}$, where V is free over K of rank at least as large as the number of parts of λ.

PROOF. Let $\mathcal{D}_\lambda = \{d(T)\}$ as T varies over the set of row standard tableaux of shape λ. This is a set of right coset representatives of $\mathfrak{S}_\lambda$ in $\mathfrak{S}_r$. The map $d \to x_\lambda d$ gives the isomorphism (a), in light of Lemma 3.2(i) of [**5**]. The isomorphism (b) works as follows. Given $d \in \mathcal{D}_\lambda$, write $d = d(T)$ for some (unique) row standard tableau T of shape λ. Use T to construct a simple tensor $v_{i_1} \otimes \cdots \otimes v_{i_r}$ of weight λ, by letting i_j be the (unique) row number in T in which j is found. This map is well defined, and is a bijection since there is an obvious inverse map. $\square$

We recall that compositions are partially ordered by *dominance*, defined as follows. Given two compositions λ, μ of r, write $\lambda \trianglerighteq \mu$ (λ dominates μ) if $\sum_{i \leqslant j} \lambda_i \geqslant \sum_{i \leqslant j} \mu_i$ for all j. One writes $\lambda \triangleright \mu$ (λ strictly dominates μ) if $\lambda \trianglerighteq \mu$ and the inequality $\sum_{i \leqslant j} \lambda_i \geqslant \sum_{i \leqslant j} \mu_i$ is strict for at least one j.

The dominance order on compositions extends to the set of row standard tableaux, as follows. Let T be a row standard λ-tableau, where λ is a composition of r. For any $s < r$ denote by $T_{\downarrow s}$ the row standard tableau that results from throwing away all boxes of T containing a number bigger than s. Let $[T_{\downarrow s}]$ be the corresponding composition of s (the composition defining the shape of $T_{\downarrow s}$). Given row standard tableaux S, T with the same number r of boxes, define

(3.2)
$$S \trianglerighteq T \text{ if for each } s \leqslant r, [S_{\downarrow s}] \trianglerighteq [T_{\downarrow s}];$$
$$S \triangleright T \text{ if for each } s \leqslant r, [S_{\downarrow s}] \triangleright [T_{\downarrow s}].$$

Note that if S, T are standard tableaux, respectively of shape λ, μ where λ and μ are partitions of r, then $S \trianglerighteq T$ if and only if $T' \trianglerighteq S'$. Here T' denotes the transposed tableau of T, obtained from T by writing its rows as columns.

Let $*$ be the K-linear anti-involution on $A = K\mathfrak{S}_r$ given by

$$\left(\sum_{w \in \mathfrak{S}_r} b_w w\right)^* \to \sum_{w \in \mathfrak{S}_r} b_w w^{-1}$$

for any $b_w \in K$. An easy calculation with the definitions shows that

(3.3)
$$x^*_{ST} = x_{TS}; \quad y^*_{ST} = y_{TS}$$

for any pair S, T of row standard λ-tableaux.

We write $c \in \{x, y\}$ in order to describe the cell structure of $A = K\mathfrak{S}_r$ relative to both bases simultaneously.

THEOREM 8 (Murphy, [**19**, Theorem 4.18]). *Let $c \in \{x, y\}$. Let λ be a partition of r. The K-module $A[\trianglerighteq \lambda] = \sum K c_{ST}$, the sum taken over all pairs (S, T) of standard μ-tableaux such that $\mu \trianglerighteq \lambda$, is a two-sided ideal of A, as is $A[\triangleright \lambda] =$*

$\sum K c_{ST}$, the sum taken over all pairs (S,T) of standard μ-tableaux such that $\mu \rhd \lambda$. For any $a \in A$ and any pair (S,T) of λ-tableaux, we have

$$(3.4) \qquad c_{ST}\, a = \sum_U r_a(T,U)\, c_{SU} \quad \mathrm{mod}\ A[\rhd\lambda]$$

where $r_a(T,U) \in K$ is independent of S, and in the sum U varies over the set of standard λ-tableaux.

In the language of cellular algebras, introduced by Graham and Lehrer [9], for $c \in \{x,y\}$ the basis $\{c_{ST}\}$ is a cellular basis of A. Note that by applying the anti-involution $*$ to (3.4) we obtain by (3.3) the equivalent condition

$$(3.5) \qquad a^*\, c_{TS} = \sum_U r_a(T,U)\, c_{US} \quad \mathrm{mod}\ A[\rhd\lambda]$$

for any $a \in A$ and any pair (S,T) of λ-tableaux.

Now fix n and r, and let P be the set of partitions λ of r such that $\lambda_1 > n$. Note that P is empty if $n \geqslant r$. Set $A[P] = \sum K y_{ST}$, where the sum is taken over the set of pairs (S,T) of standard tableaux of shape λ, for all $\lambda \in P$. It follows from (3.4), (3.5) that $A[P]$ is a two-sided ideal of A because P satisfies the property: $\lambda \in P$, $\mu \rhd \lambda \implies \mu \in P$ for any partition μ of r. Note that $A[P]$ is the zero ideal if $n \geqslant r$.

LEMMA 9. *The kernel of Φ contains $A[P]$.*

PROOF. If $n \geqslant r$ then P is empty and there is nothing to prove, so we may assume that $n < r$.

We first observe that y_λ acts as zero on any simple tensor $v_{i_1} \otimes \cdots \otimes v_{i_r} \in V^{\otimes r}$, for any $\lambda \in P$. This is because any such tensor has at most n distinct tensor factors, and thus is annihilated by the alternating sum $\alpha = \sum_{w \in \mathfrak{S}_{\{1,\dots,\lambda_1\}}} (\mathrm{sgn}\, w)w$. (Recall that $\lambda_1 > n$.) The alternating sum α is a factor of y_λ, i.e., we have $y_\lambda = \alpha\beta$ for some $\beta \in K\mathfrak{S}_r$, so y_λ acts as zero as well. Since $V^{\otimes r}$ is spanned by such simple tensors, it follows that y_λ acts as zero on $V^{\otimes r}$.

It follows immediately that every $y_{ST} = d(S)^{-1} y_\lambda d(T)$, for $\lambda \in P$, acts as zero on $V^{\otimes r}$, for any λ-tableaux S, T, since $d(S)^{-1}$ simply permutes the entries in the tensor, and then y_λ annihilates it. Since $A[P]$ is spanned by such y_{ST}, it follows that $A[P]$ is contained in the kernel of Φ.   □

We will use a lemma of Murphy to establish the opposite inclusion. Let (S,T) be a pair of λ-tableaux, where λ is a composition of r. The pair is row standard if both S, T are row standard; similarly the pair is standard if both S, T are standard. The dominance order on tableaux defined in (3.2) extends naturally to pairs of tableaux, by defining:

$$(3.6) \qquad (S,T) \unrhd (U,V) \text{ if } S \unrhd U \text{ and } T \unrhd V.$$

For $a, b \in A$ let (a,b) denote the coefficient of 1 in the expression $ab^* = \sum_{w \in \mathfrak{S}_r} c_w\, w$, where $c_w \in K$. Then $(\ ,\ )$ is a non-degenerate symmetric bilinear form on $A = K\mathfrak{S}_r$. It is straightforward to check that this bilinear form satisfies the properties

$$(3.7) \qquad (a, bd) = (ad^*, b); \quad (a, db) = (d^*a, b)$$

for any $a, b, d \in A$.

LEMMA 10 (Murphy, [20, Lemma 4.16]). *Let (S,T) be a row standard pair of μ-tableaux and (U,V) a standard pair of λ-tableaux, where μ is a given composition of r and λ a partition of r. Then:*

(a) $(x_{ST}, y_{U'V'}) = 0$ unless $(U, V) \trianglerighteq (S, T)$;

(b) $(x_{UV}, y_{U'V'}) = \pm 1$

where T' denotes the transpose of a tableau T.

This is used in proving the following result, which in particular shows that the rank (over K) of the annihilator of the symmetric group action on $V^{\otimes r}$ is independent of the characteristic of K.

PROPOSITION 11 (Härterich, [**12**, Lemma 3]). *The kernel of Φ, i.e., the annihilator* $\mathrm{ann}_{K\mathfrak{S}_r} V^{\otimes r}$, *is the cell ideal $A[P]$.*

PROOF. By Lemma 9, the kernel of Φ contains $A[P]$, so we only need to prove the reverse containment. Let

$$a = \textstyle\sum_{(S,T)} a_{ST} y_{ST} \in \ker \Phi$$

where $a_{ST} \in K$, and the sum over all pairs (S, T) of standard tableaux of shape λ, where λ is a partition of r. It suffices to prove: $(*)$ $a_{ST} = 0$ for all pairs (S, T) of standard tableaux of shape $\mu \in P^c$, where P^c is the complement of P in the set of all partitions of r.

We note that P^c is the set of conjugates λ' of partitions λ in $\Lambda(n, r)$. Write $\Lambda^+(n, r)$ for the set of partitions in $\Lambda(n, r)$; this is the set of partitions of r into not more than n parts.

We proceed by contradiction. Suppose $(*)$ is not true. Since by Lemma 9 we have $\sum_{\mathrm{shape}(S,T) \in P} a_{ST} y_{ST} \in \ker(\Phi)$, it follows that

$$b = \textstyle\sum_{\mathrm{shape}(S,T) \in P^c} a_{ST} y_{ST}$$

is also in the kernel of Φ; i.e., the element b annihilates $V^{\otimes r}$. Under the assumption we have $b \neq 0$. Let (S_0, T_0) be a minimal pair (with respect to $\trianglerighteq$) with $\mathrm{shape}(S_0, T_0) \in P^c$ such that $a_{S_0 T_0} \neq 0$. So $a_{ST} = 0$ for all pairs (S, T) with $(S_0, T_0) \rhd (S, T)$. Let λ_0 be the shape of T_0' (same as shape of S_0'). Then $\lambda_0 \in \Lambda^+(n, r)$, and we have

$$(x_{\lambda_0 S_0'} b, d(T_0')) = (x_{\lambda_0 S_0'} \textstyle\sum a_{ST} y_{ST}, d(T_0'))$$

$$= \textstyle\sum a_{ST} (d(T_0')^{-1} x_{\lambda_0 S_0'}, y_{ST}^*)$$

$$= \textstyle\sum a_{ST} (x_{T_0' S_0'}, y_{TS})$$

where all sums are taken over the set of (S, T) of shape some member of P^c. Here, we write $x_{\mu T}$ shorthand for $x_{T^\mu T}$, where (as before) T^μ is the μ-tableau in which the numbers $1, \ldots, r$ have been inserted in the boxes in order from left to right along rows, read from top to bottom.

By Lemma 10(a) all the terms in the last sum are zero unless $(S_0, T_0) \trianglerighteq (S, T)$, in other words $(x_{T_0' S_0'}, y_{TS}) = 0$ for all pairs (S, T) which are strictly more dominant than (S_0, T_0). By assumption, $a_{ST} = 0$ for all pairs (S, T) strictly less dominant than (S_0, T_0). Thus, the above sum collapses to a single term $a_{S_0 T_0}(x_{T_0' S_0'}, y_{T_0 S_0})$, and by our assumption and Lemma 10(b) this is nonzero.

This proves that $x_{\lambda_0 S_0'} b \neq 0$. Thus b does not annihilate the permutation module $M^{\lambda_0} \simeq x_{\lambda_0} A$. Since $\lambda_0 \in \Lambda^+(n, r)$ as noted above, and thus M^{λ_0} is isomorphic to a direct summand of $V^{\otimes r}$, we have arrived at a contradiction. This proves the result. $\square$

COROLLARY 12. *For any commutative integral domain K, the K-module $\Phi(K\mathfrak{S}_r)$ is free over K, of rank $r! - \sum_{\lambda \in P} N(\lambda)^2$, where $N(\lambda)$ is the number of standard tableaux of shape λ. In particular, the K-rank of $\Phi(K\mathfrak{S}_r)$ is independent of K.*

PROOF. By the preceding proposition, $\Phi(K\mathfrak{S}_r) \simeq A/A[P]$. This is free over K because it is a submodule of the free K-module $\operatorname{End}_K(V^{\otimes r})$. By definition, $A[P]$ is free over K of rank $\sum_{\lambda \in P} N(\lambda)^2$, so the result follows. $\square$

Now we return to the assumption that K is an infinite field, and consider why $\dim_K \operatorname{End}_\Gamma(V^{\otimes r})$ is independent of K. This involves facts about the representation theory of algebraic groups that are less elementary than facts used so far. We identify the group $\Gamma = \mathsf{GL}_n(K)$, the group of K-rational points in the algebraic group $\mathsf{GL}_n(\overline{K})$, where $\overline{K}$ is an algebraic closure of K, with the group scheme $\mathbf{GL}_n$ over K.

For $\Gamma = \mathsf{GL}_n(K)$ we let T be the maximal torus consisting of all diagonal elements of Γ. Regard an element $\lambda \in \mathbb{Z}^n$ as a character on T (via $\operatorname{diag}(a_1, \ldots, a_n) \to a_1^{\lambda_1} \cdots a_n^{\lambda_n}$ for $a_i \in K^\times$). Consider the Borel subgroup B consisting of the lower triangular matrices in Γ, and let $\nabla(\lambda)$ be the induced module (see [**14**, Part I, §3.3]):

$$\operatorname{ind}_B^\Gamma(K_\lambda) = \{f \in K[\Gamma] : f(gb) = b^{-1}f(g), \text{ all } b \in B, g \in G\}$$

for any $\lambda \in \mathbb{Z}^n$, where K_λ is the one dimensional T-module with character λ, regarded as a B-module by making the unipotent radical of B act trivially.

The dual space $M^* = \operatorname{Hom}_K(M, K)$ of a given rational $K\Gamma$-module M is again a rational $K\Gamma$-module, in two different ways:

    (i) $(g \cdot f)(m) = f(g^{-1}m)$;

    (ii) $(g \cdot f)(m) = f(g^t m)$ (g^T is the matrix transpose of g)

for $g \in \Gamma$, $f \in M^*$, $m \in M$. Denote the first dual by M^* and the second by M^T. Let $\Delta(\lambda) = \nabla(\lambda)^\mathsf{T}$. It is known that $\Delta(\lambda) \simeq \nabla(-w_0\lambda)^*$ where w_0 is the longest element in the Weyl group W. The modules $\nabla(\lambda)$, $\Delta(\lambda)$ are known as "dual Weyl modules" and "Weyl modules", respectively.[6] The most important property these modules satisfy, for our purposes, is the following

$$(3.8) \qquad \operatorname{Ext}_\Gamma^j(\Delta(\lambda), \nabla(\mu)) \simeq \begin{cases} K & \text{if } j = 0 \text{ and } \lambda = \mu \\ 0 & \text{otherwise.} \end{cases}$$

This is a special case of [**14**, Part II, Proposition 4.13].

Say that a Γ-module M has a ∇-filtration (respectively, Δ-filtration) if it has an ascending chain of submodules

$$0 = M_0 \subseteq M_1 \subseteq \cdots \subseteq M_{t-1} \subseteq M_t = M$$

such that each successive quotient M_i/M_{i-1} is isomorphic with $\nabla(\lambda^i)$ (respectively, $\Delta(\lambda^i)$) for some $\lambda^i \in \mathbb{Z}^n$. Another fact we need goes back to [**24**, Theorem B, page 164]:

$$(3.9) \qquad \Delta(\lambda) \otimes \Delta(\mu) \text{ has a } \Delta\text{-filtration}$$

---

[6]Weyl and dual Weyl modules for $\mathsf{GL}_n(K)$ are studied in [**10**, Chapters 4, 5], where they are respectively denoted by $D_{\lambda,K}$ and $V_{\lambda,K}$.

for any $\lambda, \mu \in \mathbb{Z}^n$. (Note that this fundamental result has been extended in [6], which in turn was extended in [18].) From (3.9) it follows immediately by taking duals that

$$(3.10) \qquad \nabla(\lambda) \otimes \nabla(\mu) \text{ has a } \nabla\text{-filtration}$$

for any $\lambda, \mu \in \mathbb{Z}^n$. The following result, which says that $V^{\otimes r}$ is a "tilting" module for Γ, is now easy to prove.

LEMMA 13. $V^{\otimes r}$ has both ∇- and Δ-filtrations.

PROOF. One has $V = \nabla(\varepsilon_1) = \Delta(\varepsilon_1)$ where $\varepsilon_1 = (1, 0, \ldots, 0)$. The result then follows from (3.9) and (3.10) by induction on r. □

For the next argument we will need the notion of formal characters. Any rational $K\Gamma$-module M has a weight space decomposition $M = \bigoplus_{\lambda \in \mathbb{Z}^n} M_\lambda$ where

$$M_\lambda = \{m \in M : tm = t^\lambda m, \text{ for all } t \in T\}.$$

Here $t^\lambda = a_1^{\lambda_1} \cdots a_n^{\lambda_n}$ where $t = \mathrm{diag}(a_1, \ldots a_n)$ as previously defined, just before (2.5). Set $X = \mathbb{Z}^n$ and let $\mathbb{Z}[X]$ be the free $\mathbb{Z}$-module on X with basis consisting of all symbols $e(\lambda)$ for $\lambda \in X$, with a multiplication given by $e(\lambda)e(\mu) = e(\lambda + \mu)$, for $\lambda, \mu \in X$. If M is finite dimensional, the formal character $\mathrm{ch}\, M \in \mathbb{Z}[X]$ of M is defined by

$$\mathrm{ch}\, M = \sum_{\lambda \in X} (\dim_K M_\lambda)\, e(\lambda).$$

The formal character of $\Delta(\lambda)$, which is the same as $\mathrm{ch}\, \nabla(\lambda)$ since the maximal torus T is fixed pointwise by the matrix transpose, is given by Weyl's character formula [14, Part II, Proposition 5.10].[7]

PROPOSITION 14. For any infinite field K, $\dim_K \mathrm{End}_\Gamma(V^{\otimes r})$ is independent of K.

PROOF. Let $0 = N_0 \subseteq N_1 \subseteq \cdots \subseteq N_{s-1} \subseteq N_s = V^{\otimes r}$ be a ∇-filtration and $0 = M_0 \subseteq M_1 \subseteq \cdots \subseteq M_{t-1} \subseteq M_t = V^{\otimes r}$ a Δ-filtration. Write $(V^{\otimes r} : \nabla(\lambda))$ for the number of successive subquotients N_i/N_{i-1} which are isomorphic to $\nabla(\lambda)$, and similarly write $(V^{\otimes r} : \Delta(\lambda))$ for the number of successive subquotients M_i/M_{i-1} which are isomorphic to $\Delta(\lambda)$. Since characters are additive on short exact sequences, we have

$$\mathrm{ch}\, V^{\otimes r} = \sum_{\lambda \in X} (V^{\otimes r} : \nabla(\lambda))\, \mathrm{ch}\, \nabla(\lambda) = \sum_{\lambda \in X} (V^{\otimes r} : \Delta(\lambda))\, \mathrm{ch}\, \Delta(\lambda).$$

Since $V^{\otimes r}$ is self-dual (under the transpose dual) we may choose the filtration (N_*) to be dual to the filtration (M_*). It follows that $s = t$ and $(V^{\otimes r} : \nabla(\lambda)) = (V^{\otimes r} : \Delta(\lambda))$ for all λ.

Now one applies (3.8) and a double induction through the filtrations. The argument is standard homological algebra, safely left at this point as an exercise for the reader. At the end one finds that

$$\dim_K \mathrm{End}_\Gamma(V^{\otimes r}) = \sum_{\lambda \in \mathbb{Z}^n} (V^{\otimes r} : \nabla(\lambda))^2$$

where the number of nonzero terms in the sum is finite. The result follows. □

---

[7]The computation of the $\mathrm{ch}\, \Delta(\lambda)$ for $\mathsf{GL}_n(\mathbb{C})$ goes back to Schur's 1901 dissertation. Thus, these characters are sometimes called *Schur functions*. See [17] or [23, Chapter 7] for exhaustive accounts of their many properties.

Now we are ready to prove the second half of Schur–Weyl duality in positive characteristic. We remind the reader that we assume the validity of Theorem 1 in case $K = \mathbb{C}$.

PROPOSITION 15. *For any infinite field K, the image $\Phi(K\mathfrak{S}_r)$ of the representation Φ is equal to the centralizer algebra $\mathrm{End}_\Gamma(V^{\otimes r})$, so the map $\overline{\Phi}$ in (1.3) is surjective.*

PROOF. The argument is essentially the same as the proof of Proposition 5. By Corollary 12 and Proposition 14 we have equalities

$$\dim_K \Phi(K\mathfrak{S}_r) = \dim_\mathbb{C} \Phi(\mathbb{C}\mathfrak{S}_r),$$

$$\dim_K \mathrm{End}_{\mathsf{GL}_n(K)}((K^n)^{\otimes r}) = \dim_\mathbb{C} \mathrm{End}_{\mathsf{GL}_{(n}(\mathbb{C})}((\mathbb{C}^n)^{\otimes r})$$

for any infinite field K. Since $\Phi(\mathbb{C}\mathfrak{S}_r) = \mathrm{End}_{\mathsf{GL}_n(\mathbb{C})}((\mathbb{C}^n)^{\otimes r})$ it follows that

$$\dim_K \Phi(K\mathfrak{S}_r) = \dim_K \mathrm{End}_{\mathsf{GL}_n(K)}((K^n)^{\otimes r})$$

for any infinite field K, and thus by comparison of dimensions the second inclusion in (1.2) must be an equality. Equivalently, the map $\overline{\Phi}$ in (1.3) is surjective.   □

By putting together Propositions 5 and 15 we have now established Theorem 1 in positive characteristic, assuming its validity for $K = \mathbb{C}$.

REMARK 16. (a) Let K be an arbitrary infinite field. Lemma 3 gives the equality $\dim_K(K\Gamma) = \binom{n^2+r-1}{r}$, and the proof of Lemma 4 in light of [**13**, Theorem 13.19] gives the equality $\dim_K \mathrm{End}_{\mathfrak{S}_r}(V^{\otimes r}) = \sum_{\lambda,\mu \in \Lambda(n,r)} N(\lambda^+, \mu^+)$, where $N(\lambda^+, \mu^+)$ counts the number of "semistandard" tableaux of shape λ^+ and weight μ^+. Corollary 12 says that $\dim_K \Phi(K\mathfrak{S}_r) = r! - \sum_{\lambda \in P} N(\lambda)^2$, where $N(\lambda)$ is the number of standard tableaux of shape λ, and the proof of Proposition 14 shows that $\dim_K \mathrm{End}_\Gamma(V^{\otimes r}) = \sum_{\lambda \in \Lambda^+(n,r)} (V^{\otimes r} : \nabla(\lambda))^2$. Thus, in order to obtain a proof of Theorem 1 in full generality (without assuming its validity for $K = \mathbb{C}$) from the methods of this paper, one only needs to demonstrate the combinatorial identities

$$(3.11) \qquad \binom{n^2+r-1}{r} = \sum_{\lambda,\mu \in \Lambda(n,r)} N(\lambda^+, \mu^+);$$

$$(3.12) \qquad r! - \sum_{\lambda \in P} N(\lambda)^2 = \sum_{\lambda \in \Lambda^+(n,r)} (V^{\otimes r} : \nabla(\lambda))^2.$$

The author has not attempted to construct a combinatorial proof of these identities. If one assumes the validity of Theorem 1 in the case $K = \mathbb{C}$, then these identities follow from the results in this paper. Alternatively, if one can find an independent proof of the identities, then one would have a new proof of Theorem 1 in full generality, including the case $K = \mathbb{C}$.

(b) There is a variant of Theorem 1 worth noting. One may twist the action of $\mathfrak{S}_r$ on $V^{\otimes r}$ by letting $w \in \mathfrak{S}_r$ act as $(\mathrm{sgn}\, w)w$ (so $\mathfrak{S}_r$ acts by "signed" place permutations). This action also commutes with the action of $\Gamma = \mathsf{GL}(V)$, and Theorem 1 holds for this action as well. This may be proved the same way. In the course of carrying out the argument, one needs to replace permutation modules by "signed" permutation modules, and interchange the role of Murphy's two bases $\{x_{ST}\}, \{y_{ST}\}$.

(c) There is also a q-analogue of Theorem 1, in which one replaces $\mathsf{GL}_n(K)$ by the quantized enveloping algebra corresponding to the Lie algebra $\mathfrak{gl}_n$, and replaces $K\mathfrak{S}_r$ by the Iwahori–Hecke algebra $\mathbf{H}(\mathfrak{S}_r)$. The generic case ($q$ not a root of unity) of this theorem was first observed in Jimbo [15], and the root of unity case was treated in Du, Parshall, and Scott [8]. Alternatively, one may derive the result in the root of unity case from Jimbo's generic version, using arguments along the lines of those sketched here.

References

[1] R. Carter and G. Lusztig, On the modular representations of general linear and symmetric groups, *Math. Zeit.* **136** (1974), 193–242.

[2] C. Curtis and I. Reiner, *Representation theory of finite groups and associative algebras*, Interscience (Wiley) 1962; reprinted by AMS Chelsea Publishing, Providence, RI, 2006.

[3] C. de Concini and C. Procesi, A characteristic free approach to invariant theory, *Advances in Math.* **21** (1976), 330–354.

[4] R. Dipper and S. Doty, The rational Schur algebra, *Represent. Theory* **12** (2008), 58–82 (electronic).

[5] R. Dipper and G. James, Representations of Hecke algebras of general linear groups, *Proc. London Math. Soc.* (3) **52** (1986), 20–52.

[6] S. Donkin, *Rational Representations of Algebraic Groups: Tensor Products and Filtrations*, Lecture Notes in Math. **1140**, Springer-Verlag, Berlin 1985.

[7] S. Doty, K. Erdmann, and A. Henke, A generic algebra associated to certain Hecke algebras, *J. Algebra* **278** (2004), 502–531.

[8] J. Du, B. Parshall, and L. Scott, Quantum Weyl reciprocity and tilting modules, *Comm. Math. Phys.* **195** (1998), 321–352.

[9] J.J. Graham and G.I. Lehrer, Cellular algebras, *Invent. Math.* **123** (1996), 1–34.

[10] J.A. Green, *Polynomial Representations of* GL_n, (Lecture Notes in Math. **830**), Springer-Verlag, Berlin 1980; Second edition 2007.

[11] R. Goodman and N.R. Wallach, *Representations and invariants of the classical groups*, Encyclopedia of Mathematics and its Applications, **68**, Cambridge University Press, Cambridge, 1998.

[12] M. Härterich, Murphy bases of generalized Temperley–Lieb algebras, *Archiv Math.* **72** (1999), 337–345.

[13] G.D. James, *The Representation Theory of the Symmetric Groups*, (Lecture Notes in Math. **682**), Springer-Verlag, Berlin 1978.

[14] J.C. Jantzen, *Representations of algebraic groups*, Second edition. Mathematical Surveys and Monographs, 107, American Mathematical Society, Providence, RI, 2003.

[15] M. Jimbo, A q-analogue of $U(\mathfrak{gl}(N+1))$, Hecke algebras, and the Yang-Baxter equation, *Lett. Math. Phys.* **11** (1986), 247–252.

[16] S. König, I.H. Slungård, and C.C. Xi, Double centralizer properties, dominant dimension, and tilting modules, *J. Algebra* **240** (2001), 393–412.

[17] I.G. Macdonald, *Symmetric functions and Hall polynomials*, Second edition, Oxford Mathematical Monographs, Oxford Science Publications, The Clarendon Press, Oxford University Press, New York, 1995.

[18] O. Mathieu, Filtrations of G-modules, *Ann. Sci. École Norm. Sup.* (4) **23** (1990), 625–644.

[19] G.E. Murphy, On the representation theory of the symmetric groups and associated Hecke algebras, *J. Algebra* **152** (1992), 492–513.

[20] G.E. Murphy, The representations of Hecke algebras in type A_n, *J. Algebra* **173** (1995), 97–121.

[21] I. Schur, Über die rationalen Darstellungen der allgemeinen linearen Gruppe (1927); in I. Schur, Gesammelte Abhandlungen III, 65–85, Springer Berlin 1973.

[22] J.-P. Serre, *Linear representations of finite groups*, Graduate Texts in Mathematics, Vol. 42, Springer-Verlag, New York-Heidelberg, 1977.

[23] R.P. Stanley, *Enumerative combinatorics, Vol. 2*, Cambridge Studies in Advanced Mathematics, **62**, Cambridge University Press, Cambridge, 1999.

[24] J.P. Wang, Sheaf cohomology on G/B and tensor products of Weyl modules, *J. Algebra* **77** (1982), 162–185.

[25] H. Weyl, *The Classical Groups, Their Invariants and Representations*, 2nd ed., Princeton Univ. Press, 1946.

MATHEMATICS AND STATISTICS, LOYOLA UNIVERSITY CHICAGO, CHICAGO, ILLINOIS 60626 U.S.A.

E-mail address: sdoty@luc.edu

Contemporary Mathematics
Volume **478**, 2009

The centers of Iwahori-Hecke algebras are filtered

Andrew Francis and Weiqiang Wang

ABSTRACT. We show that the center of the Iwahori–Hecke algebra of the symmetric group S_n carries a natural filtered algebra structure, and that the structure constants of the associated graded algebra are independent of n. A series of conjectures and open problems are also included.

1. Introduction

1.1. The main results. The class elements introduced by Geck-Rouquier [**GR**] form a basis for the center $\mathcal{Z}(\mathcal{H}_n)$ of the Iwahori-Hecke algebra $\mathcal{H}_n$ of type A over the ring $\mathbb{Z}[\xi]$, where the indeterminate ξ is related to the familiar one q by $\xi = q - q^{-1}$. In this paper, we shall parameterize these class elements $\Gamma_\lambda(n)$ by partitions λ satisfying $|\lambda| + \ell(\lambda) \leq n$ which are the so-called modified cycle types, just as Macdonald [**Mac**, pp.131] does for the usual class sums of the symmetric group S_n. Write the multiplication in $\mathcal{Z}(\mathcal{H}_n)$ as

$$(1.1) \qquad \Gamma_\lambda(n)\Gamma_\mu(n) = \sum_\nu k_{\lambda\mu}^\nu(n)\, \Gamma_\nu(n).$$

The main result of this Note is the following theorem on these structure constants $k_{\lambda\mu}^\nu(n)$.

THEOREM 1.1.   (1) *For any n, $k_{\lambda\mu}^\nu(n)$ is a polynomial in ξ with non-negative integral coefficients. Moreover, $k_{\lambda\mu}^\nu(n)$ is an even (resp., odd) polynomial in ξ if and only if $|\lambda| + |\mu| - |\nu|$ is even (resp., odd).*
  (2) *We have $k_{\lambda\mu}^\nu(n) = 0$ unless $|\nu| \leq |\lambda| + |\mu|$.*
  (3) *If $|\nu| = |\lambda| + |\mu|$, then $k_{\lambda\mu}^\nu(n)$ is independent of n.*

It follows from (2) and (3) that the center $\mathcal{Z}(\mathcal{H}_n)$ is naturally a filtered algebra and the structure constants of the associated graded algebra are independent of n. We in addition formulate several conjectures, including Conjecture 3.1 which simply states that $k_{\lambda\mu}^\nu(n)$ are polynomials in n, and further implications on the algebra generators of $\mathcal{Z}(\mathcal{H}_n)$.

2000 *Mathematics Subject Classification.* Primary: 20C08; Secondary: 16W70.
W.W. is partially supported by an NSF grant.

1.2. Motivations and connections. Our motivation comes from the original work of Farahat-Higman [**FH**] on the structures of the centers of the *integral* symmetric group algebras $\mathbb{Z}S_n$. Indeed, Theorem 1.1 specializes at $\xi = 0$ to some classical results of *loc. cit.*. In addition, Conjecture 3.1 in the specialization at $\xi = 0$ (which is a theorem in *loc. cit.*) when combined with the specialization of Theorem 1.1 allowed them to define a universal algebra $\mathcal{K}$ governing the structures of the centers $\mathcal{Z}(\mathbb{Z}S_n)$ for all n simultaneously. Farahat-Higman further developed this approach to establish a distinguished set of generators for $\mathcal{Z}(\mathbb{Z}S_n)$, which is now identified as the first n elementary symmetric polynomials in the Jucys-Murphy elements. This has applications to blocks of modular representations of S_n.

This Note arose from the hope that the results of Farahat–Higman might be generalized to the Iwahori-Hecke algebra setup and in particular it would provide a new conceptual proof of the Dipper-James conjecture. Recently, built on the earlier work of Mathas [**Mat**], the first author and Graham [**FrG**] obtained a first complete combinatorial proof of the Dipper-James conjecture that the center of $\mathcal{H}_n$ is the set of symmetric polynomials in Jucys–Murphy elements of $\mathcal{H}_n$ (cf. [**DJ**] for an earlier proof of a weaker version). A basic difficulty in completing the Farahat-Higman approach for Iwahori-Hecke algebras is that no compact explicit expression is known for the Geck-Rouquier elements $\Gamma_\lambda(n)$ (see however [**Fra**] for a useful characterization). Our Theorem 1.1 is a first positive step along the new line.

In another direction, the results of Farahat-Higman have been partially generalized by the second author [**W**] to the centers of the group algebras of wreath products $G \wr S_n$ for an arbitrary finite group G (e.g. a cyclic group $\mathbb{Z}_r$), and these centers are closely related to the cohomology ring structures of Hilbert schemes of points on the minimal resolutions. It will be interesting to develop the Farahat-Higman type results for the centers of the Iwahori-Hecke algebra of type B or more generally of the cyclotomic Hecke algebras which are q-deformation of the group algebra $\mathbb{Z}(\mathbb{Z}_r \wr S_n)$.

1.3. This Note is organized as follows. In Section 2, we prove the three parts of our main Theorem 1.1 in Propositions 2.2, 2.4 and 2.6, respectively. In Section 3, we formulate several conjectures, which are Iwahori-Hecke algebra analogues of some results of Farahat-Higman. We conclude this Note with a list of open questions.

2. Proof of the main theorem

2.1. The preliminaries. Let S_n be the symmetric group in n letters generated by the simple transpositions $s_i = (i, i+1)$, $i = 1, \ldots, n-1$.

Let ξ be an indeterminate. The *Iwahori-Hecke algebra* $\mathcal{H}_n$ is the unital $\mathbb{Z}[\xi]$-algebra generated by T_i for $i = 1, \ldots, n-1$, satisfying the relations

$$
\begin{aligned}
T_i T_{i+1} T_i &= T_{i+1} T_i T_{i+1} \\
T_i T_j &= T_j T_i, \quad |i-j| > 1 \\
T_i^2 &= 1 + \xi T_i.
\end{aligned}
$$

(2.1)

The order relation (2.1) comes from the more familiar $(T_i - q)(T_i + q^{-1}) = 0$ via the identification $\xi = q - q^{-1}$. If $w = s_{i_1} \cdots s_{i_r} \in S_n$ is a reduced expression (where r will be referred to as the *length* of w in this case), then define $T_w := T_{i_1} \cdots T_{i_r}$. This definition is independent of the reduced expression for w. It is well known that the Iwahori-Hecke algebra is a free $\mathbb{Z}[\xi]$-module with basis $\{T_w \mid w \in S_n\}$, and it is a deformation of the integral group algebra $\mathbb{Z}S_n$.

The *Jucys-Murphy elements* L_i $(1 \leq i \leq n)$ of $\mathcal{H}_n$ are defined to be $L_1 = 0$ and

$$L_i = \sum_{1 \leq k < i} T_{(k,i)},$$

for $i \geq 2$.

Given $w \in S_n$ with cycle-type $\rho = (\rho_1, \ldots, \rho_t, 1, \ldots, 1)$ for $\rho_i > 1$, we define the *modified cycle-type* of w to be $\tilde{\rho} = (\rho_1 - 1, \ldots, \rho_t - 1)$, following Macdonald [**Mac**, pp.131]. Given a partition λ, let $C_\lambda(n)$ denote the conjugacy class of S_n containing all elements of modified type λ if $|\lambda| + \ell(\lambda) \leq n$. Accordingly, let $c_\lambda(n)$ denote the class sum of $C_\lambda(n)$ if $|\lambda| + \ell(\lambda) \leq n$, and denote $c_\lambda(n) = 0$ otherwise.

The center $\mathcal{Z}(\mathcal{H}_n)$ of the Iwahori-Hecke algebra is free over $\mathbb{Z}[\xi]$ of rank equal to the number of partitions of n; that is, it has a basis indexed by the conjugacy classes of S_n (see [**GR**]). In this paper, we shall parameterize these Geck-Rouquier class elements $\Gamma_\lambda(n)$ by the modified cycle types λ. The elements $\Gamma_\lambda(n)$ for $|\lambda| + \ell(\lambda) \leq n$ are characterized by the following two properties among the central elements of $\mathcal{H}_n$ [**Fra**]:

(i) The $\Gamma_\lambda(n)$ specializes at $\xi = 0$ to the class sum $c_\lambda(n)$;

(ii) The difference $\Gamma_\lambda(n) - \sum_{w \in C_\lambda(n)} T_w$ contains no minimal length elements of any conjugacy class.

In addition, we set $\Gamma_\lambda(n) = 0$ if $|\lambda| + \ell(\lambda) > n$.

2.2. The structure constants as positive integral polynomials.

By inspection of the defining relations, $\mathcal{H}_n$ as a $\mathbb{Z}$-algebra is $\mathbb{Z}_2$-graded by declaring that ξ and T_i $(1 \leq i \leq n-1)$ have $\mathbb{Z}_2$-degree (or parity) 1 and each integer has $\mathbb{Z}_2$-degree 0.

LEMMA 2.1. *Every $\Gamma_\lambda(n)$ is homogeneous in the above $\mathbb{Z}_2$-grading with $\mathbb{Z}_2$-degree equal to $|\lambda| \bmod 2$.*

PROOF. There is a constructive algorithm [**Fra**, pp.14] for producing the elements $\Gamma_\lambda(n)$. This finite algorithm begins with the sum of T_w with minimal length elements w from the conjugacy class $C_\lambda(n)$, then at each repeat of this algorithm, the only additions are of form (i) $T_w \to T_w + T_{s_i w s_i}$, (ii) $T_{s_i w} \to T_{w s_i} + T_{s_i w} + \xi T_{s_i w s_i}$, or (iii) $T_{w s_i} \to T_{w s_i} + T_{s_i w s_i} + \xi T_{s_i w s_i}$, and the algorithm eventually ends up with the element $\Gamma_\lambda(n)$. Each of the three type of additions clearly preserves the $\mathbb{Z}_2$-degree. As the minimal length elements have the same parity as $|\lambda|$, this proves the lemma. □

Denote by $\mathbb{N}$ the set of non-negative integers.

PROPOSITION 2.2. *For any given n, $k^\nu_{\lambda\mu}(n)$ is a polynomial in ξ with non-negative integral coefficients. Moreover, $k^\nu_{\lambda\mu}(n)$ is an even (respectively, odd) polynomial in ξ if and only if $|\lambda| + |\mu| - |\nu|$ is even (respectively, odd).*

PROOF. As is seen in the proof of Lemma 2.1, the class elements are in the positive cone $\mathcal{Z}(\mathcal{H}_n)^+ := \mathcal{Z}(\mathcal{H}_n) \cap \sum_{w \in S_n} \mathbb{N}[\xi] T_w$. Because of the positive coefficients in the order relation (2.1), the positive cone is closed under additions and products. Since the class elements contain minimal length elements from exactly one conjugacy class and contains those minimal length elements with coefficient 1

[**Fra**] (see Sect. 2.1 above), the coefficient of a class element in a central element C is precisely the coefficient of the corresponding minimal length element in the T_w expansion of C. This shows that $k^\nu_{\lambda\mu}(n) \in \mathbb{N}[\xi]$.

The more refined statement on when $k^\nu_{\lambda\mu}(n)$ is an even or odd polynomial in ξ follows now from Lemma 2.1. □

2.3. The filtered algebra structure on the center. Let $m_\mu(n)$ be the monomial symmetric polynomial in the (commutative) Jucys-Murphy elements $L_1, \ldots, L_n$, parameterized by a partition μ. It is known that $m_\mu(n) \in \mathcal{Z}(\mathcal{H}_n)$. Some relations between the class elements and the monomial symmetric polynomials in Jucys–Murphy elements are summarized as follows (see [**Mat, FrG, FrJ**]).

LEMMA 2.3.   (1) *For any partition λ, we can express $m_\lambda(n)$ in terms of the class elements $\Gamma_\mu(n)$ as*

$$m_\lambda(n) = \sum_{|\mu| \le |\lambda|} b_{\lambda\mu}(n)\,\Gamma_\mu(n) \quad \text{for } b_{\lambda\mu}(n) \in \mathbb{Z}[\xi].$$

(We set $b_{\lambda\mu}(n) = 0$ if $|\mu| + \ell(\mu) > n$, or equivalently if $\Gamma_\mu(n) = 0$.)
(2) *For any λ, the coefficients $b_{\lambda\mu}(n)$ with $|\mu| = |\lambda|$ are independent of n.*
(3) *Let λ be a partition with $|\lambda| + \ell(\lambda) \le n$. Then, each $\Gamma_\lambda(n)$ is equal to $m_\lambda(n)$ plus a $\mathbb{Z}[\xi]$-linear combination of $m_\mu(n)$ with $|\mu| < |\lambda|$.*

PROOF. Part (1) is a consequence of [**Mat**, Theorems 2.7 and 2.26].

By [**FrG**, Lemma 5.2], the coefficient of T_w (for an increasing $w \in S_n$ of the right length) in a so-called quasi-symmetric polynomial in Jucys–Murphy elements is independent of n. Monomial symmetric polynomials are just sums of the corresponding quasi-symmetric polynomials, independently of n. This proves (2).

Part (3) can be read off from the proof of [**FrJ**, Theorem 4.1]. □

PROPOSITION 2.4. *We have $k^\nu_{\lambda\mu}(n) = 0$ unless $|\nu| \le |\lambda| + |\mu|$.*

PROOF. By Lemma 2.3 (3), the product $\Gamma_\lambda(n)\Gamma_\mu(n)$ is equal to $m_\lambda(n)m_\mu(n)$ plus a linear combination of products $m_{\lambda'}(n)m_{\mu'}(n)$ satisfying $|\lambda'| + |\mu'| < |\lambda| + |\mu|$. A product of monomial symmetric polynomials $m_\alpha(n)m_\beta(n)$ is a sum of monomial symmetric polynomials $m_\gamma(n)$ satisfying $|\alpha| + |\beta| = |\gamma|$. Consequently, $\Gamma_\lambda(n)\Gamma_\mu(n)$ is a sum of $m_\gamma(n)$ with partitions γ of size at most $|\lambda| + |\mu|$. The proposition follows now by Lemma 2.3 (1). □

Assign degree $|\lambda|$ to the basis element $\Gamma_\lambda(n)$ and let $\mathcal{Z}(\mathcal{H}_n)_m$ be the $\mathbb{Z}[\xi]$-span of $\Gamma_\lambda(n)$ of degree at most m, for $m \ge 0$. Then, Proposition 2.4 provides a filtered algebra structure on the center $\mathcal{Z}(\mathcal{H}_n) = \cup_m \mathcal{Z}(\mathcal{H}_n)_m$.

REMARK 2.5. Denote by A_n the (commutative) $\mathbb{Z}[\xi]$-subalgebra of $\mathcal{H}_n$ generated by the Jucys-Murphy elements $L_1, \ldots, L_n$. The algebra A_n is filtered by the subspaces $A_n^{(m)}$ ($m \ge 0$) spanned by all products $L_{i_1} \cdots L_{i_m}$ of m Jucys-Murphy elements. The filtrations on A_n and $\mathcal{Z}(\mathcal{H}_n)$ are compatible with each other by the inclusion $\mathcal{Z}(\mathcal{H}_n) \subset A_n$. However the algebra $\mathcal{H}_n$ does not seem to admit a natural filtration which is compatible with the one on A_n by inclusion $A_n \subset \mathcal{H}_n$. This is very different from the symmetric group algebra $\mathbb{Z}S_n$, which admits such a filtration by assigning degree 1 to every transposition (i, j).

2.4. The graded algebra $\mathrm{gr}\mathcal{Z}(\mathcal{H}_n)$.

PROPOSITION 2.6. *If $|\nu| = |\lambda| + |\mu|$, then $k_{\lambda\mu}^{\nu}(n)$ is independent of n. (In this case, we shall write $k_{\lambda\mu}^{\nu}(n)$ as $k_{\lambda\mu}^{\nu}$.)*

PROOF. By the definition of $k_{\lambda\mu}^{\nu}(n)$, we can assume without loss of generality that $|\lambda| + \ell(\lambda) \le n$ and $|\mu| + \ell(\mu) \le n$.

By Lemma 2.3 (3), $\Gamma_\lambda(n)\Gamma_\mu(n) = m_\lambda(n)m_\mu(n) + X$, where X is a linear combination of products of monomials whose combined partition size is less than $|\lambda| + |\mu|$. The monomials appearing in X correspond to partitions of size less than $|\nu| = |\lambda| + |\mu|$, and thus will not contribute to $k_{\lambda\mu}^{\nu}(n)$ by Lemma 2.3 (1). The product $m_\lambda(n)m_\mu(n)$ is a sum of monomials $m_\alpha(n)$ satisfying $|\alpha| = |\lambda| + |\mu|$ with coefficients independent of n; the contribution of each such $m_\alpha(n)$ to $k_{\lambda\mu}^{\nu}(n)$ is independent of n by Lemma 2.3 (2). Summing all these contributions produces $k_{\lambda\mu}^{\nu}(n)$ which is independent of n. $\square$

Proposition 2.6 is equivalent to the statement that all the structure constants of the graded algebra $\mathrm{gr}\mathcal{Z}(\mathcal{H}_n)$ associated to the filtered algebra $\mathcal{Z}(\mathcal{H}_n)$ are independent of n.

2.5. Examples. We provide some explicit calculations of the structure constants $k_{\lambda\mu}^{\nu}(n)$ for the multiplication between $\Gamma_\lambda(n)$, with $n = 3, 4, 5$. For the sake of notational simplicity, we will write $\Gamma_\lambda(n)$ as Γ_λ with n dropped in the following examples. We also drop parentheses in the subscripts of class elements, denoting $\Gamma_{(\lambda_1,\ldots,\lambda_k)}$ by $\Gamma_{\lambda_1,\ldots,\lambda_k}$. The square brackets denote the top-degree parts of each product. The compatibility of these examples with Theorem 1.1 is manifest.

(1) Let $n = 3$. In $\mathcal{Z}(\mathcal{H}_3)$, we have
$$\Gamma_1\Gamma_1 = \left[(\xi^2 + 3)\Gamma_2\right] + 2\xi\Gamma_1 + 3\Gamma_\emptyset.$$

(2) Let $n = 4$. In $\mathcal{Z}(\mathcal{H}_4)$, we have
$$\Gamma_1\Gamma_1 = \left[(\xi^2 + 3)\Gamma_2 + (\xi^2 + 2)\Gamma_{1,1}\right] + 3\xi\Gamma_1 + 6\Gamma_\emptyset,$$
$$\Gamma_1\Gamma_2 = \left[(\xi^4 + 4\xi^2 + 4)\Gamma_3\right] + (2\xi^3 + 6\xi)\Gamma_2 + (2\xi^3 + 4\xi)\Gamma_{1,1}$$
$$+ (3\xi^2 + 4)\Gamma_1 + 4\xi\Gamma_\emptyset,$$
$$\Gamma_1\Gamma_{1,1} = \left[(\xi^2 + 2)\Gamma_3\right] + 2\xi\Gamma_2 + \xi\Gamma_{1,1} + \Gamma_1.$$

(3) Let $n = 5$. In $\mathcal{Z}(\mathcal{H}_5)$, we have
$$\Gamma_1\Gamma_1 = \left[(\xi^2 + 3)\Gamma_2 + (\xi^2 + 2)\Gamma_{1,1}\right] + 4\xi\Gamma_1 + 10\Gamma_\emptyset$$
$$\Gamma_1\Gamma_2 = \left[(\xi^4 + 4\xi^2 + 4)\Gamma_3 + (\xi^4 + 2\xi^2 + 1)\Gamma_{2,1}\right]$$
$$+ (3\xi^3 + 8\xi)\Gamma_2 + (3\xi^3 + 4\xi)\Gamma_{1,1} + (6\xi^2 + 6)\Gamma_1 + 10\xi\Gamma_\emptyset$$
$$\Gamma_1\Gamma_{1,1} = \left[(\xi^2 + 2)\Gamma_3 + (2\xi^2 + 3)\Gamma_{2,1}\right] + 2\xi\Gamma_2 + 4\xi\Gamma_{1,1} + 3\Gamma_1$$
$$\Gamma_1\Gamma_3 = \left[(\xi^6 + 6\xi^4 + 10\xi^2 + 5)\Gamma_4\right]$$
$$+ (2\xi^5 + 10\xi^3 + 13\xi)\Gamma_3 + (2\xi^5 + 8\xi^3 + 7\xi)\Gamma_{2,1}$$
$$+ (3\xi^4 + 10\xi^2 + 6)\Gamma_2 + (3\xi^4 + 8\xi^2 + 4)\Gamma_{1,1} + (4\xi^3 + 6\xi)\Gamma_1 + 5\xi^2\Gamma_\emptyset$$
$$\Gamma_2\Gamma_2 = \left[(\xi^8 + 7\xi^6 + 16\xi^4 + 15\xi^2 + 5)\Gamma_4\right]$$
$$+ (2\xi^7 + 14\xi^5 + 29\xi^3 + 19\xi)\Gamma_3 + (2\xi^7 + 13\xi^5 + 22\xi^3 + 11\xi)\Gamma_{2,1}$$

$$+ (3\xi^6 + 20\xi^4 + 32\xi^2 + 7)\Gamma_2 + (3\xi^6 + 19\xi^4 + 26\xi^2 + 8)\Gamma_{1,1}$$
$$+ (4\xi^5 + 25\xi^3 + 27\xi)\Gamma_1 + (5\xi^4 + 30\xi^2 + 20)\Gamma_\emptyset$$
$$\Gamma_2\Gamma_{1,1} = \left[(\xi^6 + 6\xi^4 + 10\xi^2 + 5)\Gamma_4\right]$$
$$+ (2\xi^5 + 10\xi^3 + 11\xi)\Gamma_3 + (2\xi^5 + 9\xi^3 + 9\xi)\Gamma_{2,1}$$
$$+ (3\xi^4 + 11\xi^2 + 6)\Gamma_2 + (3\xi^4 + 9\xi^2 + 4)\Gamma_{1,1} + (4\xi^3 + 7\xi)\Gamma_1 + 5\xi^2\Gamma_\emptyset.$$

2.6. A universal graded algebra. Introduce a graded $\mathbb{Z}[\xi]$-algebra $\mathcal{G}$ with a basis given by the symbols Γ_λ, where λ runs over all partitions, and with multiplication given by

$$\Gamma_\lambda\Gamma_\mu = \sum_{|\nu|=|\lambda|+|\mu|} k_{\lambda\mu}^\nu \Gamma_\nu.$$

By Propositions 2.2 and 2.6, the structure constants $k_{\lambda\mu}^\nu$ are independent of n and actually lie in $\mathbb{N}[\xi^2]$. Furthermore, we have surjective homomorphisms $\mathcal{G} \to \operatorname{gr}\mathcal{Z}[\mathcal{H}_n]$ for all n, which send each Γ_λ to $\Gamma_\lambda(n)$. The following proposition is immediate.

PROPOSITION 2.7. *The $\mathbb{Z}[\xi]$-algebra $\mathcal{G}$ is commutative and associative.*

Below for the one-row partition (m), we shall write $\Gamma_{(m)}$ simply as Γ_m.

THEOREM 2.8. *The $\mathbb{Q}(\xi)$-algebra $\mathbb{Q}(\xi) \otimes_{\mathbb{Z}[\xi]} \mathcal{G}$ is a polynomial algebra with generators Γ_m, $m = 1, 2, \ldots$.*

PROOF. Given a partition $\lambda = (\lambda_1, \lambda_2, \ldots)$, the product is of the form

$$\Gamma_{\lambda_1}\Gamma_{\lambda_2} \cdots = \sum_\mu d_{\lambda\mu}(\xi)\, \Gamma_\mu$$

for $d_{\lambda\mu}(\xi) \in \mathbb{N}[\xi]$. As ξ goes to 0, Γ_μ goes to the class sum c_μ and $d_{\lambda\mu}(\xi)$ specifies to the structure constant $d_{\lambda\mu}$ as defined in [**Mac**, pp.132] which we recall:

$$c_{\lambda_1}c_{\lambda_2} \cdots = \sum_\mu d_{\lambda\mu}c_\mu.$$

It is known therein that the (integral) matrix $[d_{\lambda\mu}]$ for $|\lambda| = |\mu| = k$ with any k is triangular with respect to the dominance ordering of partitions and all its diagonal entries are nonzero, thus the matrix $[d_{\lambda\mu}]$ is invertible over $\mathbb{Q}$. This forces the matrix $[d_{\lambda\mu}(\xi)]$ invertible over the field $\mathbb{Q}(\xi)$. Thus each Γ_μ is generated by $\Gamma_1, \Gamma_2, \ldots$ over $\mathbb{Q}(\xi)$. By definition, the elements Γ_μ for all partitions μ are linearly independent. Thus the theorem follows by comparing the graded dimensions of the algebra $\mathbb{Q}(\xi) \otimes_{\mathbb{Z}[\xi]} \mathcal{G}$ and the polynomial algebra in Γ_m, $m = 1, 2, \cdots$. $\square$

It is not clear whether the matrix $[d_{\lambda\mu}(\xi)]$ remains triangular with respect to the dominance order when $\xi \neq 0$. A similar phenomenon with a negative answer appears with the matrix $\widetilde{M_3}$ in Mathas [**Mat**, p.310] ($\widetilde{M_k}$ gives the coefficients of class elements Γ_λ in monomial symmetric polynomials in Jucys-Murphy elements m_μ, for $\lambda, \mu \vdash k$).

3. Conjectures and discussions

3.1. Several conjectures. We expect the following conjecture to hold.

CONJECTURE 3.1. Given partitions λ, μ and ν, there exists a polynomial $f^\nu_{\lambda\mu}$ in one variable with coefficients in $\mathbb{Q}[\xi]$, such that $f^\nu_{\lambda\mu}(n) = k^\nu_{\lambda\mu}(n)$ for all n.

Recall $b_{\lambda\mu}(n)$ from Lemma 2.3 (1). Similarly, we conjecture that there exists polynomials $g_{\lambda\mu}(x)$ with coefficients in $\mathbb{Q}[\xi]$, such that $g_{\lambda\mu}(n) = b_{\lambda\mu}(n)$ for all n.

The specialization at $\xi = 0$ of Conjecture 3.1 is a result in [**FH**]. *Below we shall assume that Conjecture 3.1 holds.*

Set $\mathbb{B}$ to be the ring of polynomials in $\mathbb{Q}[x]$ which take integer values at integers. We can define a $\mathbb{B}[\xi]$-algebra $\mathcal{K}$ with a basis given by the symbols Γ_λ, where λ runs over all partitions, and the multiplication given by

$$\Gamma_\lambda \Gamma_\mu = \sum_\nu f^\nu_{\lambda\mu} \Gamma_\nu.$$

Since $f^\nu_{\lambda\mu} = 0$ unless $|\nu| \leq |\lambda| + |\mu|$, the algebra $\mathcal{K}$ is an algebra filtered by $\mathcal{K}_m$ ($m \geq 0$) which is the $\mathbb{B}[\xi]$-span of Γ_λ with $|\lambda| \leq m$. We have a natural surjective homomorphism of filtered algebras

$$p_n : \mathcal{K} \longrightarrow \mathcal{Z}(\mathcal{H}_n)$$

given by

$$p_n\left(\sum f_\lambda \Gamma_\lambda\right) = \sum_\lambda f_\lambda(n)\, \Gamma_\lambda(n).$$

The algebra $\mathcal{G}$ introduced earlier becomes the associated graded algebra for the filtered algebra $\mathcal{K}$ up to a base ring change, i.e., $\mathrm{gr}\mathcal{K} = \mathbb{B}[\xi] \otimes_{\mathbb{Z}[\xi]} \mathcal{G}$. For $r \geq 1$, set

$$E_r := \sum_{|\lambda|=r} \Gamma_\lambda.$$

CONJECTURE 3.2. The $\mathbb{B}[\xi]$-algebra $\mathcal{K}$ is generated by E_r, $r = 1, 2, \cdots$.

Conjecture 3.2 in the specialization with $\xi = 0$ is a main theorem of Farahat-Higman [**FH**].

Note (cf. e.g. [**FrG**]) that

$$E_r(n) := \sum_{|\lambda|=r} \Gamma_\lambda(n) \in \mathcal{Z}(\mathcal{H}_n)$$

can be interpreted as the r-th elementary symmetric function in the n Jucys–Murphy elements. If Conjecture 3.2 holds, then the surjectivity of the homomorphism $p_n : \mathcal{K} \to \mathcal{Z}(\mathcal{H}_n)$ implies that the center $\mathcal{Z}(\mathcal{H}_n)$ is generated by $E_r(n)$, $1 \leq r \leq n - 1$. That would provide a new and conceptual proof of the Dipper-James conjecture [**DJ**, **FrG**] along with additional results of independent interest.

3.2. Open questions and discussions. A fundamental difficulty in pursuing the approach of [**FH**] for the Iwahori-Hecke algebras is present in the following.

QUESTION 3.3. Find an explicit expression for the elements $\Gamma_\lambda(n)$ for all λ.

The more challenging Conjecture 3.2 is likely to follow from an affirmative answer to Question 3.4 below, as a similar calculation in the case of symmetric groups plays a key role in the original approach of [**FH**].

QUESTION 3.4. Calculate the structure constants $k^{\nu}_{\lambda\,(m)}$ with $|\nu| = |\lambda| + m$.

Recall the positivity and integrality from Theorem 1.1 (1).

QUESTION 3.5. Find a combinatorial or geometric interpretation of the positivity and integrality of the structure constants $k^{\nu}_{\lambda\mu}(n)$ as polynomials in ξ.

With the connections between results of Farahat-Higman and cohomology rings of Hilbert schemes of n points on the affine plane in mind (cf. [**W**]), we post the following.

QUESTION 3.6. Are there any connections between the center $\mathcal{Z}(\mathcal{H}_n)$ and the equivariant K-group of Hilbert schemes of n points on the affine plane?

Let W be an arbitrary finite Coxeter group. The group algebra $\mathbb{Z}W$ is naturally filtered by assigning degree 1 to each reflection (not just simple reflection) and degree r to any element $w \in W$ with a reduced expression in terms of reflections of minimal length r. This induces a filtered algebra structure on the center $\mathcal{Z}(\mathbb{Z}W)$ as elements of a conjugacy class have the same degree. In the cases of types A and B, this definition agrees with the notion of degree for general wreath products introduced in [**W**]. (In the case of symmetric groups, the degree of $c_{\lambda}(n)$ coincides with $|\lambda|$.) The Geck-Rouquier basis has been defined for centers of the integral Iwahori-Hecke algebras $\mathcal{H}_W$ associated to any such W [**GR**], and its characterization (as in Sect. 2.1) holds in this generality [**Fra**]. Note that the generalization of Theorem 1.1 (1) to all such $\mathcal{H}_W$ holds with the same proof. We ask for a generalization of Theorem 1.1 (2) as follows.

QUESTION 3.7. Let W be an arbitrary finite Coxeter group. If we apply the notion of degree above to the Geck-Rouquier elements in the center $\mathcal{Z}(\mathcal{H}_W)$, does it provide an algebra filtration on $\mathcal{Z}(\mathcal{H}_W)$?

We expect that the answer to Question 3.7, at least for Iwahori-Hecke algebras of type B, is positive. More generally, we ask the following.

QUESTION 3.8. Establish and characterize an appropriate basis of class elements for the centers of the integral cyclotomic Hecke algebras associated to the complex reflection groups $\mathbb{Z}_r \wr S_n$. Furthermore, generalize the results of this Note and [**W**] to the cyclotomic Hecke algebra setup. Are there any connections between these centers and equivariant K-groups of Hilbert schemes of points on the minimal resolution $\widetilde{\mathbb{C}^2/\mathbb{Z}_r}$?

It will be already nontrivial and of considerable interest to answer the question for the Iwahori-Hecke algebras of type B corresponding to $r = 2$.

Acknowledgment. The authors thank the organizers for the high level International Conference on Representation Theory in Lhasa, Tibet, China in July 2007, where this collaboration was initiated.

References

[DJ]   R. Dipper and G. James, Blocks and idempotents of Hecke algebras of general linear groups, *Proc. London Math. Soc. (3) (1)* **54** (1987), 57–82.

[FH]   H. Farahat and G. Higman, *The centres of symmetric group rings*, Proc. Roy. Soc. (A) **250** (1959), 212–221.

[Fra]   A. Francis, *The minimal basis for the centre of an Iwahori-Hecke algebra*, J. Algebra, **221** (1999), 1-28.

[FrJ]   A. Francis and L. Jones, *A new integral basis for the centre of the Hecke algebra of type A*, Preprint 2007, arXiv:0705.1581.

[FrG]   A. Francis and J. J. Graham, *Centres of Hecke algebras: the Dipper-James conjecture*, J. Algebra **306** (2006), 244–267.

[GR]   M. Geck and R. Rouquier, *Centers and simple modules for Iwahori-Hecke algebras*, In: M. Cabanes (Eds.), Finite reductive groups (Luminy, 1994), 251–272, Progr. Math., **141**, Birkhäuser Boston, 1997.

[Mac]   I. G. Macdonald, *Symmetric functions and Hall polynomials*, Second Ed., Clarendon Press, Oxford, 1995.

[Mat]   A. Mathas, *Murphy operators and the centre of Iwahori-Hecke algebras of type A*, J. Algebr. Combin. **9** (1999), 295–313.

[W]   W. Wang, *The Farahat-Higman ring of wreath products and Hilbert schemes*, Adv. in Math. **187** (2004), 417–446.

SCHOOL OF COMPUTING AND MATHEMATICS, UNIVERSITY OF WESTERN SYDNEY, NSW 1797, AUSTRALIA

E-mail address: `a.francis@uws.edu.au`

DEPARTMENT OF MATHEMATICS, UNIVERSITY OF VIRGINIA, CHARLOTTESVILLE, VA 22904, U.S.A.

E-mail address: `ww9c@virginia.edu`

Contemporary Mathematics
Volume **478**, 2009

On Kostant's Theorem for Lie Algebra Cohomology

UNIVERSITY OF GEORGIA VIGRE ALGEBRA GROUP[1]

1. Introduction

1.1. In 1961, Kostant proved a celebrated result which computes the ordinary Lie algebra cohomology for the nilradical of the Borel subalgebra of a complex simple Lie algebra $\mathfrak{g}$ with coefficients in a finite-dimensional simple $\mathfrak{g}$-module. Over the last forty years other proofs have been discovered. One such proof uses the properties of the Casimir operator on cohomology described by the Casselman-Osborne theorem (cf. [GW, §7.3] for details). Another proof uses the construction of BGG resolutions for simple finite-dimensional $\mathfrak{g}$-modules [Ro]. Recently, Polo and Tilouine [PT] constructed BGG resolutions over $\mathbb{Z}_{(p)}$ for finite-dimensional irreducible G-modules where G is a semisimple algebraic group with high weights in the bottom alcove as long as $p \geq h-1$ (h is the Coxeter number for the underlying root system). One can then use a base change argument to show that Kostant's theorem holds for these modules over algebraically closed fields of characteristic p when $p \geq h - 1$. It should be noted that Friedlander and Parshall had earlier obtained a slightly weaker formulation of this result (cf. [FP1, §2])

The aim of this paper is to investigate and compare the cohomology of the unipotent radical of parabolic subalgebras over $\mathbb{C}$ and $\overline{\mathbb{F}}_p$. We present a new proof of Kostant's theorem and Polo-Tilouine's extension in Sections 2–4. Our proof employs known linkage results in Category $\mathcal{O}_J$ and the graded G_1T category for the first Frobenius kernel G_1. There are several advantages to our approach. Our proofs of these cohomology formulas are self-contained and our approach is presented in a conceptual manner. This enables us to identify key issues in attempting to compute these cohomology groups for small primes.

In Section 5, we prove that when $p < h - 1$, there are always additional cohomology classes in $\mathrm{H}^{\bullet}(\mathfrak{u}, \overline{\mathbb{F}}_p)$ beyond those given by Kostant's formula. The proof of this result relies heavily on the modular representation theory of reductive algebraic groups. Furthermore, we exhibit natural classes that arise in $\mathrm{H}^{2p-1}(\mathfrak{u}, \overline{\mathbb{F}}_p)$ when $\Phi = A_{p+1}$ which do not arise over fields of characteristic zero. In Section 6, we examine several low rank examples of $\mathrm{H}^{\bullet}(\mathfrak{u}_J, \overline{\mathbb{F}}_p)$ which were generated using

---

[1]The members of the UGA VIGRE Algebra Group are Irfan Bagci, Brian D. Boe, Leonard Chastkofsky, Benjamin Connell, Bobbe J. Cooper, Mee Seong Im, Tyler Kelly, Jonathan R. Kujawa, Wenjing Li, Daniel K. Nakano, Kenyon J. Platt, Emilie Wiesner, Caroline B. Wright and Benjamin Wyser.

MAGMA. These examples suggest interesting phenomena which lead us to pose several open questions in Section 7.

1.2. Notation. The notation and conventions of this paper will follow those given in [Jan]. Let k be an algebraically closed field, and G a simple algebraic group defined over k with T a maximal torus of G. The root system associated to the pair (G, T) is denoted by Φ. Let Φ^+ be a set of positive roots and Φ^- be the corresponding set of negative roots. The set of simple roots determined by Φ^+ is $\Delta = \{\alpha_1, \ldots, \alpha_l\}$. We will use throughout this paper the ordering of simple roots given in [Hum1] following Bourbaki. Given a subalgebra $\mathfrak{a} \subset \mathfrak{g}$ which is a sum of root spaces, let $\Phi(\mathfrak{a})$ denote the corresponding set of roots. Let B be the Borel subgroup relative to (G, T) given by the set of negative roots and let U be the unipotent radical of B. More generally, if $J \subseteq \Delta$, let P_J be the parabolic subgroup relative to $-J$ and let U_J be the unipotent radical and L_J the Levi factor of P_J. Let Φ_J be the root subsystem in Φ generated by the simple roots in J, with positive subset $\Phi_J^+ = \Phi_J \cap \Phi^+$. Set $\mathfrak{g} = \operatorname{Lie} G$, $\mathfrak{b} = \operatorname{Lie} B$, $\mathfrak{u} = \operatorname{Lie} U$, $\mathfrak{p}_J = \operatorname{Lie} P_J$, $\mathfrak{l}_J = \operatorname{Lie} L_J$, and $\mathfrak{u}_J = \operatorname{Lie} U_J$.

Let $\mathbb{E}$ be the Euclidean space associated with Φ, and denote the inner product on $\mathbb{E}$ by $\langle \ , \ \rangle$. Let $\check{\alpha}$ be the coroot corresponding to $\alpha \in \Phi$. Set α_0 to be the highest short root. Let ρ be the half sum of positive roots. The Coxeter number associated to Φ is $h = \langle \rho, \check{\alpha}_0 \rangle + 1$.

Let $X := X(T)$ be the integral weight lattice spanned by the fundamental weights $\{\omega_1, \ldots, \omega_l\}$. Let M be a finite-dimensional T-module and $M = \oplus_{\lambda \in X} M_\lambda$ be its weight space decomposition. The character of M, denoted by $\operatorname{ch} M = \sum_{\lambda \in X} (\dim M_\lambda) e^\lambda \in \mathbb{Z}[X(T)]$. If M and N are T-modules such that $\dim M_\lambda \leq \dim N_\lambda$ for all λ then we say that $\operatorname{ch} M \leq \operatorname{ch} N$. The set X has a partial ordering defined as follows: $\lambda \geq \mu$ if and only if $\lambda - \mu \in \sum_{\alpha \in \Delta} \mathbb{Z}_{\geq 0} \, \alpha$. The set of dominant integral weights is denoted by $X^+ = X(T)_+$ and the set of p^r-restricted weights is $X_r = X_r(T)$. For $J \subseteq \Delta$, the set of J-dominant weights is

$$X_J^+ := \{\, \mu \in X \mid \langle \mu, \check{\alpha} \rangle \in \mathbb{Z}_{\geq 0} \text{ for all } \alpha \in \Phi_J^+ \,\}.$$

and denote the p-restricted J-weights by $(X_J)_1$. The bottom alcove $\overline{C}_{\mathbb{Z}}$ is defined as

$$\overline{C}_{\mathbb{Z}} := \{\lambda \in X \mid 0 \leq \langle \lambda + \rho, \check{\alpha}_0 \rangle \leq p\}.$$

Set $H^0(\lambda) = \operatorname{ind}_B^G \lambda$ where λ is the one-dimensional B-module obtained from the character $\lambda \in X^+$ by letting U act trivially. The Weyl group corresponding to Φ is W and acts on X via the dot action $w \cdot \lambda = w(\lambda + \rho) - \rho$ where $w \in W$, $\lambda \in X$.

2.   Cohomology and Composition Factors

2.1.   For this section, let $R = \mathbb{Z}$, $\mathbb{C}$ or $\overline{\mathbb{F}}_p$, and let $J \subseteq \Delta$. Then $\mathfrak{u}_J$ has a basis consisting of root vectors where the structure constants are in R. In order to construct such a basis one can take an appropriate subset of the Chevalley basis for $\mathfrak{g}$. The standard complex on $\Lambda^\bullet(\mathfrak{u}_J^*)$ has differentials which are R-linear maps and we will denote the cohomology of this complex by $\operatorname{H}^\bullet(\mathfrak{u}_J, R)$. Moreover, the torus T acts on the standard complex $\Lambda^\bullet(\mathfrak{u}_J^*)$. The differentials respect the T-action so it suffices to look at the smaller complexes $(\Lambda^\bullet(\mathfrak{u}_J^*))_\lambda$. The cohomology of this complex will be denoted by $\operatorname{H}^\bullet(\mathfrak{u}_J, R)_\lambda$. For each n, $(\Lambda^n(\mathfrak{u}_J^*))_\lambda$ is a free R-module of finite rank, so the cohomology $\operatorname{H}^n(\mathfrak{u}_J, R)_\lambda$ is a finitely generated R-module.

One can use the arguments given in Knapp [Kna, Theorem 6.10] to show that the cohomology groups when $R = \mathbb{C}$ or $\overline{\mathbb{F}}_p$ satisfy Poincaré Duality:

$$(2.1.1) \qquad \mathrm{H}^n(\mathfrak{u}_J, R) \cong \mathrm{H}^{N-n}(\mathfrak{u}_J, R)^* \otimes \Lambda^N(\mathfrak{u}_J^*)$$

as T-modules where $N = \dim \mathfrak{u}_J$. The Universal Coefficient Theorem (UCT) (cf. [R, Theorem 8.26]) can be used to relate the cohomology over $\mathbb{Z}$ to the cohomology over $\mathbb{C}$ and $\overline{\mathbb{F}}_p$. The $\mathbb{Z}$-module $\mathbb{C}$ is divisible, so from the UCT (cf. [R, Corollary 8.28]) we have

$$(2.1.2) \qquad \mathrm{H}^n(\mathfrak{u}_J, \mathbb{C})_\lambda \cong \mathrm{H}^n(\mathfrak{u}_J, \mathbb{Z})_\lambda \otimes_{\mathbb{Z}} \mathbb{C}.$$

On the other hand, when $k = \overline{\mathbb{F}}_p$, the UCT shows that

$$(2.1.3) \qquad \mathrm{H}^n(\mathfrak{u}_J, \overline{\mathbb{F}}_p)_\lambda \cong (\mathrm{H}^n(\mathfrak{u}_J, \mathbb{Z})_\lambda \otimes_{\mathbb{Z}} \overline{\mathbb{F}}_p) \oplus \mathrm{Ext}^1_{\mathbb{Z}}(\overline{\mathbb{F}}_p, \mathrm{H}^{n-1}(\mathfrak{u}_J, \mathbb{Z})_\lambda).$$

For every n, the formulas (2.1.2) and (2.1.3) demonstrate that

$$\dim \mathrm{H}^n(\mathfrak{u}_J, \mathbb{C})_\lambda \leq \dim \mathrm{H}^n(\mathfrak{u}_J, \overline{\mathbb{F}}_p)_\lambda.$$

In particular, ch $\mathrm{H}^n(\mathfrak{u}_J, \mathbb{C}) \leq$ ch $\mathrm{H}^n(\mathfrak{u}_J, \overline{\mathbb{F}}_p)$. One should observe that additional cohomology classes in $\mathrm{H}^n(\mathfrak{u}_J, \overline{\mathbb{F}}_p)_\lambda$ can arise from either the first or second summand in (2.1.3) because of p-torsion in $\mathrm{H}^\bullet(\mathfrak{u}_J, \mathbb{Z})_\lambda$.

For a $\mathfrak{u}_J$-module, one can define $\mathrm{H}^n(\mathfrak{u}_J, M)$ using a complex involving $\Lambda^\bullet(\mathfrak{u}_J^*) \otimes M$ [Jan, I 9.17]. If M is a $\mathfrak{p}_J$-module then $\mathrm{H}^n(\mathfrak{u}_J, M)$ is a $\mathfrak{l}_J$-module. If M, N are arbitrary $\mathfrak{u}_J$-modules then $\mathrm{Ext}^n_{\mathfrak{u}}(M, N) = \mathrm{H}^n(\mathfrak{u}_J, M^* \otimes N)$ for $n \geq 0$.

2.2. Category $\mathcal{O}_J$. For this section, $k = \mathbb{C}$. Fix $J \subseteq \Delta$. Denote the Weyl group of Φ_J by W_J, viewed as a subgroup of W. Let $\mathcal{U}(\mathfrak{g})$ denote the universal enveloping algebra of $\mathfrak{g}$.

DEFINITION 2.2.1. Let $\mathcal{O}_J$ be the full subcategory of the category of $\mathcal{U}(\mathfrak{g})$-modules consisting of modules V which satisfy the following conditions:

(i) The module V is a finitely generated $\mathcal{U}(\mathfrak{g})$-module.
(ii) As a $\mathcal{U}(\mathfrak{l}_J)$-module, V is the direct sum of finite-dimensional $\mathcal{U}(\mathfrak{l}_J)$-modules.
(iii) If $v \in V$, then $\dim_{\mathbb{C}} \mathcal{U}(\mathfrak{u}_J)v < \infty$.

Let Z be the center of $\mathcal{U}(\mathfrak{g})$ and denote the set of algebra homomorphisms $Z \to \mathbb{C}$ by $Z^\sharp$. We say that $\chi \in Z^\sharp$ is a *central character* of $V \in \mathcal{O}_J$ if $zv = \chi(z)v$ for all $z \in Z$ and all $v \in V$. For each $\chi \in Z^\sharp$, let $\mathcal{O}_J^\chi$ be the full subcategory of $\mathcal{O}_J$ consisting of modules $V \in \mathcal{O}_J$ such that for all $z \in Z$ and $v \in V$, v is annihilated by some power of $z - \chi(z)$. We have the decomposition

$$\mathcal{O}_J = \bigoplus_{\chi \in Z^\sharp} \mathcal{O}_J^\chi.$$

We call $\mathcal{O}_J^\chi$ an *infinitesimal block* of category $\mathcal{O}_J$.

For the purpose of this paper we will only need to apply information about the integral blocks so we can assume that the weights which arise are in X. The key objects in integral blocks of $\mathcal{O}_J$ are the parabolic Verma modules, which are defined as follows. For a finite-dimensional irreducible $\mathfrak{l}_J$-module $L_J(\mu)$ with highest weight $\mu \in X_J^+$ extend $L_J(\mu)$ to a $\mathfrak{p}_J$-module by letting $\mathfrak{u}_J^+$ act trivially. The induced module

$$Z_J(\mu) = \mathcal{U}(\mathfrak{g}) \otimes_{\mathcal{U}(\mathfrak{p}_J)} L_J(\mu)$$

is a *parabolic Verma module*, which we will abbreviate as PVM.

The module $Z_J(\mu)$ has a unique maximal submodule and hence a unique simple quotient module, which we denote by $L(\mu)$; $L(\mu)$ is also the unique simple quotient of the ordinary Verma module $Z(\mu) := U(\mathfrak{g}) \otimes_{U(\mathfrak{b})} \mu$. All simple modules in the integral blocks of $\mathcal{O}_J$ are isomorphic to some $L(\mu)$. For each $\mu \in X$, the ordinary Verma module $Z(\mu)$ (and any quotient thereof, such as $Z_J(\mu)$ or $L(\mu)$ if $\mu \in X_J^+$) has a central character which we will denote by $\chi_\mu \in Z^\sharp$. If $\chi = \chi_\mu$, write $\mathcal{O}_J^\mu := \mathcal{O}_J^{\chi_\mu}$. The Harish-Chandra linkage principle yields

$$\chi_\mu = \chi_\nu \quad \Leftrightarrow \quad \nu \in W \cdot \mu.$$

This implies that the simple modules (and hence the PVM's and projective indecomposable modules) in $\mathcal{O}_J^\mu$ can be indexed by $\{w \in W \mid w \cdot \mu \in X_J^+\}$ (by identifying repetitions).

For $\mu \in X$, let

$$\Phi_\mu = \{\alpha \in \Phi \mid \langle \mu + \rho, \check{\alpha} \rangle = 0\}.$$

If $\Phi_\mu = \varnothing$, then we say that μ is a *regular weight*; otherwise, it is a *singular weight*. If μ and ν are both regular weights, then $\mathcal{O}_J^\mu$ is equivalent to $\mathcal{O}_J^\nu$ by the Jantzen-Zuckerman translation principle.

For each $\alpha \in \Phi$, let $s_\alpha \in W$ denote the reflection in $\mathbb{E}$ about the hyperplane orthogonal to α. If μ is a regular dominant weight, then $\{w \in W \mid w \cdot \mu \in X_J^+\}$ is the set

(2.2.1)
$$^J W = \{w \in W \mid l(s_\alpha w) = l(w) + 1 \text{ for all } \alpha \in J\} = \{w \in W \mid w^{-1}(\Phi_J^+) \subseteq \Phi^+\}$$

which is the set of minimal length right coset representatives of W_J in W. Let w_0 (resp. w_J, $^J w$) denote the longest element in W (resp. W_J, $^J W$). Then $w_0 = w_J \, ^J w$.

2.3.   The following theorem provides information about the L_J composition factors in $\mathrm{H}^\bullet(\mathfrak{u}_J, L(\mu))$ when $k = \mathbb{C}$. For V a finite dimensional semisimple L_J-module, write $[V : L_J(\sigma)]_{L_J}$ for the multiplicity of $L_J(\sigma)$ as an L_J-composition factor of V.

THEOREM 2.3.1. *Let $k = \mathbb{C}$, $V \in \mathcal{O}_J$ and $\lambda \in X$.*

(a) $\mathrm{Ext}^i_{\mathcal{O}_J}(Z_J(\lambda), V) \cong \mathrm{Hom}_{\mathfrak{l}_J}(L_J(\lambda), \mathrm{H}^i(\mathfrak{u}_J, V))$

(b) *If $[\mathrm{H}^i(\mathfrak{u}_J, L(\mu)) : L_J(\sigma)]_{L_J} \neq 0$ where $\mu \in X_+$ then $\sigma = w \cdot \mu$ where $w \in {}^J W$.*

PROOF. (a) First observe that $\mathrm{Ext}^i_{\mathcal{O}_J}(Z_J(\lambda), V) \cong \mathrm{Ext}^i_{(\mathfrak{g}, \mathfrak{l}_J)}(Z_J(\lambda), V)$ (relative Lie algebra cohomology) and by Frobenius reciprocity we have

$$\mathrm{Ext}^i_{(\mathfrak{g}, \mathfrak{l}_J)}(Z_J(\lambda), V) \cong \mathrm{Ext}^i_{(\mathfrak{p}_J, \mathfrak{l}_J)}(L_J(\lambda), V) \cong \mathrm{H}^i(\mathfrak{p}_J, \mathfrak{l}_J; L_J(\lambda)^* \otimes V).$$

Since $\mathfrak{u}_J \trianglelefteq \mathfrak{p}_J$, one can use the Grothendieck spectral sequence construction given in [Jan, I Proposition 4.1] to obtain a spectral sequence,

$$E_2^{i,j} = \mathrm{H}^i(\mathfrak{p}_J/\mathfrak{u}_J, \mathfrak{l}_J/(\mathfrak{l}_J \cap \mathfrak{u}_J)); \mathrm{H}^j(\mathfrak{u}_J, 0; L_J(\lambda)^* \otimes V) \Rightarrow \mathrm{H}^{i+j}(\mathfrak{p}_J, \mathfrak{l}_J; L_J(\lambda)^* \otimes V).$$

However, $E_2^{i,j} \cong \mathrm{H}^i(\mathfrak{l}_J, \mathfrak{l}_J; \mathrm{H}^j(\mathfrak{u}_J, 0; L_J(\lambda)^* \otimes V)) = 0$ for $i > 0$, so the spectral sequence collapses and yields

$$\mathrm{Hom}_{\mathfrak{l}_J}(L_J(\lambda), \mathrm{H}^j(\mathfrak{u}_J, V)) \cong \mathrm{H}^0(\mathfrak{l}_J, \mathfrak{l}_J; \mathrm{H}^j(\mathfrak{u}_J, L_J(\lambda)^* \otimes V)) \cong \mathrm{H}^j(\mathfrak{p}_J, \mathfrak{l}_J; L_J(\lambda)^* \otimes V).$$

(b) Suppose that $[\mathrm{H}^i(\mathfrak{u}_J, L(\mu)) : L_J(\sigma)]_{L_J} \neq 0$. Then from part (a),

$$[\mathrm{H}^i(\mathfrak{u}_J, L(\mu)) : L_J(\sigma)]_{L_J} = \dim \mathrm{Hom}_{\mathfrak{l}_J}(L_J(\sigma), \mathrm{H}^i(\mathfrak{u}_J, L(\mu)))$$

$$= \dim \mathrm{Ext}^i_{\mathcal{O}_J}(Z_J(\sigma), L(\mu)).$$

But, $\mathrm{Ext}^i_{\mathcal{O}_J}(Z_J(\sigma), L(\mu)) \neq 0$ implies by linkage that $\sigma = w \cdot \mu$ where $w \in {}^J W$. $\quad\square$

2.4. Now let us assume that $k = \overline{\mathbb{F}}_p$. Let W_p be the affine Weyl group and $\widehat{W}_p$ be the extended affine Weyl group. In this setting we regard G as an affine reductive group scheme with $F: G \to G$ denoting the Frobenius morphism. Let F^r be this morphism composed with itself r times and set $G_r T = (F_r)^{-1}(T)$. The category of $G_r T$-modules has a well developed representation theory (cf. [Jan, II Chapter 9]). Group schemes analogous to $G_r T$ can be defined similarly using the Frobenius morphism for L_J, P_J, B, U, etc.

The following theorem provides information about the composition factors in the $\mathfrak{u}_J$-cohomology for $p \geq 3$.

THEOREM 2.4.1. *Let $k = \mathbb{F}_p$ with $p \geq 3$.*
  (a) *If $[\mathrm{H}^i(\mathfrak{u}_J, L(\mu)) : L_J(\sigma)]_{L_J} \neq 0$ where $\mu \in X^+$ then $\mu = w \cdot \sigma$ where $w \in \widehat{W}_p$.*
  (b) *If $[\mathrm{H}^i(\mathfrak{u}_J, L(\mu)) : L_J(\sigma)]_{L_J} \neq 0$ where $\mu \in X_1$ and $\sigma \in (X_J)_1$ then $\mu = w \cdot \sigma$ where $w \in W_p$.*

PROOF. (a) Suppose that $[\mathrm{H}^i(\mathfrak{u}_J, L(\mu)) : L_J(\sigma)]_{L_J} \neq 0$. From the Steinberg tensor product theorem, we can write $L_J(\sigma) = L_J(\sigma_0) \otimes L_J(\sigma_1)^{(1)}$ where $\sigma_0 \in (X_J)_1$ and $\sigma_1 \in X_J^+$. Therefore, $[\mathrm{H}^i(\mathfrak{u}_J, L(\mu)) : L_J(\sigma_0) \otimes p\gamma_1]_{(L_J)_1 T} \neq 0$ for some $\gamma_1 \in X$. One can also express $\mu = \mu_0 + p\mu_1$ where $\mu_0 \in X_1$ and $\mu_1 \in X^+$ so that

$$\mathrm{H}^i(\mathfrak{u}_J, L(\mu)) \cong \mathrm{H}^i(\mathfrak{u}_J, L(\mu_0)) \otimes L(\mu_1)^{(1)}.$$

Therefore, $[\mathrm{H}^i(\mathfrak{u}_J, L(\mu)) : L_J(\sigma_0) \otimes p\gamma_1]_{(L_J)_1 T} \neq 0$ implies that $[\mathrm{H}^i(\mathfrak{u}_J, L(\mu_0)) \otimes p\gamma_2 : L_J(\sigma_0) \otimes p\gamma_1]_{(L_J)_1 T} \neq 0$ for some $\gamma_2 \in X$, thus $[\mathrm{H}^i(\mathfrak{u}_J, L(\mu_0)) : L_J(\sigma_0) \otimes p\gamma]_{(L_J)_1 T} \neq 0$ for some $\gamma \in X$ (where $\gamma = \gamma_1 - \gamma_2$).

Observe that
(2.4.1)
$$[\mathrm{H}^i(\mathfrak{u}_J, L(\mu_0)) : L_J(\sigma_0) \otimes p\gamma]_{(L_J)_1 T} = \dim \mathrm{Hom}_{(L_J)_1 T}(P_J(\sigma_0) \otimes p\gamma, \mathrm{H}^i(\mathfrak{u}_J, L(\mu_0))).$$

where $P_J(\sigma_0) \otimes p\gamma$ is the $(L_J)_1 T$ projective cover of $L_J(\sigma_0) \otimes p\gamma$.

Next consider the composition factor multiplicities for the cohomology of $L(\mu_0)$ over the Frobenius kernel $(U_J)_1$,

$$[\mathrm{H}^i((U_J)_1, L(\mu_0)) : L_J(\sigma_0) \otimes p\gamma]_{(L_J)_1 T}$$
$$= \dim \mathrm{Hom}_{(L_J)_1 T}(P_J(\sigma_0) \otimes p\gamma, \mathrm{H}^i((U_J)_1, L(\mu_0))).$$

We can also give another interpretation of this composition factor multiplicity. First, let us apply the Lyndon-Hochschild-Serre spectral sequence for $(U_J)_1 \trianglelefteq (P_J)_1 T$, $(P_J)_1 T / (U_J)_1 \cong (L_J)_1 T$:
(2.4.2)
$$E_2^{i,j} = \mathrm{Ext}^i_{(L_J)_1 T}(P_J(\sigma_0) \otimes p\gamma, \mathrm{H}^j((U_J)_1, L(\mu_0))) \Rightarrow \mathrm{Ext}^{i+j}_{(P_J)_1 T}(P_J(\sigma_0) \otimes p\gamma, L(\mu_0)).$$

Since $P := P_J(\sigma_0) \otimes p\gamma$ is projective as an $(L_J)_1 T$-module, the spectral sequence collapses and we have

$$\mathrm{Hom}_{(L_J)_1 T}(P, \mathrm{H}^i((U_J)_1, L(\mu_0)) \cong \mathrm{Ext}^i_{(P_J)_1 T}(P, L(\mu_0))$$
$$\cong \mathrm{Ext}^i_{G_1 T}(\mathrm{coind}^{G_1 T}_{(P_J)_1 T} P, L(\mu_0)).$$

For $p \geq 3$, there exists another first quadrant spectral sequence which can be used to relate these two different composition factor multiplicities [FP2, (1.3) Proposition]:

$$E_2^{2i,j} = S^i(\mathfrak{u}_J^*)^{(1)} \otimes \mathrm{H}^j(\mathfrak{u}_J, L(\mu_0)) \Rightarrow \mathrm{H}^{2i+j}((U_J)_1, L(\mu_0)).$$

Since the functor $\mathrm{Hom}_{(L_J)_1 T}(P, -)$ is exact, we can compose it with the spectral sequence above to get another spectral sequence:

$$(2.4.3) \quad E_2^{2i,j} = S^i(\mathfrak{u}_J^*)^{(1)} \otimes \mathrm{Hom}_{(L_J)_1 T}(P, \mathrm{H}^j(\mathfrak{u}_J, L(\mu_0)))$$

$$\Rightarrow \mathrm{Hom}_{(L_J)_1 T}(P, \mathrm{H}^{2i+j}((U_J)_1, L(\mu_0))).$$

Suppose that $\sigma_0 + p\gamma \notin W_p \cdot \mu_0$. Then by the linkage principle for $G_1 T$:

$$\mathrm{Hom}_{(L_J)_1 T}(P_J(\sigma_0) \otimes p\gamma, \mathrm{H}^i((U_J)_1, L(\mu_0)))$$

$$\cong \mathrm{Ext}^i_{G_1 T}(\mathrm{coind}_{(P_J)_1 T}^{G_1 T} P_J(\sigma_0) \otimes p\gamma, L(\mu_0)) = 0$$

for all $i \geq 0$. Therefore, the spectral sequence (2.4.3) abuts to zero. The differential d_2 in the spectral sequence maps $E_2^{0,j}$ to $E_2^{2,j-1}$. Note that $E_2^{2i,j} = S^i(\mathfrak{u}_J^*)^{(1)} \otimes E_2^{0,j}$ for all $i, j \geq 0$. Since $0 = E_0 = E_2^{0,0}$, it follows that $E_2^{2i,0} = 0$ for $i \geq 0$. Therefore, $E_2^{0,1} = 0$, thus $E_2^{2i,1} = 0$ for $i \geq 0$. Continuing in this fashion, we have $E_2^{2i,j} = 0$ for all i, j. In particular, using (2.4.1) and (2.4.3), $[\mathrm{H}^j(\mathfrak{u}_J, L(\mu_0)) : L_J(\sigma_0) \otimes p\gamma]_{(L_J)_1 T} = \dim \mathrm{Hom}_{(L_J)_1 T}(P, \mathrm{H}^j(\mathfrak{u}_J, L(\mu_0))) = \dim E_2^{0,j} = 0$ for all j which is a contradiction. This implies that μ_0 and σ_0 are in the same orbit under $\widehat{W}_p$, thus $\mu = w \cdot \sigma$ where $w \in \widehat{W}_p$.

(b) Under the hypotheses, we can apply the above argument with $0 = \gamma_1 = \gamma_2 = \gamma$. Therefore, $\mu = w \cdot \sigma$ where $w \in W_p$. $\square$

2.5.   We present the following proposition which allows one to compare composition factors of the cohomology with coefficients in a module to the cohomology with trivial coefficients. Note that this proposition is independent of the characteristic of the field k.

PROPOSITION 2.5.1. *Let $J \subseteq \Delta$ and V be a finite-dimensional P_J-module. If $[\mathrm{H}^i(\mathfrak{u}_J, V) : L_J(\sigma)]_{L_J} \neq 0$ for $\sigma \in X_J^+$ then $[\mathrm{H}^i(\mathfrak{u}_J, k) \otimes V : L_J(\sigma)]_{L_J} \neq 0$.*

PROOF. The simple finite-dimensional P_J-modules are the simple finite-dimensional L_J-modules inflated to P_J by making U_J act trivially. We will prove the proposition by induction on the composition length n of V. For $n = 1$, this is clear because V is simple and U_J acts trivially so

$$\mathrm{H}^i(\mathfrak{u}_J, V) \cong \mathrm{H}^i(\mathfrak{u}_J, k) \otimes V.$$

Now assume that the proposition holds for modules of composition length n, and let V have composition length $n + 1$. There exists a short exact sequence

$$0 \to V' \to V \to L \to 0$$

where V' has composition length n and L is a simple P_J-module. We have a long exact sequence in cohomology which shows that if $[\mathrm{H}^i(\mathfrak{u}_J, V) : L_J(\sigma)]_{L_J} \neq 0$ then either $[\mathrm{H}^i(\mathfrak{u}_J, V') : L_J(\sigma)]_{L_J} \neq 0$ or $[\mathrm{H}^i(\mathfrak{u}_J, L) : L_J(\sigma)]_{L_J} \neq 0$. By the induction hypothesis, this implies $[\mathrm{H}^i(\mathfrak{u}_J, k) \otimes V' : L_J(\sigma)]_{L_J} \neq 0$ or $[\mathrm{H}^i(\mathfrak{u}_J, k) \otimes L : L_J(\sigma)]_{L_J} \neq 0$.

The short exact sequence above can be tensored by $\mathrm{H}^i(\mathfrak{u}_J, k)$ to obtain a short exact sequence:

$$0 \to \mathrm{H}^i(\mathfrak{u}_J, k) \otimes V' \to \mathrm{H}^i(\mathfrak{u}_J, k) \otimes V \to \mathrm{H}^i(\mathfrak{u}_J, k) \otimes L \to 0.$$

The result now follows because one of the terms on the end has an L_J composition factor of the form $L_J(\sigma)$ by the induction hypothesis, so the middle term has to have a composition factor of this form. $\qquad\square$

3. Parabolic Computations

In this section we prove several elementary results which will be ingredients in our proof of Kostant's Theorem and its generalization to prime characteristic in Section 4.

3.1. Given $\Psi \subset \Phi^+$, write

$$\langle \Psi \rangle = \sum_{\beta \in \Psi} \beta.$$

For $w \in W$ put

(3.1.1) $$\Phi(w) = -(w\Phi^+ \cap \Phi^-) = w\Phi^- \cap \Phi^+ \subset \Phi^+.$$

We recall some basic facts about $\Phi(w)$.

LEMMA 3.1.1. *Let $w \in W$.*
  (a) $|\Phi(w)| = l(w)$.
  (b) $w \cdot 0 = -\langle \Phi(w) \rangle$.
  (c) *If $w = s_{j_1} \ldots s_{j_t}$ is a reduced expression, then*

$$\Phi(w^{-1}) = \{\alpha_{j_t}, s_{j_t}\alpha_{j_{t-1}}, s_{j_t}s_{j_{t-1}}\alpha_{j_{t-2}}, \ldots, s_{j_t}\ldots s_{j_2}\alpha_{j_1}\}.$$

PROOF. (a) [Hum1, Lemma 10.3A], (b) [Kna, Proposition 3.19], (c) [Hum2, Exercise 5.6.1] $\qquad\square$

LEMMA 3.1.2. *Let $J \subseteq \Delta$ and $w \in W$.*
  (a) $\Phi(w) \subset \Phi^+ \setminus \Phi_J^+ = \Phi(\mathfrak{u}_J)$ *if and only if $w \in {}^J W$.*
  (b) *If $w \cdot 0 = -\langle \Psi \rangle$ for some $\Psi \subset \Phi^+$ then $\Psi = \Phi(w)$.*

PROOF. (a) Assume $w \in {}^J W$. Let $\beta \in \Phi(w)$. Then $\beta \in \Phi^+$, and $\beta \in w\Phi^-$ whence $w^{-1}\beta \in \Phi^-$. Thus $\beta \notin \Phi_J^+$ by the second characterization of ${}^J W$ in (2.2.1).

Conversely, assume $w \notin {}^J W$. Then by the first characterization of ${}^J W$ in (2.2.1), w has a reduced expression beginning with s_α for some $\alpha \in J$ (by the Exchange Condition, for instance). Then by Lemma 3.1.1(c), $\alpha \in \Phi(w)$; but $\alpha \in \Phi_J^+$ so $\Phi(w) \not\subset \Phi^+ \setminus \Phi_J^+$.

(b) We prove this by induction on $l(w)$. If $l(w) = 0$ then $w = 1$ and $w \cdot 0 = 0$, so clearly the only possible Ψ is $\Psi = \varnothing = \Phi(1)$.

Given w with $l(w) > 0$, write $w = s_\alpha w'$ with $\alpha \in \Delta$ and $l(w') = l(w) - 1$. Then $\alpha \in \Phi(w)$ and $\alpha \notin \Phi(w') = s_\alpha(\Phi(w) \setminus \{\alpha\})$; cf. the proof of [Hum2, Lemma 1.6]. Suppose $w \cdot 0 = -(\gamma_1 + \cdots + \gamma_m)$ for distinct $\gamma_1, \ldots, \gamma_m \in \Phi^+$. Then

$$w' \cdot 0 = s_\alpha \cdot (w \cdot 0) = s_\alpha(w \cdot 0) + s_\alpha\rho - \rho = -(s_\alpha\gamma_1 + \cdots + s_\alpha\gamma_m + \alpha).$$

There are two cases.

Case 1: No $\gamma_i = \alpha$. Then $s_\alpha\gamma_1, \ldots, s_\alpha\gamma_m, \alpha$ are distinct positive roots: s_α permutes the positive roots other than α, and no $s_\alpha\gamma_i = \alpha$ because $s_\alpha(-\alpha) = \alpha$. But then by induction, $\{s_\alpha\gamma_1, \ldots, s_\alpha\gamma_m, \alpha\} = \Phi(w')$, and this contradicts $\alpha \notin \Phi(w')$.

Case 2: Some $\gamma_i = \alpha$. Say $\gamma_m = \alpha$. Then $s_\alpha(\gamma_m) = -\alpha$, so $w' \cdot 0 = -(s_\alpha \gamma_1 + \cdots + s_\alpha \gamma_{m-1})$. By induction, $\Phi(w') = \{s_\alpha \gamma_1, \ldots, s_\alpha \gamma_{m-1}\}$. Hence $\Phi(w) = s_\alpha \Phi(w') \cup \{\alpha\} = \{\gamma_1, \ldots, \gamma_m\}$ as required. $\qquad \square$

3.2. Saturation. Lemma 3.1.2 guarantees that, for $w \in {}^J W$, $w \cdot 0 = -\langle \Phi(w) \rangle$ is a weight in $\Lambda^n(\mathfrak{u}_J^*)$, where $n = l(w)$. Specifically, if $\Phi(w) = \{\beta_1, \ldots, \beta_n\}$ then the vector $f_{\Phi(w)} := f_{\beta_1} \wedge \cdots \wedge f_{\beta_n}$ has the desired weight, where $\{f_\beta \mid \beta \in \Phi(\mathfrak{u}_J)\}$ is the basis for $\mathfrak{u}_J^*$ dual to a fixed basis of weight vectors $\{x_\beta \mid \beta \in \Phi(\mathfrak{u}_J)\}$ for $\mathfrak{u}_J$. Lemmas 3.1.1 and 3.1.2 guarantee that the weight $w \cdot 0$ occurs with multiplicity one in $\Lambda^\bullet(\mathfrak{u}_J^*)$. In particular, since the differentials in the complex $0 \to \Lambda^\bullet(\mathfrak{u}_J^*)$ preserve weights, we see that $f_{\Phi(w)}$ descends to an element of $\mathrm{H}^n(\mathfrak{u}_J, k)$ of weight $w \cdot 0$, and n is the only degree in which this weight occurs in $\mathrm{H}^\bullet(\mathfrak{u}_J, k)$ (where $k = \mathbb{C}$ or $\mathbb{F}_p$).

In order to prove that $f_{\Phi(w)}$ generates an L_J-submodule of $\mathrm{H}^\bullet(\mathfrak{u}_J, k)$ of highest weight $w \cdot 0$, we need the following condition, which could be described by saying that $\Phi(w)$ is "saturated" with respect to Φ_J^+.

PROPOSITION 3.2.1. *Let $w \in {}^J W$. If $\beta \in \Phi(w)$, $\gamma \in \Phi_J^+$, and $\delta = \beta - \gamma \in \Phi$, then $\delta \in \Phi(w)$.*

PROOF. We prove this by induction on $l(w)$. If $w = 1$ then $\Phi(w) = \varnothing$ and the statement is vacuously true. So assume $l(w) > 0$. Write $w = w' s_\alpha$ with $\alpha \in \Delta$ and $l(w') = l(w) - 1$; then necessarily $w' \in {}^J W$. To see this, note that $w\alpha < 0$, so $(w')^{-1}(\Phi_J^+) = s_\alpha w^{-1}(\Phi_J^+) \subset s_\alpha(\Phi^+ \smallsetminus \{\alpha\}) \subset \Phi^+$. Now

$$\begin{aligned}
\Phi(w) &= \Phi^+ \cap w' s_\alpha \Phi^- \\
&= \Phi^+ \cap w'(\Phi^- \smallsetminus \{-\alpha\} \cup \{\alpha\}) \\
&= (\Phi^+ \cap w' \Phi^-) \cup \{w'\alpha\} \\
&= \Phi(w') \cup \{w'\alpha\},
\end{aligned}$$

where in the third equality we have used the fact that $w'\alpha > 0$. By induction, $\Phi(w')$ is saturated with respect to Φ_J^+. So it remains to check the condition of the lemma when $\beta = w'\alpha$.

Let $\beta = w'\alpha$ and suppose $\delta = \beta - \gamma \in \Phi$ for some $\gamma \in \Phi_J^+$. Since $\beta \in \Phi^+ \smallsetminus \Phi_J^+$ by Lemma 3.1.2, and $\gamma \in \Phi_J^+$, necessarily $\delta \in \Phi^+ \smallsetminus \Phi_J^+$. Consider the root $(w')^{-1}\delta = (w')^{-1}(w'\alpha - \gamma) = \alpha - (w')^{-1}(\gamma)$. Since $w' \in {}^J W$ and $\gamma \in \Phi_J^+$, we know $(w')^{-1}(\gamma) > 0$. Since α is simple, $(w')^{-1}\delta < 0$. That is, $\delta \in w'(\Phi^-)$. Thus, $\delta \in \Phi(w') \subset \Phi(w)$, as required. $\qquad \square$

3.3. Prime characteristic. In the prime characteristic setting we will need to work harder than in characteristic zero, because our control over the composition factors in cohomology in Theorem 2.4.1 is much weaker than in Theorem 2.3.1. We begin by recording two simple technical facts which will be needed later.

PROPOSITION 3.3.1.   (a) *Let $\lambda, \mu \in X$ and suppose $\lambda = w\mu$ where $w = s_{j_1} \ldots s_{j_t}$ with t minimal. Then $\langle \alpha_{j_r}, s_{j_{r+1}} \ldots s_{j_t} \mu \rangle \neq 0$ for $1 \leq r \leq t - 1$.*
      (b) *Suppose $\widetilde{\alpha} \in \Phi^+$ has maximal height in its W-orbit. Then $\langle \beta, \widetilde{\alpha} \rangle \geq 0$ for all $\beta \in \Phi^+$.*

PROOF. (a) Since t is minimal,

$$s_{j_{r+1}} \ldots s_{j_t} \mu \neq s_{j_r} \ldots s_{j_t} \mu = s_{j_{r+1}} \ldots s_{j_t} \mu - \langle s_{j_{r+1}} \ldots s_{j_t} \mu, \check{\alpha}_{j_r} \rangle \alpha_{j_r}.$$

This implies the desired inequality.

(b) Otherwise, $s_\beta(\tilde{\alpha}) = \tilde{\alpha} - \langle \tilde{\alpha}, \check{\beta} \rangle \beta$ would be a root of the same length as $\tilde{\alpha}$, but higher, contradicting the hypothesis. □

We will be able to cut down the possible weights in cohomology when $p \geq h - 1$. The proof will make use of certain special sums of positive roots. For $1 \leq i \leq l$ set

$$\Phi_i = \{ \alpha \in \Phi^+ \mid \langle \omega_i, \alpha \rangle > 0 \}$$

$$= \{ \alpha \in \Phi^+ \mid \alpha = \sum r_j \alpha_j \text{ with } r_i > 0 \}$$

(3.3.1)
$$= \Phi^+ \setminus \Phi_J^+, \text{ where}$$

$$J = J_i = \Delta \setminus \{ \alpha_i \},$$

$$\Phi_i' = \{ \alpha \in \Phi_i \mid \langle \alpha, \alpha_i \rangle \geq 0 \},$$

and define

(3.3.2)
$$\delta_i = \langle \Phi_i \rangle, \qquad \delta_i' = \langle \Phi_i' \rangle.$$

We begin by collecting some elementary properties of δ_i.

PROPOSITION 3.3.2.   (a) $w(\Phi_i) = \Phi_i$ for all $w \in W_J$.
(b) $\delta_i = c\omega_i$ for some $c \in \mathbb{Z}$.
(c) $-\delta_i = {}^J w \cdot 0$ where $J = \Delta \setminus \{ \alpha_i \}$ (recall ${}^J w$ is the longest element of ${}^J W$).

PROOF. (a) For $j \neq i$ and α a positive root involving α_i, $s_j(\alpha)$ is again a positive root involving α_i. Thus s_j permutes Φ_i. Since the s_j with $j \neq i$ generate W_J, the result follows.

(b) From (a), for $j \neq i$, $s_j(\delta_i) = \delta_i$. Thus when δ_i is written as a linear combination of fundamental dominant weights, the coefficient of ω_j is 0. That is, $\delta_i = c\omega_i$ for some scalar c. Since $\delta_i \in \mathbb{Z}\Phi$, $c \in \mathbb{Z}$.

(c) Write

$$2\rho = \sum_{\substack{\alpha \in \Phi^+ \\ \langle \omega_i, \alpha \rangle > 0}} \alpha + \sum_{\substack{\alpha \in \Phi^+ \\ \langle \omega_i, \alpha \rangle = 0}} \alpha = \delta_i + 2\rho_J.$$

Apply the longest element w_J of W_J, and use the first computation in (a):

$$2w_J\rho = w_J\delta_i - 2\rho_J = \delta_i - 2\rho_J.$$

Thus

$$w_J\rho = \tfrac{1}{2}\delta_i - \rho_J = \tfrac{1}{2}\delta_i - (\rho - \tfrac{1}{2}\delta_i) = \delta_i - \rho,$$

and so

$$-\delta_i = -w_J\rho - \rho = w_J w_0 \rho - \rho = {}^J w \cdot \rho.$$

□

3.4.   The crucial property of δ_i' is that $\langle \delta_i', \check{\alpha}_i \rangle \leq h$. The proof will require a few steps. First, put $J = \Delta \setminus \{ \alpha_i \}$ as before, and recall that w_J denotes the longest element of the parabolic subgroup W_J. Let $w_i \in W$ be an element of shortest possible length such that

(3.4.1)
$$w_i w_J \alpha_i = \tilde{\alpha}, \text{ the highest root in } W\alpha_i.$$

PROPOSITION 3.4.1. Let i, J, w_i and $\tilde{\alpha}$ be as above.
(a) $w_J(\Phi_i \setminus \Phi_i') = \Phi(w_i^{-1})$.
(b) $w_J(\delta_i - \delta_i') = \rho - w_i^{-1}\rho$.
(c) $\langle \delta_i', \check{\alpha}_i \rangle = 1 + \langle \rho, \tilde{\alpha}^\vee \rangle$.

(d) $\langle \delta_i', \check{\alpha}_i \rangle \le h$.

PROOF. (a) Observe that $\beta \in w_J(\Phi_i \setminus \Phi_i')$ if and only if $\beta = w_J\alpha$ with $\alpha \in \Phi_i$ and $\langle \alpha, \alpha_i \rangle < 0$; equivalently (using Proposition 3.3.2(a)), $\langle \beta, w_J\alpha_i \rangle < 0$ and $\beta \in \Phi_i$. Thus

$$(3.4.2) \qquad \beta \in w_J(\Phi_i \setminus \Phi_i') \iff \beta \in \Phi_i \text{ and } \langle w_i\beta, \tilde{\alpha} \rangle < 0.$$

Assuming $\beta \in w_J(\Phi_i \setminus \Phi_i')$, then $\beta \in \Phi^+$ and $w_i\beta \in \Phi^-$ (by Proposition 3.3.1(b)); equivalently $\beta \in \Phi(w_i^{-1})$ (by (3.1.1)).

To prove the reverse inclusion, assume that $\beta \in \Phi(w_i^{-1})$; i.e., $\beta \in \Phi^+$ and $w_i\beta \in \Phi^-$. We claim it is enough to show that $\langle w_i\beta, \tilde{\alpha} \rangle < 0$ (the second condition of (3.4.2)). For if $\beta \notin \Phi_i$ then $\beta \in \Phi_J^+$, hence $w_J\beta \in \Phi_J^-$, and thus $\langle w_i\beta, \tilde{\alpha} \rangle = \langle \beta, w_J\alpha_i \rangle = \langle w_J\beta, \alpha_i \rangle \ge 0$, since $\langle \alpha_j, \alpha_i \rangle \le 0$ for $j \ne i$.

It remains to show $\langle w_i\beta, \tilde{\alpha} \rangle < 0$, or, equivalently, $\langle w_i\beta, \tilde{\alpha} \rangle \ne 0$, since $w_i\beta \in \Phi^-$ (recall Proposition 3.3.1(b)). Write $w_i = s_{j_1} \dots s_{j_t}$ with t minimal. By Lemma 3.1.1(c) we have $\beta = s_{j_t} \dots s_{j_{r+1}}\alpha_{j_r}$ for some $1 \le r \le t$. Put $\mu = w_J\alpha_i$ and $\lambda = \tilde{\alpha}$ in Proposition 3.3.1(a) to obtain

$$\langle w_i\beta, \tilde{\alpha} \rangle = \langle s_{j_1} \dots s_{j_r}\alpha_{j_r}, s_{j_1} \dots s_{j_t} w_J\alpha_i \rangle = \langle \alpha_{j_r}, s_{j_{r+1}} \dots s_{j_t} w_J\alpha_i \rangle \ne 0.$$

(b) Using (a) and Lemma 3.1.1(b),

$$w_J(\delta_i - \delta_i') = \langle w_J(\Phi_i \setminus \Phi_i') \rangle = \langle \Phi(w_i^{-1}) \rangle = -w_i^{-1} \cdot 0 = \rho - w_i^{-1}\rho.$$

(c) Using (b) and the idea of the proof of Proposition 3.3.2(c),

$$\delta_i - \delta_i' = w_J(\rho - w_i^{-1}\rho) = w_J[(\rho - \tfrac{1}{2}\delta_i) + \tfrac{1}{2}\delta_i] - w_Jw_i^{-1}\rho$$
$$= -(\rho - \tfrac{1}{2}\delta_i) + \tfrac{1}{2}\delta_i - w_Jw_i^{-1}\rho = \delta_i - \rho - w_Jw_i^{-1}\rho.$$

Thus $\delta_i' = \rho + w_Jw_i^{-1}\rho$ and so

$$\langle \delta_i', \check{\alpha}_i \rangle = \langle \rho + w_Jw_i^{-1}\rho, \check{\alpha}_i \rangle = 1 + \langle \rho, w_iw_J\check{\alpha}_i \rangle = 1 + \langle \rho, \tilde{\alpha}^\vee \rangle.$$

(d) Combine (c) with the inequality $\langle \rho, \tilde{\alpha}^\vee \rangle \le \langle \rho, \check{\alpha}_0 \rangle = h - 1$. □

3.5. The next proposition is the key to our proof of Kostant's Theorem in characteristic $p \ge h - 1$.

PROPOSITION 3.5.1. *Assume $p \ge h - 1$. Suppose $\sigma = w \cdot 0 + p\mu$ is a weight of $\Lambda^\bullet(\mathfrak{u}^*)$ where $w \in W$ and $\mu \in X$. Then $\sigma = x \cdot 0$ for some $x \in W$.*

PROOF. The proof is again by induction on $l(w)$. Assume $w = 1$ so that $p\mu$ is a sum of distinct negative roots. Set $\nu = -\mu$ so that $p\nu = \langle \Psi \rangle$ for some $\Psi \subset \Phi^+$. For any $1 \le i \le l$,

$$\langle \langle \Psi \rangle, \check{\alpha}_i \rangle \le \langle \delta_i, \check{\alpha}_i \rangle \le \langle \delta_i', \check{\alpha}_i \rangle.$$

The first inequality follows because $\langle \alpha_j, \check{\alpha}_i \rangle \le 0$ if $j \ne i$ whereas $\langle \alpha_i, \check{\alpha}_i \rangle = 2$, so including only positive roots that involve α_i can only make the inner product bigger. The second inequality follows similarly: including only those positive roots α with $\langle \alpha, \check{\alpha}_i \rangle \ge 0$ obviously can only increase the inner product. Writing $\langle \Psi \rangle = 2\rho - \langle \Psi^c \rangle$, where $\Psi^c = \Phi^+ \setminus \Psi$, applying the same inequality for Ψ^c, and using the fact that $\langle \rho, \check{\alpha}_i \rangle = 1$, we obtain

$$2 - \langle \delta_i', \check{\alpha}_i \rangle \le \langle \langle \Psi \rangle, \check{\alpha}_i \rangle \le \langle \delta_i', \check{\alpha}_i \rangle.$$

But we also have $\langle \delta_i', \check{\alpha}_i \rangle \leq h$ by Proposition 3.4.1(d). Thus

$$(3.5.1) \qquad\qquad 2 - h \leq p\langle \nu, \check{\alpha}_i \rangle \leq h.$$

Since $p \geq h - 1$ and $\langle \nu, \check{\alpha}_i \rangle \in \mathbb{Z}$, the first inequality implies $\langle \nu, \check{\alpha}_i \rangle \geq 0$ for all i. That is, ν is dominant. If $p > h$, the second inequality implies that $\langle \nu, \check{\alpha}_i \rangle = 0$ for all i, and thus $\nu = 0$. This completes the proof in the case $w = 1$ when $p > h$.

From Proposition 3.3.1(b), it follows that

$$p\langle \nu, \check{\alpha}_0 \rangle = \langle \langle \Psi \rangle, \check{\alpha}_0 \rangle \leq \langle 2\rho, \check{\alpha}_0 \rangle = 2(h - 1).$$

Since $p \geq h - 1$, we deduce that $\langle \nu, \check{\alpha}_0 \rangle = 0$, 1 or 2. Suppose for the moment that we handle the case $\langle \nu, \check{\alpha}_0 \rangle = 2$; this case does not arise if $p = h$. Recall also that we know ν is dominant. If $\langle \nu, \check{\alpha}_0 \rangle = 0$ then $\nu = 0$; this can be seen since $\check{\alpha}_0$ is the highest root of the dual root system, and thus involves every dual simple root $\check{\alpha}_i$ with positive coefficient [Hum1, Lemma 10.4A]. So the coefficient of ω_i in ν must be 0 for every i. Suppose $\langle \nu, \check{\alpha}_0 \rangle = 1$. Then ν is a minuscule dominant weight. Also $p\nu = \langle \Psi \rangle$ must belong to the root lattice.

When $p = h - 1$, one can check for each irreducible root system that p does not divide the index of connection f (the index of the weight lattice in the root lattice); cf. [Hum1, p. 68]. Thus ν itself must lie in the root lattice. However, a case-by-case check using the list of minuscule weights (e.g., [Hum1, Exercise 13.13 and Table 13.1]) shows that this never happens.

Assume $p = h$. The Coxeter number is prime only in type A_l. In this case every fundamental dominant weight ω_i is minuscule, and $h = f = l + 1$ so $p\omega_i$ is in the root lattice. Suppose $\nu = \omega_i$. Recall from Proposition 3.3.2(b) that $\delta_i = c\omega_i$; we compute

$$c = \langle c\omega_i, \check{\alpha}_i \rangle = \left\langle \sum_{\substack{\alpha \in \Phi^+ \\ \langle \omega_i, \alpha \rangle > 0}} \alpha, \check{\alpha}_i \right\rangle = 2 + (l - 1) = l + 1 = h,$$

where we have used the fact that $\langle \alpha_i, \check{\alpha}_i \rangle = 2$, $\langle \alpha_j + \cdots + \alpha_i, \check{\alpha}_i \rangle = \langle \alpha_i + \cdots + \alpha_k, \check{\alpha}_i \rangle = 1$ for $1 \leq j < i$ and $i < k \leq l$, and $\langle \alpha, \check{\alpha}_i \rangle = 0$ for all other positive roots in type A_l which involve α_i. Thus $p\mu = -h\omega_i = -\delta_i = x \cdot 0$ for some $x \in W$ by Proposition 3.3.2(c), as required.

To complete the proof for $w = 1$, there remains to handle the case $\langle \nu, \check{\alpha}_0 \rangle = 2$ when $p = h - 1$. Set $\Psi_0 = \{ \alpha \in \Phi^+ \mid \langle \alpha, \check{\alpha}_0 \rangle > 0 \}$ and $\gamma = \langle \Psi_0 \rangle$. We claim that $\gamma = (h - 1)\alpha_0$. To see this, note that $s_{\alpha_0} \Psi_0 = -\Psi_0$ (recall that $\langle \alpha, \check{\alpha}_0 \rangle \geq 0$ for $\alpha \in \Phi^+$). So $s_{\alpha_0} \gamma = -\gamma$. Substituting this into the formula for $s_{\alpha_0} \gamma$ gives $\gamma = \frac{1}{2}\langle \gamma, \check{\alpha}_0 \rangle \alpha_0$. But $\langle \gamma, \check{\alpha}_0 \rangle = \langle 2\rho, \check{\alpha}_0 \rangle = 2(h - 1)$, and this proves the claim.

Now assume $p = h - 1$, $\langle \nu, \check{\alpha}_0 \rangle = 2$, and $(h - 1)\nu = \langle \Psi \rangle$ for some $\Psi \subset \Phi^+$. Then

$$2(h - 1) = (h - 1)\langle \nu, \check{\alpha}_0 \rangle = \langle \langle \Psi \rangle, \check{\alpha}_0 \rangle \leq \langle 2\rho, \check{\alpha}_0 \rangle = 2(h - 1),$$

so we must have equality at the third step. It follows from the definition of Ψ_0 above, and the fact that $\langle \gamma, \check{\alpha}_0 \rangle = 2(h-1)$, that $\Psi_0 \subset \Psi$. But then $\langle \Psi_0 \setminus \Psi \rangle = (h-1)(\alpha_0 - \nu)$, so $\alpha_0 - \nu$ is a dominant weight (by the argument given for $\langle \Psi \rangle$ at the beginning of this proof), and $\langle \alpha_0 - \nu, \check{\alpha}_0 \rangle = 0$ by the definition of Ψ_0. As mentioned earlier, this implies $\alpha_0 - \nu = 0$. Thus $\sigma = p\mu = -p\nu = -(h-1)\alpha_0 = -\langle \rho, \check{\alpha}_0 \rangle \alpha_0 = s_{\alpha_0} \cdot 0$. This completes the case $w = 1$.

The induction step is almost identical to that in Lemma 3.1.2(b). Write $w = s_\alpha w'$ as in that proof, and suppose as before that $w \cdot 0 + p\mu = -(\gamma_1 + \cdots + \gamma_m)$ for distinct $\gamma_1, \ldots, \gamma_m \in \Phi^+$. Then

$$w' \cdot 0 + p s_\alpha \mu = -(s_\alpha \gamma_1 + \cdots + s_\alpha \gamma_m + \alpha).$$

This is a sum of $m \pm 1$ distinct negative roots (according to whether or not some $\gamma_i = \alpha$). By induction, $w' \cdot 0 + p s_\alpha \mu = x' \cdot 0$ for some $x' \in W$. Apply $s_\alpha \cdot$ to get the result. $\qquad \square$

3.6.   In this section we prove results about complete reducibility of modules that will be later used in our cohomology calculations.

PROPOSITION 3.6.1. *Let $p \geq h - 1$, $w \in {}^J W$, and $\lambda \in \overline{C}_{\mathbb{Z}} \cap X^+$. Then*

(a) *$L_J(w \cdot 0)$ is in the bottom alcove for L_J;*
(b) *$L_J(w \cdot 0) \otimes L(\lambda)$ is completely reducible as an L_J-module.*

PROOF. (a) First decompose $J := J_1 \cup J_2 \cdots \cup J_t$ into indecomposable components, and let β_0 be the highest short root of one of the components $J_i =: K$. Observe that for $w \in {}^J W$,

$$\langle w \cdot 0 + \rho_K, \check{\beta}_0 \rangle = \langle w\rho - \rho + \rho_K, \check{\beta}_0 \rangle = \langle w\rho, \check{\beta}_0 \rangle = \langle \rho, w^{-1} \check{\beta}_0 \rangle$$

where in the second equality we have used that both ρ and ρ_K have inner product 1 with each simple coroot appearing in the decomposition of $\check{\beta}_0$. Now since $w \in {}^J W$ and $\beta_0 \in \Phi_J^+$, $w^{-1}\beta_0 \in \Phi^+$, and thus $0 \leq \langle \rho, w^{-1}\check{\beta}_0 \rangle \leq h - 1 \leq p$. Hence, $w \cdot 0$ belongs to the closure of the bottom L_J alcove.

(b) Suppose that $L_J(\nu + \mu)$ is an L_J composition factor of $L_J(w \cdot 0) \otimes L(\lambda)$ where $\nu + \mu$ is J-dominant and ν is a weight of $L_J(w \cdot 0)$ and μ is a weight of $L(\lambda)$. We will show that $\nu + \mu$ belongs to the closure of the bottom L_J alcove. First observe that $\langle \mu, \check{\alpha} \rangle \leq \langle \lambda, \check{\alpha}_0 \rangle$ for all $\alpha \in \Phi$. Indeed, we can choose $w \in W$ such that $w\mu$ is dominant and since μ is a weight of $L(\lambda)$, $w\mu \leq \lambda$. Therefore,

$$\langle \mu, w^{-1}\check{\beta} \rangle = \langle w\mu, \check{\beta} \rangle \leq \langle w\mu, \check{\alpha}_0 \rangle \leq \langle \lambda, \check{\alpha}_0 \rangle$$

for all $\beta \in \Phi$.

Using the notation and results in (a), in addition to the fact that $\lambda \in \overline{C}_{\mathbb{Z}}$, we have

$$
\begin{aligned}
\langle \nu + \mu + \rho_K, \check{\beta}_0 \rangle &= \langle \nu + \rho_K, \check{\beta}_0 \rangle + \langle \mu, \check{\beta}_0 \rangle \\
&\leq \langle w \cdot 0 + \rho_K, \check{\beta}_0 \rangle + \langle \mu, \check{\beta}_0 \rangle \\
&\leq (h - 1) + \langle \lambda, \check{\alpha}_0 \rangle \\
&= \langle \rho, \check{\alpha}_0 \rangle + \langle \lambda, \check{\alpha}_0 \rangle \\
&= \langle \lambda + \rho, \check{\alpha}_0 \rangle \\
&\leq p.
\end{aligned}
$$

The complete reducibility assertion follows by the Strong Linkage Principle [Jan, Proposition 6.13] because all the composition factors of $L_J(w \cdot 0) \otimes L(\lambda)$ are in the bottom L_J alcove. $\qquad \square$

4. Kostant's Theorem and Generalizations

4.1. In this section we will prove Kostant's theorem, and its extension to characteristic p by Friedlander-Parshall ($p \geq h$) [FP1] and by Polo-Tilouine ($p \geq h - 1$) [PT], for dominant highest weights in the closure of the bottom alcove. We begin by proving the result for trivial coefficients, and then use our tensor product results to prove it in the more general setting.

THEOREM 4.1.1. *Let $J \subseteq \Delta$. Assume $k = \mathbb{C}$ or $k = \overline{\mathbb{F}}_p$ with $p \geq h - 1$. Then as an L_J-module*

$$\mathrm{H}^n(\mathfrak{u}_J, k) \cong \bigoplus_{\substack{w \in {}^J W \\ l(w) = n}} L_J(w \cdot 0).$$

PROOF. First observe that when $p = 2$ the condition that $p \geq h - 1$ implies that $\Phi = A_1$ or A_2. For these cases the theorem can easily be verified directly. So assume that $p \geq 3$.

We first prove that every irreducible L_J-module in the sum on the right side is a composition factor of the left side. By the remarks at the beginning of Section 3.2, we have for each $w \in {}^J W$ with $l(w) = n$ the vector $f_{\Phi(w)} \in \mathrm{H}^n(\mathfrak{u}_J, k)$, where $\Phi(w) = \{\beta_1, \ldots, \beta_n\}$. To show that $f_{\Phi(w)}$ is a maximal vector for the Levi subalgebra $\mathfrak{l}_J$, fix $\gamma \in \Phi_J^+$. Then

$$(4.1.1) \qquad x_\gamma f_{\Phi(w)} = \sum_{i=1}^m f_{\beta_1} \wedge \cdots \wedge x_\gamma f_{\beta_i} \wedge \cdots \wedge f_{\beta_m}.$$

Fix $\beta = \beta_i$ for some $1 \leq i \leq m$. For any root vector x_δ,

$$(x_\gamma f_\beta)(x_\delta) = -f_\beta([x_\gamma, x_\delta])$$

is nonzero if and only if $0 \neq [x_\gamma, x_\delta] \in \mathfrak{g}_\beta$, if and only if $\beta = \gamma + \delta$ (since root spaces are one-dimensional). Assume $x_\gamma f_\beta$ is nonzero; then it is a scalar multiple of f_δ where $\delta = \beta - \gamma$ is a root. Since $\beta \in \Phi(w)$, Proposition 3.2.1 implies that $\delta \in \Phi(w)$; that is, $\delta = \beta_j$ for some $j \neq i$. Thus $x_\gamma f_\beta = f_{\beta_j}$ already occurs in the wedge product in (4.1.1). So every term on the right hand side of (4.1.1) is 0, proving that $f_{\Phi(w)}$ is the highest weight vector of a L_J (resp. $(L_J)_1$) composition factor of $\mathrm{H}^n(\mathfrak{u}_J, k)$ when $k = \mathbb{C}$ (resp. $k = \overline{\mathbb{F}}_p$). But, the high weight is in the bottom L_J-alcove so we can conclude in general that this high weight corresponds to a L_J composition factor isomorphic to $L_J(w \cdot 0)$.

We now prove that all composition factors in cohomology appear in Kostant's formula. By Theorem 2.3.1 when $k = \mathbb{C}$, and by Theorem 2.4.1, Proposition 3.5.1, and Lemma 3.1.2 when $k = \overline{\mathbb{F}}_p$, any L_J composition factor of $\mathrm{H}^n(\mathfrak{u}_J, k)$ is an $L_J(w \cdot 0)$ for $w \in {}^J W$. By Lemmas 3.1.1(a) and 3.1.2(b), $l(w) = n$ and $L_J(w \cdot 0)$ occurs with multiplicity one in cohomology.

Moreover, when $k = \overline{\mathbb{F}}_p$, by Proposition 3.6.1 all the composition factors $L_J(w \cdot 0)$ lie in the bottom L_J alcove. By the Strong Linkage Principle, there are no nontrivial extensions between these irreducible L_J modules. So in either case, $\mathrm{H}^n(\mathfrak{u}_J, k)$ is completely reducible and given by Kostant's formula. $\square$

We remark that the largest weight in $\Lambda^\bullet(\mathfrak{u}^*)$ is 2ρ. Moreover, $\langle 2\rho + \rho, \check{\alpha}_0 \rangle = 3\langle \rho, \check{\alpha}_0 \rangle = 3(h-1)$. This weight is not in the bottom alcove unless $p \geq 3(h-1)$. This necessitates a more delicate argument for the complete reducibility of the cohomology when $p \geq h - 1$.

4.2.   We can now use the previous theorem to compute the cohomology of $\mathfrak{u}_J$ with coefficients in a finite-dimensional simple $\mathfrak{g}$-module.

THEOREM 4.2.1. *Let $J \subseteq \Delta$ and $\mu \in X^+$. Assume that either $k = \mathbb{C}$, or $k = \overline{\mathbb{F}}_p$ with $\langle \mu + \rho, \check{\beta} \rangle \leq p$ for all $\beta \in \Phi^+$. Then as an L_J-module,*

$$\mathrm{H}^n(\mathfrak{u}_J, L(\mu)) \cong \bigoplus_{\substack{w \in {}^J W \\ l(w) = n}} L_J(w \cdot \mu).$$

PROOF. Observe that the conditions on μ imply $p \geq h - 1$. Namely, we have

(4.2.1) $p \geq \langle \mu + \rho, \check{\alpha}_0 \rangle = h - 1 + \langle \mu, \check{\alpha}_0 \rangle \geq h - 1.$

For $p = 2$, the only case that remains to be checked is the case when $\Phi = A_1$ and $L(\mu) = L(1)$ is the two dimensional natural representation. This can be easily verified using the definition of cocycles and differentials in Lie algebra cohomology. So assume that $p \geq 3$.

First consider the case $k = \overline{\mathbb{F}}_p$ with $p = h - 1$. Then the inequalities in (4.2.1) must all be equalities, whence $\langle \mu, \check{\alpha}_0 \rangle = 0$. Since $\mu \in X^+$, it follows that $\mu = 0$. But now we are back to the setting of Theorem 4.1.1, where the result is proved. Thus for the rest of this proof we may assume $k = \mathbb{C}$ or $k = \overline{\mathbb{F}}_p$ with $p \geq h$.

We first prove that every L_J composition factor of the cohomology occurs in the direct sum on the right side. Let $L_J(\sigma)$ be an L_J composition factor of $\mathrm{H}^n(\mathfrak{u}_J, L(\mu))$. By Proposition 2.5.1 and Theorem 4.1.1, we have that $L_J(\sigma)$ is an L_J composition factor of $L_J(w \cdot 0) \otimes L(\mu)$ for some $w \in {}^J W$ with $l(w) = n$. Moreover, by definition $\mu \in X_1(T)$ and by the proof of Proposition 3.6.1(b), $\sigma \in (X_J)_1$. Hence, by Theorem 2.3.1 or 2.4.1, $\sigma = y \cdot \mu$ for some $y \in W_p$ (when $k = \mathbb{C}$ we set $W_p = W$).

According to Proposition 3.6.1, $L_J(w \cdot 0) = H_J^0(w \cdot 0)$, and $L_J(w \cdot 0) \otimes L(\mu)$ is completely reducible. Therefore, by using Frobenius reciprocity

$$0 \neq \mathrm{Hom}_{L_J}(L_J(\sigma), L_J(w \cdot 0) \otimes L(\mu)) \cong \mathrm{Hom}_{B_{L_J}}(L_J(\sigma), w \cdot 0 \otimes L(\mu)).$$

From this statement, one can see that

$$\sigma = y \cdot \mu = w \cdot 0 + \tilde{\nu}$$

for some weight $\tilde{\nu}$ of $L(\mu)$.

Choose $x \in W$ such that $\tilde{\nu} = x\nu$ with ν dominant. Note that ν is still a weight of $L(\mu)$, so in particular $\nu \leq \mu$. Rewriting the previous equation gives

$$(w^{-1}y) \cdot \mu = w^{-1}x\nu.$$

Applying [Jan, Lemma II.7.7(a)] with $\lambda = 0$, $\nu_1 = \mu \in X(T)_+ \cap W(\mu - \lambda)$, we conclude that $\nu = \mu$. Now apply [Jan, Lemma II.7.7(b)] to conclude that there exists $w_1 \in W_p$ such that

$$w_1 \cdot 0 = 0 \quad \text{and} \quad w_1 \cdot \mu = w^{-1}x\mu.$$

But since $p \geq h$, ρ lies in the interior of the bottom alcove, so the stabilizer of 0 under the dot action of W_p is trivial; i.e., $w_1 = 1$. Thus $\mu = w^{-1}x\mu$, or equivalently, $w \cdot \mu = w \cdot 0 + x\mu = w \cdot 0 + \tilde{\nu} = y \cdot \mu = \sigma$. Since $w \in {}^J W$ and $l(w) = n$, this proves that every composition factor in cohomology occurs in Kostant's formula (possibly with multiplicity greater than one).

We now prove that every L_J irreducible on the right side occurs as a composition factor in cohomology, with multiplicity one. Let $\sigma = w \cdot \mu$ for $w \in {}^J W$ with

$l(w) = n$. The σ weight space of $C^\bullet = \Lambda^\bullet(\mathfrak{u}_J^*) \otimes L(\mu)$ contains at least the one dimensional space

$$\Lambda^n(\mathfrak{u}_J^*)_{w \cdot 0} \otimes L(\mu)_{w\mu}$$

since $w \cdot \mu = w \cdot 0 + w\mu = -\langle \Phi(w) \rangle + w\mu$. To see that this is the entire σ weight space of $C^\bullet$ we use a simple argument of Cartier [Cart], which we reproduce here for the reader's convenience.

Note first that there is a bijection between subsets $\Psi \subset \Phi^+$ and subsets $\widetilde{\Psi} \subset \Phi$ satisfying

$$\Phi = \widetilde{\Psi} \amalg -\widetilde{\Psi},$$

namely

$$\Psi = \widetilde{\Psi} \cap \Phi^+ \quad \text{and} \quad \widetilde{\Psi} = \Psi \cup -(\Phi^+ \smallsetminus \Psi).$$

Note that the collection of sets of the form $\widetilde{\Psi}$ is invariant under the ordinary action of W. It is easy to check that for such pairs,

(4.2.2) $$\rho - \langle \Psi \rangle = -\tfrac{1}{2}\langle \widetilde{\Psi} \rangle.$$

Suppose $\sigma = -\langle \Psi \rangle + \nu$ for some $\Psi \subset \Phi^+$ and some weight ν of $L(\mu)$. It suffices to show $\Psi = \Phi(w)$ and $\nu = w\mu$. We have $\sigma + \rho = \rho - \langle \Psi \rangle + \nu = -\tfrac{1}{2}\langle \widetilde{\Psi} \rangle + \nu$.

Thus

$$\mu + \rho = w^{-1}(\sigma + \rho) = w^{-1}\nu - \tfrac{1}{2}\langle w^{-1}\widetilde{\Psi} \rangle = w^{-1}\nu - \langle \Gamma \rangle + \rho,$$

where we have applied (4.2.2) to $w^{-1}\widetilde{\Psi}$ and set $\Gamma = w^{-1}\widetilde{\Psi} \cap \Phi^+$.

But since $w^{-1}\nu$ is a weight of $L(\mu)$ we can write $w^{-1}\nu = \mu - \sum_i m_i \alpha_i$ with $m_i \in \mathbb{Z}_{\geq 0}$. So

$$\mu = \mu - \sum_i m_i \alpha_i - \langle \Gamma \rangle.$$

We conclude that all $m_i = 0$, so $w^{-1}\nu = \mu$ and $\nu = w\mu$. Also,

$$\Gamma = \varnothing \implies w^{-1}\widetilde{\Psi} = \Phi^- \implies \widetilde{\Psi} = w\Phi^- \implies \Psi = w\Phi^- \cap \Phi^+ = \Phi(w).$$

This is what we wanted to show.

Since the $w \cdot \mu$ weight space in the chain complex $C^\bullet$ is one dimensional and occurs in C^n, we conclude, as in the case of trivial coefficients, that $w \cdot \mu$ is a weight in the cohomology $\mathrm{H}^n(\mathfrak{u}_J, L(\mu))$. A corresponding weight vector in C^n is

$$v = f_{\Phi(w)} \otimes v_{w\mu},$$

where $f_{\Phi(w)}$ is as in the proof of Theorem 4.1.1 and $0 \neq v_{w\mu} \in L(\mu)_{w\mu}$. Fix $\gamma \in \Phi_J^+$; then $x_\gamma v = x_\gamma f_{\Phi(w)} \otimes v_{w\mu} + f_{\Phi(w)} \otimes x_\gamma v_{w\mu}$. We know from the proof of Theorem 4.1.1 that $x_\gamma f_{\Phi(w)} = 0$. Suppose $x_\gamma v_{w\mu}$ were not zero. Then it would be a weight vector in $L(\mu)$ of weight $w\mu + \gamma$. By W-invariance, $\mu + w^{-1}\gamma$ would be a weight of $L(\mu)$. But $w \in {}^J W$ and $\gamma \in \Phi_J^+$ imply $w^{-1}\gamma \in \Phi^+$, and this contradicts that μ is the highest weight of $L(\mu)$. Therefore v is annihilated by the nilradical of the Levi subalgebra, and hence its image in cohomology generates an L_J composition factor of $\mathrm{H}^n(\mathfrak{u}_J, L(\mu))$ isomorphic to $L_J(w \cdot \mu)$. Note also that our argument proves that this composition factor occurs with multiplicity one.

The L_J highest weights are in the closure of the bottom L_J alcove by Propositions 2.5.1 and 3.6.1, and thus the cohomology is completely reducible as an L_J-module. $\qquad \square$

5. The Converse of Kostant's Theorem

5.1. Existence of extra cohomology. The following theorem shows that there are extra cohomology classes (beyond those given by Kostant's formula) that arise in $\mathrm{H}^\bullet(\mathfrak{u}, k)$ when char $k = p$ and $p < h - 1$. This can be viewed as a converse to Theorem 4.1.1 in the case when $J = \varnothing$. Examples in Section 6 will indicate that the situation is much more subtle for $J \neq \varnothing$ (i.e., extra cohomology classes may or may not arise depending on the size of J relative to the rank).

THEOREM 5.1.1. *Let $k = \overline{\mathbb{F}}_p$ with $p < h - 1$. Then $\operatorname{ch} \mathrm{H}^\bullet(\mathfrak{u}, k) \neq \operatorname{ch} \mathrm{H}^\bullet(\mathfrak{u}, \mathbb{C})$.*

PROOF. Fix a simple root α and let $J = \{\alpha\}$; shortly we will choose α more precisely. There exists a Lyndon-Hochschild-Serre spectral sequence

$$E_2^{i,j} = \mathrm{H}^i(\mathfrak{u}/\mathfrak{u}_J, \mathrm{H}^j(\mathfrak{u}_J, k)) \Rightarrow \mathrm{H}^{i+j}(\mathfrak{u}, k).$$

Since $\dim \mathfrak{u}/\mathfrak{u}_J = 1$, $E_2^{i,j} = 0$ for $i \neq 0, 1$. Therefore, the spectral sequence collapses, yielding

(5.1.1) $$\mathrm{H}^n(\mathfrak{u}, k) \cong \mathrm{H}^n(\mathfrak{u}_J, k)^{\mathfrak{u}/\mathfrak{u}_J} \oplus \mathrm{H}^1(\mathfrak{u}/\mathfrak{u}_J, \mathrm{H}^{n-1}(\mathfrak{u}_J, k)).$$

By the remarks at the beginning of Section 3.2, we can find explicit cocycles such that, as a T-module,

$$\bigoplus_{w \in W} w \cdot 0 \hookrightarrow \mathrm{H}^\bullet(\mathfrak{u}, k)$$

whereas by Lemmas 3.1.1 and 3.1.2, the only weights in $\mathrm{H}^\bullet(\mathfrak{u}_J, k)$ (or even in $\Lambda^\bullet(\mathfrak{u}_J^*)$) of the form $w \cdot 0$ with $w \in W$ occur when $w \in {}^J W$. So we must have

$$\bigoplus_{w \in W \smallsetminus {}^J W} w \cdot 0 \hookrightarrow \mathrm{H}^1(\mathfrak{u}/\mathfrak{u}_J, \mathrm{H}^\bullet(\mathfrak{u}_J, k)).$$

Thus it suffices to find "extra" cohomology in the first term on the right hand side of (5.1.1), meaning a cohomology class in characteristic p which does not have an analog in characteristic zero.

Since $\mathfrak{u}/\mathfrak{u}_J$ is isomorphic to the nilradical of the Levi subalgebra $\mathfrak{l}_J$, the first part of the proof of Theorem 4.1.1 shows that for $w \in {}^J W$ with $l(w) = n$, we have an explicit invariant vector of weight $w \cdot 0$ in $\mathrm{H}^n(\mathfrak{u}_J, k)^{\mathfrak{u}/\mathfrak{u}_J}$. Thus we get an inclusion

$$\bigoplus_{\substack{w \in {}^J W \\ l(w) = n}} w \cdot 0 \hookrightarrow \mathrm{H}^n(\mathfrak{u}_J, k)^{\mathfrak{u}/\mathfrak{u}_J} \subset \mathrm{H}^n(\mathfrak{u}_J, k).$$

By [Jan, Lemma 2.13] this induces an L_J-homomorphism from a sum of Weyl modules (for L_J)

(5.1.2) $$\phi \colon S = \bigoplus_{\substack{w \in {}^J W \\ l(w) = n}} V_J(w \cdot 0) \to \mathrm{H}^n(\mathfrak{u}_J, k)$$

which is injective on the direct sum of the highest weight spaces. Next we claim that

(5.1.3) $$\mathrm{Hd}_{L_J} \phi(S) = \bigoplus_{\substack{w \in {}^J W \\ l(w) = n}} L_J(w \cdot 0).$$

To see this, note first that $\phi(S) \cong S/\mathrm{Ker}\,\phi$, and $\mathrm{Ker}\,\phi \subset \mathrm{Rad}_{L_J} S$ because of the injectivity of ϕ on the highest weight spaces of the indecomposable direct summands $V_J(w \cdot 0)$ of S. This means that

$$\mathrm{Rad}_{L_J}\,\phi(S) \cong \mathrm{Rad}_{L_J}(S/\mathrm{Ker}\,\phi) = (\mathrm{Rad}_{L_J} S)/\mathrm{Ker}\,\phi.$$

Thus

$$\begin{aligned}
\mathrm{Hd}_{L_J}\,\phi(S) &= \phi(S)/\mathrm{Rad}_{L_J}\,\phi(S) \\
&\cong (S/\mathrm{Ker}\,\phi)/((\mathrm{Rad}_{L_J} S)/\mathrm{Ker}\,\phi) \\
&\cong S/\mathrm{Rad}_{L_J} S \\
&\cong \bigoplus_{\substack{w \in {}^J W \\ l(w)=n}} L_J(w \cdot 0)
\end{aligned}$$

as claimed.

Now choose α to be a short simple root, and fix $\widetilde{w} \in W$ such that $\widetilde{w}^{-1}\alpha = \alpha_0$, the highest short root. Then $\widetilde{w} \in {}^J W$ and

$$\langle \widetilde{w} \cdot 0 + \rho, \check{\alpha} \rangle = \langle \widetilde{w}\rho, \check{\alpha} \rangle = \langle \rho, \widetilde{w}^{-1}\check{\alpha} \rangle = \langle \rho, \check{\alpha}_0 \rangle = h - 1 > p.$$

Thus $\lambda := \widetilde{w} \cdot 0$ is not in the restricted region for L_J. Write $\lambda = \lambda_0 + p\lambda_1$ with $\lambda_0 \in (X_J)_1$ and $0 \neq \lambda_1 \in X_J^+$. There are two cases, according to whether or not $\phi(V_J(\lambda))$ is a simple L_J-module.

Case 1: $\phi(V_J(\lambda)) \cong L_J(\lambda)$. By Steinberg's tensor product theorem, $L_J(\lambda) \cong L_J(\lambda_0) \otimes L_J(\lambda_1)^{(1)}$. Since $\lambda_1 \neq 0$ (on J), $L_J(\lambda_1)^{(1)}$ has dimension at least two, and $\mathfrak{u}/\mathfrak{u}_J$ acts trivially on it. So this produces at least a two-dimensional space of vectors in $\mathrm{H}^n(\mathfrak{u}_J, k)^{\mathfrak{u}/\mathfrak{u}_J}$ arising from $L_J(\lambda)$ which produces "extra" cohomology.

Case 2: $N := \mathrm{Rad}_{L_J}\,\phi(V_J(\lambda)) \neq 0$. Then $N \subset \mathrm{Rad}_{L_J}\,\phi(S)$ and

$$0 \neq N^{\mathfrak{u}/\mathfrak{u}_J} \subset \phi(S)^{\mathfrak{u}/\mathfrak{u}_J} \subset \mathrm{H}^n(\mathfrak{u}_J, k)^{\mathfrak{u}/\mathfrak{u}_J}.$$

Since by (5.1.3) all the "characteristic zero" cohomology in $\mathrm{H}^n(\mathfrak{u}_J, k)^{\mathfrak{u}/\mathfrak{u}_J}$ has already been accounted for in $\mathrm{Hd}_{L_J}\,\phi(S)$, the vectors in $N^{\mathfrak{u}/\mathfrak{u}_J} \subset \mathrm{Rad}_{L_J}\,\phi(S)$ must be "extra" cohomology in characteristic p. $\square$

5.2. Explicit extra cohomology.

In this section we exhibit additional cohomology that arises in $\mathrm{H}^\bullet(\mathfrak{u}, k)$ where $k = \overline{\mathbb{F}}_p$ in case $\Phi = A_n$.

THEOREM 5.2.1. *Let p be prime and Φ be of type A_n where $n = p + 1$. Then the vector*

$$\sum_{i=1}^p f_{-\alpha_0} \wedge \gamma_1 \wedge \gamma_2 \wedge \cdots \wedge \widehat{\gamma_i} \wedge \cdots \wedge \gamma_p$$

appears as extra cohomology in $\mathrm{H}^{2p-1}(\mathfrak{u}, k)$, where $\gamma_i = f_{-(\alpha_1+\cdots+\alpha_i)} \wedge f_{-(\alpha_{i+1}+\cdots+\alpha_n)}$.

PROOF. Let $E = \sum_{i=1}^p f_{-\alpha_0} \wedge \gamma_1 \wedge \gamma_2 \wedge \cdots \wedge \widehat{\gamma_i} \wedge \cdots \wedge \gamma_p$. Consider the vector

$$f_{-\alpha_0} \wedge [d(f_{-\alpha_0})]^{p-1} := f_{-\alpha_0} \wedge \underbrace{d(f_{-\alpha_0}) \wedge d(f_{-\alpha_0}) \wedge \cdots \wedge d(f_{-\alpha_0})}_{p-1 \text{ times}},$$

with $d(f_{-\alpha_0}) = \gamma_1 + \gamma_2 + \cdots + \gamma_p \in \Lambda^2(\mathfrak{u}^*)$. First note by direct calculation one has $\gamma_i \wedge \gamma_j = \gamma_j \wedge \gamma_i$ and $\gamma_i \wedge \gamma_i = 0$. We can now apply the multinomial theorem for $[d(f_{-\alpha_0})]^m = [\gamma_1 + \gamma_2 + \cdots + \gamma_p]^m$ for $m \geq 2$:

$$[d(f_{-\alpha_0})]^m = [\gamma_1 + \gamma_2 + \cdots + \gamma_p]^m$$

$$= \sum_{r_1, \ldots, r_p} \binom{m}{r_1, \ldots, r_p} [\gamma_1]^{r_1} \wedge [\gamma_2]^{r_2} \wedge \cdots \wedge [\gamma_p]^{r_p}$$

where $\sum_{i=1}^{p} r_i = m$ and $\binom{m}{r_1, \ldots, r_p} = \frac{m!}{r_1! r_2! \ldots r_p!}$. Consider the case when $m = p - 1$. Since $[\gamma_i]^{r_i} = 0$ for $r_i \geq 2$, the only nonzero terms occur where $r_i = 0$ for some i, and $r_j = 1$ for all $j \neq i$. We have

$$[d(f_{-\alpha_0})]^{p-1} = \sum_{i=1}^{p} \binom{p-1}{0, 1, \ldots, 1} (\gamma_1 \wedge \cdots \wedge \widehat{\gamma_i} \wedge \cdots \wedge \gamma_p)$$

$$= (p-1)! \sum_{i=1}^{p} (\gamma_1 \wedge \cdots \wedge \widehat{\gamma_i} \wedge \cdots \wedge \gamma_p).$$

So we have that $f_{-\alpha_0} \wedge [d(f_{-\alpha_0})]^{p-1} = (p-1)! \, E$. Since the terms in the above sum are linearly independent, this shows that $E \neq 0$. To prove that $E \in \mathrm{Ker}\, d$, we look at $f_{-\alpha_0} \wedge [d(f_{-\alpha_0})]^{p-1}$. Since d is a differential, $d(d(f_{-\alpha_0})) = 0$. Also note we can apply the multinomial theorem again to get $[d(f_{-\alpha_0})]^p = p!(\gamma_1 \wedge \gamma_2 \wedge \cdots \wedge \gamma_i \wedge \cdots \wedge \gamma_p)$. Consequently,

$$d\left(f_{-\alpha_0} \wedge [d(f_{-\alpha_0})]^{p-1}\right) = d(f_{-\alpha_0}) \wedge [d(f_{-\alpha_0})]^{p-1} - f_{-\alpha_0} \wedge d\left([d(f_{-\alpha_0})]^{p-1}\right)$$

$$= [d(f_{-\alpha_0})]^p - f_{-\alpha_0} \wedge \left(\sum_{i=1}^{p-1} d(d(f_{-\alpha_0})) \wedge [d(f_{-\alpha_0})]^{p-2}\right)$$

$$= p! \, (\gamma_1 \wedge \gamma_2 \wedge \cdots \wedge \gamma_p).$$

It now follows that

$$d(E) = d\left(\frac{1}{(p-1)!} (f_{-\alpha_0} \wedge [d(f_{-\alpha_0})]^{p-1})\right)$$

$$= \frac{p!}{(p-1)!} (\gamma_1 \wedge \gamma_2 \wedge \cdots \wedge \gamma_p)$$

$$= p \, (\gamma_1 \wedge \gamma_2 \wedge \cdots \wedge \gamma_p).$$

Thus $d(E) = 0$ in characteristic p (but not in characteristic 0).

We need to verify that E is not in the image of the previous differential. This will follow by demonstrating that $\Lambda^{2p-2}(\mathfrak{u}^*)_{-p\alpha_0} = 0$ because the differentials respect weight spaces. Any weight in $\Lambda^{2p-2}(\mathfrak{u}^*)$ is of the form $\beta_1 + \beta_2 + \cdots + \beta_{2p-2}$ where the β_i are distinct negative roots. Observe that $\langle \beta_1 + \beta_2 + \cdots + \beta_{2p-2}, \check{\alpha}_0 \rangle \geq -2p + 1$. One can deduce this because for each i, $\langle \beta_i, \check{\alpha}_0 \rangle = 0, \pm 1, \pm 2$ and is equal to -2 if and only if $\beta_i = -\alpha_0$. On the other hand, $\langle -p\alpha_0, \check{\alpha}_0 \rangle = -2p$, thus $\Lambda^{2p-2}(\mathfrak{u}^*)_{-p\alpha_0} = 0$. $\qquad \square$

6. Examples for $\mathbf{H}^\bullet(\mathfrak{u}_J, \overline{\mathbb{F}}_p)$

The following low rank examples were calculated using our computer package developed in MAGMA [BC, BCP]. Recall that the cohomology has a palindromic behavior so in the A_4 table the degrees are only listed up to half the dimension of $\mathfrak{u}_J$. Set $H^n = \dim H^n(\mathfrak{u}_J, k)$.

Type A_3, $h - 1 = 3$

J	p	H^0	H^1	H^2	H^3	H^4	H^5	H^6
$\emptyset$	0	1	3	5	6	5	3	1
	2	1	3	6	8	6	3	1
$\{1\}$ or $\{3\}$	0, 2	1	3	6	6	3	1	
$\{2\}$	0, 2	1	4	5	5	4	1	
$\{1,3\}$ or $\{2,4\}$	0, 2	1	4	6	4	1		
$\{1,2\}$ or $\{2,3\}$	0, 2	1	3	3	1			

Type A_4, $h - 1 = 4$

J	p	H^0	H^1	H^2	H^3	H^4	H^5	H^6	H^7
$\emptyset$	0	1	4	9	15	20	22	...	...
	2	1	4	11	25	38	42	...	...
	3	1	4	9	17	25	28	...	...
$\{1\}$ or $\{4\}$	0	1	4	10	19	26	...	...	...
	2	1	4	12	25	32	...	...	...
	3	1	4	10	20	27	...	...	...
$\{2\}$ or $\{3\}$	0	1	5	12	19	23	...	...	...
	2	1	5	12	23	33	...	...	...
	3	1	5	12	20	24	...	...	...
$\{1,3\}$ or $\{2,4\}$	0, 2, 3	1	6	13	23	30	...	...	...
$\{1,4\}$	0, 2, 3	1	4	14	25	28	...	...	...
$\{1,2\}$ or $\{3,4\}$	0, 3	1	4	12	18	...	...	...	...
	2	1	4	12	19	...	...	...	...
$\{2,3\}$	0, 3	1	6	14	14	...	...	...	...
	2	1	6	14	15	...	...	...	...
$\{1,3,4\}$ or $\{1,2,4\}$	0, 2, 3	1	6	15	20	...	...	...	...
$\{1,2,3\}$ or $\{2,3,4\}$	0, 2, 3	1	4	6	...	...	...	...	...

Type G_2, $h - 1 = 5$

J	p	H^0	H^1	H^2	H^3	H^4	H^5	H^6
$\varnothing$	0	1	2	2	2	2	2	1
	2, 3	1	3	6	8	6	3	1
$\{1\}$	0, 2	1	4	5	5	4	1	
	3	1	4	7	7	4	1	
$\{2\}$	0	1	2	3	3	2	1	
	2	1	3	6	6	3	1	
	3	1	4	7	7	4	1	

7.   Further Questions

The results in the preceding sections and our low rank examples naturally suggest the following open questions which are worthy of further study.

(7.1) Let G be a simple algebraic group over $\overline{\mathbb{F}}_p$ and $\mathfrak{g} = \operatorname{Lie} G$. Determine a maximal $c(J, p) > 0$ such that

$$\operatorname{ch} H^n(\mathfrak{u}_J, \mathbb{C}) = \operatorname{ch} H^n(\mathfrak{u}_J, \overline{\mathbb{F}}_p)$$

for $0 \le n \le c(J, p)$.

(7.2) Let $\Phi = A_n$ with $|\Delta| = n$.

a) Does $|\Delta - J| > p$ imply that $\operatorname{ch} H^\bullet(\mathfrak{u}_J, \mathbb{C}) \ne \operatorname{ch} H^\bullet(\mathfrak{u}_J, \overline{\mathbb{F}}_p)$?

b) Does $|\Delta - J| < p$ imply that $\operatorname{ch} H^\bullet(\mathfrak{u}_J, \mathbb{C}) = \operatorname{ch} H^\bullet(\mathfrak{u}_J, \overline{\mathbb{F}}_p)$?

We have seen that when $|\Delta - J| = p$ either conclusion can hold in the example where $\Phi = A_4$ and $|\Delta - J| = p = 2$.

c) What is the appropriate formulation of parts (a) and (b) when Φ is of arbitrary type?

(7.3) Let G be a simple algebraic group over $\overline{\mathbb{F}}_p$ and $\mathfrak{g} = \operatorname{Lie} G$. Assume that p is a good prime. Let $\mathcal{N}_1(\mathfrak{g}) = \{x \in \mathfrak{g} : x^{[p]} = 0\}$ (restricted nullcone). From work of Nakano, Parshall and Vella [NPV], there exists $J \subseteq \Delta$ such that $\mathcal{N}_1(\mathfrak{g}) = G \cdot \mathfrak{u}_J$ (i.e., closure of a Richardson orbit).

Does there exist $J \subseteq \Delta$ with $\mathcal{N}_1(\mathfrak{g}) = G \cdot \mathfrak{u}_J$ such that

$$\operatorname{ch} H^\bullet(\mathfrak{u}_J, \mathbb{C}) = \operatorname{ch} H^\bullet(\mathfrak{u}_J, \overline{\mathbb{F}}_p)?$$

(7.4) Let G be a simple algebraic group over $\overline{\mathbb{F}}_p$ and $\mathfrak{g} = \operatorname{Lie} G$. Compute

$$\operatorname{ch} H^n(\mathfrak{u}_J, \overline{\mathbb{F}}_p)$$

for all p. It would be even better to describe the L_J-module structure.

Solving (7.4) would complete the analog of Kostant's theorem for the trivial module for all characteristics. One might be able to use (7.3) as a stepping stone to perform this computation. Moreover, this calculation would have major implications in determining cohomology for Frobenius kernels and algebraic groups (cf. [BNP]).

8. VIGRE Algebra Group at the University of Georgia

This project was initiated during Fall Semester 2006 under the Vertical Integration of Research and Education (VIGRE) Program sponsored by the National Science Foundation (NSF) at the Department of Mathematics at the University of Georgia (UGA). We would like to acknowledge the NSF grant DMS-0089927 for its financial support of this project. The VIGRE Algebra Group consists of 3 faculty members, 2 postdoctoral fellows, 8 graduate students, and 1 undergraduate. The group is led by Brian D. Boe, Leonard Chastkofsky and Daniel K. Nakano. The email addresses of the members of the group are given below.

Faculty:

Brian D. Boe	brian@math.uga.edu
Leonard Chastkofsky	lenny@math.uga.edu
Daniel K. Nakano	nakano@math.uga.edu

Postdoctoral Fellows:

Jonathan R. Kujawa	kujawa@math.ou.edu
Emilie Wiesner	ewiesner@ithaca.edu

Graduate Students:

Irfan Bagci	bagci@math.uga.edu
Benjamin Connell	bconnell@math.uga.edu
Bobbe J. Cooper	bcooper@math.uga.edu
Mee Seong Im	msi@math.uga.edu
Wenjing Li	wli@math.uga.edu
Kenyon J. Platt	platt@math.uga.edu
Caroline B. Wright	cwright@math.uga.edu
Benjamin Wyser	bwyser@math.uga.edu

Undergraduate Student:

Tyler Kelly	tlkelly@uga.edu

9. Acknowledgements

We would like to thank the referee for providing useful suggestions.

References

[BNP]   C.P. Bendel, D.K. Nakano, C. Pillen, Second cohomology for Frobenius kernels and related structures, *Advances in Math.* **209** (2007), 162–197.

[BC]   W. Bosma, J. Cannon, *Handbook on Magma Functions*, Sydney University, 1996.

[BCP]   W. Bosma, J. Cannon, C. Playoust, The Magma Algebra System I: The User Language, *J. Symbolic Computation* **24** (1997), 235–265.

[Cart] P. Cartier, Remarks on "Lie algebra cohomology and the generalized Borel-Weil theorem", by B. Kostant, *Ann. of Math.* (2) **74** (1961), 388–390.

[FP1]   E.M. Friedlander and B.J. Parshall, Cohomology of infinitesimal and discrete groups, *Math. Ann.* **273** (1986), no. 3, 353–374.

[FP2]   ———, Geometry of *p*-unipotent Lie algebras, *J. Algebra* **109** (1986), 25–45.

[GW]   R. Goodman, N. R. Wallach, *Representations and Invariants of the Classical Groups*, Encyclopedia of Mathematics and its Applications 68, Cambridge University Press, 1998.

[Hum1]   J.E. Humphreys, *Introduction to Lie Algebras and Representation Theory*, Springer-Verlag, New York, 1972.

[Hum2]   ———, *Conjugacy Classes in Semisimple Algebraic Groups*, Math. Surveys and Monographs, AMS, vol. 43, 1995.

[Jan]   J.C. Jantzen, *Representations of Algebraic Groups*, Academic Press, 1987.

[Kna]   Anthony W. Knapp, *Lie Groups, Lie Algebras, and Cohomology*, Mathematical Notes **34**, Princeton University Press, 1988.

[NPV]   D.K. Nakano, B.J. Parshall, D.C. Vella, Support varieties for algebraic groups, *J. Reine Angew. Math.* **547** (2002), 15–49.

[PT]   P. Polo and J. Tilouine, Bernstein-Gelfand-Gelfand complexes and cohomology of nilpotent groups over $\mathbb{Z}_{(p)}$ for representations with p-small weights, *Astérisque* (2002), no. 280, 97–135, Cohomology of Siegel varieties.

[Ro]   A. Rocha-Caridi, Splitting criteria for $\mathfrak{g}$-modules induced form a parabolic and the Bernstein-Gelfand-Gelfand resolution of a finite dimensional irreducible $\mathfrak{g}$-module, *Trans. AMS* **262** (1980), 335–366.

[R]   J. Rotman, *An Introduction to Homological Algebra*, Academic Press, 1979.

DEPARTMENT OF MATHEMATICS, UNIVERSITY OF GEORGIA, ATHENS, GEORGIA 30602

Contemporary Mathematics
Volume **478**, 2009

G-stable pieces and partial flag varieties

Xuhua He

ABSTRACT. We will use the combinatorics of the G-stable pieces to describe the closure relation of the partition of partial flag varieties in [**L3**, section 4].

Introduction

In 1977, Lusztig introduced a finite partition of a (partial) flag variety Y. In the case where Y is the full flag variety, this partition is the partition into Deligne-Lusztig varieties (see [**DL**]). In this case, it follows easily from the Bruhat decomposition that the closure of a Deligne-Lusztig variety is the union of some other Deligne-Lusztig varieties and the closure relation is given by the Bruhat order on the Weyl group.

In this paper, we will use some combinatorial technique in [**H4**] to study the partition on a partial flag variety. We show that the partition is a stratification and the closure relation is given by the partial order introduced in [**H2**, 5.4] and [**H3**, 3.8 & 3.9]. We also study some other properties of the locally closed subvarieties that appear in the partition.

1. Some combinatorics

1.1. Let **k** be an algebraic closure of the finite field $\mathbf{F}_q$ and G be a connected reductive algebraic group defined over $\mathbf{F}_q$ with Frobenius map $F : G \to G$. We fix an F-stable Borel subgroup B of G and an F-stable maximal torus $T \subset B$. Let I be the set of simple roots determined by B and T. Then F induces an automorphism on the Weyl group W which we deonte by δ. The autmorphism restricts to a bijection on the set I of simple roots. By abusion notations, we also denote the bijection by δ.

For any $J \subset I$, let P_J be the standard parabolic subgroup corresponding to J and $\mathcal{P}_J$ be the set of parabolic subgroups that are G-conjugate to P_J. We simply write $\mathcal{P}_\emptyset$ as $\mathcal{B}$. Let L_J be the Levi subgroup of P_J that contains T.

For any parabolic subgroup P, let U_P be the unipotent radical of P. We simply write U for U_B.

2000 *Mathematics Subject Classification.* 14M15, 20G40.
The author is partially supported by NSF grant DMS-0700589.

For $J \subset I$, we denote by W_J the standard parabolic subgroup of W generated by J and by W^J (resp. $^J W$) the set of minimal coset representatives in W/W_J (resp. $W_J \backslash W$). For $J, K \subset I$, we simply write $W^J \cap {}^K W$ as $^K W^J$.

For $P \in \mathcal{P}_J$ and $Q \in \mathcal{P}_K$, we write $\mathrm{pos}(P, Q) = w$ if $w \in {}^J W^K$ and there exists $g \in G$ such that $P = g P_J g^{-1}$, $Q = g \dot{w} P_K \dot{w}^{-1} g^{-1}$, where $\dot{w}$ is a representative of w in $N(T)$.

For $g \in G$ and $H \subset G$, we write $^g H$ for gHg^{-1}.

We first recall some combinatorial results.

1.2. For $J \subset I$, let $\mathcal{T}(J, \delta)$ be the set of sequences $(J_n, w_n)_{n \geq 0}$ such that
(a) $J_0 = J$,
(b) $J_n = J_{n-1} \cap \mathrm{Ad}(w_{n-1}) \delta(J_{n-1})$ for $n \geq 1$,
(c) $w_n \in {}^{J_n} W^{\delta(J_n)}$ for $n \geq 0$,
(d) $w_n \in W_{J_n} w_{n-1} W_{\delta(J_{n-1})}$ for $n \geq 1$.
Then for any sequence $(J_n, w_n)_{n \geq 0} \in \mathcal{T}(J, \delta)$, we have that $w_n = w_{n+1} = \cdots$ and $J_n = J_{n+1} = \cdots$ for $n \gg 0$. By [**Be**], the assignment $(J_n, w_n)_{n \geq 0} \mapsto w_m^{-1}$ for $m \gg 0$ defines a bijection $\mathcal{T}(J, \delta) \to W^J$.

Now we prove some result that will be used in the proof of Lemma 2.5.

LEMMA 1.1. *Let $(J_n, w_n)_{n \geq 0} \in \mathcal{T}(J, \delta)$ be the element that corresponds to w. Then*

(1) $w(L_J \cap U_{P_{J_1}}) w^{-1} \subset U_{P_{\delta(J)}}$.
(2) $w(L_{J_i} \cap U_{P_{J_{i+1}}}) w^{-1} \subset L_{\delta(J_{i-1})} \cap U_{P_{\delta(J_i)}}$ for $i \geq 1$.

PROOF. We only prove part (1). Part (2) can be proved in the same way.

Assume that part (1) is not true. Then there exists $\alpha \in \Phi_J^+ - \Phi_{J_1}^+$ such that $wa \in \Phi_{\delta(J)}^+$. Let $i \in J - J_1$ with $\alpha_i \leq \alpha$. Since $w \in W^J$, we have that $w\alpha_i \in \Phi_{\delta(J)}^+$. By definition, $w^{-1} = w_1 v$ for some $v \in W_{\delta(J)}$. Then $\alpha_i \in w^{-1}\Phi_{\delta(J)}^+ = w_1 v \Phi_{\delta(J)}^+ = w_1 \Phi_{\delta(J)}$. Since $w_1 \in W^{\delta(J)}$, we must have $\alpha_i = w_1 \alpha_j$ for some $j \in \delta(J)$. Hence $i \in J_1$, which is a contradiction. Part (1) is proved. $\square$

1.3. Define a W_J-action on W by $x \cdot y = \delta(x) y x^{-1}$. For $w \in W^J$, set
$$I(J, \delta; w) = \max\{K \subset J; \mathrm{Ad}(w)(K) = \delta(K)\}$$
and $[w]_J = W_J \cdot (w W_{I(J, \delta; w)})$. Then $W = \sqcup_{w \in W^J} [w]_J$. See [**H4**, Corollary 2.6].

Given $w, w' \in W$ and $j \in J$, we write $w \xrightarrow{s_j}_\delta w'$ if $w' = s_{\delta(j)} w s_j$ and $l(w') \leq l(w)$. If $w = w_0, w_1, \cdots, w_n = w'$ is a sequence of elements in W such that for all k, we have $w_{k-1} \xrightarrow{s_j}_\delta w_k$ for some $j \in J$, then we write $w \to_{J, \delta} w'$.

We call $w, w' \in W$ *elementarily strongly (J, δ)-conjugate* if $l(w) = l(w')$ and there exists $x \in W_J$ such that $w' = \delta(x) w x^{-1}$ and either $l(\delta(x)w) = l(x) + l(w)$ or $l(wx^{-1}) = l(x) + l(w)$. We call w, w' *strongly (J, δ)-conjugate* if there is a sequence $w = w_0, w_1, \cdots, w_n = w'$ such that w_{i-1} is elementarily strongly (J, δ)-conjugate to w_i for all i. We will write $w \sim_{J, \delta} w'$ if w and w' are strongly (J, δ)-conjugate. If $w \sim_{J, \delta} w'$ and $w \to_{J, \delta} w'$, then we say that w and w' are in the same (J, δ)-cyclic shift and write $w \approx_{J, \delta} w'$. Then it is easy to see that $w \approx_{J, \delta} w'$ if and only if $w \to_{J, \delta} w'$ and $w' \to_{J, \delta} w$.

By [**H4**, Proposition 3.4], we have the following properties:
(a) for any $w \in W$, there exists $w_1 \in W^J$ and $v \in W_{I(J, \delta; w_1)}$ such that $w \to_{J, \delta} w_1 v$.

(b) if w, w' are in the same W_J-orbit $\mathcal{O}$ of W and w, w' are of minimal length in $\mathcal{O}$, then $w \sim_{J,\delta} w'$. If moreover, $\mathcal{O} \cap W^J \neq \emptyset$, then $w \approx_{J,\delta} w'$.

1.4. By [**H4**, Corollary 4.5], for any W_J-orbit $\mathcal{O}$ and $v \in \mathcal{O}$, the following conditions are equivalent:

(1) v is a minimal element in $\mathcal{O}$ with respect to the restriction to $\mathcal{O}$ of the Bruhat order on W.

(2) v is an element of minimal length in $\mathcal{O}$.

We denote by $\mathcal{O}_{\min}$ the set of elements in $\mathcal{O}$ satisfy the above conditions. The elements in $(W_J \cdot w)_{\min}$ for some $w \in W^J$ are called *distinguished elements* (with respect to J and δ).

As in [**H4**, 4.7], we have a natural partial order $\leq_{J,\delta}$ on W^J defined as follows:

Let $w, w' \in W^J$. Then $w \leq_{J,\delta} w'$ if for some (or equivalently, any) $v' \in (W_J \cdot w')_{\min}$, there exists $v \in (W_J \cdot w)_{\min}$ such that $v \leq v'$.

In general, for $w \in W^J$ and $w' \in W$, we write $w \leq_{J,\delta} w'$ if there exists $v \in (W_J \cdot w)_{\min}$ such that $v \leq w'$.

2. G_F-stable pieces

2.1. For $J \subset I$, set $Z_J = \{(P, gU_P); P \in \mathcal{P}_J, g \in G\}$ with the $G \times G$-action defined by

$$(g_1, g_2) \cdot (P, gU_P) = ({}^{g_2}P, g_1 g U_P g_2^{-1}).$$

Set $h_J = (P_J, U_{P_J})$. Then the isotropic subgroup R_J of h_J is $\{(lu_1, lu_2); l \in L_J, u_1, u_2 \in U_{P_J}\}$. It is easy to see that

$$Z_J \cong (G \times G)/R_J.$$

Set $G_F = \{(g, F(g)); g \in G\} \subset G \times G$. For $w \in W^J$, set

$$Z_{J,F;w} = G_F(B, BwB) \cdot h_J.$$

We call $Z_{J,F;w}$ a G_F-stable piece of Z_J.

LEMMA 2.1. *Let $w, w' \in W$.*
(1) If $w \rightarrow_{J,\delta} w'$, then

$$G_F(B, BwB) \cdot h_J \subset G_F(B, Bw'B) \cdot h_J \cup \cup_{v < w} G_F(B, BvB) \cdot h_J.$$

(2) If $w \approx_{J,\delta} w'$, then

$$G_F(B, BwB) \cdot h_J = G_F(B, Bw'B) \cdot h_J.$$

PROOF. It suffices to prove the case where $w \xrightarrow{s_j}_\delta w'$ for some $j \in J$. Notice that $F(Bs_iB) = Bs_{\delta(i)}B$ for $i \in I$.
If $ws_j < w$, then

$$\begin{aligned}
(B, BwB) \cdot h_J &= (B, Bws_jB)(B, Bs_jB) \cdot h_J \\
&= (Bs_jB, Bws_jB) \cdot h_J \\
&\subset G_F(B, Bs_{\delta(j)}Bws_jB, B) \cdot h_J \\
&\subset G_F(B, Bw'B) \cdot h_J \cup G_F(B, Bws_jB) \cdot h_J.
\end{aligned}$$

If moreover, $l(w') = l(w)$, then $Bs_{\delta(j)}Bws_jB = Bw'B$ and $G_F(B, BwB) \cdot h_J = G_F(B, Bw'B) \cdot h_J$.

If $s_{\delta(j)}w < w$, then

$$\begin{aligned}
(B, BwB) \cdot h_J &= (B, Bs_{\delta(j)}B)(B, Bs_{\delta(j)}wB) \cdot h_J \\
&\subset G_F(Bs_jB, Bs_{\delta(j)}wB) \cdot h_J \\
&= G_F(B, Bs_{\delta(j)}wBs_jB) \cdot h_J \\
&\subset G_F(B, Bw'B) \cdot h_J \cup G_F(B, Bs_{\delta(j)}wB) \cdot h_J.
\end{aligned}$$

If moreover, $l(w') = l(w)$, then $Bs_{\delta(j)}wBs_jB = Bw'B$ and $G_F(B, BwB) \cdot h_J = G_F(B, Bw'B) \cdot h_J$.

If $ws_{\delta(j)} > w$ and $s_jw > w$, then $l(w') = l(w)$. By [**L1**, Proposition 1.10], $w' = w$. The statements automatically hold in this case. $\qquad\square$

LEMMA 2.2. *We have that* $Z_J = \cup_{w \in W^J} Z_{J,F;w}$.

REMARK. We will see in subsection 2.3 that Z_J is the disjoint union of $Z_{J,F;w}$ for $w \in W^J$.

PROOF. Let $z \in Z_J$. Since $G \times G$ acts transitively on Z_J, z is contained in the G-orbit of an element $(1, g) \cdot h_J$ for some $g \in G$. By the Bruhat decomposition of G, we have that $z \in G_F(1, Bw_1B) \cdot h_J$ for some $w_1 \in W$. We may assume furthermore that w_1 is of minimal length among all the Weyl group elements w_1' with $z \in G_F(1, Bw_1'B) \cdot h_J$.

By part (1) of the previous lemma and 1.3 (a),

$$z \in G_F(B, BwvB) \cdot h_J \cup \cup_{l(w') < l(w_1)} G_F(B, Bw'B) \cdot h_J$$

for some $w \in W^J$ and $v \in W_{I(J,\delta;w)}$. By our assumption on w_1, we have that $z \in G_F(B, BwvB) \cdot h_J$ and $l(wv) = l(w_1)$. In particular, z is contained in the G_F-orbit of an element $(1, g'l) \cdot h_J$ for some $l \in L_K$ and $g' \in U_{P_{\delta(K)}}wU_{P_K}$, where $K = I(J, \delta; w)$.

Set $F' : L_K \to L_K$ by $F'(l_1) = w^{-1}F(l_1)w$. By Lang's theorem for F', we can find $l_1 \in L_K$ such that $F'(l_1)ll_1^{-1} = 1$. Then

$$\begin{aligned}
(l_1, F(l_1))(1, g'l) \cdot h_J &= (1, F(l_1)g'll_1^{-1}) \cdot h_J \\
&\in (1, U_{P_{\delta(K)}}wF'(l_1)ll_1^{-1}U_{P_K}) \cdot h_J \subset (1, BwB) \cdot h_J
\end{aligned}$$

and $z \in Z_{J,F;w}$. $\qquad\square$

2.2. For any parabolic subgroups P and Q of G, we set $P^Q = (P \cap Q)U_P$. It is known that P^Q is a parabolic subgroup of G. The following properties are easy to check.

(1) For any $g \in G$, $({}^gP)^{({}^gQ)} = {}^g(P^Q)$.

(2) If $P \in \mathcal{B}$, then $P^Q = P$ for any parabolic subgroup Q.

LEMMA 2.3. *Let* $J, K \subset I$ *and* $w \in {}^JW$. *Set* $J_1 = J \cap \mathrm{Ad}(w_1)K$, *where* $w_1 = \min(wW_K)$. *Then for* $g \in BwB$, *we have that* $P_J^{({}^gP_K)} = P_{J_1}$.

PROOF. By 2.2 (1), it suffices to prove the case where $g = \dot{w}$. Now

$$\begin{aligned}
(P_J)^{({}^{\dot w}P_K)} &= (P_J)^{({}^{\dot w_1}P_K)} = (L_J \cap {}^{\dot w_1}L_K)(L_J \cap {}^{\dot w_1}U_{P_K})(U_{P_J} \cap {}^{\dot w_1}P_K)U_{P_J} \\
&= (L_J \cap {}^{\dot w_1}L_K)((L_J \cap U) \cap {}^{\dot w_1}L_K)(L_J \cap {}^{\dot w_1}U_{P_K})(U_{P_J} \cap {}^{\dot w_1}P_K)U_{P_J}.
\end{aligned}$$

Since $w \in {}^J W$ and $w_1 = \min(w W_K)$, we have that $w_1 \in {}^J W^K$. Therefore $L_J \cap {}^{\dot{w}_1} L_K = L_{J_1}$ and

$$
\begin{aligned}
((L_J &\cap U) \cap {}^{\dot{w}_1} L_K)(L_J \cap {}^{\dot{w}_1} U_{P_K}) \\
&= ((L_J \cap U) \cap {}^{\dot{w}_1} L_K)((L_J \cap U) \cap {}^{\dot{w}_1} U_{P_K}) \\
&= (L_J \cap U) \cap {}^{\dot{w}_1} P_K = L_J \cap U.
\end{aligned}
$$

So $(P_J)^{({}^{\dot{w}} P_K)} = L_{J_1}(L_J \cap U) U_{P_J} = L_{J_1} U = P_{J_1}$. $\qquad\square$

LEMMA 2.4. *To each $(P, g U_P) \in Z_J$, we associate a sequence $(P^n, J_n, w_n)_{n \geq 0}$ as follows*

$$
P^0 = P, \quad P^n = (P^{n-1})^{F({}^g P^{n-1})} \qquad \text{for } n \geq 1,
$$
$$
J_n \subset I \text{ with } P^n \in \mathcal{P}_{J_n}, \quad w_n = \operatorname{pos}(P^n, F({}^g P^n)) \qquad \text{for } n \geq 0.
$$

Let $w \in W^J$. Let $(P, g U_P) \in Z_{J,F;w}$ and $(P^n, J_n, w_n)_{n \geq 0}$ be the sequence associated to $(P, g U_P)$. Then $(J_n, w_n)_{n \geq 0} \in \mathcal{T}(J, \delta)$ and $w_m^{-1} = w$ for $m \gg 0$.

PROOF. Using 2.2 (1), it is easy to see by induction on n that the sequence associated to $({}^{F(h)} P, h g U_P F(h)^{-1})$ is $({}^{F(h)} P^n, J_n, w_n)_{n \geq 0}$. Then it suffices to prove the case where $(P, g U_P) = (P_J, k U_{P_J})$ for some $k \in B \delta^{-1}(w)^{-1} B$.

Let $(J_n', w_n')_{n \geq 0} \in \mathcal{T}(J, \delta)$ be the element that corresponds to w. Then $w_n' = \min(w^{-1} W_{\delta(J_n)})$ for $n \geq 0$. By the previous lemma, we can show by induction on n that $P^n = P_{J_n'}$ for all $n \geq 0$. Then $J_n = J_n'$ for $n \geq 0$. Moreover, $w_n = \operatorname{pos}(P^n, F({}^k P^n)) = \operatorname{pos}(P_{J_n}, {}^{F(k)} P_{\delta(J_n)}) = w_n'$ since $k \in B \delta^{-1}(w)^{-1} B$. $\qquad\square$

(A similar result with a similar proof appears in [**H1**, Lemma 2.3].)

2.3. We can now define a map $\beta : Z_J \to W^J$ by $\beta(P, g U_P) = w_m^{-1}$ for $m \gg 0$, where $(P^n, J_n, w_n)_{n \geq 0}$ is the sequence associated to $(P, g U_P)$. Then $Z_J = \sqcup_{w \in W^J} \beta^{-1}(w)$ is a partition of Z_J into locally closed subvarieties. Since $Z_{J,F;w} \subset \beta^{-1}(w)$ and $Z_J = \cup_{w \in W^J} Z_{J,F;w}$, we have that $Z_{J,F;w} = \beta^{-1}(w)$ and

$$
Z_J = \sqcup_{w \in W^J} Z_{J,F;w}.
$$

Fix $w \in W^J$ and let $(J_n, w_n)_{n \geq 0}$ be the element in $\mathcal{T}(J, \delta)$ that corresponds to w. Clearly, the map $(P, g U_P) \mapsto P^m$ for $m \gg 0$ is a morphism $\vartheta : Z_{J,F;w} \to \mathcal{P}_{I(J, \delta; w)}$.

LEMMA 2.5. *Let $w \in W^J$. Set $x = \delta^{-1}(w)^{-1}$ and $K = \delta^{-1} I(J, \delta; w)$. Then*

$$
(U_{P_K})_F(x, 1) \cdot h_J = (U_{P_K} x, U_{P_{\delta(K)}}) \cdot h_J.
$$

PROOF. Notice that

$$
\begin{aligned}
(U_{P_K} x, U_{P_{\delta(K)}}) \cdot h_J &= (U_{P_K})_F(x, U_{P_{\delta(K)}}) \cdot h_J \\
&= (U_{P_K})_F(x, U_{P_{\delta(K)}} \cap L_J) \cdot h_J \\
&= (U_{P_K})_F(x(U_{P_{\delta(K)}} \cap L_J), 1) \cdot h_J.
\end{aligned}
$$

So it suffices to show that for any $v \in U_{P_{\delta(K)}} \cap L_J$, there exists $u \in U_{P_K} \cap L_{\delta^{-1}(J)}$ such that $x^{-1} u x F(u)^{-1} \in v U_{P_J}$.

Let $(J_n, w_n)_{n \geq 0} \in \mathcal{T}(J, \delta)$ be the element that corresponds to w. By Lemma 1.1,

$$x^{-1}(L_{\delta^{-1}(J)} \cap U_{P_{\delta^{-1}(J_1)}})x \subset U_{P_J},$$

$$x^{-1}(L_{\delta^{-1}(J_i)} \cap U_{P_{\delta^{-1}(J_{i+1})}})x \subset L_{J_{i-1}} \cap U_{P_{J_i}} \text{ for } i \geq 1.$$

We have that $\delta(K) = J_m$ for some $m \in \mathbb{N}$. Now $v = v_m v_{m-1} \cdots v_0$ for some $v_i \in L_{J_i} \cap U_{P_{J_{i+1}}}$. We define $u_i \in L_{\delta^{-1}(J_i)} \cap U_{P_{\delta^{-1}(J_{i+1})}}$ as follows:

Let $u_m = 1$. Assume that $k < m$ and that $u_i \in L_{\delta^{-1}(J_i)} \cap U_{P_{\delta^{-1}(J_{i+1})}}$ are already defined for $k < i \leq m$ and that

$$(x^{-1}(u_m u_{m-1} \cdots u_{k+2})^{-1}x)F(u_m u_{m-1} \cdots u_{k+1})^{-1} = v_m v_{m-1} \cdots v_{k+1}.$$

Let u_k be the element with

$$F(u_k)^{-1} = (x^{-1}(u_m u_{m-1} \cdots u_{k+1})x)v_m v_{m-1} \cdots v_k F(u_m u_{m-1} \cdots u_{k+1})$$

$$= (x^{-1}u_{k+1}x)F(u_m u_{m-1} \cdots u_{k+1})^{-1}v_k F(u_m u_{m-1} \cdots u_{k+1})$$

$$\in L_{J_K} \cap U_{P_{J_{k+1}}}.$$

Thus $u_{k+1} \in L_{\delta^{-1}(J_k)} \cap U_{P_{\delta^{-1}(J_{k+1})}}$ and that

$$(x^{-1}(u_m u_{m-1} \cdots u_{k+1})x)F(u_m u_{m-1} \cdots u_k)^{-1} = v_m v_{m-1} \cdots v_k.$$

This completes the inductive definition.

Now set $u = u_m u_{m-1} \cdots u_0$. Then

$$(x^{-1}ux)F(u)^{-1}$$

$$= (x^{-1}(u_m u_{m-1} \cdots u_1)x)F(u)^{-1}(F(u)(x^{-1}u_0 x)F(u)^{-1})$$

$$\in vU_{P_{\delta(J)}}.$$

The lemma is proved.   □

By the proof of Lemma 2.2,

$$Z_{J,F;w} = G_F(U_{P_{\delta^{-1}I(J,\delta;w)}}\delta^{-1}(w)^{-1}U_{P_{\delta(I(J,\delta;w))}}, 1) \cdot h_J.$$

Then we have the following consequence.

COROLLARY 2.6. *Let $w \in W^J$. Then G_F acts transitively on $Z_{J,F;w}$.*

REMARK. Therefore there are only finitely many G_F-orbits on Z_J and they are indexed by W^J. This is quite different from the set of G_Δ-orbits on Z_J.

PROPOSITION 2.7. *Let $w \in W$. Then*

$$\overline{G_F(B, Bw) \cdot h_J} = \sqcup_{w' \in W^J, w' \leq_{J, \delta} w} Z_{J, F; w'}.$$

REMARK. Similar results appear in [H3, Proposition 4.6], [H2, Corollary 5.5] and [H4, Proposition 5.8]. The following proof is similar to the proof of [H4, Proposition 5.8].

PROOF. We prove by induction on $l(w)$.

Using the proper map $p : G_F \times_{B_F} Z_J \to Z_J$ defined by $(g, z) \mapsto (g, F(g)) \cdot z$, one can prove that

$$\overline{G_F(B, Bw) \cdot h_J} = G_F \overline{(B, Bw) \cdot h_J} = \cup_{v \leq w} G_F(B, Bv) \cdot h_J.$$

By 1.3 (a), $w \to_{J,\delta} w_1 v$ for some $w_1 \in W^J$ and $v \in W_{I(J,\delta;w)}$. By Lemma 2.1,

$$\overline{G_F(B, Bw) \cdot h_J} = G_F(B, Bw) \cdot h_J \cup \cup_{w' < w} G_F(B, Bw') \cdot h_J$$
$$= G_F(B, Bw_1 v) \cdot h_J \cup \cup_{w' < w} G_F(B, Bw') \cdot h_J.$$

By the proof of Lemma 2.2, $G_F(B, Bw_1 v) \cdot h_J \subset G_F(B, Bw_1) \cdot h_J$. Thus by induction hypothesis,

$$\overline{G_F(B, Bw) \cdot h_J} \subset \cup_{w' \in W^J, w' \leq_{J,\delta} w} Z_{J,F;w'}.$$

On the other hand, if $w' \in W^J$ with $w' \leq_{J,\delta} w$, then there exists $w'' \approx_{J,\delta} w'$ with $w'' \leq w$. Then by Lemma 2.1,

$$Z_{J,F;w'} = G_F(B, Bw'') \cdot h_J \subset \overline{G_F(B, Bw) \cdot h_J}.$$

Therefore $\overline{G_F(Bw, B) \cdot h_J} = \cup_{w' \in W^J, w' \leq_{J,\delta} w} Z_{J,F;w'}$. By 2.3, $Z_{J,F;w_1} \cap Z_{J,F;w_2} = \emptyset$ if $w_1, w_2 \in W^J$ and $w_1 \neq w_2$. Thus $\overline{G_F(Bw, B) \cdot h_J} = \sqcup_{w' \in W^J, w' \leq_{J,\delta} w} Z_{J,F;w'}$. The proposition is proved. $\square$

3. A stratification of partial flag varieties

3.1. It is easy to see that there is a canonical bijection between the G_F-orbits on Z_J and the R_J-orbits on $(G \times G)/G_F$ which sends $G_F(1, w) \cdot h_J$ to $R_J(1, w^{-1}) G_F/G_F$. Notice that the map $(g_1, g_2) \mapsto g_2 F(g_1)^{-1}$ gives an isomorphism of R_J-varieties $(G \times G)/G_F \cong G$, where the R_J-action on G is defined by

$$(lu_1, lu_2) \cdot g = lu_2 g F(lu_1)^{-1}.$$

Using the results of G_F-orbits on Z_J above, we have the following results.

(1) For $w \in {}^J W$, $R_J \cdot w = R_J \cdot (BwB)$. If moreover, $w' \in (W_J \cdot w)_{\min}$, then $R_J \cdot (Bw'B) = R_J \cdot (BwB)$.

(2) $G = \sqcup_{w \in {}^J W} R_J \cdot w$.

(3) For $w \in W$, $\overline{R_J \cdot w} = \sqcup_{w' \in {}^J W, (w')^{-1} \leq_{J,\delta} w^{-1}} R_J \cdot w'$.

Notice that if $J = \emptyset$, part (2) above follows easily from Bruhat decomposition. One may regard (2) as an extension of Bruhat's Lemma. We will also discuss a variation of (2) in section 4.

3.2. Now we review the partition on $\mathcal{P}_J$ introduced by Lusztig in [**L3**, section 4].

To each $P \in \mathcal{P}_J$, we associate a sequence $(P^n, J_n, w_n)_{n \geq 0}$ as follows

$$P^0 = P, \quad P^n = (P^{n-1})^{F(P^{n-1})} \text{ for } n \geq 1,$$

$$J_n \subset I \text{ with } P^n \in \mathcal{P}_{J_n}, \quad w_n = \text{pos}(P^n, F(P^n)) \qquad \text{for } n \geq 0.$$

By [**L3**, 4.2], $(J_n, w_n)_{n \geq 0} \in \mathcal{T}(J)$. Thus we have a map $i : \mathcal{P}_J \to {}^J W$. For $w \in {}^J W$, let

$$\mathcal{P}_{J,w} = \{P \in \mathcal{P}_J; w_m = w \text{ for } m \gg 0\}.$$

Then $\mathcal{P}_J = \sqcup_{w \in {}^J W} \mathcal{P}_{J,w}$.

It is easy to see that $\mathcal{P}_{J,w} = \{P \in \mathcal{P}_J; (P, U_P) \in Z_{J,F;w^{-1}}\}$.

Notice that $\text{Lie}(G_\Delta) + \text{Lie}(G_F) = \text{Lie}(G) \oplus \text{Lie}(G)$. Then for any $x \in Z_J$, $G_\Delta \cdot x$ and $G_F \cdot x$ intersects transversally at x. In particular, $\mathcal{P}_{J,w}$ is the transversal intersection of $G_\Delta \cdot h_J$ and $Z_{J,F;w^{-1}}$.

We simply write $\mathcal{P}_{\emptyset,w}$ as $\mathcal{B}_w$. By 3.2 (3),

$$\mathcal{B}_w = \{B_1 \in \mathcal{B}; \text{pos}(B_1, F(B_1)) = w\} = \{{}^g B; g^{-1} F(g) \in B\dot{w}B\}.$$

Since the Lang isogeny $g^{-1}F(g)$ is an isomorphism $G^F \backslash G \to G$, we have that

(a)
$$\overline{\mathcal{B}_w} = \sqcup_{v \leq w} \mathcal{B}_v.$$

Now we can prove our main theorem.

THEOREM 3.1. *Let $p : \mathcal{B} \to \mathcal{P}_J$ be the morphism which sends a Borel subgroup B' to the unique parabolic subgroup in $\mathcal{P}_J$ that contains B'. Then*
(1) For $w \in {}^J W$, $p(\mathcal{B}_w) = \mathcal{P}_{J,w}$. If moreover, $v \in (W_J \cdot w)_{\min}$, then $p(\mathcal{B}_v) = p(\mathcal{B}_w) = \mathcal{P}_{J,w}$.
(2) For $w \in W$, $\overline{\mathcal{P}_{J,w}} = p(\overline{\mathcal{B}_w}) = \sqcup_{w' \in {}^J W, (w')^{-1} \leq_{J,\delta} w^{-1}} \mathcal{P}_{J,w'}$.

REMARK. The closure relation of $\mathcal{P}_{J,w}$ was conjectured by G. Lusztig in private conversation.

PROOF. (1) Let $w \in {}^J W$ and $g \in G$ with ${}^g B \in \mathcal{B}_w$. Then $g^{-1}F(g) \in BwB$. Thus

$$({}^g P_J, U_{P_J}) = (g,g) \cdot h_J = (g, F(g))(1, F(g)^{-1}g) \cdot h_J$$
$$\in G_F(B, Bw^{-1}) \cdot h_J = Z_{J,F;w^{-1}}$$

and $p(\mathcal{B}_w) \subset \mathcal{P}_{J,w}$.

By 3.1, for any $g \in G$, there exists $l \in L_J$ such that $(gl)^{-1}F(gl) \in B\dot{w}'B$ for some $w' \in {}^J W$. Hence

(a)
$$p(\mathcal{B}) = \cup_{g \in G} p({}^g B) = \cup_{w' \in {}^J W} \cup_{g^{-1}F(g) \in B\dot{w}'B} p({}^g B)$$
$$= \cup_{w' \in {}^J W} p(\mathcal{B}_{w'}) \subset \sqcup_{w' \in {}^J W} \mathcal{P}_{J,w} = \mathcal{P}_J.$$

Since p is proper, we have that $p(\mathcal{B}) = \mathcal{P}_J$. Thus the inequality in (a) is actually an equality and $p(\mathcal{B}_{w'}) = \mathcal{P}_{J,w'}$ for all $w' \in {}^J W$.

If moreover, $v \in (W_J \cdot w)_{\min}$, then by 3.1, there exists $l \in L_J$ such that $(gl)^{-1}F(gl) = l^{-1}g^{-1}F(g)F(l) \in B\dot{w}B$. Thus $p({}^g B) = {}^g P_J = {}^{gl} P_J \in p(\mathcal{B}_w)$ and $p(\mathcal{B}_v) \subset p(\mathcal{B}_w)$. Similarly, we have that $p(\mathcal{B}_w) \subset p(\mathcal{B}_v)$. Then $p(\mathcal{B}_v) = p(\mathcal{B}_w)$.

Part (1) is proved.

(2) Since p is proper, we have that $\overline{\mathcal{P}_{J,w}} = p(\overline{\mathcal{B}_w})$. By 3.2 (a), $\overline{\mathcal{B}_w} = \sqcup_{v \leq w} \mathcal{B}_v$ and $p(\overline{\mathcal{B}_w}) = \cup_{v \leq w} p(\mathcal{B}_v) = \cup_{g \in G, g^{-1}F(g) \in \overline{B\dot{v}B}} p({}^g B)$. By 3.1 (3),

$$p(\overline{\mathcal{B}_w}) = \cup_{w' \in {}^J W, (w')^{-1} \leq_{J,\delta} w^{-1}} \cup_{g^{-1}F(g) \in B\dot{w}'B} p({}^g B)$$
$$= \cup_{w' \in {}^J W, (w')^{-1} \leq_{J,\delta} w^{-1}} p(\mathcal{B}'_w)$$
$$= \cup_{w' \in {}^J W, (w')^{-1} \leq_{J,\delta} w^{-1}} \mathcal{P}_{J,w'}.$$

Part (2) is proved.   □

Let us discuss some other properties of $\mathcal{P}_{J,w}$.

PROPOSITION 3.2. *Assume that G is quasi-simple and $J \neq I$. Then $\mathcal{P}_{J,w}$ is irreducible if and only if $\operatorname{supp}_\delta(w) = I$.*

PROOF. By [L3, 4.2 (d)], $\mathcal{P}_{J,w}$ is isomorphic to $\mathcal{P}_{K,w}$, where $K = I(J, \delta; w)$. By [BR, Theorem 2], $\mathcal{P}_{K,w}$ is irreducible if and only if wW_K is not contained in $W_{J'}$ for any δ-stable proper subset J' of I.

Let J' be the minimal δ-stable subset of I with $wW_K \subset W_{J'}$. It is easy to see that if $\operatorname{supp}_\delta(w) = I$, then $J' = I$. On the other hand, suppose that $\operatorname{supp}_\delta(w) \neq I$ and $J' = I$. Then for any $i \in K - \operatorname{supp}_\delta(w)$, we have that $w\alpha_i \in \delta(K)$. Since

$w\alpha_i \in \alpha_i + \sum_{j \in \mathrm{supp}(w)} \mathbb{Z}\alpha_j$, we must have that $w\alpha_i = \alpha_i$ and $i \in \delta(K)$. In particular, $K - \mathrm{supp}_\delta(w)$ is δ-stable, $w\alpha_i = \alpha_i$ for all $i \in K - \mathrm{supp}_\delta(w)$ and $K - \mathrm{supp}_\delta(w) = I - \mathrm{supp}_\delta(w)$. Since G is quasi-simple, there exists $i \in K - \mathrm{supp}_\delta(w)$ such that $(\alpha_i, \alpha_j^\vee) < 0$ for some $j \in \mathrm{supp}(w)$. Now assume that $w = s_{j_1} s_{j_2} \cdots s_{j_m}$ is a reduced expression and $m' = \max\{n; (\alpha_i, \alpha_{j_n}^\vee) \neq 0\}$. Then $s_{j_1} s_{j_2} \cdots s_{j_{m'}} \alpha_i = s_{j_1} s_{j_2} \cdots s_{j_m} \alpha_i = \alpha_i$. Thus

$$0 > (\alpha_i, \alpha_{j_{m'}}^\vee) = (s_{j_1} \cdots s_{j_{m'}} \alpha_i, s_{j_1} \cdots s_{j_{m'}} \alpha_{j_{m'}}^\vee) = (\alpha_i, s_{j_1} \cdots s_{j_{m'}} \alpha_{j_{m'}}^\vee).$$

However, $s_{j_1} \cdots s_{j_{m'}} \alpha_{j_{m'}}^\vee$ is a negative coroot. Thus

$$(\alpha_i, s_{j_1} \cdots s_{j_{m'}} \alpha_{j_{m'}}^\vee) \geq 0,$$

which is a contradiction. Therefore if $\mathrm{supp}_\delta(w) \neq I$, then $J' \neq I$. The proposition is proved. $\qquad\square$

3.3. By [**L3**, 4.2 (d)], $\mathcal{P}_{J,w}$ is isomorphic to $\mathcal{P}_{K,w}$, where $K = I(J, \delta; w)$. Similar to [**DL**, 1.11], we have that

$$\mathcal{P}_{K,w} = \{g \in G; g^{-1}F(g) \in P_K \dot{w} P_K\}/P_K$$
$$= \{g \in G; g^{-1}F(g) \in \dot{w} P_K\}/P_K \cap {}^{\dot{w}}P_K$$
$$= \{g \in G; g^{-1}F(g) \in \dot{w} U_{P_K}\}/L_K^{\mathrm{Ad}(\dot{w}) \circ F}(U_{P_K} \cap {}^{\dot{w}}U_{P_K}).$$

Let $P \in \mathcal{P}_{K,w}$ such that there exists a F-stable Levi subgroup L of P. Then similar to [**DL**, 1.17], we have that

$$\mathcal{P}_{K,w} = \{g \in G; g^{-1}F(g) \in PF(P)\}/P$$
$$= \{g \in G; g^{-1}F(G) \in F(P)\}/P \cap F(P)$$
$$= \{g \in G; g^{-1}F(g) \in F(U_P)\}/L^F(U_P \cap F(U_P)).$$

4. An extension of Bruhat decomposition

After the paper was submitted, I learned from A. Vasiu about his conjecture in [**Va**, 2.2.1]. We state it in the following slightly stronger version.

COROLLARY 4.1. *Let P be a parabolic subgroup of G of type J with a Levi subgroup L. Let $R = \{(lu, lu'); l \in L, u, u' \in U_P\}$ and define the action of R on G by $(lu, lu') \cdot g = lug F(lu')^{-1}$. Then*

(1) There are only finitely many R-orbits on G, indexed by ${}^J W$.

(2) If moreover, there exists a maximal torus $T' \subset P$ such that $F(T') = T'$, then each R-orbit contains an element in $N_G(T')$.

PROOF. We may assume that $P = {}^g P_J$ and $L = {}^g L_J$. For any $w \in {}^J W$, set $w^* = gwF(g)^{-1}$. Then it is to see that $R \cdot w^* = g(R_J \cdot w)F(g)^{-1}$. Now part (1) follows from 3.1 (2).

If moreover, $T' = {}^g T \subset P$ is F-stable, then we have that $g^{-1}F(g) \in N_G(T)$. Thus $w^* = gwF(g)^{-1} = g(wF(g)^{-1}g)g^{-1}$ and $wF(g)^{-1}g \in N_G(T)$. So $w^* \in N_G(T')$ and part (2) is proved. $\qquad\square$

Acknowledgements

We thank G. Lusztig for suggesting the problem and some helpful discussions. We also thank Z. Lin for explaining to me a conjecture by A. Vasiu and some helpful discussions. After the paper was submitted, I learned from T. A. Springer that he also obtained some similar results about the G_F-stable pieces in a different way using the approach in [**Sp**].

References

[Be] R. Bédard, *On the Brauer liftings for modular representations*, J. Algebra 93 (1985), no. 2, 332-353.

[BR] C. Bonnafé and R. Rouquier, *On the irreducibility of Deligne-Lusztig varieties*, Comptes Rendus Math. Acad. Sci. Paris, 343 (2006), 37–39.

[DL] P. Deligne and G. Lusztig, *Representations of reductive groups over finite fields*, Ann. of Math. (2) 103 (1976), no. 1, 103–161.

[H1] X. He, *Unipotent variety in the group compactification*, Adv. in Math. 203 (2006), 109-131.

[H2] X. He, *The character sheaves on the group compactification*, Adv. in Math., 207 (2006), 805-827.

[H3] X. He, *The G-stable pieces of the wonderful compactification*, Trans. Amer. Math. Soc., 359 (2007), 3005-3024.

[H4] X. He, *Minimal length elements in some double cosets of Coxeter groups*, Adv. in Math. 215 (2007), 469-503.

[L1] G. Lusztig, *Hecke algebras with unequal parameters*, CRM Monograph Series 18, American Mathematical Society, 2003.

[L2] G. Lusztig, *Parabolic character sheaves*, I, II, Mosc. Math. J. **4** (2004), no. 1, 153–179; no. 4, 869–896.

[L3] G. Lusztig, *A class of perverse sheaves on a partial flag manifold*, Represent. Theory 11 (2007), 122-171.

[Sp] T. A. Springer, *An extension of Bruhat's Lemma*, J. Alg. **313** (2007), 417–427.

[Va] A. Vasiu, *Mod p classification of Shimura F-crystals*, arXiv:math/0304030.

DEPARTMENT OF MATHEMATICS, STONY BROOK UNIVERSITY, STONY BROOK, NY 11794
E-mail address: hugo@math.sunysb.edu

Contemporary Mathematics
Volume **478**, 200 9

Steinberg representations and duality properties of arithmetic groups, mapping class groups, and outer automorphism groups of free groups

Lizhen Ji

ABSTRACT. An important cohomological property of a countable group Γ concerns its duality property of its homology and cohomology groups with coefficients. In this paper, we review the duality property of arithmetic subgroups of linear algebraic groups and the identification of the dualizing module with the Steinberg representation (or module) of the arithmetic subgroups, discuss duality property of the mapping class groups of surfaces (for example, they are duality groups but not Poincaré duality groups) and also the identification of the dualizing module with a generalized Steinberg module, and prove that the outer automorphism group $Out(F_n)$ of free groups F_n is also a duality group but not a Poincaré duality group. Based on the discussions in this paper, we also propose a conjecture that the outer automorphism group of a duality group is a virtual duality group. Hope that the discussions in this paper will add to the list of amazing applications to many different fields of the Steinberg representations.

CONTENTS

1991 *Mathematics Subject Classification.* Primary 57P10, 11F75, 30F60, 20F65.
Key words and phrases. Steinberg representation, arithmetic group, Tits building, mapping class group, Teichmüller space, outer automorphism group, duality group, Poincaré duality.
Partially supported by NSF grant DMS 0604878.

1. Introduction

There are many similarities between finite groups of Lie type (or finite Chevalley groups) and arithmetic subgroups of linear semisimple algebraic groups defined over $\mathbb{Q}$ (or over other global fields). For example, they are both naturally associated with algebraic groups over $\mathbb{Z}$ as in the case of $SL(n, \mathbb{Z}/p\mathbb{Z})$ (a finite group) and $SL(n, \mathbb{Z})$ (an arithmetic subgroup of $SL(n, \mathbb{Q})$) arising from the algebraic group $SL(n) = \{g \in GL(n, \mathbb{C}) \mid \det g = 1\}$, which is defined over $\mathbb{Z}$.

In the study of representation theory of finite groups of Lie type, the Steinberg representation and its variants have played a fundamental role. In fact, it has been used crucially in both the ordinary representations over $\mathbb{C}$ and modular representations of such finite groups, representations of algebraic groups over algebraically closed fields of positive characteristic, and representations of Lie algebras, and determination of the number of unipotent elements of finite Chevalley groups. See the comments and notes [St, pp. 580-586], the survey paper [Hu] and the book [Ca] for extended discussions about the Steinberg representations and their applications mentioned above. See also the book [Be] for the generalized Steinberg module for finites groups at primes p which divide the order of the groups, which is the Lefschetz module of the canonical action of the groups on the simplicial complex associated with the poset of p-subgroups.

There are several constructions of the Steinberg representation for finite Chevalley groups. One method is to obtain it as the top dimensional non-vanishing homology group of the Tits building of the associated algebraic group (or a BN-pair) in [So]. This method is the most useful one for the purpose of defining the generalized Steinberg modules (or representations) and for studying duality properties of groups considered in this paper. One reason is that the geometry at infinity of Lie groups and symmetric spaces is often described by the Tits building (or rather parabolic subgroups) of the Lie groups.

In fact, in the study of cohomology groups of arithmetic subgroups Γ of linear semisimple algebraic groups $\mathbf{G}$ defined over the field $\mathbb{Q}$ of rational numbers (or more generally over number fields), the Tits building $\Delta_{\mathbb{Q}}(\mathbf{G})$ of $\mathbf{G}$ also occurs naturally and an analogue of the Steinberg representation (or module) of $\mathbf{G}(\mathbb{Q})$ was used by Borel and Serre [BS1] to show that arithmetic subgroups Γ are virtual duality groups, but not virtual Poincaré duality groups if they are non-uniform, i.e., the quotient $\Gamma \backslash \mathbf{G}(\mathbb{R})$ is non-compact, which is equivalent to that the the $\mathbb{Q}$-rank r of $\mathbf{G}$ is positive (Theorem 6.1). In fact, the symmetric space X associated with its real locus $G = \mathbf{G}(\mathbb{R})$ admits the Borel-Serre partial compactification $\overline{X}^{BS}$ whose boundary is homotopy equivalent to the Tits building $\Delta_{\mathbb{Q}}(\mathbf{G})$. Various cohomological finiteness properties of torsion-free arithmetic subgroups Γ can be derived from the action of Γ on X and $\overline{X}^{BS}$, and the duality properties of Γ follow from the Poincaré duality for manifolds with boundary applied to $\overline{X}^{BS}$ and the Solomon-Tits theorem on the homotopy type of $\Delta_{\mathbb{Q}}(\mathbf{G})$ (or rather the induced Steinberg representation of Γ on $H_{r-1}(\Delta_{\mathbb{Q}}(\mathbf{G}))$). A particularly important connection with the previous case of finite Chevalley groups is that the dualizing module of arithmetic subgroups is equal to the generalized Steinberg module. In this paper, we also give a proof that non-uniform arithmetic subgroups are not Poincaré duality groups without using the Borel-Serre compactification and related results on Tits building, in the case the symmetric space $X = \mathbf{G}(\mathbb{R})/K$ associated with $\mathbf{G}$ is linear, i.e., a homothety section of a symmetric cone.

A natural generalization of arithmetic subgroups is the class of S-arithmetic subgroups such as $SL(n, \mathbb{Z}[\frac{1}{p_1}, \cdots, \frac{1}{p_m}])$, where $p_1, \cdots, p_m$ are prime numbers. The same results were proved for S-arithmetic subgroups of $\mathbf{G}(\mathbb{Q})$ in [BS2] (see Theorem 6.6 below). For this purpose, we need to use Bruhat-Tits buildings of $\mathbf{G}(\mathbb{Q}_{p_i})$ as replacement of symmetric spaces for the group $\mathbf{G}(\mathbb{Q}_{p_i})$ to act on properly. A similar result holds for S-arithmetic subgroups of reductive algebraic groups of rank 0 over function fields (Theorem 6.7).

Another important class of groups closely related to the class of arithmetic subgroups consists of the mapping class group $\Gamma_{g,p}$ of surfaces of genus g with p punctures. In [Ha1] (see also [Ha2] [I2]), Harer proved that $\Gamma_{g,p}$ is a virtual duality group. In a joint work with Ivanov [IJ], it was proved that $\Gamma_{g,p}$ is not a virtual Poincaré duality group. In this case, the dualizing module is also given by a generalized Steinberg module, which is the homology group of a natural simplicial complex, the curve complex $\mathcal{C}(S_{g,p})$. In the proofs, actions of $\Gamma_{g,p}$ on the Teichmüller space $T_{g,p}$, the Borel-Serre partial compactification $\overline{T_{g,p}}^{BS}$ and the curve complex $\mathcal{C}(S_{g,p})$, and a weak analogue of the Solomon-Tits theorem on the homotopy type of $\mathcal{C}(S_{g,p})$ (Theorem 8.4) are all used. Once the non-Poincaré duality of $\Gamma_{g,p}$ is proved, a stronger analogue of the Solomon-Tits theorem on the homotopy type of $\mathcal{C}(S_{g,p})$ can be derived (Theorem 8.5).

Yet another class of groups closely related to arithmetic groups consists of the outer automorphism group $Out(F_n)$ of free groups F_n on n generators, $n \geq 2$. The analogue of the symmetric spaces and Teichmüller spaces for $Out(F_n)$ is the outer space X_n consisting of suitable marked metric graphs where $Out(F_n)$ acts by changing the marking, introduced by Culler and Vogtmann in [CV]. In [BF], Bestvina and Feighn proved that $Out(F_n)$ is a virtual duality group (Theorem 10.1) by using an analogue of the Borel-Serre partial compactification of X_n. In this paper, we show that $Out(F_n)$ is not a virtual Poincaré duality group (Theorem 10.2). In this case, the dualizing module has not been identified, though there is some candidate for an analogue of the Steinberg module [HV].

The plan of the rest of this paper is as follows. In §2, we explain different Poincaré dualities of homology and cohomology groups of various kinds of manifolds: compact closed orientable manifolds, noncompact manifolds, and nonorientable manifolds, and also homology and cohomology with local coefficient systems. In §3, we introduce the homology and cohomology of groups with coefficients, and establish their relations with homology and cohomology with local coefficient systems of manifolds. In §4, motivated by the Poincaré dualities of manifolds, we define duality groups and Poincaré duality groups. We also discuss several criterions for duality groups and Poincaré duality groups. In §5, we introduce the spherical Tits building of a reductive algebraic group $\mathbf{G}$ over the field $\mathbb{Q}$, the Solomon-Tits theorem, and the Steinberg module of $\mathbf{G}(\mathbb{Q})$. We also remark that similar constructions and results hold for reductive algebraic groups over general fields k, in particular, finite fields and p-adic fields. In §6, we construct a finite CW-complex

$B\Gamma$-space for arithmetic subgroups Γ, explain the relation between its boundary and the spherical Tits building, and show that arithmetic subgroups are virtual duality groups whose dualizing module is given by the Steinberg module. We also prove that non-uniform arithmetic subgroups acting on linear symmetric spaces are not virtual Poincaré duality groups without using the Borel-Serre compactification and the Tits building. We conclude this section by explaining that S-arithmetic subgroups are also duality groups but not Poincaré duality groups in general. In §7, we introduce mapping class groups $Mod(S)$ of a surface S and the Teichmüller space $T(S)$ associated with S, and a realization of the Borel-Serre type compactification of $T(S)$ in terms of truncated submanifolds. In §8, we show that the curve complex $\mathcal{C}(S)$ plays the same role in describing the geometry at infinity of Teichmüller space as the Tits building in describing that of symmetric spaces, and explain how it can be used to prove that $Mod(S)$ is a duality group, and the non-Poincaré duality of $Mod(S)$ implies a stronger analogue of the Solomon-Tits theorem for $\mathcal{C}(S)$. In §9, we introduce the outer automorphism group $Out(F_n)$ and the associated outer space X_n. In §10, we first recall the result of Bestvina-Feighn that $Out(F_n)$ is a virtual duality group, and then prove that $Out(F_n)$ is not a virtual Poincaré duality group. In §11, we raise a conjecture that if Γ is a duality group, then $Out(\Gamma)$ is a virtual duality group and discuss cases where this conjecture is true.

By studying side-by-side duality properties of these three classes of groups: arithmetic groups, mapping class groups, and outer automorphism groups, we hope that we could bring out further their similarities and emphasize the role of the generalized Steinberg modules in studying them.

Acknowledgments.

I would like to thank K.Brown and M.Goresky for very helpful and informative correspondences about duality properties of homology and cohomology groups, and references. This paper is based on a talk at *The 4th International Conference on Representation Theory* held in Lhasa, Tibet, from July 16–20, 2007. I would like to thank the organizers for providing a stimulating environment and their hospitality during the conference. I would also like to thank an anonymous referee for reading the paper carefully and for simplifying the proof of Theorem 10.2.

2. Poincaré duality of closed manifolds and generalizations

In this section, we introduce many kinds of Poincaré dualities for manifolds, which could be noncompact, have nonempty boundary, or could be non-orientable. These versions will be needed in studying homology of groups and their duality properties in later sections. Besides the first standard version of the Poincaré duality, other versions do not seem to be so well-known to non-experts. In this paper, all manifolds are connected.

The basic Poincaré duality for oriented manifolds.

An important property of the cohomology and homology groups of a compact oriented manifold M^n is the well-known *Poincaré duality*: for every $i = 0, \cdots, n$, there is a canonical isomorphism

(2.1) $H^i(M, \mathbb{Z}) \cong H_{n-i}(M, \mathbb{Z}),$

which is obtained by the cap product with the fundamental class $[M]$ of M.

As a consequence, the Betti numbers b_i of M, which is equal to the rank of $H_i(M, \mathbb{Z})$, or the dimension of the real vector space $H_i(M, \mathbb{R})$, satisfy the equality: for every i,

$$b_i = b_{n-i}.$$

This is the original form of the duality as stated by Poincaré.

Another form of the Poincaré duality states that there is a well-defined intersection pairing

(2.2) $$H_i(M, \mathbb{Z}) \times H_{n-i}(M, \mathbb{Z}) \to H_0(M, \mathbb{Z}) \cong \mathbb{Z}$$

and that the pairing becomes nondegenerate when it is tensored with $\mathbb{Q}$:

(2.3) $$H_i(M, \mathbb{Q}) \times H_{n-i}(M, \mathbb{Q}) \to H_0(M, \mathbb{Q}) \cong \mathbb{Q}.$$

(Note that the tensor product with $\mathbb{Q}$ is important. Otherwise, the pairing in Equation (2.2) is not nondegenerate in general if the homology groups contain torsion elements.)

Poincaré-Lefschetz duality for manifolds with boundary.

The Poincaré duality in Equation 2.1 does not hold for noncompact manifolds or manifolds with nonempty boundary.

On the other hand, a modified version holds. Suppose that M^n is an oriented manifold with possible nonempty boundary ∂M. M is not necessarily compact. Let $H_c^i(M, \mathbb{Z})$ be the cohomology group of X with compact support. Then the *Poincaré duality* for *noncompact* M is the following isomorphism: for every $i \leq n$,

(2.4) $$H_c^i(M, \mathbb{Z}) \cong H_{n-i}(M, \partial M, \mathbb{Z}).$$

This duality is often called the *Poincaré-Lefschetz duality*. See [Ma, p. 343] [BS1, p. p. 473] [Br1, p. 211] for some discussions and references for this version of Poincaré duality. For later applications, we note that this Poincaré duality also holds for manifolds with corners, since the corners of the manifold can be smoothed out without changing its topology. We will apply it to $E\Gamma$, the universal covering space of a classifying space $B\Gamma$ of Γ when $B\Gamma$ is a compact manifold with boundary or corners.

Poincaré duality for closed non-orientable manifolds.

If M^n is a non-orientable closed manifold, then the Poincaré duality for M is usually stated in terms of homology and cohomology groups with the coefficient $\mathbb{Z}/2\mathbb{Z}$. For example, if M^n is a closed connected non-orientable manifold, then for every $i = 0, \cdots, n$, there is a canonical isomorphism

(2.5) $$H^i(M, \mathbb{Z}/2\mathbb{Z}) \cong H_{n-i}(M, \mathbb{Z}/2\mathbb{Z}).$$

See [Mu, p. 383] for example. On the other hand, in order to connect better with the non-orientable Poincaré duality groups defined in the next section, it is helpful to state a stronger version of this Poincaré duality in Equation (2.5) in terms of the orientation sheaf.

Specifically, let ω_M be the orientation sheaf of M. It can be defined in several different but equivalent ways. First, for any open subset $U \subset M$, assign the abelian group $\text{Hom}_{\mathbb{Z}}(H_c^n(U, \mathbb{Z}), \mathbb{Z})$, where $H_c^n(U, \mathbb{Z})$ is the cohomology with compact support. Then ω_M is the associated sheaf and is locally constant [Ive]. Alternatively, let $\tilde{M}$ be the universal covering space of M. Then $\tilde{M}$ is orientable. Let Ω be the orientation module of $\tilde{M}$, which is an infinite cyclic group whose two generators correspond to the two orientations of $\tilde{M}$. The fundamental group $\pi_1(M)$ acts on $\tilde{M}$ and also on Ω such that $\gamma \in \pi_1(M)$ acts on Ω as ± 1 depending on whether γ is orientation preserving or not. Then the associated locally system $\tilde{M} \times_{\pi_1(M)} \Omega$ on M is equal to the locally constant orientation sheaf ω.

In terms of the orientation sheaf ω, the manifold M is orientable if and only if ω is a constant sheaf with the stalk equal to $\mathbb{Z}$.

Then the general Poincaré duality for a closed manifold M^n, independent of whether it orientable or not, states that for every $i = 0, \cdots, n$, there is a canonical isomorphism

$$(2.6) \qquad H^i(M, \mathbb{Z}) \cong H_{n-i}(M, \omega \otimes \mathbb{Z}).$$

In the above equation, the group on the right hand, $H_{n-i}(M, \omega \otimes \mathbb{Z})$, is the homology of M with local coefficient system $\omega \otimes \mathbb{Z}$.

Since $\pi_1(M)$ acts trivially on $\Omega/2\Omega \cong \mathbb{Z}/2\mathbb{Z}$, the above duality in Equation (2.6) implies the Poincaré duality in Equation (2.5) for non-orientable closed manifolds.

Poincaré duality for closed manifolds with local coefficient systems.

More generally, for any local coefficient system E on M, and for every $i = 0, \cdots, n$, there is a canonical isomorphism

$$(2.7) \qquad H^i(M, E) \cong H_{n-i}(M, \omega \otimes E).$$

Suppose that M^n is a closed oriented manifold. Then the orientation sheaf is trivial, and for any local coefficient system E on M, we have the Poincaré duality:

$$(2.8) \qquad H^i(M, E) \cong H_{n-i}(M, E).$$

See [S] and [Ei] for definition and other details on homology and cohomology with local coefficient systems, and intersection theory of cycles and Poincaré duality for non-orientable manifolds.

There are also other versions of Poincaré duality for not necessarily orientable noncompact manifolds with boundary and with local coefficient systems, but they are not needed in this paper.

REMARK 2.1. If M is a compact topological space that is not a smooth manifold, then the Poincaré duality, in particular the version in Equations (2.1) or (2.3) does not hold for M. For some stratified spaces, there is a family of intersection homology groups, depending on some parameters, called perversities, which satisfy an analogue of the duality in Equation (2.3). See [GM] for details.

3. Homology and cohomology of groups

Given a group Γ with discrete topology, there exists a topological space (or rather a CW-complex) $B\Gamma$ which satisfies the conditions:

$$\pi_1(B\Gamma) = \Gamma, \quad \pi_i(B\Gamma) = 0, \ i \geq 2.$$

The latter vanishing conditions $\pi_i(B\Gamma) = 0$, $i \geq 2$, are equivalent to that the universal covering space $E\Gamma = \widetilde{B\Gamma}$ is contractible.

The space $B\Gamma$ is unique up to homotopy equivalence and is called the classifying space of Γ. (It is also the $K(\Gamma, 1)$-space in the family of $K(\Gamma, n)$-spaces, $n \geq 1$.) The space $E\Gamma$ admits a proper and fixed point free action of Γ and is a universal space for proper and fixed-point free actions of Γ. In fact, it is characterized by the following properties:

(1) $E\Gamma$ is contractible.

(2) $E\Gamma$ is a Γ-CW-complex such that Γ acts properly and fixed point freely.

In practice, an effective method to find good models of $B\Gamma$ is to find good models of $E\Gamma$ first, and then take the quotients $\Gamma\backslash E\Gamma$ as $B\Gamma$-spaces.

Since $B\Gamma$ depends on Γ up to homotopy equivalence, its homology groups $H_i(B\Gamma, \mathbb{Z})$ and cohomology groups $H^i(B\Gamma, \mathbb{Z})$ only depend on Γ. They give rise to homology and cohomology groups of Γ with the trivial coefficient $\mathbb{Z}$:

$$(3.1) \qquad H_i(\Gamma, \mathbb{Z}) = H_i(B\Gamma, \mathbb{Z}), \quad H^i(\Gamma, \mathbb{Z}) = H^i(B\Gamma, \mathbb{Z}).$$

For many applications to groups Γ, we also need to consider homology and cohomology groups $H_i(\Gamma, E)$, $H^i(\Gamma, E)$ of Γ with coefficients in $\mathbb{Z}\Gamma$-modules E (or more generally $R\Gamma$-modules, where R is a ring).

Usually they are defined in terms of a projective resolution $\mathbf{F} : \cdots \to F_2 \to F_1 \to F_0 \to E$. The homology is defined by

$$(3.2) \qquad H_i(\Gamma, E) = H_i(\mathbf{F} \otimes_\Gamma E),$$

where $\mathbf{F} \otimes_\Gamma E$ is the complex given by $F_i \otimes_\Gamma E = F_i \otimes E/\sim$, and the relation $\sim$ is defined by $f \otimes m \sim \gamma f \otimes \gamma m$, for $f \in F_i, m \in E, \gamma \in \Gamma$. The cohomology is defined by

$$(3.3) \qquad H^i(\Gamma, E) = H^i(\mathcal{H}om_\Gamma(F, E)).$$

See [Br1, Chap. III] for more detail.

If $E = \mathbb{Z}$, the trivial $\mathbb{Z}\Gamma$-module, then the chain complex of $E\Gamma = \widetilde{B\Gamma}$ gives a free and hence projective resolution of $\mathbb{Z}$, and the identification in Equation (3.1) follows. (Note that $B\Gamma$ has a CW-complex structure, and its lift to $E\Gamma$ is equivariant with respect to Γ).

If E is a nontrivial $\mathbb{Z}\Gamma$-module, then it gives a nontrivial local system on $B\Gamma$: $E\Gamma \times_\Gamma E$, still denoted by E, and similar geometric interpretations of the homology and cohomology groups of Γ hold:

$$(3.4) \qquad H^i(\Gamma, E) = H^i(B\Gamma, E), \quad H_i(\Gamma, E) = H_i(B\Gamma, E).$$

See [Ei, Theorem 28.1] for details.

An important cohomological invariant of a group Γ is its *cohomological dimension*, denoted by cd Γ, which is defined by:

$$(3.5) \qquad \text{cd } \Gamma = \sup\{i \in \mathbb{Z} \mid H^i(\Gamma, E) \neq 0, \text{ for some } \mathbb{Z}\Gamma\text{-module } E\}.$$

It is known that if cd $\Gamma < +\infty$, then Γ is torsion-free. If a group Γ contains nontrivial torsion elements but contains some torsion-free subgroups Γ' of finite index, then cd (Γ') is independent of the choice of Γ and is called the *virtual cohomological dimension* of Γ, denoted by vcd Γ.

Then the geometric interpretations in Equation 3.4 immediately implies the following.

PROPOSITION 3.1. *For any group Γ and any $B\Gamma$-space, cd $\Gamma \leq \dim B\Gamma$. In particular, if there exists a $B\Gamma$-space with $\dim B\Gamma < +\infty$, then cd $\Gamma < +\infty$.*

A group Γ is said to be of *type FP* if the trivial $\mathbb{Z}\Gamma$-module $\mathbb{Z}$ admits a finite length resolution by finitely generated projective modules. It follows immediately from the definition that if Γ is of type FP, then cd $\Gamma < +\infty$, and the homology and cohomology groups $H_i(\Gamma, \mathbb{Z})$ and $H^i(\Gamma, \mathbb{Z})$ are finitely generated in every degree i.

A group Γ is said to be of *type FP_∞* if the trivial $\mathbb{Z}\Gamma$-module $\mathbb{Z}$ admits a resolution by finitely generated projective modules. It also follows immediately from the definition that for every group Γ of type FP_∞, the homology and cohomology groups $H_i(\Gamma, \mathbb{Z})$ and $H^i(\Gamma, \mathbb{Z})$ are finitely generated in every degree i.

An effective method to prove that a group Γ is of type FP or FP_∞ is to construct good models of $B\Gamma$-spaces.

PROPOSITION 3.2. *If Γ admits a finite CW-complex $B\Gamma$-space, then Γ is of type FP. If Γ contains a subgroup Γ' of finite index which is of type FP, for example, if it admits a finite CW-complex $B\Gamma'$-space, then Γ is of type FP_∞.*

Proof. The first statement follows from the projective resolution of Γ induced from the chain complex of $E\Gamma$, and the second statement follows from the first and [Br1, Proposition 5.1, p. 197].

4. Duality groups, Poincaré duality groups

In this section, we first introduce the notion of duality groups, Poincaré duality groups, virtual duality groups, and virtual Poincaré duality groups. Then we give a geometric interpretation of the dualizing module (Proposition 4.4), and several useful criterions for duality properties (Propositions 4.5 and 4.8, and Corollary 4.9).

Let M be an oriented closed manifold. Suppose that M is aspherical, i.e., $\pi_i(M) = \{e\}$ for $i \geq 2$. Denote the fundamental group $\pi_1(M)$ by Γ. Then the Poincaré duality for M and the identifications $H^i(M, \mathbb{Z}) = H^i(\Gamma, \mathbb{Z})$ and $H_i(M, \mathbb{Z}) = H_i(\Gamma, \mathbb{Z})$ in Equation (3.1) implies the following duality property:

$$(4.1) \qquad\qquad H^i(\Gamma, \mathbb{Z}) \cong H_{n-i}(\Gamma, \mathbb{Z}).$$

By the Poincaré duality for homology groups with local coefficient systems in Equation (2.8) and the realization in Equation (3.4), it follows that for any Γ- (or $\mathbb{Z}\Gamma$-) module E, and every $i \geq 0$, there is a canonical isomorphism:

$$(4.2) \qquad\qquad H^i(\Gamma, E) \cong H_{n-i}(\Gamma, E).$$

Now let Γ be any group with discrete topology. If the group Γ satisfies the above condition in Equation (4.2) for all $\mathbb{Z}\Gamma$-modules E, then Γ is called *an orientable Poincaré duality group* of dimension n.

Assume that M is a closed non-orientable aspherical manifold. Its universal covering space $\tilde{M}$ is clearly orientable, and its orientation sheaf is trivial. Identify its orientation module Ω with $\mathbb{Z}$ after a choice of an orientation of M. Then $\Gamma = \pi_1(M)$ acts on $\mathbb{Z}$ nontrivially. The Poincaré duality for non-orientable manifolds in Equation (2.7) and the realization in Equation (3.4) imply that for every $\mathbb{Z}\Gamma$-module E, and every $i \geq 1$,

$$(4.3) \qquad\qquad H^i(\Gamma, E) \cong H_{n-i}(\Gamma, \mathbb{Z} \otimes E).$$

In general, if a group Γ acts on $\mathbb{Z}$, i.e., $\mathbb{Z}$ is a $\mathbb{Z}\Gamma$-module, and satisfies the above condition in Equation (4.3) for all $\mathbb{Z}\Gamma$-module E, then Γ is called *a Poincaré duality group* of dimension n. If the action of Γ on $\mathbb{Z}$ is trivial, then $\mathbb{Z} \otimes E \cong E$, and Γ

is an *orientable* Poincaré duality group. Otherwise, Γ is a *non-orientable* Poincaré duality group.

More generally, a group Γ is called *a duality group* (or *a generalized Poincaré duality group*) of dimension n if there exists a $\mathbb{Z}\Gamma$-module D such that for every $i \geq 0$ and every $\mathbb{Z}\Gamma$-module E, there exists an isomorphism

$$(4.4) \qquad H^i(\Gamma, E) \cong H_{n-i}(\Gamma, D \otimes E).$$

In this case, the module D is called the *dualizing module* of Γ. It is known [Fa, Theorem 3] that under the assumption that Γ is of type FP, if D is finitely generated abelian group, then D must be isomorphic to $\mathbb{Z}$ which is regarded as a Γ-module, and hence Γ is a Poincaré duality group of dimension n.

A group Γ is called a *virtual Poincaré duality group* if there exists a subgroup of finite index Γ' which is a Poincaré duality group. The notion of *virtual duality groups* can be defined similarly.

PROPOSITION 4.1. *If Γ is a duality group of dimension n, then Γ is of type FP, and the cohomological dimension of Γ is equal to n; in particular, Γ is torsion-free. If Γ is a virtual duality group of dimension n, then the virtual cohomological dimension of Γ is equal to n.*

For the first statement, see [Bi, Theorem 9.2], and for the second, see [Br1, p. 220, Theorem 10.1(iv)]. Therefore, duality groups enjoy various cohomological finiteness properties.

A useful algebraic criterion for duality is the following one [Br1, p. 220] [Fa, Theorem 3] (see also [Bi]).

PROPOSITION 4.2. *Let Γ be a group of type FP. Then Γ is a duality group of dimension n if and only if one of the following equivalent conditions holds:*
  (1) *$H^i(\Gamma, \mathbb{Z}\Gamma) = 0$ for $i \neq n$, and $H^n(\Gamma, \mathbb{Z}\Gamma)$ is a torsion-free abelian group.*
  (2) *for every abelian group E, the cohomology group $H^i(\Gamma, \mathbb{Z}\Gamma \otimes E) = 0$ for $i \neq n$.*

If the above conditions are satisfied, then $H^n(\Gamma, \mathbb{Z}\Gamma)$ is the dualizing module D of Γ. As an abelian group, D is either isomorphic to $\mathbb{Z}$ or infinitely generated.

REMARK 4.3. In the above proposition, if Γ is a nontrivial duality group, then the module $H^n(\Gamma, \mathbb{Z}\Gamma)$ is automatically nontrivial. In fact, by [Br, Proposition 6.7, p. 202], under the assumption that Γ is of type FP, the equality

$$(4.5) \qquad \operatorname{cd} \Gamma = \max\{i \mid H^i(\Gamma, \mathbb{Z}\Gamma) \neq 0\}$$

holds. Then the conditions in the proposition imply that $\operatorname{cd}\Gamma$ is finite. This implies that Γ is torsion-free and hence the non-triviality of Γ implies that $\Gamma \supseteq \mathbb{Z}$. By the monotonicity of the cohomological dimension, it follows that $\operatorname{cd}\Gamma \geq \operatorname{cd}\mathbb{Z} = 1$. Then Equation 4.5 again implies that $H^n(\Gamma, \mathbb{Z}\Gamma)$ is nontrivial.

Examples of Poincaré duality groups.

The existence of a good model of $B\Gamma$ is often the key to show that Γ is a duality group. As in the situation at the beginning of this section, if a group Γ admits a $B\Gamma$-space given by a *compact* aspherical manifold M, then Γ is a Poincaré duality group.

For example, it is clear that the free abelian group $\mathbb{Z}^n$ admits a $B\Gamma$-space given by the closed orientable aspherical manifold $\mathbb{Z}^n \backslash \mathbb{R}^n$ and is hence an orientable Poincaré duality group of dimension n.

More generally, every torsion-free, finitely generated nilpotent group Γ is also an orientable Poincaré duality group. In fact, by a known result of Malcev [M], such a group Γ can be embedded into a simply connected nilpotent Lie group N as a lattice. Since every lattice Γ in N is automatically uniform, i.e., $\Gamma\backslash N$ is compact, and N is diffeomorphic to $\mathbb{R}^n$, $n = \dim N$, the closed orientable manifold $\Gamma\backslash N$ gives a desired $B\Gamma$-space.

If G is a simply connected solvable Lie group and Γ is a lattice subgroup of G. Then Γ is torsion-free, and $\Gamma\backslash G$ is a closed manifold and gives a $B\Gamma$-space (see [Ra1]). In particular, Γ is a Poincaré duality group of dimension n, where $n = \dim G$.

BΓ-space given by manifolds with nonempty boundary.

If the *compactness* condition on $B\Gamma$ is dropped, or $B\Gamma$ is given by a compact manifold with *nonempty boundary*, then it is not obvious if Γ is a Poincaré duality group or even a duality group. For example, if Γ is a torsion-free non-uniform arithmetic subgroup of a semisimple linear algebraic group, such as a torsion-free subgroup of $SL(2,\mathbb{Z})$ of finite index, then there are $B\Gamma$-spaces given by compact manifold with boundary but Γ is not a Poincaré duality group.

On the other hand, the following result is very useful in studying the duality of Γ.

PROPOSITION 4.4. *Suppose that a $B\Gamma$-space is given by a finite CW-complex. Let $E\Gamma = \widetilde{B\Gamma}$ be the universal covering space of $B\Gamma$. Then*

$$H^i(\Gamma, \mathbb{Z}\Gamma) \cong H_c^i(E\Gamma, \mathbb{Z}).$$

In particular, if Γ is a Poincare duality group of dimension n, then its dualizing module is equal to $H_c^n(E\Gamma, \mathbb{Z})$.

Suppose that a $B\Gamma$-space is given by an n-dimensional smooth compact manifold with nonempty boundary. Then its universal covering space $E\Gamma$ is a contractible manifold with boundary. Fix an orientation of $E\Gamma$.

PROPOSITION 4.5. *Assume that Γ admits a $B\Gamma$-space given by an n-dimensional smooth compact manifold with nonempty boundary as above. If the boundary $\partial E\Gamma$ is homotopy equivalent to a bouquet of spheres S^{r-1} of dimension $r-1$ (i.e., a wedge product of S^{r-1}), then Γ is a duality group of dimension $n-r$. Furthermore, if the bouquet contains at least one but only finitely many spheres S^{r-1}, then it must contain exactly one sphere, and Γ is a Poincaré duality group in this case; otherwise, the bouquet contains infinitely many spheres and Γ is not a Poincaré duality group.*

Proof. By the Poincaré duality for manifolds with boundary in Equation (2.4),

$$H_c^i(E\Gamma, \mathbb{Z}) \cong H_{n-i}(E\Gamma, \partial E\Gamma, \mathbb{Z}).$$

By the long exact sequence for the pair $E\Gamma, \partial E\Gamma$, we obtain that

$$H_c^i(E\Gamma, \mathbb{Z}) \cong \tilde{H}_{n-i-1}(\partial E\Gamma, \mathbb{Z}).$$

Then the assumption on the homotopy type of $E\Gamma$ implies that $H_c^i(E\Gamma, \mathbb{Z})$ is equal to 0 for $i \neq n-r$ and is free for $n-r$, and hence Γ is a duality group of dimension $n-r$ by Proposition 4.2.

In the above proposition, if $B\Gamma$ is a closed manifold, the boundary $\partial E\Gamma$ is empty and hence of dimension -1, i.e., $r = 0$. This implies that Γ is a duality group of dimension n.

REMARK 4.6. Note that every smooth compact manifold with corners is homeomorphic to a smooth manifold with boundary by smoothing out the corners, and we can also assume the $B\Gamma$-space in the above proposition is a compact manifold with corners.

A related result on cohomological dimension is the following [Br1, Proposition 8.1, p. 210].

PROPOSITION 4.7. *Suppose that a $B\Gamma$-space is a compact manifold with nonempty boundary. Then* cd $\Gamma \le \dim B\Gamma - 1$.

If M is a closed aspherical manifold with $\pi_1(M) = \Gamma$, then it is a $B\Gamma$-space. On the other hand, the product $M \times [-1, 1]$ is a manifold with nonempty boundary and gives another $B\Gamma$-space. In general, if a group Γ admits a $B\Gamma$-space given by a manifold which could be noncompact or with boundary, it seems difficult to determine the minimal dimension in which a manifold exists, and it seems also difficult to decide whether a duality group Γ is a Poincaré duality group or not by using the condition whether such a $B\Gamma$-manifold has non-empty boundary or not.

On the other hand, the following algebraic criterion for non-Poincaré duality is useful for some groups [Str].

PROPOSITION 4.8. *Suppose Γ is of type FP and its cohomological dimension,* cd Γ, *is equal to n. If $H^n(\Gamma, \mathbb{Z}\Gamma)$ is finitely generated as an abelian group, then for every subgroup Γ' of Γ of infinite index,* cd $\Gamma' <$ cd Γ.

COROLLARY 4.9. *If Γ is a group of type FP and there exists a subgroup Γ' of infinite index such that* cd $\Gamma' =$ cd Γ, *then Γ is not a Poincaré duality group.*

Proof. If not, then by definition, $H^n(\Gamma, \mathbb{Z}\Gamma)$ is isomorphic to $\mathbb{Z}$. Then the above proposition implies that cd $\Gamma' <$ cd Γ. This is a contradiction.

5. Algebraic groups, Tits building, Steinberg representations, arithmetic subgroups

In this section, we first introduce the notion of spherical Tits building $\Delta_{\mathbb{Q}}(\mathbf{G})$ of a reductive algebraic group $\mathbf{G}$ defined over $\mathbb{Q}$. Then we recall the Solomon-Tits theorem on the homotopy type of the building $\Delta_{\mathbb{Q}}(\mathbf{G})$, and obtain the Steinberg module of $\mathbf{G}(\mathbb{Q})$. In Remark 5.4, we introduce the Steinberg module of reductive algebraic groups over general fields, which will be used in studying duality properties of S-arithmetic subgroups over both number fields and function fields of curves over finite fields. Then we define arithmetic subgroups and point out some references on applications of the Steinberg module $St_{\mathbb{Q}}(\mathbf{G})$ of arithmetic subgroups in number theory, in particular, modular symbols.

Let $\mathbf{G} \subset GL(n, \mathbb{C})$ be a connected linear algebraic group defined over $\mathbb{Q}$. It is often obtained as subgroups preserving some structures on $\mathbb{C}^n$, for example, a volume form, which leads to $SL(n)$, or a skew-symmetric form, which leads to $Sp(n)$.

For simplicity, we assume in this section that $\mathbf{G}$ is a reductive algebraic group. An important object associated with $\mathbf{G}$ is the spherical Tits building $\Delta_{\mathbb{Q}}(\mathbf{G})$, which is a simplicial complex whose simplexes are parametrized by $\mathbb{Q}$-parabolic subgroups of $\mathbf{G}$.

Recall that a subgroup $\mathbf{P}$ of $\mathbf{G}$ is called a parabolic subgroup if the quotient $\mathbf{P}\backslash\mathbf{G}$ is a projective variety. If $\mathbf{P}$ is defined over $\mathbb{Q}$, then $\mathbf{P}$ is called a $\mathbb{Q}$-parabolic subgroup.

For each proper $\mathbb{Q}$-parabolic subgroup $\mathbf{P}$ of $\mathbf{G}$, denote the corresponding simplex of $\Delta_{\mathbb{Q}}(\mathbf{G})$ by $\sigma_{\mathbf{P}}$. Then these simplexes satisfy the following compatibility and normalizing conditions:

(1) If $\mathbf{P}$ is a proper maximal $\mathbb{Q}$-parabolic subgroup of $\mathbf{G}$, then its simplex $\sigma_{\mathbf{P}}$ is zero dimensional, i.e., a point.

(2) For every pair of $\mathbb{Q}$-parabolic subgroups $\mathbf{P}_1, \mathbf{P}_2$, $\sigma_{\mathbf{P}_1}$ is a face of $\sigma_{\mathbf{P}_2}$ if and only if $\mathbf{P}_1$ contains $\mathbf{P}_2$. In particular, for every simplex $\sigma_{\mathbf{P}}$, its vertices correspond to the set of all proper maximal $\mathbb{Q}$-parabolic subgroups that contain $\mathbf{P}$.

Let r be the $\mathbb{Q}$-rank of $\mathbf{G}$. Then $r > 0$ if and only if $\mathbf{G}$ contains a proper $\mathbb{Q}$-parabolic subgroup. Assume that $r > 0$. From the above description and the structure of $\mathbb{Q}$-parabolic subgroups (in particular, a minimal $\mathbb{Q}$-parabolic subgroup is contained in exactly r different proper maximal $\mathbb{Q}$-parabolic subgroups), it follows that the dimension of $\Delta_{\mathbb{Q}}(\mathbf{G})$ is equal to $r - 1$.

Clearly, the $\mathbb{Q}$-locus $\mathbf{G}(\mathbb{Q})$ of $\mathbf{G}$ acts on the set of $\mathbb{Q}$-parabolic subgroups of $\mathbf{G}$ by conjugation and hence acts on $\Delta_{\mathbb{Q}}(\mathbf{G})$ by simplicial automorphisms.

A known result of Solomon and Tits is the following [So] [Br2].

PROPOSITION 5.1. *The Tits building $\Delta_{\mathbb{Q}}(\mathbf{G})$ is homotopy equivalent to a bouquet of countably infinitely many spheres S^{r-1}. In particular, $H_{r-1}(\Delta_{\mathbb{Q}}(\mathbf{G}), \mathbb{Z})$ is a free abelian group on infinitely many generators.*

The action of $\mathbf{G}(\mathbb{Q})$ on $\Delta_{\mathbb{Q}}(\mathbf{G})$ induces an action on $H_{r-1}(\Delta_{\mathbb{Q}}(\mathbf{G}), \mathbb{Z})$, and hence $H_{r-1}(\Delta_{\mathbb{Q}}(\mathbf{G}), \mathbb{Z})$ becomes a $\mathbf{G}(\mathbb{Q})$-module. This is called the *Steinberg module* of $\mathbf{G}(\mathbb{Q})$ (or rather of $\mathbb{Z}\mathbf{G}(\mathbb{Q})$), and denoted by $St_{\mathbb{Q}}(\mathbf{G})$ or $St(\mathbf{G})$. If the rank $r = 0$, we call the trivial $\mathbb{Z}\mathbf{G}(\mathbb{Q})$-module $\mathbb{Z}$ the Steinberg module of $\mathbf{G}(\mathbb{Q})$.

REMARK 5.2. The spheres S^{r-1} in the bouquet can be parametrized as follows. The building $\Delta_{\mathbb{Q}}(\mathbf{G})$ contains sub-complexes, which are all isomorphic to the Coxeter complex associated with the Weyl group of $\mathbf{G}$ over $\mathbb{Q}$. These sub-complexes are called apartments. Each apartment gives a triangulation of the sphere S^{r-1} and contains finitely many simplexes, and the top dimensional simplexes of apartments are called chambers of the building. The building $\Delta_{\mathbb{Q}}(\mathbf{G})$ is the union of such apartments. In fact, for any fixed chamber C of $\Delta_{\mathbb{Q}}(\mathbf{G})$, the union of the collection $\mathcal{A}$ of the apartments which contain this chamber is equal to $\Delta_{\mathbb{Q}}(\mathbf{G})$. Then each apartment in $\mathcal{A}$ corresponds to a sphere in the bouquet in the above proposition.

REMARK 5.3. Since $\mathbf{G}(\mathbb{Q})$ acts transitively on the set of minimal $\mathbb{Q}$-parabolic subgroups, it follows that as a $\mathbb{Z}\mathbf{G}(\mathbb{Q})$-module, the Steinberg module $St_{\mathbb{Q}}(\mathbf{G})$ is generated by one element. Since the action is not simply transitive, it is not a free $\mathbb{Z}\mathbf{G}(\mathbb{Q})$-module of rank 1.

Let $\mathbf{P}$ be a minimal $\mathbb{Q}$-parabolic subgroup $\mathbf{G}$, and $\mathbf{U}_{\mathbf{P}}$ be the unipotent radical of $\mathbf{P}$. Then $\mathbf{U}_{\mathbf{P}}(\mathbb{Q})$ acts simply transitively on Σ. This implies that as a $\mathbb{Z}\mathbf{U}_{\mathbf{P}}(\mathbb{Q})$-module, $H_{r-1}(\Delta_{\mathbb{Q}}(\mathbf{G}), \mathbb{Z})$ is free of rank 1. See [BS1, Remark 8.6.8] for more detail.

REMARK 5.4. Instead of the field $\mathbb{Q}$, we can consider any field k, such as a finite field $\mathbb{F}_q$ (for example, $\mathbb{Z}/p\mathbb{Z}$, where p is a prime number), the field $\mathbb{Q}_p$ of p-adic numbers, or the function field of a curve over a finite field $\mathbb{F}_q$, for example, $\mathbb{F}_q(t)$, where t is a free variable. Then for any reductive algebraic group $\mathbf{G}$ over k, there is also a spherical Tits building $\Delta_k(\mathbf{G})$ whose simplexes are parametrized by k-parabolic subgroups of $\mathbf{G}$. There is also a Solomon-Tits theorem stating that $\Delta_k(\mathbf{G})$ is homotopy equivalent to a bouquet of sphere S^{r-1}, where r is the k-rank of $\mathbf{G}$. The non-vanishing homology group $H_{r-1}(\Delta_k(\mathbf{G}), \mathbb{Z})$ is a $\mathbb{Z}\mathbf{G}(k)$-module, called the *Steinberg module* of $\mathbf{G}(k)$, and denoted by $St_k(\mathbf{G})$. The building $\Delta_k(\mathbf{G})$ is a finite simplicial complex if and only if k is a finite field, and hence $St_k(\mathbf{G})$ is a finite abelian group if and only if k is a finite field.

When $\mathbf{G}$ is a Chevalley group and k is finite field, the module $St_k(\mathbf{G})$ is the original Steinberg module [So]. As pointed out in the introduction, this module and generalizations have played a fundamental role in representation theories of algebraic groups and many other applications (see [St, pp. 580-586] [Ca] [Hu] [Be]).

REMARK 5.5. Since $St_k(\mathbf{G})$ is generated over $\mathbb{Z}\mathbf{G}(k)$ by one element, there is a surjective map $\mathbb{Z}\mathbf{G}(k) \to St_k(\mathbf{G})$. See [To] for descriptions of the kernel of this map.

DEFINITION 5.6. A subgroup Γ of $\mathbf{G}(\mathbb{Q})$ is called an arithmetic subgroup if it is commensurable with $\mathbf{G}(\mathbb{Z}) = \mathbf{G}(\mathbb{Q}) \cap GL(n, \mathbb{Z})$, i.e., $\Gamma \cap \mathbf{G}(\mathbb{Z})$ is of finite index in both Γ and $\mathbf{G}(\mathbb{Z})$.

Clearly, an arithmetic subgroup Γ acts on the set of $\mathbb{Q}$-parabolic subgroups and also on the Tits building $\Delta_{\mathbb{Q}}(\mathbf{G})$. The action of Γ on $H_{r-1}(\Delta_{\mathbb{Q}}(\mathbf{G}), \mathbb{Z})$ gives the latter a $\mathbb{Z}\Gamma$-module structure. The module is also called the *Steinberg module* of Γ (or $\mathbb{Z}\Gamma$).

Besides the application in the next section that the Steinberg module of a torsion-free arithmetic subgroup Γ is its dualizing module, this Steinberg module is also important in number theory. For example, when $\mathbf{G} = SL(n)$ and $\Gamma = SL(n, \mathbb{Z})$, then Γ acts transitively on $\Delta_{\mathbb{Q}}(\mathbf{G})$, and it is reasonable to expect that the Steinberg module is generated over $\mathbb{Z}\Gamma$ by one element [AP, Theorem 4.1]. This has important applications in number theory through the so-called modular symbols. See the references of [To] and [GuM] for other related results.

6. Borel-Serre compactification and duality properties of arithmetic subgroups

In this section, we first use the Borel-Serre compactification of locally symmetric spaces to construct finite CW-complex $B\Gamma$-spaces for torsion-free arithmetic subgroups of reductive algebraic groups $\mathbf{G}$, and hence show that arithmetic subgroups are of type FP_∞ and of type FP if they are torsion-free. Then we use the relation between the boundary of the Borel-Serre partial compactification of the symmetric space and the spherical Tits building to show that arithmetic subgroups are duality groups. For non-uniform arithmetic groups acting on linear symmetric spaces, we give a proof without using the Borel-Serre compactification that they are not Poincaré duality groups (Theorem 6.4). We conclude this section with duality properties of S-arithmetic subgroups.

Let $\mathbf{G}$ be a linear algebraic group as in the previous section. Let $G = \mathbf{G}(\mathbb{R})$ be the real locus of $\mathbf{G}$. Then every arithmetic subgroup $\Gamma \subset \mathbf{G}(\mathbb{Q})$ is a discrete subgroup of G.

In this section, we assume that Γ is a lattice of G, i.e., $\Gamma\backslash G$ has finite volume. This is not always true, for example, if $\mathbf{G} = GL(n)$ and $\Gamma = GL(n, \mathbb{Z})$. On the other hand, if $\mathbf{G}$ is semisimple or $\mathbf{G}$ has no nontrivial character defined over $\mathbb{Q}$, then this condition on Γ is satisfied.

Let $K \subset G$ be a maximal compact subgroup of G. Then $X = G/K$ is diffeomorphic to $\mathbb{R}^n$, where $n = \dim X$. If G is reductive, then with respect to an invariant metric, X is a symmetric space not containing any compact factors and hence is of non-positive sectional curvature. Let r be the $\mathbb{Q}$-rank of $\mathbf{G}$ as above. Then it is known that $\Gamma\backslash X$ is compact if and only if $r = 0$.

A basic and important result of [BS1] is the following.

THEOREM 6.1. *Let $\Gamma \subset \mathbf{G}(\mathbb{Q})$ be an arithmetic subgroup as above. Then Γ is a virtual duality group of dimension $n - r$, where $n = \dim X$, and the dualizing module is given by the Steinberg module; and Γ is a virtual Poincaré duality group if and only if the $\mathbb{Q}$-rank $r = 0$, i.e., Γ is a uniform lattice.*

To prove this theorem, we need some preparations, in particular, the Borel-Serre compactification in [BS1]. Since Γ is a discrete subgroup of G, it acts properly on X. If Γ is torsion-free, then Γ acts fixed point freely. This implies that $\Gamma\backslash X$ is a $B\Gamma$-space and X is a $E\Gamma$-space.

When $r = 0$ and Γ is torsion-free, then $\Gamma\backslash X$ is a closed aspherical manifold with $\pi_1 = \Gamma$. This implies that Γ is a Poincaré duality group, and Theorem 6.1 is proved in this case.

On the other hand, when $r > 0$, $\Gamma\backslash X$ is not finite CW-complex, and Proposition 4.5 can not be applied. One way to obtain a $B\Gamma$-space given by a finite CW-complex is to construct a compactification $\overline{\Gamma\backslash X}$ of $\Gamma\backslash X$ such that the inclusion

$$\Gamma\backslash X \hookrightarrow \overline{\Gamma\backslash X}$$

is a homotopy equivalence.

This is achieved by the Borel-Serre compactification $\overline{\Gamma\backslash X}^{BS}$ of $\Gamma\backslash X$ in [BS1]. Briefly, it is constructed in two steps:

(1) Construct a partial compactification $\overline{X}^{BS}$ by attaching a boundary component $e_{\mathbf{P}}$ for every proper $\mathbb{Q}$-parabolic subgroup $\mathbf{P}$ of $\mathbf{G}$:

(6.1) $$\overline{X}^{BS} = X \cup \coprod_{\mathbf{P}} e_{\mathbf{P}}.$$

Each boundary component $e_{\mathbf{P}}$ is contractible and the $\overline{X}^{BS}$ is a real analytic manifold with corners with the interior equal to X.

(2) Show that the Γ-action on X extends to a proper action on $\overline{X}^{BS}$ with a compact quotient, which gives $\overline{\Gamma\backslash X}^{BS}$.

Since a manifold with corners is clearly homotopy equivalent to its interior, the inclusion $\Gamma\backslash X \hookrightarrow \overline{\Gamma\backslash X}^{BS}$ is a homotopy equivalence when Γ is torsion-free. Consequently, $\overline{\Gamma\backslash X}^{BS}$ is a compact $B\Gamma$-space when Γ is torsion-free. A simple example to keep in mind is the case when X is the Poincaré upper half-plane, and $\Gamma\backslash X$ is a hyperbolic (Riemann) surface. In this case, the Borel-Serre partial compactification $\overline{X}^{BS}$ is obtained by adding a copy of $\mathbb{R}$ at every rational boundary point, i.e., a point in $\mathbb{Q} \cup \{i\infty\} \subset \mathbb{R} \cup \{i\infty\} = \partial X$, and every cusp of $\Gamma\backslash X$ is compactified by adding a circle.

An immediate corollary is the following.

PROPOSITION 6.2. *If Γ is a torsion-free arithmetic subgroup, then Γ is of type FP. In general, every arithmetic subgroup Γ, which might contain nontrivial torsion elements, is of type FP_∞. In particular, in every degree i, $H_i(\Gamma, \mathbb{Z})$ and $H^i(\Gamma, \mathbb{Z})$ are finitely generated.*

This follows from Proposition 3.2 and the fact that every arithmetic subgroup contains a torsion-free subgroup of finite index.

In the Borel-Serre partial compactification $\overline{\Gamma\backslash X}^{BS}$, the boundary components $e_{\mathbf{P}}$ are contractible and satisfy the incidence relation:

$$\mathbf{P}_1 \subset \mathbf{P}_2 \text{ if and only if } e_{\mathbf{P}_1} \text{ is contained in the closure of } e_{\mathbf{P}_2}.$$

Together with the fact that every boundary component $e_{\mathbf{P}}$ is contractible and the nerve-cover principle, this implies the following result [BS1].

PROPOSITION 6.3. *The boundary $\partial \overline{X}^{BS}$ is homotopy equivariant to the Tits building $\Delta_{\mathbb{Q}}(\mathbf{G})$, and hence is homotopy equivalent to a bouquet of countably infinitely many spheres S^{r-1}, when the $\mathbb{Q}$-rank r of $\mathbf{G}$ is positive.*

Proof of Theorem 6.1.

The case $r = 0$ was proved earlier. Assume that $r > 0$. When Γ is a torsion-free, $\overline{X}^{BS}$ is a $E\Gamma$-space with a compact quotient. Together with Proposition 4.5, Proposition 6.3 implies that Γ is a duality group of dimension $\dim X - r$ with the dualizing module equal to the Steinberg module $St_{\mathbb{Q}}(\mathbf{G}) \cong H_{r-1}(\Delta_{\mathbb{Q}}(\mathbf{G})) = H_{r-1}(\partial \overline{X}^{BS})$.

To motivate results in the later section, we describe a realization of $\overline{\Gamma\backslash X}^{BS}$ in terms of a submanifold $\Gamma\backslash X_T$ of $\Gamma\backslash X$ such that the inclusion $\Gamma\backslash X_T \hookrightarrow \Gamma\backslash X$ is a homotopy equivalence. When $\Gamma\backslash X$ is a hyperbolic Riemann surface, it is clear that $\Gamma\backslash X_T$ can be obtained by cutting off each cusp end. Since each cusp neighborhood is a topological cylinder, the whole space $\Gamma\backslash X$ can be deformed retracted to $\Gamma\backslash X_T$. This is equivalent to the following steps:

(1) Remove Γ-equivariant horoballs at rational boundary points from the Poincaré upper half-plane to get a truncated submanifold X_T.
(2) Show that X can be deformation retracted to X_T.
(3) The Γ-action preserves X_T, and the quotient $\Gamma\backslash X_T$ is the desired submanifold $\Gamma\backslash X_T$.

For a general symmetric space X and arithmetic subgroup Γ, the construction is similar. But the notion of neighborhoods of points at infinity, i.e., the analogues of the removed horoballs here, is more complicated. See the paper [Sa] for details and references.

A natural question is whether X_T can be further deformed so that there exists a subspace Y of X satisfying the following conditions:

(1) Y is an equivariant deformation retract of X with respect to an arithmetic subgroup Γ,
(2) the quotient $\Gamma\backslash Y$ is compact,

(3) and the dimension of Y is equal to cd Γ when Γ is torsion-free.

Certainly, the dimension on Y is the minimal one allowed. It turns out that when X is a linear symmetric space, i.e., a homothety section of a symmetric cone, such a subspace Y and deformation retraction exist [As]. Using this, we can give a different proof of that fact that Γ is not a Poincaré duality group when Γ is a torsion-free arithmetic subgroup.

The most basic example of linear symmetric spaces is $X = SL(n, \mathbb{R})/SO(n)$, which is a homothety section of the symmetric cone $\Pi_n(\mathbb{R})$ of positive definite matrices of rank n over $\mathbb{R}$. Besides this family of symmetric cones, there are three other families:

(1) $\Pi_n(\mathbb{C})$, the space of positive definite matrices over the complex numbers $\mathbb{C}$,
(2) $\Pi_n(\mathbb{H})$, the space of positive definite matrices over the quaternions $\mathbb{H}$,
(3) L_n, the Lorentz cone,
(4) an exceptional symmetric cone $\Pi_3(\mathbb{O})$, the cone of all 3×3 positive definite matrices over the algebra $\mathbb{O}$ of octonions.

See the book [FK] for more details.

THEOREM 6.4. *Assume that $X = G/K$ is a linear symmetric space and $\Gamma \subset \mathbf{G}(\mathbb{Q})$ is a torsion-free non-uniform arithmetic subgroup, for example, when $\Gamma \subset SL(n, \mathbb{Z})$ is a torsion-free subgroup of finite index. Then the following results hold:*

(1) *There exists a Γ-equivariant deformation retraction of X to a simplicial complex Y which is of dimension equal to $\dim X - r$ such that $\Gamma \backslash Y$ is compact, i.e, a finite CW-complex. In particular, cd $\Gamma \leq \dim X - r$.*
(2) *There exists a subgroup Γ' of Γ such that cd $\Gamma' = \dim X - r$. This implies that*
$$\text{cd } \Gamma = \dim X - r.$$
(3) *The index $[\Gamma : \Gamma']$ is equal to infinity, and hence Γ is not a Poincaré duality group.*

Proof. The existence of Y and an equivariant deformation of X were proved in [As]. To prove (2), let $\mathbf{P}$ be a minimal $\mathbb{Q}$-parabolic subgroup of $\mathbf{G}$, and $\mathbf{N_P}$ its unipotent radical. Then its real locus by $P = \mathbf{P}(\mathbb{R})$ admits a $\mathbb{Q}$-Langlands decomposition
$$P = N_\mathbf{P} A_\mathbf{P} M_\mathbf{P} \cong N_\mathbf{P}(\mathbb{R}) \times A_\mathbf{P} \times M_\mathbf{P},$$
where $N_\mathbf{P} = N_\mathbf{P}(\mathbb{R})$, $A_\mathbf{P}$ is the identity component of the real locus of a maximal $\mathbb{Q}$-split torus of $\mathbf{P}$, and $M_\mathbf{P}$ is a reductive Lie group. Then $\Gamma_P = \Gamma \cap P$ is a discrete subgroup and is contained in the subgroup $N_\mathbf{P} M_\mathbf{P}$ with a compact quotient. Let K_M be a maximal compact subgroup of $M_\mathbf{P}$. Then $N_\mathbf{P} M_\mathbf{P}/K_M$ is diffeomorphic to $\mathbb{R}^{\dim X - r}$. In fact, the Langlands decomposition of P induces a horospherical decomposition of X:
$$X = N_\mathbf{P} \times A_\mathbf{P} \times M_\mathbf{P}/K_M,$$
where $M_\mathbf{P}/K_M$ is a symmetric space not containing any compact factors, and hence
$$N_\mathbf{P} M_\mathbf{P}/K_M \cong N_\mathbf{P} \times M_\mathbf{P}/K_M \cong \mathbb{R}^{\dim X - r}$$
and $\dim A_\mathbf{P} = r$. Since $\mathbf{P}$ is a minimal $\mathbb{Q}$-parabolic subgroup, $\Gamma_P \backslash N_\mathbf{P} M_\mathbf{P}/K_M$ is a closed aspherical manifold, and consequently Γ_P is a Poincaré duality group of dimension $\dim X - r$. This implies that cd $\Gamma_P = \dim X - r$. This proves (2). Since

the volume of $\Gamma_P \backslash X \cong \Gamma_P \backslash N_{\mathbf{P}} M_{\mathbf{P}} / K_M \times A_{\mathbf{P}}$ is infinite (think of the case when X is the Poincaré upper half plane and $\Gamma_P \backslash X$ is a cylinder with an exponentially growing end), the index $[\Gamma : \Gamma_P] = \infty$. Then (3) follows from Corollary 4.9 and the equality cd Γ_P = cd Γ.

REMARK 6.5. If X is not a linear symmetric space but the $\mathbb{Q}$-rank r of $\mathbf{G}$ is equal to 1, we can also show directly, without using the Borel-Serre compact-ification, that every torsion-free arithmetic subgroup Γ is not a Poincaré duality group.

The argument goes as follows. Each end of $\Gamma \backslash X$ is a topological cylinder and truncating off these ends gives a compact submanifold $\Gamma \backslash X_T$ with boundary which is homotopy equivalent to $\Gamma \backslash X$. By Proposition 4.7, cd $\Gamma \leq \dim X - 1$. On the other hand, as in the proof of the above theorem, for any proper $\mathbb{Q}$-parabolic subgroup $\mathbf{P}$, the intersection $\Gamma_P = \Gamma \cap P$, where $P = \mathbf{P}(\mathbb{R})$, is a subgroup of infinite index with cd $\Gamma' = \dim X - 1$. This implies that

$$\text{cd } \Gamma = \dim X - 1 = \text{ cd } \Gamma_P.$$

Since $[\Gamma : \Gamma_P] = \infty$, Corollary 4.9 implies that Γ is not a Poincaré duality group. (For truncating a general locally symmetric space to obtain a compact submanifold with boundary which is a deformation retraction of the whole space, see [Ra2].)

The statement that Γ is a duality group but not a Poincare duality group can also be proved using the truncated submanifold $\Gamma \backslash X_T$ as follows. It is clear that the universal covering X_T of $\Gamma \backslash X_T$ has a nonempty boundary ∂X_T which is homotopy equivalent to a bouquet of infinitely many spheres S^0 (note S^0 consists of two points and ∂X_T has infinitely contractible connected components). By Proposition 4.5, Γ is a duality group but not a Poincaré duality group.

S-arithmetic subgroups and duality properties.

An important generalization of arithmetic arithmetic subgroups is the class of S-arithmetic subgroups. Let $p_1, \cdots, p_k$ be a finite set of prime numbers and $S = \{\infty, p_1, \cdots, p_k\}$. The ring $\mathbb{Z}_S = \mathbb{Z}[\frac{1}{p_1}, \cdots, \frac{1}{p_k}]$ of S-integers consist of rational numbers whose denominators are only divisible by primes $p_1, \cdots, p_k$.

Then a subgroup Γ of $\mathbf{G}(\mathbb{Q})$ is called an S-arithmetic subgroup if it is commen-surable with $\mathbf{G}(\mathbb{Q}) \cap GL(n, \mathbb{Z}_S)$. An important example of S-arithmetic subgroups is $SL(n, \mathbb{Z}_S)$.

For each prime p_i, denote the $\mathbb{Q}_{p_i}$-rank of $\mathbf{G}$ by r_i. Denote the Steinberg module of $\mathbf{G}(\mathbb{Q}_{p_i})$ (see below) by St_i. Then an important result of Borel and Serre in [BS2] is the following.

THEOREM 6.6. *Let $\mathbf{G}$ be a reductive algebraic group defined over $\mathbb{Q}$, and $\Gamma \subset \mathbf{G}(\mathbb{Q})$ be an S-arithmetic subgroup. Then Γ is a virtual duality group of dimension $\dim X + \sum_{i=1}^{k} r_i - r$, where r is the $\mathbb{Q}$-rank of $\mathbf{G}$, and the dualizing module is the tensor product of local Steinberg modules $St_{\mathbb{Q}}(\mathbf{G}) \otimes \prod_{i=1}^{k} St_i$. If either $r > 0$ or S contains at least one prime number, then Γ is not a virtual Poincaré duality group.*

The idea is as follows. For each prime p_i, there is a Bruhat-Tits building X_i associated with $\mathbf{G}(\mathbb{Q}_{p_i})$, which is an infinite simplicial complex of dimension r_i and on which $\mathbf{G}(\mathbb{Q}_{p_i})$ acts simplicially and properly. Endowed with the Tits metric, X_i is also a so-called CAT(0) and hence contractible (see [Br2] and [Ji]). It can be compactified by adding the spherical Tits building $\Delta(\mathbf{G}(\mathbb{Q}_{p_i}))$ of the algebraic group $\mathbf{G}$ over $\mathbb{Q}_{p_i}$. Then it follows that

$$H_c^i(X_i, \mathbb{Z}) \cong H^i(X_i \cup \Delta(\mathbf{G}(\mathbb{Q}_{p_i})), \Delta(\mathbf{G}(\mathbb{Q}_{p_i})), \mathbb{Z}) \cong H^{i-1}(\Delta(\mathbf{G}(\mathbb{Q}_{p_i})), \mathbb{Z})$$

is non-zero only in dimension r_i (note that $\Delta(\mathbf{G}(\mathbb{Q}_{p_i}))$ is a simplicial complex of dimension $r_i - 1$). The action of $\mathbf{G}(\mathbb{Q}_{p_i})$ on $H_c^{r_i}(X_i, \mathbb{Z})$ is the *Steinberg module* St_i. As an abelian group, $H_c^{r_i}(X_i, \mathbb{Z})$ is infinitely generated. (Note that X_i is not a manifold and hence there is no version of Poincaré duality which can be used to compute $H_c^i(X_i, \mathbb{Z})$ as in the previous cases.)

Since an S-arithmetic subgroup Γ is embedded in the product $G \times \prod_{i=1}^k \mathbf{G}(\mathbb{Q}_{p_i})$ as a discrete subgroup, it follows that Γ acts properly on the space

$$X_S = X \times \prod_{i=1}^k X_i.$$

If Γ is torsion-free, then $\Gamma \backslash X_S$ is a $B\Gamma$-space.

If $r > 0$, $\Gamma \backslash X_S$ is noncompact and is not a finite CW-compex. On the other hand, replacing X by the Borel-Serre partial compactification $\overline{X}^{BS}$ gives a finite CW-complex $B\Gamma$-space $\Gamma \backslash \overline{X_S}^{BS}$, where

$$\overline{X_S}^{BS} = \overline{X}^{BS} \times \prod_{i=1}^k X_i.$$

The Kunneth formula implies that when $j = \dim X + \sum_{i=1}^k r_i - r$,

$$H_c^j(\overline{X_S}^{BS}, \mathbb{Z}) = H_c^{\dim X - r}(\overline{X}^{BS}, \mathbb{Z}) \otimes \prod_{i=1}^k H_c^{r_i}(X_i, \mathbb{Z}) = St_{\mathbb{Q}}(\mathbf{G}) \otimes \prod_{i=1}^k St_i,$$

and vanishes for other values of j.

This implies that if Γ is torsion-free, then it is a duality group of dimension equal to $\dim X + \sum_{i=1}^k r_i - r$ with the dualizing module given by the tensor product $St_{\mathbb{Q}}(\mathbf{G}) \otimes \prod_{i=1}^k St_i$. In general, Γ is a virtual duality group.

As pointed out before, for each i, St_i is not finitely generated as an abelian group. This implies that if either $r > 0$ or S contains at least one prime p_i, then the dualizing module of Γ is not finitely generated and hence Γ is not a Poincaré duality group.

S-arithmetic subgroups over function fields.

Global fields consist of two types: number fields, which are finite extensions of $\mathbb{Q}$, and function fields of curves over finite fields $\mathbb{F}_q$, which are finite separable extensions of $\mathbb{F}_q(t)$, where t is a free variable.

For a function field k and a reductive algebraic group $\mathbf{G}$ defined over k, we can also define S-arithmetic subgroups. By methods similar to those described in the previous paragraphs, in particular, the Bruhat-Tits building associated with $\mathbf{G}(k_{\mathfrak{p}})$ for any place $\mathfrak{p}$ of k, the following result is also true [BS2, Theorem 6.2].

THEOREM 6.7. *Suppose $\mathbf{G}$ is a reductive algebraic group defined over a function field k and the k-rank of $\mathbf{G}$ is equal to 0. Then every S-arithmetic subgroup of $\mathbf{G}(k)$ is a virtual duality group, but not a virtual Poincaré duality group unless the S-arithmetic subgroup is finite.*

The assumption that the k-rank of $\mathbf{G}$ is equal to 0 in the above theorem is crucial. Otherwise, S-arithmetic subgroups of algebraic groups of positive rank over function fields always contain nontrivial torsion elements and hence their virtual cohomology dimension is equal to infinity, which implies that they are not virtual duality groups. In fact, a stronger result holds: they are not of type FP_∞. See [BuW] and references there.

7. Mapping class groups and Teichmüller spaces

Let S be an orientable surface. Let $\mathrm{Diff}(S)$ be the group of all diffeomorphisms of S, and $\mathrm{Diff}^0(S)$ its identity component. Then $Mod_S = \mathrm{Diff}(S)/\mathrm{Diff}^0(S)$ is the *extended mapping class group* (or also called the *extended Teichmüller modular group*) of S. Let $\mathrm{Diff}^+(S)$ be the subgroup of orientation preserving diffeomorphisms of S. Then $Mod^+(S) = \mathrm{Diff}^+(S)/\mathrm{Diff}^0(S)$ is the *mapping class group* of S.

When $S = \mathbb{Z}^2\backslash\mathbb{R}^2$ is the torus, i.e., S is a closed surface of genus 1, then $Mod^+(S)$ can be identified with $SL(2,\mathbb{Z})$, and $Mod(S)$ can be identified with $GL(2,\mathbb{Z})$. Therefore, if S is a closed surface of genus $g \geq 2$, or more generally an oriented surface of negative Euler characteristic $\chi(S)$, then $Mod(S)$ and $Mod^+(S)$ are natural generalizations of $GL(2,\mathbb{Z})$ and $SL(2,\mathbb{Z})$.

For simplicity, in the following we assume that S is a closed orientable surface of genus $g \geq 2$, and Mod_S is also denoted by Mod_g.

The analogue of the symmetric spaces for the mapping class groups is the Teichmüller space T_g which consists of equivalence classes marked hyperbolic metrics on S, where a marked a hyperbolic metric on S is a hyperbolic surface Σ together with a homotopy class of diffeomorphisms $\varphi : S \to \Sigma$, and denoted by $(\Sigma, [\varphi])$. Two marked hyperbolic surfaces $(\Sigma_1, [\varphi_1])$ and $(\Sigma_2, [\varphi_2])$ are equivalent if there is an isometry between Σ_1 and Σ_2 which commutes with the two markings. Therefore,

$$T_g = \{(\Sigma, [\varphi]) \mid \varphi : S \to \Sigma \text{ is a diffeomorphism}\}/ \sim .$$

(As abstracted hyperbolic surfaces, if there exists an isometry between Σ_1 and Σ_2, then they are the same. Two equivalent marked hyperbolic surfaces $(\Sigma_1, [\varphi_1])$ and $(\Sigma_2, [\varphi_2])$ are also the same marked abstract hyperbolic surface. Therefore, T_g is *the space of all marked hyperbolic structures* on S. Because of this, we also write

$$T_g = \{(\Sigma, [\varphi]) \mid \varphi : S \to \Sigma \text{ is a diffeomorphism}\}.$$

Another way to define T_g is as follows. Let $\mathcal{H}(S)$ be the space of all hyperbolic metrics on S. Then the group of all diffeomorphisms of S, $\mathrm{Diff}(S)$, acts on $\mathcal{H}(S)$, and the quotient $\mathrm{Diff}^0(S)\backslash\mathcal{H}(S)$ under the identity component $\mathrm{Diff}^0(S)$ can be identified with T_g.)

It is known that T_g is a manifold diffeomorphic to $\mathbb{R}^{6g-6}$ and Mod_g acts on T_g by changing the marking: for any $[\psi] \in Mod_g$, where $\psi \in \mathrm{Diff}(S)$, and a marked hyperbolic surface $(\Sigma, [\varphi])$, then

$$[\psi] \cdot (\Sigma, [\varphi]) = (\Sigma, [\varphi \circ \psi]).$$

It is also known that this action of Mod_g on T_g is proper. Therefore, for any torsion-free subgroup Γ of Mod_g of finite index, $\Gamma\backslash T_g$ is a $B\Gamma$-space.

By definition, the quotient $Mod_g^+\backslash T_g$ is the moduli space of Riemann surfaces (or algebraic curves) of genus g and is hence noncompact. The reason for the noncompactness is that starting with any hyperbolic surface, we can pinch a closed geodesic and get a sequence of hyperbolic surfaces which has no convergent subsequences.

To get a finite CW-complex $B\Gamma$-space for a torsion-free subgroup $\Gamma \subset Mod_g$ of finite index, we can use with a Borel-Serre type compactification $\overline{\Gamma\backslash T_g}^{BS}$ or a truncated subspace $\Gamma\backslash T_g(\varepsilon)$ such that the inclusions below are homotopy equivalence:

$$\Gamma\backslash T_g(\varepsilon) \hookrightarrow \Gamma\backslash T_g \hookrightarrow \overline{\Gamma\backslash T_g}^{BS}.$$

The Borel-Serre type compactification $\overline{\Gamma\backslash T_g}^{BS}$ has been constructed (or outlined) by Harvey [Hav] and finished by Ivanov [I1] (see also the references of [IJ]). On the other hand, it is easier to construct the truncated subspace $\Gamma\backslash T_g(\varepsilon)$.

For every point $x = (\Sigma, [\varphi])$ of T_g, and closed geodesic c in Σ, denote the length of c with respect to the hyperbolic metric by $\ell_x(c)$. Then it is known that there exists a positive constant ε_g depending only on g such that every two closed geodesics c_1 and c_2 of Σ with $\ell_x(c_1), \ell_x(c_2) \le \varepsilon_g$ are disjoint.

For any $\varepsilon \le \varepsilon_g$, define a truncated subspace $T_g(\varepsilon)$ of T_g by

$$T_g(\varepsilon) = \{x = (\Sigma, [\varphi]) \in T_g \mid \text{ for every closed geodesic } c \subset \Sigma, \quad \ell_x(c) \ge \varepsilon\}.$$

Then $T_g(\varepsilon)$ is a real analytic manifold with corners. By the Mumford criterion, the quotient $\Gamma\backslash T_g(\varepsilon)$ is compact if $[Mod_g : \Gamma] < +\infty$ and is a real analytic with corners if Γ is also torsion-free.

It is also known that $T_g(\varepsilon)$ is a deformation retract of T_g and hence is contractible. Together with Proposition 3.2, this implies the following result (see [I1] for detail and references).

PROPOSITION 7.1. *For every torsion-free subgroup Γ of finite index of Mod_g, $\Gamma\backslash T_g(\varepsilon)$ is a finite CW-complex $B\Gamma$-space, and hence Γ is of type FP. Consequently, Mod_g is of type FP_∞.*

8. Curve complex, duality properties of mapping class groups

As pointed in the previous section, Mod_g is an analogue of arithmetic subgroups. It is naturally expected that they share many properties. One important result in [Ha2, Theorem 4.1] is the following.

THEOREM 8.1. *For every torsion-free subgroup $\Gamma \subset Mod_g$ of finite index, Γ is a duality group of dimension $4g - 5$. Therefore, Mod_g is a virtual duality group of virtual cohomological dimension $4g - 5$.*

The dualizing module is given by an analogue of the Steinberg module and will be described later.

Since $Mod_g\backslash T_g$ is noncompact, Mod_g is an analogue of non-uniform arithmetic subgroups, and the following result is naturally expected [IJ, Theorem 1.5].

THEOREM 8.2. *For every torsion-free subgroup $\Gamma \subset Mod_g$ of finite index, Γ is not a Poincaré duality group. Therefore, Mod_g is not a virtual Poincaré duality group.*

A natural method to prove the above results is to understand the homotopy type of the boundary $\partial T_g(\varepsilon)$ and apply the criterion in Proposition 4.5. For this purpose, we need to introduce the curve complex $\mathcal{C}(S)$ of the surface S, which is an analogue of the spherical Tits building $\Delta_{\mathbb{Q}}(\mathbf{G})$ associated with algebraic groups earlier and was introduced by Harvey in [Hav].

By definition, the vertices of the curve complex $\mathcal{C}(S)$ are free homotopy classes $[c]$ of simple closed curves c in S. Vertices $[c_1], \ldots, [c_{k+1}]$ form the vertices of a k-simplex if and only if they are all different and every two curves c_{i_1} and c_{i_2} for $1 \leq i_1 < i_2 \leq k+1$ are isotopic to disjoint curves. It is well known and easy to see that $\mathcal{C}(S)$ is a simplicial complex of dimension $3g - 4$. Clearly, Mod_g acts simplicially on $\mathcal{C}(S)$.

For every homotopy class of simple closed curves c of S and every point $x = (\Sigma, [\varphi]) \in T_g$, the homotopy class $[\varphi(c)]$ contains a unique simple closed geodesics, still denoted by c, and its length, denoted by $\ell_x(c)$, defines a length function $\ell(c) : T_g \to \mathbb{R}_{>0}$.

Then for $\varepsilon \leq \varepsilon_g$, the boundary faces of $T_g(\varepsilon)$ are defined by these length functions: $\ell(c) = \varepsilon$. Clearly, only when the geodesics $c_1, \cdots, c_{k+1}$ are disjoint, there are hyperbolic surfaces $x = (\Sigma, [\varphi]) \in T_g$ such that $\ell_x(c_i) = \varepsilon$, for $i = 1, \cdots, k+1$.

From this, it is clear that the boundary faces of $T_g(\varepsilon)$ are parametrized the simplexes of $\mathcal{C}(S)$. In fact, the following result is true.

PROPOSITION 8.3. *The space $T_g(\varepsilon)$ is a contractible manifold with corners. Its boundary faces when $T_g(\varepsilon)$ is considered as a manifold with corners are contractible and parametrized by the simplexes of the curve complex $\mathcal{C}(S)$, and the whole boundary $\partial T_g(\varepsilon)$ is homotopy equivalent to $\mathcal{C}(S)$.*

Each boundary face of $\partial T_g(\varepsilon)$ is the product of a truncated Teichmüller space with some $\mathbb{R}^m$, and hence is contractible [I1], and the inclusion relation between these faces is the opposite of the inclusion of the simplexes of $\mathcal{C}(S)$. Therefore, $\mathcal{C}(S)$ plays the role of spherical Tits building.

A natural and important problem is to prove an analogue of the Solomon-Tits theorem. In [**Ha2**] (see also [**Ha1**, Chap. 4, §1], and [**I4**, §3], or [**I1**, Theorem 3.3.A] for a different proof), Harer proved the following weak analogue of the Solomon-Tits theorem.

THEOREM 8.4. *The curve complex $\mathcal{C}(S)$ is homotopy equivalent to a bouquet of spheres $\vee S^n$, where the dimension $n = 2g - 2$.*

On the other hand, it is not obvious that the bouquet contains at least one sphere, i.e., $\mathcal{C}(S)$ is not contractible. In fact, this question has been raised by several people (see the introduction of [IJ] for more details about this question), and can be answered positively as follows: The above Theorem 8.4, and Propositions 4.5 and 4.2 together with a fact that the cohomological dimension of a group is greater than or equal to the cohomological dimension of any subgroup imply that Mod_g is a virtual duality group of dimension $4g - 5$, and then the bouquet for the homotopy type of $\mathcal{C}(S)$ contains at least one sphere. The dualizing module of Mod_g (or rather its torsion-free subgroups of finite index) is $H_{2g-2}(\mathcal{C}(S), \mathbb{Z})$ with the natural action of Mod_g. See [IJ] for more details.

In [IJ, Theorem 1.4], the following result was proved and gives a stronger analogue of the Solomon-Tits theorem.

THEOREM 8.5. *The curve complex $\mathcal{C}(S)$ is homotopy equivalent to the bouquet of countably infinitely many spheres S^n, where $n = 2g - 2$.*

Unfortunately, the proof is not direct as in the case of spherical Tits buildings. Indeed, in [IJ], the idea was to show that Mod_g contains a subgroup, a so-called Mess subgroup Γ', of infinite index whose virtual cohomological dimension is equal to that of Mod_g, i.e., $4g - 5$. Then by Corollary 4.9, Mod_g is not a virtual Poincaré duality group. Applying Proposition 4.5 to $E\Gamma = T_g(\varepsilon)$ implies that the homotopy type of the boundary $\partial T_g(\varepsilon)$, or equivalently the curve complex $\mathcal{C}(S)$, contains infinitely many spheres.

It will be desirable to give a more direct proof of Theorem 8.5. See [Bro] for results on $H_{2g-2}(\mathcal{C}(S))$ as a $\mathbb{Z}Mod_g$-module.

9. Outer automorphism groups of free groups, outer space

In this section, we first recall the reduced outer space X_n and its spine K_n, then the determination of the virtual cohomological dimension of $Out(F_n)$.

Let F_n, $n \geq 2$, be the free group on n generators, and

$$Out(F_n) = Aut(F_n)/Inn(F_n)$$

be the group of outer aumorphisms of F_n. When $n = 2$, $Out(F_n) = GL(2, \mathbb{Z})$. It is known that there is a surjective homomorphism

$$\pi : Out(F_n) \to GL(n, \mathbb{Z})$$

induced by the map $F_n \to \mathbb{Z}_n = \mathbb{Z} \times \cdots \times \mathbb{Z}$. In particular, $Out(F_n)$ contains a distinguished subgroup $\pi^{-1}(SL(n, \mathbb{Z}))$, which is usually denoted by $SOut(F_n)$.

The outer automorphism group $Out(F_n)$ plays an important role in the theory of combinatorial group theory. See [V1-2] [Be] for surveys on various aspects of this group.

One important reason for considering $Out(F_n)$ together with arithmetic subgroups of linear algebraic groups and mapping class groups is that they share many similar properties. The results in this section provide more evidence for this point. In view of all these, it is natural that $Out(F_n)$ can be considered as a a natural variation of arithmetic subgroups of semisimple Lie groups and mapping class groups of surfaces.

As discussed in the previous section, symmetric spaces and actions of arithmetic subgroups on them are fundamental to understand arithmetic subgroups and Teichmüller spaces are also crucial in studying the mapping class groups.

The analogous space for $Out(F_n)$ is the so-called reduced outer space X_n. Briefly, the outer space was introduced by Culler and Vogtmann in [CV] and was defined to be the space of marked metric graphs with the fundamental group equal to F_n and the total length of all edges equal to 1.

Specifically, let R_n be the rose with n petals, i.e., the wedge product of n circles S^1 (or the bouquet of n circles S^1). Then $\pi_1(R_n) = F_n$, and R_n is a $B\Gamma$-space for $\Gamma = F_n$. (We can identify $\pi_1(R_n)$ with F_n by sending the homotopy class of each petal to a generator of F_n). We consider only graphs G that do not contain vertices of valency 1 or 2. Then a marked graph G with $\pi_1(G) = F_n$ is a graph together with a homotopy class of homotopy equivalence $\varphi : R_n \to G$.

A metric graph is a graph G with an assignment of nonnegative edge lengths $\ell(e)$ to edges e of G such that the sum of edges in every nontrivial loop in G is is strictly positive (but some edge lengths could be zero). We scale the edge lengths so that the total sum of all edge lengths is equal to 1. A metric graph is denoted by (G, ℓ), and a marked metric graph is denoted by $(G, \ell, [\varphi])$. Two marked metric graphs $(G_1, \ell_1, [\varphi_1])$ and $(G_2, \ell_2, [\varphi_2])$ are defined equivalent if there is an isometry between (G_1, ℓ_1) and (G_2, ℓ_2) that commutes with the markings $[\varphi_1]$ and $[\varphi_2]$. (In other words, an equivalence class of marked metric graphs is one marked *abstract* metric graph, or an abstract metric graph together with a specified isomorphism $\pi_1 \cong F_n$.)

Then the set of equivalence classes of such marked metric graphs with $\pi_1 = F_n$ is called the *outer space* associated with F_n. If we only consider the sub-collection of graphs which do not contain any separating edge (i.e., its complement is disconnected), we get the *reduced outer space* X_n.

Let $Htp(R_n)$ be the group of all homotopy equivalences of R_n, and $Htp(R_n)^0$ the identity component of $Htp(R_n)$. Then the quotient $Htp(R_n)/Htp(R_n)^0$ is canonically isomorphic to $Out(F_n)$. Specifically, every automorphism $\varphi : F_n \to F_n$ corresponds to an homotopy equivalence between R_n and R_n, which maps a petal representing a generator x of F_n to the loop representing the $\varphi(x)$, and the automorphism of F_n is inner if and only if the homotopy class contains the identity map. It follows that $Out(F_n)$ acts on X_n by changing the marking of the marked metric graphs in X_n as in the case of the action of Mod_g on the Teichmüller space.

It is known that the outer space X_n is an infinite simplicial complex of dimension $3n - 4$, and the action of the group $Out(F_n)$ on X_n is proper. The quotient $Out(F_n)\backslash X_n$ is the moduli space of normalized metric graphs without containing any separating edges. When $n = 2$, X_n can be canonically realized with an ideal triangulation of the upper half plane $\mathbb{H}^2 = \{x + iy \mid x \in \mathbb{R}, y > 0\}$ (or the unit disc in $\mathbb{C}$) [V1-2].

An important result in [CV, Theorem, p. 93] is the following.

THEOREM 9.1. *The reduced outer space X_n is contractible.*

This implies that for any torsion-free subgroup $\Gamma \subset Out(F_n)$, $\Gamma\backslash X_n$ is a $B\Gamma$-space, and hence cd $\Gamma \leq 3n - 4$. Consequently, we have an upper bound on the virtual cohomological dimension of $Out_n(F_n)$:

$$\text{vcd } Out_n(F_n) \leq \dim X_n = 3n - 4.$$

Since the total sum of some nontrivial loops could become arbitrarily small and go to 0, the quotient $Out(F_n)\backslash X_n$ is noncompact. Therefore, $Out(F_n)$ is an analogue of non-uniform arithmetic subgroups, and the above upper bound on vcd $Out(F_n)$ is unlikely to be sharp.

Indeed, let K_n be the spine of X_n, i.e., the geometric realization of the partially ordered set of open simplices of X_n, which can be canonically realized as a subset of X_n. It turns out to be a simplicial complex of dimension $2n - 3$, and $Out(F_n)$ leaves K_n stable and acts properly on it with the compact quotient $Out(F_n)\backslash K_n$ [CV].

Another important result in [CV, Theorem 6.1.1] is the following.

THEOREM 9.2. *The spine K_n is an equivariant deformation retract of X_n and hence is contractible.*

Together with Proposition 3.2, it implies the following corollary.

PROPOSITION 9.3. *For every torsion-free subgroup $\Gamma \subset Out(F_n)$, $\Gamma\backslash K_n$ is a finite CW-complex $B\Gamma$-space, and hence Γ is of type FP and cd $\Gamma \leq 2n - 3$. Consequently, $Out(F_n)$ is of type FP_∞, and vcd $Out(F_n) \leq 2n - 3$.*

The following observation of Gersten [CV, p. 93] shows that the upper bound on vcd $Out(F_n)$ in the above theorem is sharp.

PROPOSITION 9.4. *For every $n \geq 2$, $Out(F_n)$ contains a subgroup isomorphic to $\mathbb{Z}^{2n-3}$, which implies $\mathrm{vcd}\, Out(F_n) \geq \mathrm{vcd}\, \mathbb{Z}^{2n-3} = 2n - 3$. Consequently,*

$$\mathrm{vcd}\, Out(F_n) = 2n - 3.$$

This subgroup can be written down explicitly [CV, p. 93].

10. Duality properties of outer automorphism groups

Given the duality results for arithmetic subgroups and mapping class groups, it is natural to expect that similar duality results hold for $Out(F_n)$.

In [BF, Theorems 1.1 and 1.4], by applying a version of Morse theory to a partial compactification of the outer space X_n, which is an analogue of the Borel-Serre partial compactification $\overline{X}^{BS}$ of symmetric spaces X, Bestvina and Feighn proved the following result, using the algebraic criterion in Proposition 4.2.

THEOREM 10.1. *The outer group $Out(F_n)$ is $(2n-5)$-connected at infinity, and hence for any torsion-free subgroup Γ of $Out(F_n)$ of finite index, $H^i(\Gamma, \mathbb{Z}\Gamma) = 0$ for $i < 2n - 3$; for $i = 2n - 3$, $H^{2n-3}(\Gamma, \mathbb{Z})$ is a free abelian group. Consequently, $Out(F_n)$ is a virtual duality group of dimension $2n - 3$.*

Since $Out(F_n) \backslash X_n$ is noncompact and $Out(F_n)$ is similar to a non-uniform arithmetic subgroup, the following result is naturally expected.

THEOREM 10.2. *For every $n \geq 2$, the group $Out(F_n)$ is not a virtual Poincaré duality group.*

Proof. Suppose that the opposite is true, i.e., that $Out(F_n)$ is a virtual Poincaré duality group. Then there exists a torsion-free subgroup Γ of $Out(F_n)$ of finite index which is a Poincaré duality group, i.e., $H^{2n-3}(\Gamma, \mathbb{Z}\Gamma) \cong \mathbb{Z}$.

By Corollary 4.9, it suffices to find an infinite index subgroup Γ' of Γ such that

$$\mathrm{cd}\, \Gamma' = 2n - 3 = \mathrm{cd}\, \Gamma.$$

Indeed, this will lead to a contradiction with the assumption that Γ is a Poincaré duality group.

By Proposition 9.4, $Out(F_n)$ contains a subgroup Λ isomorphic to $\mathbb{Z}^{2n-3}$. We claim that Γ also contains a free abelian subgroup of rank $2n - 3$, which clearly has the cohomological dimension equal to $2n - 3$. To prove this, consider the left-multiplication of Λ on the finite set $Out(F_n)/\Gamma$. The stabilizer in Λ of the identity coset Γ is equal to $\Gamma \cap \Lambda$ and clearly has finite index in Λ (note that the Λ-orbit through the coset Γ is finite). This implies that $\Gamma \cap \Lambda$ is also a free abelian group of rank $2n - 3$.

REMARK 10.3. In the previous cases of arithmetic subgroups of linear algebraic groups and mapping class groups, the dualizing modules can be identified with some Steinberg modules, which are defined in terms of the top non-vanishing homology group of natural simplicial complexes, and the duality property follows from the Poincaré duality for manifolds with boundary. In the case of $Out(F_n)$, the dualizing module does not have such a concrete interpretation yet, and the duality is also proved using the algebraic criterion in Proposition 4.2. Since X_n is not a manifold, probably one could not appeal to the Poincaré duality for manifolds. On the other hand, it is an important and natural problem to find an analogue of the spherical Tits building $\Delta_{\mathbb{Q}}(\mathbf{G})$ and the curve complex $\mathcal{C}(S)$ which describes the geometry at infinity of X_n and whose homology group gives the dualizing module. One such candidate is the factor complex in [HV].

REMARK 10.4. There is no doubt that Theorem 10.2 was expected by many people and should be known to many experts. On the other hand, it has not been written down explicitly before. It was observed by the author while trying to list similar properties of the three classes of groups discussed in this paper. From such listing, this result is natural.

Let $x_1, \cdots, x_n$ be a free basis of F_n. The symmetric automorphism subgroup Σ_n of $Aut(F_n)$ consists of those automorphisms that send each generator x_i to a conjugate of some x_j. The pure symmetric automorphism group, denoted by $P\Sigma_n$, is the subgroup of symmetric automorphisms that send each x_i to a conjugate of x_i. The image of $P\Sigma_n$ in $Out(F_n)$ is called the outer pure symmetric automorphism group and denoted by $OP\Sigma_n$. After a preliminary version of this paper was submitted, we learnt the following result [BMM].

THEOREM 10.5. *The group $OP\Sigma_n$ is a duality group of dimension $n-2$, and the group $P\Sigma_n$ is a duality group of dimension $n-1$.*

By [Co, Theorem 5.1], the group $OP\Sigma_n$ contains a free abelian subgroup of rank $n-2$, and the group $P\Sigma_n$ contains a free abelian subgroup of rank $n-1$. Then by the same argument as in the proof of Theorem 10.2, we can prove the following result.

THEOREM 10.6. *For $n \geq 3$, the groups $OP\Sigma_n$ and $P\Sigma_n$ are not Poincaré duality groups.*

REMARK 10.7. The above arguments show that if a virtually non-abelian group Γ is of type FP and of cohomological dimension d but contains a free abelian subgroup of rank d, then Γ is not a virtual Poincaré duality group.

Besides the examples discussed above, other groups satisfying this condition includes the partial symmetric outer automorphism group $P\Sigma(n, k)$, which consists of automorphisms which send the first k generators of $\{x_1, \cdots, x_n\}$ to conjugates of themselves (see [BCV] for details). Therefore, it follows that $P\Sigma(n, k)$ is not a virtual Poincare duality group either.

11. Comments and conjectures

In the above sections, we have shown that three important classes of groups are duality groups. It is natural to consider which group operations preserve the class of duality groups.

Clearly, if Γ_1 and Γ_2 are both duality groups which admit $E\Gamma_1$ and $E\Gamma_2$ that are manifolds satisfying the conditions in Propisition 4.5, then $E\Gamma_1 \times E\Gamma_2$ is a $E(\Gamma_1 \times \Gamma_2)$-space satisfying the conditions of Proposition 4.5 as well. This implies that $\Gamma_1 \times \Gamma_2$ is also a duality group. (Note that $E\Gamma_1 \times E\Gamma_2$ is a manifold with corners. But the corners can be smoothed out as pointed out in Remark 4.6.)

In fact, a stronger result is true. Specifically, by [Bi, Theorem 9.10], the class of duality groups is closed under extension. Specifically, assume that Γ_1, Γ_2 are duality groups of dimensions n_1, n_2, and that Γ fits into an exact sequence:

$$1 \to \Gamma_1 \to \Gamma \to \Gamma_2 \to 1,$$

then Γ is a duality group of dimension $n_1 + n_2$.

An important procedure to construct a new group from a given group Γ is to consider the outer automorphism group

$$Out(\Gamma) = Aut(\Gamma)/Inn(\Gamma).$$

Since $Inn(\Gamma)$ is isomorphic to $\Gamma/Z(\Gamma)$, where $Z(\Gamma)$ is the center of Γ, it is reasonable to divide out $Inn(\Gamma)$ in order to get a new group.

Based on the examples discussed in this paper, it is reasonable to formulate the following general conjecture.

CONJECTURE 11.1. *If Γ is a duality group, then $Out(\Gamma)$ is also a duality group. Consequently, if $Z(\Gamma)$ is trivial, then $Aut(\Gamma)$ is also a duality group.*

In fact, if $Z(\Gamma)$ is trivial, then $Aut(\Gamma)$ fits into an exact sequence $1 \to \Gamma \to Aut(\Gamma) \to Out(\Gamma) \to 1$, and the duality of $Out(\Gamma)$ implies the duality of $Aut(\Gamma)$.

REMARK 11.2. It might be worthwhile to point out that when Γ is a duality group, subgroups and quotient subgroups of Γ are usually not virtual duality groups. For example, the free group F_n contains subgroups and are also mapped onto some quotient groups which are not finitely generated. Since duality groups are of type FP (Proposition 4.1) and hence finitely generated, the above subgroups and quotient groups are not duality groups.

Now we examine cases where the above conjecture holds. It is known that the free group F_n is a duality group [Br1, p. 223, Example 5]. (Another geometric way to understand this is to note that if $\Gamma \subset SL(2, \mathbb{R})$ is a torsion-free non-uniform lattice subgroup, then Γ is a free group on finitely many generators and the surface $\Gamma \backslash SL(2, \mathbb{R})/SO(2)$ admits a Borel-Serre compactification, which implies that Γ is a duality group of dimension 1 as in §6.) By the results in §10, $Out(F_n)$ is a virtual duality group.

Suppose that $\Gamma = \mathbb{Z}^n$, the free abelian group, then $Out(\mathbb{Z}^n) = GL(n, \mathbb{Z})$ and is a virtual duality group by §6.

Suppose that S is a compact surface of genus $g \geq 1$ and $\Gamma = \pi_1(S)$, a surface group. Then it is known that $Out(\pi_1(S)) = Mod(S) = Mod_g$ [I1, Theorem 2.9.A]. By the results of §8, $Out(\pi_1(S))$ is a virtual duality group. (As pointed out before, if S is non-compact, then $\pi_1(S)$ is a free group.)

If Γ is a finitely generated nilpotent group, it is a virtual duality group since it admits a torsion-free subgroup of finite index and then by applying either the discussions in §4 or the above result that the class of duality groups is closed under extensions, together with the fact that $\mathbb{Z}$ is a duality group. By [Se], $Out(\Gamma)$ is a finite extension of an arithmetic group. Therefore, $Out(\Gamma)$ is a virtual duality group. (By an arithmetic group, we mean a group that is isomorphic to an arithmetic subgroup of a linear algebraic group defined over $\mathbb{Q}$.)

If Γ is a polycyclic-by-finite group, then Γ is a virtual duality group by the same arguments as in the previous paragraph, i.e., the class of duality groups is closed under extension. By [BaG], $Out(\Gamma)$ is an arithmetic group, and hence by §6, $Out(\Gamma)$ is also a virtual duality group.

If Γ is an arithmetic subgroup of a linear semisimple algebraic $\mathbf{G}$ over $\mathbb{Q}$, then Γ is a virtual duality group by the results in §6. Assume that the associated symmetric space $X = G/K$ is not the Poincaré upper half-plane, then the Mostow strong rigidity implies that $Out(\Gamma) = Out(G)$ is a finite group and hence is a virtual duality group. If Γ is an arithmetic subgroup of $\mathbf{G} = SL(2)$ in the exceptional case, then $Out(\Gamma)$ is essentially a mapping class group of the surface $\Gamma \backslash SL(2,\mathbb{R})/SO(2)$ and is a virtual duality group as well. The same result holds for any lattice subgroup of a *semisimple* (or more generally a reductive) Lie group G.

Given the above discussions, it is conceivable that if G is any (virtually) connected Lie group, which is not necessarily reductive, and Γ is a lattice subgroup of G, then $Out(\Gamma)$ is a virtual duality group.

References

[As]   A.Ash, *Deformation retracts with lowest possible dimension of arithmetic quotients of self-adjoint homogeneous cones*, Math. Ann. 225 (1977) 69–76.

[AR]   A.Ash, L.Rudolph, *The modular symbol and continued fractions in higher dimensions*, Invent. Math. 55 (1979) 241–250.

[BaG]   O.Baues, F.Grunewald, *Automorphism groups of polycyclic-by-finite groups and arithmetic groups*, Publ. Math. Inst. Hautes Études Sci. No. 104 (2006) 213–268.

[BaT]   G.Baumslag, T.Taylor, *The centre of groups with one defining relator*, Math. Ann. 175 (1968) 315–319.

[B]   D.Benson, *Representations and cohomology. II. Cohomology of groups and modules*, Cambridge Studies in Advanced Mathematics, 31. Cambridge University Press, 1991. x+278 pp.

[Be]   M.Bestvina, *The topology of* Out(F_n), Proc. of International Congress of Mathematicians, Vol. II (Beijing, 2002), pp. 373–384, Higher Ed. Press, 2002.

[BF]   M.Bestvina, M.Feighn, *The topology at infinity of* Out(F_n), Invent. Math. 140 (2000) 651–692.

[Bi]   R.Bieri, *Homological dimension of discrete groups*, Second edition. Queen Mary College Mathematical Notes. Queen Mary College, London, 1981. iv+198 pp.

[Bo]   A.Borel, *Introduction aux groupes arithmétiques*, Hermann, Paris, 1969, 125 pp.

[BS1]   A.Borel, J.P.Serre, *Corners and arithmetic groups*, Comment. Math. Helv. 48 (1973) 436–491.

[BS2]   A.Borel, J.P.Serre, *Cohomologie d'immeubles et de groupes S-arithmétiques*, Topology 15 (1976) 211–232.

[BMM]   N.Brady, J.McCammond, J.Meier, A.Miller, *The pure symmetric automorphisms of a free group form a duality group*, J. Algebra 246 (2001) 881–896.

[Bro]   N.Broaddus, *Homology of the curve complex and the Steinberg module of the mapping class group*, arXiv:0711.0011.

[Br1]   K.Brown, *Cohomology of groups*, Graduate Texts in Mathematics, 87. Springer-Verlag, 1994. x+306 pp.

[Br2]   K.Brown, *Buildings*, Springer-Verlag, 1989. viii+215 pp.

[BCV]   K.Bux, R.Charney, K.Vogtmann, *Automorphism groups of RAAGs and partially symmetric automorphisms of free groups*, preprint.

[BuW]   K.Bux, K.Wortman, *Finiteness properties of arithmetic groups over function fields*, Invent. Math. 167 (2007) 355–378.

[Ca]   R.Carter, *Finite groups of Lie type. Conjugacy classes and complex characters*, Wiley Classics Library. A Wiley-Interscience Publication, 1993. xii+544 pp.

[Co]   D.Collins, *Cohomological dimension and symmetric automorphisms of a free group*, Comment. Math. Helv. 64 (1989) 44–61.

[CV]   M.Culler, K.Vogtmann, *Moduli of graphs and automorphisms of free groups*, Invent. Math. 84 (1986) 91–119.

[Ei]   S.Eilenberg, *Homology of spaces with operators. I*, Trans. Amer. Math. Soc. 61(1947) 378–417.

[FK]   J.Faraut, A.Koranyi, *Analysis on symmetric cones*, The Clarendon Press, Oxford University Press, 1994. xii+382 pp.

[Fa]   F.Farrell, *Poincaré duality and groups of type* (FP), Comment. Math. Helv. 50 (1975) 187–195.

[GM]   M.Goresky, R.MacPherson, *Intersection homology theory*, Topology 19 (1980) 135–162.

[GuM]   P.Gunnells, M.McConnell, *Hecke operators and $\mathbb{Q}$-groups associated to self-adjoint homogeneous cones*, J. Number Theory 100 (2003) 46–71.

[Ha1]   J.Harer, *The virtual cohomological dimension of the mapping class group of an orientable surface*, Invent. Math. 84 (1986) 157–176.

[Ha2]   J.Harer, *The cohomology of the moduli space of curves*, in *Theory of moduli* (Montecatini Terme, 1985), pp. 138–221, Lecture Notes in Math., 1337, 1988.

[Hav]   W.Harvey, *Boundary structure of the modular group*, in *Riemann surfaces and related topics*, pp. 245–251, Ann. of Math. Stud., 97, Princeton Univ. Press, 1981.

[HV]   A.Hatcher, K.Vogtmann, *The complex of free factors of a free group*, Quart. J. Math. Oxford Ser. 49 (1998) 459–468.

[Hu]   J.Humphreys, *The Steinberg representation*, Bull. Amer. Math. Soc. 16 (1987) 247–263.

[I1]   N.Ivanov, *Mapping class groups*, in *Handbook of geometric topology*, pp. 523–633, North-Holland, Amsterdam, 2002.

[I2]   N.Ivanov, *Complexes of curves and Teichmüller modular groups*, Russian Math Surveys, V. 42, No. 3 (1987) 55-107.

[I3]   N.Ivanov, *Attaching corners to Teichmüller space*, Leningrad Math. J. 1 (1990) 1177–1205.

[I4]   N.Ivanov, *Complexes of curves and Teichmüller spaces*, Math. Notes 49 (1991), no. 5-6, 479–484.

[IJ]   N.Ivanov, L.Ji, *Infinite topology of curve complex and non-Poincaré duality of mapping class group*, to appear in L'Enseignement Mathématique.

[Ive]   B.Iversen, *Cohomology of sheaves*, Universitext. Springer-Verlag, Berlin, 1986. xii+464 pp.

[Ji]   L.Ji, *Buildings and their applications in geometry and topology*, Asian J. Math. 10 (2006) 11–80.

[M]   A.Malcev, *On a class of homogeneous spaces*, Amer. Math. Soc. Translation 1951, no. 39, 33 pp.

[Ma]   W.Massey, *Homology and cohomology theory. An approach based on Alexander-Spanier cochains*, Monographs and Textbooks in Pure and Applied Mathematics, Vol. 46, Marcel Dekker, Inc., 1978. xiv+412 pp.

[Mu]   J.Munkres, *Elements of algebraic topology*, Addison-Wesley Publishing Company, 1984. ix+454 pp.

[Ra1]   M.S.Raghunathan, *Discrete subgroups of Lie groups*, Ergebnisse der Mathematik und ihrer Grenzgebiete, Band 68. Springer-Verlag, 1972. ix+227 pp.

[Ra2]   M.S.Raghunathan, *A note on quotients of real algebraic groups by arithmetic subgroups*, Invent. Math. 4 (1967/1968) 318–335.

[Sa]   L.Saper, *Tilings and finite energy retractions of locally symmetric spaces*, Comment. Math. Helv. 72 (1997) 167–202.

[Se]   D.Segal, *On the outer automorphism group of a polycyclic group*, in *Proceedings of the Second International Group Theory Conference*, (Bressanone, 1989), Rend. Circ. Mat. Palermo (2) Suppl. No. 23 (1990), 265–278.

[So]   L.Solomon, *The Steinberg character of a finite group with BN-pair*, in *Theory of Finite Groups*, pp. 213–221, Benjamin, New York, 1969.

[S]   N.Steenrod, *Homology with local coefficients*, Ann. of Math. 44 (1943) 610–627.

[St]   R.Steinberg, *Robert Steinberg collected papers*, American Mathematical Society, 1997. xx+599 pp.

[Str]   R.Strebel, *A remark on subgroups of infinite index in Poincaré duality groups*, Comment. Math. Helvetici. 52 (1977) 317–324.

[To]   A.Toth, *On the Steinberg module of Chevalley groups*, Manuscripta Math. 116 (2005) 277–295.

[V1]   K.Vogtmann, *Automorphisms of free groups and outer space*, Geom. Dedicata 94 (2002) 1–31.

[V2]   K.Vogtmann, *The cohomology of automorphism groups of free groups*, Proc. of International Congress of Mathematicians. Vol. II, pp. 1101–1117, Eur. Math. Soc., 2006.

DEPARTMENT OF MATHEMATICS, UNIVERSITY OF MICHIGAN, ANN ARBOR, MI 48109
E-mail address: lji@umich.edu

Contemporary Mathematics
Volume **478**, 2009

Characters of simplylaced nonconnected groups versus characters of nonsimplylaced connected groups

Shrawan Kumar, George Lusztig and Dipendra Prasad

ABSTRACT. Let G be a connected, simply-connected, almost simple semisimple group over $\mathbf{C}$ of simplylaced type and let σ be a nontrivial diagram automorphism of G. Let $G\langle\sigma\rangle$ be the (disconnected) group generated by G and σ. As a consequence of a theorem of Jantzen the character of an irreducible representation of $G\langle\sigma\rangle$ (also irreducible on G) on $G\sigma$ can be expressed in terms of a character of an irreducible representation of a certain connected simply connected semisimple group G_σ of nonsimplylaced type. We show how Jantzen's theorem can be deduced from properties of the canonical bases.

Let G be a connected, simply-connected, almost simple algebraic group of simplylaced type over $\mathbf{C}$. Let T be a maximal torus of G. Let $x_i : \mathbf{C} \to G$, $y_i : \mathbf{C} \to G$ $(i \in I)$ be homomorphisms which together with T form a pinning (épinglage) of G. We fix a nontrivial automorphism σ of G such that $\sigma(T) = T$, and such that for some permutation $i \mapsto \tilde{i}$ of I we have $\sigma(x_i(a)) = x_{\tilde{i}}(a)$, $\sigma(y_i(a)) = y_{\tilde{i}}(a)$ for all $a \in \mathbf{C}$. For $i \in I$ we write $\sigma(i) = \tilde{i}$. Let $\langle\sigma\rangle$ be the finite subgroup of the automorphism group of G generated by σ and let $G\langle\sigma\rangle$ be the semidirect product of G with $\langle\sigma\rangle$.

Let X be the group of characters $T \to \mathbf{C}^*$; let Y be the group of one parameter subgroups $\mathbf{C}^* \to T$ and let $\langle,\rangle : Y \times X \to \mathbf{Z}$ be the standard pairing. For $i \in I$ we define $\alpha_i \in X$ by $x_i(\alpha_i(t)) = tx_i(1)t^{-1}$, $y_i(\alpha_i(t)^{-1}) = ty_i(1)t^{-1}$ for all $t \in T$. This is a root of G. Let $\breve{\alpha}_i \in Y$ be the corresponding coroot. Note that

(a) $(Y, X, \langle,\rangle, \breve{\alpha}_i, \alpha_i (i \in I))$

is the root datum of G. Now σ induces automorphisms of X, Y denoted again by σ; these are compatible with $\langle,\rangle$ and we have $\sigma(\alpha_i) = \alpha_{\sigma(i)}, \sigma(\breve{\alpha}_i) = \breve{\alpha}_{\sigma(i)}$ for $i \in I$. Let $X^+ = \{\lambda \in X; \langle\breve{\alpha}_i, \lambda\rangle \in \mathbf{N}\forall i \in I\}$.

We set $Y_\sigma = Y/(\sigma-1)Y$, ${}^\sigma X = \{\lambda \in X; \sigma(\lambda) = \lambda\}$. Note that $\langle,\rangle : Y \times X \to \mathbf{Z}$ induces a perfect pairing $Y_\sigma \times {}^\sigma X \to \mathbf{Z}$ denoted again by $\langle,\rangle$. Let I_σ be the set of σ-orbits on I. For any $\mathcal{O} \in I_\sigma$ let $\breve{\alpha}_{\mathcal{O}} \in Y_\sigma$ be the image of $\breve{\alpha}_i$ under $Y \to Y_\sigma$ where i is any element of $\mathcal{O}$. Since $\{\breve{\alpha}_i; i \in I\}$ is a $\mathbf{Z}$-basis of Y we see that $\{\breve{\alpha}_{\mathcal{O}}; \mathcal{O} \in I_\sigma\}$ is a $\mathbf{Z}$-basis of Y_σ. For any $\mathcal{O} \in I_\sigma$ let $\alpha_{\mathcal{O}} = 2^h \sum_{i\in\mathcal{O}} \alpha_i \in {}^\sigma X$ where h is the number of unordered pairs (i, j) such that $i, j \in \mathcal{O}$, and $\alpha_i + \alpha_j$ is a root. Note that $h = 0$

The first author was supported in part by NSF Grant

except when G is of type A_{2n} when $h = 0$ for all $\mathcal{O}$ but one and $h = 1$ for one $\mathcal{O}$. Note that

(b) $(Y_\sigma, {}^\sigma X, \langle,\rangle, \check{a}_{\mathcal{O}}, a_{\mathcal{O}}(\mathcal{O} \in I_\sigma))$

is a root datum, see [**Ja, p.29**]. Let ${}^\sigma X^+ = \{\lambda \in {}^\sigma X; \langle \check{a}_{\mathcal{O}}, \lambda \rangle \in \mathbf{N} \forall \mathcal{O} \in I_\sigma\} = {}^\sigma X \cap X^+$. Let G_σ be the connected semisimple group over $\mathbf{C}$ with root datum (b). By definition, G_σ is provided with an épinglage $(T_\sigma, x_{\mathcal{O}}, y_{\mathcal{O}}$ $(\mathcal{O} \in I_\sigma)$ where $T_\sigma := \mathbf{C}^* \otimes Y_\sigma = T/\{\sigma(t)t^{-1}; t \in T\}$ is a maximal torus of G_σ and $x_{\mathcal{O}} : \mathbf{C} \to G_\sigma$, $y_{\mathcal{O}} : \mathbf{C} \to G_\sigma$ satisfy $x_{\mathcal{O}}(a_{\mathcal{O}}(t_1)) = t_1 x_{\mathcal{O}}(1)t_1^{-1}$, $y_{\mathcal{O}}(a_{\mathcal{O}}(t_1)^{-1}) = t_1 y_{\mathcal{O}}(1)t_1^{-1}$ for all $t_1 \in T_\sigma$. (We have ${}^\sigma X = \mathrm{Hom}(T_\sigma, \mathbf{C}^*)$ canonically.)

Note that G_σ is simply connected and that $G_\sigma \cong {}^L(((^LG)^\sigma)^0)$ where $^L()$ denotes the Langlands dual group and $(^LG)^\sigma$ denotes the fixed point set of the automorphism of ^LG induced by σ. Now G_σ is only isogenous to $^L(G^\sigma)$ where G^σ is the fixed point set of $\sigma : G \to G$.

Let $\lambda \in {}^\sigma X^+$. We can view λ both as a character of T and as a character of T_σ. Let V (resp. V') be a finite dimensional complex irreducible representation of G (resp. G_σ) with a non-zero vector η (resp. η') such that $x_i(a)\eta = 0$ for all $i \in I, a \in \mathbf{C}$ (resp. $x_{\mathcal{O}}(a)\eta' = 0$ for all $\mathcal{O} \in I_\sigma, a \in \mathbf{C}$) and $t\eta = \lambda(t)\eta$ for all $t \in T$ (resp. $t'\eta' = \lambda(t')\eta'$ for all $t' \in T_\sigma$). Now V can be regarded as a representation of $G\langle\sigma\rangle$ whose restriction to G is as above and on which the action of σ satisfies $\sigma(\eta) = \eta$.

Let $\mu \in X$. Let $V_\mu = \{x \in V; tx = \mu(t)x \quad \forall t \in T\}$. Note that $\sigma : V \to V$ permutes the weight spaces V_μ among themselves. A weight space V_μ is σ-stable if and only if $\mu \in {}^\sigma X$; in this case μ can be viewed as a character of T_σ and we set $V'_\mu = \{x' \in V'; t'x' = \mu(t')x' \forall t' \in T_\sigma\}$.

THEOREM (JANTZEN [**Ja, Satz 9**]). *For $\mu \in {}^\sigma X$ we have* $\mathrm{tr}(\sigma : V_\mu \to V_\mu) = \dim V'_\mu$.

COROLLARY. *Let $\varpi : T \to T_\sigma$ be the canonical homomorphism. For any $t \in T$ we have* $\mathrm{tr}(t\sigma : V \to V) = \mathrm{tr}(\varpi(t), V')$.

The corollary describes completely the character of V on $G\sigma$ in terms of the character of V' since any semisimple element in $G\sigma$ is G-conjugate to an element of the form $t\sigma$ with $t \in T$. Note also that there is a well defined bijection between the set of semisimple G-conjugacy classes in $G\sigma$ and the set of semisimple G_σ-conjugacy classes in G_σ which for any $t \in T$ maps the G-conjugacy class of $t\sigma$ to the G_σ-conjugacy class of $\varpi(t)$; see [**L2, 6.26**], [**Mo**].

We now show (assuming that G is not of type A_{2n}) how Jantzen's theorem can be deduced from properties of canonical bases in [**L1**]. According to [**L1**], V has a canonical basis B_λ and V' has a canonical basis B'_λ. Also, B_λ (resp. B'_λ) can be naturally viewed as a subset of $\mathbf{B}$ (resp. $\mathbf{B}'$), the canonical basis of the $+$ part of the universal enveloping algebra attached to the root datum (a) (resp. (b)). Now σ acts naturally on $\mathbf{B}$ (preserving the subset B_λ) and [**L1, Theorem 14.4.9**] provides a canonical bijection between $\mathbf{B}'$ and the fixed point set of σ on $\mathbf{B}$. (This theorem is applicable since the Cartan datum of (b) is obtained from the Cartan datum of (a) by the general "folding" procedure [**L1, 14.1**] which appplies to any simplylaced Cartan datum of not necessarily finite type together with an admissible automorphism; here we use that G is not of type A_{2n}.) This restricts to a bijection between B'_λ and the fixed point set ${}^\sigma B_\lambda$ of σ on B_λ. Next we note that B_λ (resp. B'_λ) is compatible with the decomposition of V (resp. V') into weight spaces and from the definitions we see that the bijection above carries $B'_\lambda \cap V'_\mu$

bijectively onto $^\sigma B_\lambda \cap V_\mu$. Since $B_\lambda \cap V_\mu$ is a basis of V_μ which is σ-stable we have $\mathrm{tr}(\sigma : V_\mu \to V_\mu) = \sharp(^\sigma B_\lambda \cap V_\mu)$. Using the bijection above this equals $\sharp(B'_\lambda \cap V'_\mu)$ and this is equal to $\dim V'_\mu$ since $B'_\lambda \cap V'_\mu$ is a basis of V'_μ. This gives the desired result.

We refer the reader to [**FSS, FRS, NS, N1, N2, We**] for other approaches to Jantzen's theorem. We thank S. Naito and the referee for pointing out these references to us.

The first two authors were supported in part by the National Science Foundation. The third author thanks the Institute for Advanced Study where this work was done, and gratefully acknowledges receiving support through grants to the Institute by the Friends of the Institute, and the von Neumann Fund.

References

[FSS]   J. Fuchs, B. Schellekens and G. Schweigert, *From Dynkin diagrams symmetries to fixed point structures*, Comm. Math. Phys. **180** (1996), 39-97.

[FRS]   J. Fuchs, U. Ray and G. Schweigert, *Some automorphisms of generalized Kac-Moody algebras*, J. Algebra **191** (1997), 518-590.

[Ja]   J. C. Jantzen, *Darstellungen Halbeinfacher Algebraischer Groupen,*, Bonner Math. Schriften **67** (1973).

[L1]   G. Lusztig, *Introduction to quantum groups*, Progress in Math., vol. 110, Birkhäuser, 1993.

[L2]   G. Lusztig, *Classification of unipotent representations in simple p-adic groups, II*, Represent. Theory (2002), 243-289.

[Mo]   S. Mohrdieck, *Conjugacy classes of non-connected semisimple algebraic groups*, Transfor. Groups **8** (2003), 377-395.

[N1]   S.Naito, *Twining character formulas of Borel-Weil-Bott type*, J. Math. Sci. Univ. Tokyo **9** (2002), 637-658.

[N2]   S. Naito, *Twining characters, Kostant's homology formula and the Bernstein-Gelfand-Gelfand resolution*, J. Math. Kyoto Univ. **42** (2002), 83-103.

[NS]   S. Naito and D. Sagaki, *Lakshmibai-Seshadri paths fixed by a diagram automorphism*, J. Algebra **245** (2001), 395-412.

[We]   R. Wendt, *Weyl's character formula for non connected Lie groups and orbital theory for twisted affine Lie algebras*, J. Funct. Anal. **180** (2001), 31-65.

S.K.: DEPARTMENT OF MATHEMATICS, UNIVERSITY OF NORTH CAROLINA, CHAPEL HILL, NC 27599-3250, USA

G.L.: DEPARTMENT OF MATHEMATICS, M.I.T., CAMBRIDGE, MA 02139, USA
E-mail address: gyuri@math.mit.edu

D.P.: SCHOOL OF MATHEMATICS, TATA INSTITUTE OF FUNDAMENTAL RESEARCH, COLABA, MUMBAI 400005, INDIA, AND THE INSTITUTE FOR ADVANCED STUDY, PRINCETON, NJ 08540, USA

Contemporary Mathematics
Volume **478**, 2009

Classification of Finite-dimensional Basic Hopf Algebras According to Their Representation Type

Gongxiang Liu

ABSTRACT. The main aim of this paper is to give the classification of finite-dimensional basic Hopf algebras according to their representation type. We attach every finite-dimensional basic Hopf algebra H a natural number n_H, which will help us to determine the representation type of H. The class of finite-dimensional basic Hopf algebras of finite representation type is determined completely. A complete list of local Frobenius algebras of tame type is given. By using this list, we get all possible algebraic structures of tame basic Hopf algebras.

CONTENTS

1. Introduction

1.1. Throughout this paper k denotes an algebraically closed field. All spaces are k-spaces. By an algebra we mean an associative algebra with identity element. For an algebra A, J_A denotes its Jacobson radical. We freely use the results, notation, and conventions of [**49**].

1.2. The classification of Hopf algebras is one of central problems of Hopf algebra theory. The first celebrated result on this problem is now known as the following Cartier-Kostant-Milnor-Moore theorem.

1991 *Mathematics Subject Classification.* Primary 16W30, 16G30; Secondary 16G20.
Key words and phrases. Basic Hopf algebra, representation type, Nakayama algebra, Frobenius algebra.

THEOREM 1.1. *A cocommutative Hopf algebra over an algebraically closed field* k *of characteristic 0 is a semidirect product of a group algebra and the enveloping algebra of a Lie algebra. In particular, a finite-dimensional cocommutative Hopf algebra over* k *is a group algebra.*

1.3. In recent years, some substantial classification results in the infinite-dimensional case are given. All possible cotriangular Hopf algebras were determined [24], and a class of pointed Hopf algebras with finite Gelfand-Kirillov dimension is classified [2][5]. Lu, Wu and Zhang introduced the concept of *homological integral*, which generalizes the usual integral defined for finite-dimensional Hopf algebras to a large class of infinite-dimensional Hopf algebras, and use it to research particularly noetherian affine Hopf algebras of Gelfand-Kirillov dimension 1 [43]. Although these results shed some light on the structure of infinite-dimensional Hopf algebras, it is still very hard to handle infinite-dimensional Hopf algebras in general, and the classification of finite-dimensional Hopf algebras is of more interest for us. We can use the following diagram (see [1]) to explain the general procedure to classify finite-dimensional Hopf algebras.

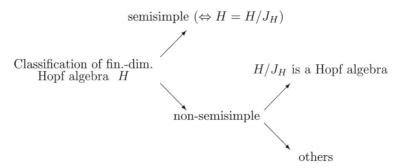

$$\text{semisimple } (\Leftrightarrow H = H/J_H)$$

Classification of fin.-dim. Hopf algebra H

H/J_H is a Hopf algebra

non-semisimple

others

Over the last decade, under various assumption, considerable progress has been made in classifying finite-dimensional Hopf algebras. To the author's knowledge, the classification of finite-dimensional Hopf algebras mainly consists of the following four aspects:

(1) Classification of semisimple cosemisimple Hopf algebras;
(2) Classification of non-semisimple Hopf algebras;
(3) Classification of all Hopf algebras of a prescribed dimension;
(4) Classification of triangular Hopf algebras.

1.4. For (1), if the characteristic of k is 0, then we know that a Hopf algebra is semisimple is equivalent to that it is cosemisimple by [39]. By a beautiful result of Etingof and Gelaki [20], problems in positive characteristic can be reduced to similar problems in characteristic 0. Therefore, we often consider the classification of semisimple Hopf algebras over an algebraically closed field of characteristic 0. Some semisimple Hopf algebras of special dimensions, particularly for dimensions p, p^2, p^3, pq, pq^2 [22][30][45][46][52][53][65] and dimensions ≤ 60 [54], were intensively studied. For example, semisimple Hopf algebras of dimension pq are shown to be trivial. That is, they are isomorphic to group algebras or dual group algebras. The class of semisimple Hopf algebras that are simple as Hopf algebras is researched recently [27][28]. Under a systematic study of of fusion categories, Etingof, Dikshych and Ostrik asked an interesting question (Question 8.45 in [25]) about the semisimple Hopf algebras: dose there exist a finite-dimensional semisimple Hopf algebra whose representation category is not group-theoretical? This question was answered affirmatively by Nikshych [59]. But, the classification of semisimple Hopf algebras is still a widely open question. We refer to the survey papers [1][50] for a more detailed exposition.

1.5. For (2), substantial results in this case are known for the class of pointed Hopf algebras over an algebraically closed field of characteristic 0. There are two different methods which were used to classify pointed Hopf algebras. One method was formulated by N.Andruskiewitsch and H.-J.Schneider. They reduced the study of pointed Hopf algebras to the study of Nichols algebras via bosonization given by Radford [61] and Majid [44]. This method gets great success. One of most remarkable properties of this method is that it allows Lie theory to enter into the picture through quantum groups [7]. Many new examples about pointed Hopf algebras were found through this way. It can also help us to give counterexamples for Kaplansky's Conjecture 10. For details see [6][7][9][31][33][34][35] [36]. Recently, Andruskiewitsch and Schneider have classified all finite-dimensional pointed Hopf algebras whose group of group-like elements $G(H)$ is abelian such that all prime divisors of the order of $G(H)$ are > 7. See [10].

Another method, mainly due to Pu Zhang and his co-workers, is to use quivers and their representation theory. This method depends heavily on one of Cibils-Rosso's conclusions [14]. One of merits of this method is that it introduces the combinatorial methods to enter into the field of the classification of pointed Hopf algebras. By using this method, the classification of so called Monomial Hopf algebras was gotten. Locally finite simple-pointed Hopf algebras can also be classified. For details see [13][60].

1.6. For (3), the starting point of this direction should be the following Y.Zhu's result [65].

THEOREM 1.2. *Let p be a prime number number and k an algebraically closed field of characteristic 0. Then a Hopf algebra of dimension p over k is necessarily semisimple and isomorphic to the group algebra of Z_p.*

By using [8] and [48], S-H. Ng [55] classified all Hopf algebras of dimension p^2 and showed that they are the group algebras and the Taft algebras. For dimension pq with $p \neq q$, a folklore conjecture says that such Hopf algebras are semisimple. If it is true, the results given in Subsection 1.4 imply that Hopf algebras of dimension pq, where p and q are distinct, are trivial. This conjecture was verified for some particular values of p and q [4][12][23][56][57][58]. There are other classification results in low dimension. All Hopf algebras of dimension ≤ 15 were classified and the most recent result in dimension 16 is [29]. See [29] and references therein.

1.7. For (4), the works of N. Andruskiewitsch, P. Etingof and S. Gelaki should be considered the most important. P. Etingof and S. Gelaki indeed show that semisimple triangular Hopf algebras are very closed to group algebras [21]. The structure of minimal triangular Hopf algebras is also given [3].

There are some nice surveys about classification of finite-dimensional Hopf algebras, see for instance [1].

1.8. According to the fundamental theorem of Drozd [**17**], the category of finite-dimensional algebras over k can be divided into disjoint classes of finite representation, tame and wild algebras. This fact stimulates us to classify finite-dimensional Hopf algebras through their representation type. In order to realize this idea, we need add some conditions on the Hopf algebra H:

(1): We assume that H/J_H is a quotient Hopf algebra. By the general procedure to classify finite-dimensional Hopf algebras, this requirement is not strange. Note that H satisfies this condition if and only if the coradical of the dual Hopf algebra H^* is a Hopf subalgebra of H^*.

(2): By 1.4, the classification of semisimple Hopf algebras is still a widely open question. This suggests us that we should consider semisimple Hopf algebra H/J_H which can be described easily.

A good candidate satisfying these conditions is the class of basic Hopf algebras. That is, as an algebra, it is basic. This implies that H/J_H is a Hopf algebra automatically (see Lemma 1.1 in [**32**]). Since the field k is algebraically closed and H/J_H is a Hopf algebra, the condition "basic" implies that $H/J_H \cong (kG)^*$ for some finite group G.

Before giving the classification of finite-dimensional basic Hopf algebras, we should give an effective way to determine their representation type at first. This is indeed what we will do in the next section. Explicitly, we can attach to every finite-dimensional basic Hopf algebra H a natural number n_H and prove that (i) H is of finite representation type if and only if $n_H = 0$ or $n_H = 1$; (ii) if H is tame, then $n_H = 2$ and (iii) if $n_H \geq 3$, then H is wild.

The dual of a basic Hopf algebra is a pointed Hopf algebra, and vice versa. So, all classification results on pointed Hopf algebras (some of them mentioned in subsection 1.5) can be applied by duality to basic Hopf algebras. But, I think, there is no possibility to give structures of all basic Hopf algebras. Inspired by the case of path algebras, it is quite natural to give the classification of finite-dimensional basic Hopf algebras of finite representation type and tame type, and Section 3 and Section 4 are devoted to classifying finite-dimensional basic Hopf algebras of finite representation type and tame type respectively. Indeed, in Section 3, the class of finite-dimensional basic Hopf algebras of finite representation type is classified completely (see Theorem 3.1). In the case of tame type, we give a list of algebras which contains all possible tame basic Hopf algebras in Section 4. Notice that it is still a problem to give the actual determination of tame basic Hopf algebras in this list (see Problem 4.1 in Section 4).

1.9. A finite-dimensional algebra A is said to be of *finite representation type* provided there are finitely many non-isomorphic indecomposable A-modules. A is of *tame type* or A is a *tame* algebra if A is not of finite representation type, whereas for any dimension $d > 0$, there are finite number of A-$k[T]$-bimodules M_i which are free as right $k[T]$-modules such that all but a finite number of indecomposable A-modules of dimension d are isomorphic to $M_i \otimes_{k[T]} k[T]/(T - \lambda)$ for $\lambda \in k$. We say that A is of *wild type* or A is a *wild* algebra if there is a finitely generated A-$k < X, Y >$-bimodule B which is free as a right $k < X, Y >$-module such that the functor $B \otimes_{k<X,Y>} -$ from mod-$k < X, Y >$, the category of finitely generated $k < X, Y >$-modules, to mod-A, the category of finitely generated A-modules, preserves indecomposability and reflects isomorphisms. See [**18**] for more details.

For other unexplained notations about representation theory of finite-dimensional algebras in this paper, see [11][18].

2. Representation type of basic Hopf algebras

In the rest of this paper, all algebras are assumed to be finite-dimensional. In this section, the definition of the covering quiver $\Gamma_G(W)$, introduced by Green and Solberg [32], is given at first. Then we observe that we can associate to this covering quiver $\Gamma_G(W)$ a natural number $n_{\Gamma_G(W)}$, which can help us to determine the representation type of the finite-dimensional algebra A whose Ext-quiver is $\Gamma_G(W)$. For a finite-dimensional basic Hopf algebra H, it is known that its Ext-quiver is a covering quiver. So the above results can be applied to the case of finite-dimensional basic Hopf algebras directly.

DEFINITION 2.1. *Let G be a finite group and let $W = (w_1, w_2, \ldots, w_n)$ be a sequence of elements of G. We say W is a* weight sequence *if, for each $g \in G$, the sequences W and $(gw_1g^{-1}, gw_2g^{-1}, \ldots, gw_ng^{-1})$ are the same up to a permutation. Define a quiver, denoted by $\Gamma_G(W)$, as follows. The vertices of $\Gamma_G(W)$ is the set $\{v_g\}_{g \in G}$ and the arrows are given by*

$$\{(a_i, g): v_{g^{-1}} \to v_{w_ig^{-1}} | i = 1, 2, \ldots, n, g \in G\}.$$

We call this quiver the covering quiver (with respect to G and W).

EXAMPLE 2.1. (1): Let $G = <g>$, $g^n = 1$ and $W = (g)$. The corresponding covering quiver is

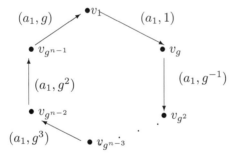

We call such quiver a *basic cycle of length n.*

(2): Let $G = K_4 = \{1, a, b, ab\}$, the Klein four group, and $W = (1)$. Then the corresponding covering quiver is

$$\bullet^1 \circlearrowleft, \quad \bullet^a \circlearrowleft, \quad \bullet^b \circlearrowleft, \quad \bullet^{ab} \circlearrowleft$$

For a covering quiver $\Gamma_G(W)$, define $n_{\Gamma_G(W)}$ to be the length of W. For an algebra A, it is Morita equivalent to a unique basic algebra $B(A)$ and for this basic algebra $B(A)$, the Gabriel's theorem says that there exists a unique quiver Q and an admissible ideal I (i.e. $J^N \subseteq I \subseteq J^2$ where J is the ideal generated by all arrows of Q) such that $B(A) \cong kQ/I$. See [11]. This quiver is called the *Ext-quiver* of A.

It is known that for a finite quiver Q, the path algebra kQ is of finite representation type if and only if the underlying graph $\overline{Q}$ of Q is one of Dynkin diagrams: A_n, D_n, E_6, E_7, E_8, and is of tame type if and only if the underlying graph $\overline{Q}$ is one of Euclidean diagrams: $\widetilde{A_n}$, $\widetilde{D_n}$, $\widetilde{E_6}$, $\widetilde{E_7}$, $\widetilde{E_8}$. For details, see [11][63]. These facts will be used freely in the proof of the following conclusion.

THEOREM 2.1. *Let $\Gamma_G(W)$ be a covering quiver, $n_{\Gamma_G(W)}$ defined as the above and assume A is an algebra with Ext-quiver $\Gamma_G(W)$. Then*

(i) *A is of finite representation type if and only if $n_{\Gamma_G(W)} = 0$ or $n_{\Gamma_G(W)} = 1$;*

(ii) *A is tame only if $n_{\Gamma_G(W)} = 2$;*

(iii) *If $n_{\Gamma_G(W)} \geq 3$, then A is wild.*

PROOF. (i): "If part: " When $n_{\Gamma_G(W)} = 0$, there is no any arrow in $n_{\Gamma_G(W)}$. This implies A is a semisimple algebra and so is of finite representation type. When $n_{\Gamma_G(W)} = 1$, $\Gamma_G(W)$ is a finite union of basic cycles. It is well known that a basic algebra is Nakayama if and only if its Ext-quiver is A_n or a basic cycle. Thus the basic algebra of A is a Nakayama algebra. Since every Nakayama algebra must be of finite representation type ([**11**], p. 197) and A is Morita equivalent to its basic algebra, A is of finite representation type.

"Only if part: " It is sufficient to prove that A is not of finite representation type if $n_{\Gamma_G(W)} \geq 2$. In order to prove this, it is enough to consider the case $n_{\Gamma_G(W)} = 2$ (in fact, we will show later that if $n_{\Gamma_G(W)} \geq 3$, then A is wild). We denote the basic algebra of A by $B(A)$. We need only to prove that $B(A)$ is of infinite representation type. By the Gabriel's theorem, $k\Gamma_G(W)/I \cong B(A)$ for an admissible ideal I. Denote the ideal generating all arrows in $k\Gamma_G(W)$ by J. By the definition of admissible ideal, we have an algebra epimorphism

$$B(A) \twoheadrightarrow k\Gamma_G(W)/J^2.$$

Thus it is enough to prove that $k\Gamma_G(W)/J^2$ is not of finite representation type. Since the Jacobson radical of $k\Gamma_G(W)/J^2$ is clearly 2-nilpotent, $k\Gamma_G(W)/J^2$ is stably equivalent to the following hereditary algebra (see Theorem 2.4 in Chapter X in [**11**]):

$$\Lambda = \begin{pmatrix} k\Gamma_G(W)/J & 0 \\ J/J^2 & k\Gamma_G(W)/J \end{pmatrix}$$

The Ext-quiver of Λ is indeed the separated quiver of $\Gamma_G(W)$ (see the proof of Theorem 2.6 in Chapter X of [**11**]).

Assume $W = (w_1, w_2)$. If $w_1 = w_2$, we can find that the separated quiver of $\Gamma_G(W)$ is a disjoint union of quivers of following form:

$$i \; \bullet \rightrightarrows \bullet \; j'$$

This means Λ is not of finite representation type since clearly above quiver is a Kronecher quiver which is not a Dynkin diagram (see also Theorem 2.6 in Chapter X of [**11**]).

If $w_1 \neq w_2$, $\Gamma_G(W)_s$ must contain the following sub-quiver:

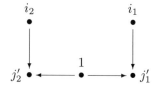

Here 1 is the identity element of G. If $i_1 = i_2$, $\Gamma_G(W)_s$ is not a Dynkin diagram and thus Λ is of infinite representation type. If it is not, $\Gamma_G(W)_s$ contains the following sub-quiver:

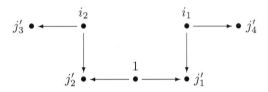

If $j'_4 = j'_l$ for $l = 1, 2, 3$, $\Gamma_G(W)_s$ is not a Dynkin diagram and thus Λ is of infinite representation type. If it is not, repeats above process and by the definition of covering quiver, there exit i_t, i_s or j'_t, j'_s satisfying $i_t = i_s$ or $j'_t = j'_s$. In a word, $\Gamma_G(W)_s$ is not a Dynkin diagram and thus Λ is of infinite representation type. A celebrated result of H.Krause [**38**] states that two stably equivalent algebras have the same representation type. Thus $k\Gamma_G(W)/J^2$ is not of finite representation type since Λ is so. Therefore $B(A)$ is not of finite representation type.

Clearly, (ii) $\Leftrightarrow$ (iii). So it is enough to prove (iii). Since $n_{\Gamma_G(W)} \geq 3$, we assume $W = (w_1, w_2, w_3, \ldots)$. Just like analysis of the "Only if part" of (i), we consider the separated quiver of $k\Gamma_G(W)/J^2$. If $w_1 = w_2$, we have the following form sub-quiver of $\Gamma_G(W)_s$:

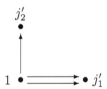

It is clearly not a Euclidean diagram and thus $k\Gamma_G(W)/J^2$ is a wild algebra. If $w_1 \neq w_2$, not loss generality, we can assume $w_i \neq w_j$ for $1 \leq i \neq j \leq 3$. This implies $\Gamma_G(W)_s$ contains the following sub-quiver

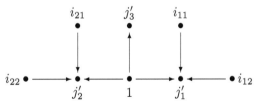

which is clearly not Euclidean diagram and thus $k\Gamma_G(W)/J^2$ is a wild algebra. Therefore $B(A)$ and thus A is a wild algebra. $\square$

The following conclusion (see Theorem 2.3 in [**32**]) states the importance of covering quivers.

LEMMA 2.2. *Let H be a finite-dimensional basic Hopf algebra over k. Then there exists a finite group G and a weight sequence $W = (w_1, w_2, \ldots, w_n)$ of G, such that $H \cong k\Gamma_G(W)/I$ for an admissible ideal I.*

This result indeed tells us that the Ext-quiver of a basic Hopf algebra H must be a covering quiver $\Gamma_G(W)$. By this, define $n_H := n_{\Gamma_G(W)}$.

COROLLARY 2.3. *Let H be a finite-dimensional basic Hopf algebra and n_H defined as above. Then*
   (i) *H is of finite representation type if and only if $n_H = 0$ or $n_H = 1$;*
   (ii) *If H is tame, then $n_H = 2$;*
   (iii) *If $n_H \geq 3$, then H is of wild type.*

We want to take this opportunity to give two applications of Theorem 2.1. The first one is to give a new proof of Theorem 3.1 in [**42**]:

COROLLARY 2.4. [Theorem 3.1 in [**42**]] *Let H be a finite-dimensional basic Hopf algebra. Then H is of finite representation type if and only if it is a Nakayama algebra.*

PROOF. It is enough to prove the necessity since every Nakayama algebra must be of finite representation type. By the Theorem 2.1, we know that H is of finite representation type if and only if $n_H = 0$ or $n_H = 1$. When $n_H = 0$, there is no arrow in $\Gamma_G(W)$. This means H is semisimple and of course Nakayama. When $n_H = 1$, $\Gamma_G(W)$ is a disjoint union of basic cycles and H is Nakayama too (see the first paragraph of the proof of Theorem 2.1). $\qquad\square$

The second one is to give an easy way to determine the representation type of a kind of Drinfeld doubles. Consider the basic cycle of length n (Example 2.1 (1)) and we denote this quiver by Z_n and by γ_i^m the path of length m starting at the vertex e_i ($i = 1, \ldots, n$).
We consider the quotient algebra $\Gamma_{n,d} := kZ_n / J^d$ with $d|n$. It is a Hopf algebra with comultiplication Δ, counit ε and antipode defined as follows. We fix a primitive d-th root of unity q.

$$\Delta(e_t) = \sum_{j+l=t} e_j \otimes e_l, \quad \Delta(\gamma_t^1) = \sum_{j+l=t} e_j \otimes \gamma_l^1 + q^l \gamma_j^1 \otimes e_l,$$

$$\varepsilon(e_t) = \delta_{t0}, \quad \varepsilon(\gamma_t^1) = 0, \quad S(e_t) = e_{-t}, \quad S(\gamma_t^1) = -q^{t+1}\gamma_{-t-1}^1.$$

As a Hopf algebra, $(\Gamma_{n,d})^{*cop}$ is isomorphic to the generalized Taft algebra $T_{nd}(q)$ [**37**] which as an associative algebra is generated by two elements g and x with relations

$$g^n = 1, \quad x^d = 0, \quad xg = qgx,$$

with comultiplication Δ, counit ε, and antipode S given by

$$\Delta(g) = g \otimes g, \quad \Delta(x) = 1 \otimes x + x \otimes g,$$

$$\varepsilon(g) = 1, \quad \varepsilon(x) = 0,$$

$$S(g) = g^{-1}, \quad S(x) = -xg^{-1}.$$

For details, see [**19**].
In [**19**], the authors studied the representation theory of the Drinfeld Double $\mathcal{D}(\Gamma_{n,d})$ and proved the following conclusion.

LEMMA 2.5. [Theorem 2.25 in [19]] *The Ext-quiver of $\mathcal{D}(\Gamma_{n,d})$ has $\frac{n^2}{d}$ isolated vertices which correspond to the simple projective modules, and $\frac{n(d-1)}{2}$ copies of the quiver*

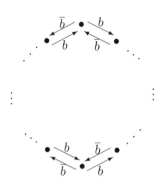

with $\frac{2n}{d}$ vertices and $\frac{4n}{d}$ arrows. The relations on this quiver are bb, $\overline{b}\overline{b}$ and $b\overline{b} - \overline{b}b$.

From this lemma, the authors of [19] find that $\mathcal{D}(\Gamma_{n,d})$ is a special biserial algebra and thus it is of finite representation type or tame type (see [19]). After listing all indecomposable modules of $\mathcal{D}(\Gamma_{n,d})$, they get that $\mathcal{D}(\Gamma_{n,d})$ is a tame algebra. Indeed, even without the complete list of indecomposable $\mathcal{D}(\Gamma_{n,d})$-modules, we also can prove that it is tame now.

COROLLARY 2.6. $\mathcal{D}(\Gamma_{n,d})$ *is a tame algebra.*

PROOF. We have known that $\mathcal{D}(\Gamma_{n,d})$ is a special biserial algebra and thus it is tame or of finite representation type. Thus in order to prove that it is tame, it is enough to show that it is not of finite representation type. Note that the above quiver is a covering quiver $\Gamma_G(W)$ by setting $G = <g|g^{\frac{2n}{d}} = 1>$ and $W = (g, g^{-1})$, and then $n_{\Gamma_G(W)} = 2$. Therefore, Theorem 2.1 gives us the desire conclusion. □

3. Classification of basic Hopf algebras of finite representation type

The classification of basic Hopf algebras of finite representation type indeed has been given by the author (with F. Li) in [42]. In [42], the conclusion is given in the language of pointed Hopf algebras. Note that the dual of pointed Hopf algebras are basic ones. For our purpose, we rewrite the result out in the language of basic Hopf algebras without proof (see Theorem 4.6 in [42]).

THEOREM 3.1. *Let H be a finite-dimensional basic Hopf algebra. Then*
(i) *H is semisimple if and only if $H \cong (kG)^*$ for some finite group G;*
(ii) *Assume the characteristic of k is zero and H is not semisimple, then H is of finite representation type if and only if $H^* \cong A(\alpha)$ for some group datum $\alpha = (G, g, \chi, \mu)$;*
(iii) *Assume the characteristic of k is p and H is not semisimple, then H is of finite representation type if and only if there exist two natural numbers $n > 0$, $r \geq$*

0, a d_0-th primitive root of unity $q \in k$ with $d_0|n$, and $d = p^r d_0 \geq 2$ such that

$$H^* \cong C_d(n) \oplus \cdots \oplus C_d(n)$$

as coalgebras and

$$H^* \cong C_d(n)\#_\sigma k(G/N)$$

as Hopf algebras, where $G = G(H)$ and $N = G(C_d(n))$.

REMARK 3.2. (i) Here a group datum (for details, see [13]) over k is defined to be a sequence $\alpha = (G, g, \chi, \mu)$ consisting of

(1) a finite group G, with an element g in its center,

(2) a one-dimensional k-representation χ of G,

(3) an element $\mu \in k$ such that $\mu = 0$ if $o(g) = o(\chi(g))$, and if $\mu \neq 0$ then $\chi^{o(\chi(g))} = 1$.

For a group datum $\alpha = (G, g, \chi, \mu)$ over k, the corresponding Hopf algebra $A(\alpha)$ was defined in [13], which is generated as an algebra by x and all $h \in G$ with relations

$$x^d = \mu(1 - g^d), \quad xh = \chi(h)hx, \quad \forall\, h \in G$$

where $d = o(\chi(g))$. Its comultiplication Δ, counit ε, and antipode S are defined by

$$\Delta(x) = g \otimes x + x \otimes 1, \quad \varepsilon(x) = 0,$$

$$\Delta(h) = h \otimes h, \quad \varepsilon(h) = 1 \quad \forall\, h \in G,$$

$$S(x) = -g^{-1}x, \quad S(h) = h^{-1}, \quad \forall\, h \in G.$$

When d is a prime, the corresponding Hopf algebra $A(\alpha)$ appeared before [13] in [16].

(ii) For any quiver Γ, we define $C_d(\Gamma) := \oplus_{i=1}^{d-1} k\Gamma(i)$ for $d \geq 2$, where $\Gamma(i)$ is the set of all paths of length i in Γ. We denote the basic cycle of length n (Example 2.1 (1)) by Z_n and denote $C_d(Z_n)$ by $C_d(n)$.

For more details about this theorem, see [42].

4. Classification of basic Hopf algebras of tame type

For the radically graded tame basic Hopf algebras, all possible structure are determined in the author's paper [41]. In this section, we determine the structure of tame basic Hopf algebras (without the assumption of radical grading) completely.

We now give a short description of our method which is a kind of generalization of the method used in [41]. Let H be a finite-dimensional basic Hopf algebra over k. Then we have a Hopf epimorphism $H \twoheadrightarrow H/J_H$ where J_H is the Jacobson radical of H. By a work of H.-J. Schneider (see [64]), we have $H \cong R_H \#_\sigma H/J_H$, where $R_H = \{a \in H | (id \otimes \pi)\Delta(a) = a \otimes 1\}$ and $\pi : H \to H/J_H$ the canonical epimorphism. We will show that R_H is a local Frobenius algebra. By [40], we know that H and R_H have the same representation type. These results help us to reduce the study of tame basic Hopf algebras to that of tame local Frobenius algebras. Fortunately, we classify all tame local Frobenius algebras and show that there are only ten classes of local algebras which are tame Frobenius (see Theorem 4.1). By this, we find one possible structure given in [41] will not happen and the detail will be given at the end of this section.

4.1. A complete list of tame local Frobenius algebras. Denote the characteristic of k by char$k$. The main result of this subsection is the following.

THEOREM 4.1. *Let Λ be a tame local Frobenius algebra. If char$k \neq 2$, then $\Lambda \cong k < x, y > /I$ where I is one of forms:*

(1): $I = (x^m - y^n, \ yx - ax^m, \ xy)$ for $a \in k$ and $m, n \geq 2$;

(2): $I = (x^2, \ y^2, \ (xy)^m - a(yx)^m)$ for $0 \neq a \in k$ and $m \geq 1$;

(3): $I = (x^2 - (yx)^m, \ y^2, \ (xy)^m + (yx)^m)$ for $m \geq 1$;

(4): $I = (x^2 - (yx)^m, \ y^2 - (xy)^m, \ (xy)^m + (yx)^m, \ (xy)^m x)$ for $m \geq 1$;

(5): $I = (x^2, \ y^2, \ (xy)^m x - (yx)^m y)$ for $m \geq 1$;

(6): $I = (x^2 - (yx)^{m-1}y - b(xy)^m, \ y^2, \ (xy)^m - a(yx)^m)$ for $a, \ b \in k$ with $a \neq 0$ and $m \geq 2$;

(7): $I = (x^2 - (yx)^{m-1}y - b(xy)^m, \ y^2 - (xy)^m, \ (xy)^m + (yx)^m, \ (xy)^m x)$ for $a, \ b \in k$ with $a \neq 0$ and $m \geq 2$;

(8): $I = (x^2 - (yx)^{m-1}y - f(xy)^m, \ y^2 - (xy)^{m-1}x - e(xy)^m, \ (xy)^m - a(yx)^m, \ (xy)^m x)$ for $a, \ e, \ f \in k$ with $a \neq 0$ and $m \geq 2$;

(9): $I = (x^2 - (yx)^m, \ y^2, \ (xy)^m x - a(yx)^m y)$ for $0 \neq a \in k$ and $m \geq 1$;

(10): $I = (x^2 - (yx)^m, \ y^2 - (xy)^m, \ (xy)^m x - a(yx)^m y, \ (xy)^{m+1})$ for $0 \neq a \in k$ and $m \geq 1$.

We want to prove Theorem 4.1 now. Some preliminaries must be given at first. It is easy to see that a local algebra is Frobenius if and only if the dimension of its socle equals to one. In this section, Λ always denotes a local Frobenius algebra and J_Λ its Jacobson radical. Recall that for any self-injective algebra Λ, we always have $soc \ _\Lambda\Lambda = soc \ \Lambda_\Lambda$ (see [**51**]). This fact will be used frequently.

Any tame local algebra A must have a quiver of the form

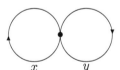

We denote this quiver by Q. By the Gabriel's Theorem, we know $A \cong k < x, y > /I$ for some ideal $J^2 \subseteq I \subseteq J^N$ where J is the ideal of $k < x, y >$ generated by x, y and $N \geq 2$. Therefore, if A is Frobenius then $dim_k A \geq 4$.

For convenience, we always denote the image of x, y in A by x, y too.

PROPOSITION 4.2. *All local algebras listed in Theorem 4.1 are tame local Frobenius algebras.*

PROOF. By checking the dimension of the socle, it is easy to see that they are Frobenius algebras. It is known that Λ and $\Lambda/soc\Lambda$ have the same representation type. Now we can find all $\Lambda/soc\Lambda$ are images of maximal tame local algebras which given by C. Ringel [**62**]. Thus they are tame or of finite representation type. But it is known that $k < x, y > /(x, y)^2$ is tame and clearly there is a natural algebra epimorphism $\Lambda \twoheadrightarrow k < x, y > /(x, y)^2$ for any Λ in Theorem 4.1. Therefore, they are all tame.   □

LEMMA 4.3. *Let $\Lambda = kQ/I$ be a local Frobenius algebra such that J_Λ^2 is generated by x^2 and y^2. Then $xy = 0$ if and only if $I = (x^m - y^n, \ yx - ax^m, xy)$ for $0 \neq a \in k$ with $m, n \geq 2$ or $xy = yx = 0$. Moreover, if $xy = yx = 0$, then $I = (x^m - y^n, \ xy, \ yx)$ for $m, n \geq 2$.*

PROOF. It is enough to prove the necessity. By assumption, we have that Λ is spanned by $1, x, x^2, \cdots, y, y^2, \cdots$. We may write $yx = x^c w + y^d z$ where w, z are units in $k[[x]]$ and $k[[y]]$ respectively and $c, d \geq 2$. Then $0 = xyx = x^{c+1}w$ and then $x^{c+1} = 0$. Since also $x^c y = 0$, it follows that $x^c \in soc\Lambda$, the socle of Λ. Moreover, $0 = yxy = y^{d+1}z$ and we deduce that $y^{d+1} = 0$. Since also $xy^d = 0$, it follows that $y^d \in soc\Lambda$. This shows that $yx \in soc\Lambda$ since $soc\Lambda$ is an ideal of Λ.

Assume $yx \neq 0$ now. Let m, n be the maximal integers such that $x^m \neq 0$, $y^n \neq 0$ and $x^{m+1} = 0$, $y^{n+1} = 0$. Clearly, $m, n \geq 2$ and x^m, $y^n \in soc\Lambda$. By $dim_k soc\Lambda = 1$, there are $a, b \in k$ with $ab \neq 0$ such that $x^m = ay^n$ and $yx = bx^m$. Let $y' = \sqrt[n]{a}y$, then $x^m = y'^n$. The last statement is clear and the lemma is proved.   □

LEMMA 4.4. *Assume that chark $\neq 2$ and Λ is a 4-dimensional local Frobenius algebra. Then Λ is isomorphic to one of the following algebras:*
(1): $kQ/(x^2 - y^2, \; yx - ax^2, xy)$ for $0 \neq a \in k$;
(2): $kQ/(x^2, \; y^2, \; xy - ayx)$ for $0 \neq a \in k$.

PROOF. Let x, y be generators of J_Λ. Since $dim_k\Lambda = 4$ and Λ is Frobenius, xy and yx belong to the socle of Λ.

(I): Assume $xy = 0$. If $yx \neq 0$, then $y^2 \neq 0$ and $x^2 \neq 0$ since $dim_k soc\Lambda = 1$. Therefore, by above lemma, we can find $m = 2$, $n = 2$ since otherwise the dimension of Λ will bigger than 4. Thus, $\Lambda \cong kQ/(x^2 - y^2, \; yx - ax^2, xy)$ for $0 \neq a \in k$.

If $yx = 0$. In this case, we know that $x^2 = ay^2$ for $0 \neq a \in k$. Let $u \in k$ with $u^2 = -a$, then, by chark $\neq 2$, $X = x + uy$, $Y = x - uy$ are generators. And, $X^2 = Y^2 = x^2 + u^2y^2 = x^2 - ay^2 = 0$, $XY = YX = x^2 - u^2y^2 = x^2 + ay^2$. Therefore, $\Lambda \cong k < X, Y > /(X^2, \; Y^2, \; XY - YX)$ which is a special case of (2).

(II): Assume $xy \neq 0 \neq yx$. Then $xy = cyx$ for $0 \neq c \in k$. By $dim_k soc\Lambda = 1$, we have $x^2 = axy$ and $y^2 = bxy$. If $a = b = 0$, then $\Lambda \cong kQ/(x^2, \; y^2, \; xy - cyx)$. Otherwise, no loss generality, assume $a \neq 0$. Let $Y = x - ay$, then $xY = 0$. Therefore we are in case (I) again.   □

LEMMA 4.5. *Let Λ be a local Frobenius algebra. Then*
(i): If Λ is tame then $dim_k J_\Lambda^2/J_\Lambda^3 \leq 2$.
(ii): If chark $\neq 2$ and $dim_k J_\Lambda^2/J_\Lambda^3 \leq 1$ then $dim_k\Lambda = 4$ or Λ is an algebra as in Theorem 4.1 (1).

PROOF. (i) If $dim_k J_\Lambda^2/J_\Lambda^3 \geq 3$, then there is a homomorphic image which is wild (see (2.1) of [62]). This implies Λ is wild which contradict the assumption that Λ is tame.

(ii) Suppose now that $dim_k J_\Lambda^2/J_\Lambda^3 \leq 1$. Then the dimension must be 1, since otherwise x, y would lie in $soc\Lambda$ and $soc\Lambda$ would not be simple. By this, we know that $dim_k\Lambda/J_\Lambda^3 = 4$.

Case (1): If Λ/J_Λ^3 is Frobenius, then by Lemma 4.4 we have

$$\Lambda/J_\Lambda^3 \cong kQ/(x^2 - y^2, \; yx - ax^2, xy) \text{ or } \Lambda/J_\Lambda^3 \cong kQ/(x^2, \; y^2, \; xy - ayx)$$

for $a \neq 0$.

If $\Lambda/J_\Lambda^3 \cong kQ/(x^2 - y^2, \; yx - ax^2, xy)$, then $xy, \; yx - ax^2, \; x^2 - y^2 \in J_\Lambda^3$. By $xy \in J_\Lambda^3$, $x^2y, yxy, xy^2, xyx \in J_\Lambda^4$. By $x^2 - y^2 \in J_\Lambda^3$, $x^3 - xy^2 \in J_\Lambda^4$ and thus $x^3 \in J_\Lambda^4$. Using $x^2 - y^2 \in J_\Lambda^3$ again, we can find $x^3 - y^2x \in J_\Lambda^4$ and thus $y^2x \in J_\Lambda^4$. Similarly, by $yx^2 - ax^3 \in J_\Lambda^4$ and $x^2y - y^3 \in J_\Lambda^4$, we have $yx^2, y^3 \in J_\Lambda^4$. Therefore, $J_\Lambda^3 \subseteq J_\Lambda^4$ and thus $J_\Lambda^3 = 0$. This implies $dim_k\Lambda = 4$.

If $\Lambda/J_\Lambda^3 \cong kQ/(x^2,\ y^2,\ xy - ayx)$, we have $x^2,\ y^2,\ xy - ayx \in J_\Lambda^3$. By this, it is easy to show that $xy^2,\ x^2y,\ yx^2,\ y^2x,\ x^3,\ y^3,\ xyx,\ yxy \in J_\Lambda^4$. Thus $J_\Lambda^3 \subseteq J_\Lambda^4$ and so $J_\Lambda^3 = 0$. This also means that $dim_k\Lambda = 4$.

Case (2): Assume now Λ/J_Λ^3 is not Frobenius. Therefore, $dim_k soc(\Lambda/J_\Lambda^3) \geq 2$. This implies $x \in soc(\Lambda/J_\Lambda^3)$ or $y \in soc(\Lambda/J_\Lambda^3)$. Not loss generality, assume $x \in soc(\Lambda/J_\Lambda^3)$. Thus we have $x^2,\ xy,\ yx \in J_\Lambda^3$. This means J_Λ^2 is generated by y^2 and thus $xy = uy^l$ where u is a unit of $k[[y]]$ and $l \geq 3$. Let $x' = x - uy^{l-1}$ and we have $x'y = 0$. By Lemma 4.3, we know that $yx' = 0$ or $I = (x'^m - y^n,\ yx' - ax'^m,\ x'y)$.

If $I = (x'^m - y^n,\ yx' - ax'^m,\ x'y)$, the proof is done. In the case of $yx' = 0$, we write $x'^2 = vy^s$ for u a unit of $k[[y]]$ and $s \geq 3$. Thus $x'^3 = vy^s x' = 0$ and so $x'^2 \in soc\Lambda$. Take m to be the maximal integer such that $y^m \neq 0$ and $y^{m+1} = 0$. Therefore, $y^m \in soc\Lambda$ and thus $x'^2 = ay^m$. Let $y' = \lambda y$. Take a suitable λ, we have $x'^2 = y'^m$ and $\Lambda \cong kQ/I$ for $I = (x'^2 - y'^m,\ x'y',\ y'x')$. This is a special case of Theorem 4.1 (1). $\qquad\square$

The following lemma is given in [18] (page 84).

LEMMA 4.6. *Let A be a tame local algebra with the quiver Q, of dimension 5, with $J^3 = 0$. Then $\Lambda \cong kQ/L$ where L is one of the following ideals:*

(1): $(xy,\ yx)$;
(2): $(yx - x^2,\ xy)$;
(3): $(yx - x^2,\ xy - ay^2)$ where $a \in k$ and $0 \neq a \neq 1$;
(4): $(x^2,\ y^2)$;
(5): $(yx - x^2,\ y^2)$.

LEMMA 4.7. *Let Λ be a local Frobenius algebra such that xy and yx lie in J_Λ^3. Assume $chark \neq 2$, then $dim_k\Lambda = 4$ or Λ is an algebra as in Theorem 4.1 (1).*

PROOF. By Lemma 4.4, we may assume $dim_k\Lambda > 4$. The algebra has a basis of the form

$$\{1,\ x, \ldots, x^s,\ y, \ldots, y^t\}.$$

If $xy = 0$, then by Lemma 4.3, Λ is of the form given in Theorem 4.1 (1).

We consider now the case when $xy \neq 0 \neq yx$. Let p be as large as possible such that xy and $yx \in J_\Lambda^p$. Clearly, $p \geq 3$. Then one of them does not lie in J_Λ^{p+1}. Write $xy \equiv x^p u + y^p v$ and $yx \equiv x^p w + y^p z$ (modulo J_Λ^{p+1}) where $u,\ v,\ w,\ z \in k$. We may replace $x,\ y$ by $x',\ y'$ where $x' = x - y^{p-1}z$ and $y' = y - x^{p-1}u$. Then we have new relations $xy \equiv cy^p$ and $yx \equiv dx^p$ for $c,\ d \in k$. Moreover, one of them, c say, is non-zero. Now, J_Λ^{p+1} is generated by x^{p+1} and y^{p+1}, and $cy^{p+1} \equiv yxy \equiv dx^p y \equiv dcx^{p-1}y^p \in J_\Lambda^{p+2}$ and already $J_\Lambda^{p+1} = (x^{p+1})$.

If $d \neq 0$, then similarly $x^{p+1} \in J_\Lambda^{p+2}$ and thus $J_\Lambda^{p+1} = 0$. So $J_\Lambda^p \subseteq soc\Lambda$. We assumed that $0 \neq xy \in J_\Lambda^p$ and it follows that $J_\Lambda^p = soc\Lambda$. We have $x^p = ay^p$ for $0 \neq a \in k$ and also $xy = cy^p$ and $yx = dx^p$ with $cd \neq 0$. Replace now $x,\ y$ by $x' = x - cy^{p-1}$ and y. Then $x'y = 0$ and thus we can use Lemma 4.3 again.

Suppose now that $d = 0$. Then $yx \in J_\Lambda^{p+1} = (x^{p+1})$. Note that $yx \neq 0$ and consequently $yx = x^m u$ for $m \geq p + 1$, where u is a unit of $k[[x]]$. Replace y by $y' = y - x^{m-1}u$, then $y'x = 0$. We also can use Lemma 4.3 again. $\qquad\square$

LEMMA 4.8. *Let Λ be a local Frobenius algebra such that $yx - x^2$ and xy lie in J_Λ^3. Then $\Lambda \cong kQ/(x^m - y^n,\ yx - ax^m,\ xy)$ for $0 \neq a \in k$.*

PROOF. Claim: J_Λ *does not have generators* x', y' *with* $(x'y')$ *and* $(y'x')$ *lying in* J_Λ^3. This claim was proved in [18] (see Lemma III.7 of [18]).

We have that $x^3 \equiv xyx$ (modulo J_Λ^4). But $xyx \in J_\Lambda^4$ and thus $x^3 \in J_\Lambda^4$. So J_Λ^3 is generated by y^3. So we may have $xy = y^s u$ where u is a unit in $k[[y]]$ and $s \geq 3$. Let $x' = x - y^{s-1}u$, then we have $x'y = 0$. By the claim, $yx' \neq 0$ and thus Lemma 4.3 is applied. $\qquad\square$

LEMMA 4.9. *Let* Λ *be a local Frobenius algebra such that* $yx - x^2$ *and* $xy - ay^2$ *lie in* J_Λ^3 *where* $0 \neq a \neq 1$. *Then* $dim_k\Lambda = 4$.

PROOF. By assumption, we have $x^3 \equiv xyx \equiv ay^2x \equiv ayx^2 \equiv ax^3$ (modulo J_Λ^4) and $x^2y \equiv axy^2 \equiv a^2y^3 \equiv ayxy \equiv ax^2y$ (modulo J_Λ^4). Since $a \neq 1$, $x^3 \in J_\Lambda^4$ and $x^2y \in J_\Lambda^4$. Since $a \neq 0$, it follows that all the other monomials occurring lie in J_Λ^4. This means that $J_\Lambda^3 \subseteq J_\Lambda^4$ and thus $J_\Lambda^3 = 0$. So, $J_\Lambda^2 \subseteq soc\Lambda$. By Λ is Frobenius, $dim_k\Lambda = 4$. $\qquad\square$

LEMMA 4.10. *Let* Λ *be a local Frobenius algebra such that* x^2 *and* y^2 *lie in* J_Λ^3. *Assume* $chark \neq 2$, *then* Λ *is isomorphic to one of algebras in Theorem 4.1.*

PROOF. By Lemma 4.4 and Lemma 4.5, we can assume that $dim_k\Lambda > 4$ and $dim_k J_\Lambda^2/J_\Lambda^3 = 2$. Thus xy and yx are generators of J_Λ^2 which are independent.

Case (1): Assume x^2 and y^2 lie in $soc\Lambda$. Let $m \geq 1$ be the integer such that $(xy)^m \neq 0$ and $(xy)^{m+1} = 0$. We claim that $soc\Lambda = ((xy)^m)$ or $soc\Lambda = ((xy)^m x)$. Indeed, since $y^2 \in soc\Lambda$, we always have $(xy)^m y = 0$. Thus if $(xy)^m x = 0$ then $0 \neq (xy)^m \in soc\Lambda$. By $dim_k soc\Lambda = 1$, $soc\Lambda = ((xy)^m)$. Otherwise, $(xy)^m x \neq 0$. By $(xy)^{m+1} = 0$ and $(xy)^m x^2 = 0$, $soc\Lambda = ((xy)^m x)$. Thus the claim is proved.

If $soc\Lambda = ((xy)^m)$, then there exists $0 \neq a \in k$ such that $(yx)^m = a(xy)^m$. By x^2 and y^2 lie in $soc\Lambda$, we have $x^2 = b(xy)^m$ and $y^2 = c(xy)^m$ for $b, c \in k$. If $a \neq -1$, let $d = \frac{b}{1+a}$, $e = \frac{c}{1+a}$ and $x' = x - dy(xy)^{m-1}$, $y' = y - e(xy)^{m-1}y$, then we have $x'^2 = 0$ and $y'^2 = 0$. Thus it is isomorphic to the one of algebras in Theorem 4.1 (2). If $a = -1$, then consider b, c. If $b = c = 0$, it is isomorphic to the one of algebras in Theorem 4.1 (2). If one of b, c is zero while the other is not zero, say $b \neq 0$ and $c = 0$. Let $x' = \lambda x$ and $y' = \mu y$ for $\lambda, \mu \in k$. By suitable choice of λ, μ, we can assume $x'^2 = (x'y')^m$ and $y'^2 = 0$. Thus the algebra has the form as in Theorem 4.1 (3). Similarly, if $bc \neq 0$ then we can show the algebra has the form as in Theorem 4.1 (4).

If $soc\Lambda = ((xy)^m x)$, then there exists $0 \neq a \in k$ such that $(yx)^m y = a(xy)^m x$. By x^2 and y^2 lie in $soc\Lambda$, we have $x^2 = b(xy)^m x$ and $y^2 = cy(xy)^m$ for $b, c \in k$. Let $x' = b(xy)^m - x$ and $y' = c(yx)^m - y$. Then we can find $x'^2 = 0$ and $y'^2 = 0$. Moreover, clearly we can take a to be 1. Thus the algebra has the form as in Theorem 4.1 (5).

Case (2): Otherwise, choose n, l such that $x^2 \in J_\Lambda^n - J_\Lambda^{n+1}$ and $y^2 \in J_\Lambda^l$ with $l \geq n$. Take also n as large as possible with respect to these conditions.

Claim: $J_\Lambda^{n+1} \subseteq soc\Lambda$. This claim was proved in [18] (see Lemma III.10 of [18]). Now we consider two possibilities: $soc\Lambda$ is an even power of J_Λ or $soc\Lambda$ is an odd power of J_Λ.

(i) If $soc\Lambda$ is an even power of J_Λ, then $soc\Lambda = J_\Lambda^{2m}$. By the hypothesis at the beginning, $J_\Lambda^n \not\subseteq soc\Lambda$. So $J_\Lambda^{n+1} \neq 0$ and thus $J_\Lambda^{n+1} = soc\Lambda$. Therefore, $n + 1 = 2m$. Clearly, $soc\Lambda = ((xy)^m)$ and $(yx)^m = a(xy)^m$ for $0 \neq a \in k$. Then $x^2 = c(xy)^{m-1}x + d(yx)^{m-1}y + f(xy)^m$ and $(c, d) \neq (0, 0)$. Without loss of generality, we can assume $c = 0$ since otherwise we can replace x by $x' = x - c(xy)^{m-1}$.

Now consider y^2. By the hypothesis, the element lies in $soc_2\Lambda$. If $y^2 \in soc\Lambda$, then we show similarly as in Case (1) that Λ is an algebra in Theorem 4.1 (6), (7). Otherwise, $y^2 = a(xy)^{m-1}x + b(yx)^{m-1}y + e(xy)^m$ and $(a, b) \neq (0, 0)$. Similarly, we can assume $b = 0$. Thus $a \neq 0$.

We have now $x^2 = d(yx)^{m-1}y + f(xy)^m$ and $y^2 = a(xy)^{m-1}x + e(xy)^m$. Let $x' = \lambda x$ and $y' = \mu y$. By a suitable choice of λ, μ, we can assume $a = 1 = d$. This is an algebra in Theorem 4.1 (8).

(ii) Otherwise, $soc\Lambda$ is an odd power of J_Λ. Similarly, we have $soc\Lambda = J_\Lambda^{n+1} = ((xy)^m x)$ and $(xy)^m x = a(yx)^m y$ for $a \neq 0$. By assumption, $x^2 = b(xy)^m + c(yx)^m + f(xy)^m x$. As before, we can assume $b = 0$. Also, we consider y^2. If $y^2 \in soc\Lambda$, then by the discussion of Case (1), the algebra is isomorphic to one of algebras in Theorem 4.1 (9). If not, we have $y^2 = d(xy)^m + e(yx)^m + g(xy)^m x$ for d, e, $g \in k$. Similarly, we can assume that $e = 0$ and $d \neq 0$. As in Case (1), we also can assume $f = g = 0$. So now we have $x^2 = c(yx)^m$ and $y^2 = d(xy)^m$. Similarly, let $x' = \lambda x$ and $y' = \mu y$ and choose suitable λ, μ, we may assume $c = d = 1$. Thus it is an algebra in Theorem 4.1 (10). □

LEMMA 4.11. *There is no local Frobenius algebra Λ such that Λ/J_Λ^3 satisfies Lemma 4.6 (5).*

PROOF. Suppose such algebra exists.

Claim: $J_\Lambda^3 = (xyx) \subseteq soc\Lambda$. We have that J_Λ^3 is generated by xyx and yxy by the given relations. Moreover, modulo J_Λ^4 we have that $xyx \equiv x^3 \equiv yxx \equiv y^2x \equiv 0$ and therefore $J_\Lambda^3 = (yxy)$. This implies $J_\Lambda^4 = ((yx)^2) \subseteq yJ_\Lambda^4 \subseteq J_\Lambda^5$. Thus $J_\Lambda^4 = 0$ as required.

We claim yxy must be zero now. Otherwise, assume $yxy \neq 0$ and thus $J_\Lambda^3 = (yxy) = soc\Lambda$. Since $J_\Lambda^4 = 0$, we know $xyx = 0$ and $xy^2 = 0$. This means $xy \in soc\Lambda$. Clearly, $xy \neq 0$ since otherwise $yxy = 0$. Since $dim_k soc\Lambda = 1$, there exists non-zero $c \in k$ such that $xy = cyxy$. So we have $xy = cyxy = c^2y^2xy = 0$. It's a contradiction. This means $yxy = 0$ and thus $J_\Lambda^3 = 0$ and $J_\Lambda^2 \subseteq soc\Lambda$. Therefor $soc\Lambda$ is not simple, which is absurd. □

Proof of Theorem 4.1: Since Λ is tame, $dim_k J_\Lambda^2/J_\Lambda^3 \leq 2$ by Lemma 4.5.

If $dim_k J_\Lambda^2/J_\Lambda^3 = 1$, Lemma 4.5 shows that Λ is one of algebras of this list.

If $dim_k J_\Lambda^2/J_\Lambda^3 = 2$, then $dim_k \Lambda/J_\Lambda^3 = 5$. This means that Λ/J_Λ^3 satisfies the conditions of Lemma 4.6. Therefore, Lemma 4.7-4.11 give our desired conclusion.

4.2. Tame basic Hopf algebras. The main aim of this subsection is to describe the structure of tame basic Hopf algebras (see Theorem 4.15).

Let H be a basic Hopf algebra and J_H be its Jacobson radical. Recall $H/J_H \cong (kG)^*$ for some finite group G. In this section, we always assume $chark \neq 2$ and $chark \nmid |G|$. Thus kG is always semisimple.

Denote H/J_H by $\overline{H}$. Now we have a Hopf algebra epimorphism

$$H \twoheadrightarrow \overline{H}.$$

By a result which given by H.-J. Schneider [64], there is an algebra R_H such that

$$H \cong R_H \#_\sigma \overline{H}.$$

LEMMA 4.12. R_H is a local algebra.

PROOF. For any finite-dimensional algebra A, we write $gr A = A/J_A \oplus J_A/J_A^2 \oplus \cdots$. By [61] and [44], there is an algebra R_{grH}, which is a graded braided Hopf algebra in $^H_H \mathcal{YD}$, such that $grH \cong R_{grH} \# \overline{H}$. Hence R_{grH} is Frobenius by [26] and is local since the degree 0 part is k. Thus R_{grH} is a local Frobenius algebra.

By Blattner-Montgomery Duality Theorem (see Section 9.4 in [49]), we have

$$(R_H \#_\sigma \overline{H}) \# (\overline{H})^* \cong M_n(R_H),$$

$$(R_{grH} \# \overline{H}) \# (\overline{H})^* \cong M_n(R_{grH})$$

where $n = dim_k \overline{H}$. Note that $(\overline{H})^*$ is a group algebra now, thus we have $J_{(R_H \#_\sigma \overline{H}) \# (\overline{H})^*} = (J_{R_H \#_\sigma \overline{H}}) \# (\overline{H})^*$. This means we have the following isomorphism

$$gr M_n(R_H) \cong (gr(R_H \#_\sigma \overline{H})) \# (\overline{H})^*.$$

Thus,

$$M_n(gr R_H) \cong gr M_n(R_H) \cong (grH) \# (\overline{H})^* \cong (R_{grH} \# \overline{H}) \# (\overline{H})^* \cong M_n(R_{grH}).$$

So we have $M_n(gr R_H) \cong M_n(R_{grH})$ and thus $gr R_H \cong R_{grH}$. By R_{grH} is local, $gr R_H$ and thus R_H is local. $\qquad\square$

LEMMA 4.13. R_H is a Frobenius algebra.

PROOF. By the Lemma 4.12, it is enough to show that R_H is self-injective since any basic self-injective algebra must be Frobenius. By Blattner-Montgomery Duality Theorem, we need only show $H \# (\overline{H})^*$ is self-injective. Let P be a projective $H \# (\overline{H})^*$-module, we need to show that P is also injective.

For $H \# (\overline{H})^*$-modules M, N, let $i: M \hookrightarrow N$ and $h: M \to P$ be two $H \# (\overline{H})^*$-module morphisms such that i is injective. In order to prove that P is injective as an $H \# (\overline{H})^*$-module, it is enough to find an $\widetilde{f} \in Hom_{H \# (\overline{H})^*}(N, P)$ satisfying $h = \widetilde{f} i$. It is known that $H \# (\overline{H})^*$ is a free H-module. Thus P is also a projective H-module. By H is Frobenius, P is injective as an H-module. Thus there exists an H-morphism f such that $h = fi$. Define $\widetilde{f}(n) = \sum S(t_1) \cdot f(t_2 \cdot n)$ for $n \in N$, where t is a non-zero right integral with $\varepsilon(t) = 1$. Then $\widetilde{f}$ is $H \# (\overline{H})^*$-linear by [15] and satisfies $h = \widetilde{f} i$. $\qquad\square$

The following lemma is proved in [40] (see Theorem 2.6 in [40]).

LEMMA 4.14. Let A be a finite-dimensional algebra and H a finite-dimensional Hopf algebra. If H and H^* are semisimple, then $A \#_\sigma H$ and A have the same representation type.

The next conclusion will give us all possible structures of tame basic Hopf algebras.

THEOREM 4.15. Let H be a basic Hopf algebra. Assume $chark \neq 2$ and $dim_k H/J_H$ is invertible in k, then H is tame if and only if $H \cong k < x, y > /I \#_\sigma (kG)^*$ for some finite group G and some ideal I which is one of the following forms:

(1): $I = (x^m - y^n, yx - ax^m, xy)$ for $a \in k$ and $m, n \geq 2$;
(2): $I = (x^2, y^2, (xy)^m - a(yx)^m)$ for $0 \neq a \in k$ and $m \geq 1$;
(3): $I = (x^2 - (yx)^m, y^2, (xy)^m + (yx)^m,)$ for $m \geq 1$;

(4): $I = (x^2 - (yx)^m, \ y^2 - (xy)^m, \ (xy)^m + (yx)^m, \ (xy)^m x)$ *for $m \geq 1$;*

(5): $I = (x^2, \ y^2, \ (xy)^m x - a(yx)^m y)$ *for $0 \neq a \in k$ and $m \geq 1$;*

(6): $I = (x^2 - (yx)^{m-1}y - b(xy)^m, \ y^2, \ (xy)^m - a(yx)^m)$ *for $a, \ b \in k$ with*
$a \neq 0$ and $m \geq 2$;

(7): $I = (x^2 - (yx)^{m-1}y - b(xy)^m, \ y^2 - (xy)^m, \ (xy)^m + (yx)^m, \ (xy)^m x)$
for $a, \ b \in k$ with $a \neq 0$ and $m \geq 2$;

(8): $I = (x^2 - (yx)^{m-1}y - f(xy)^m, \ y^2 - (xy)^{m-1}x - e(xy)^m, \ (xy)^m - a(yx)^m, \ (xy)^m x)$
for $a, \ e \ f \in k$ with $a \neq 0$ and $m \geq 2$;

(9): $I = (x^2 - (yx)^m, \ y^2, \ (xy)^m x - a(yx)^m y)$ *for $0 \neq a \in k$ and $m \geq 1$;*

(10): $I = (x^2 - (yx)^m, \ y^2 - (xy)^m, \ (xy)^m x - a(yx)^m y, \ (xy)^{m+1})$ *for*
$0 \neq a \in k$ and $m \geq 1$.

PROOF. "Only if part: " On one hand, by Lemma 4.12, Lemma 4.13 and Lemma 4.14, R_H is a tame local Frobenius algebra. On the other hand, H/J_H is a commutative semisimple Hopf algebra and thus $H/J_H \cong (kG)^*$ for some finite group. Therefore, by Theorem 4.1 and $H \cong R_H \#_\sigma H/J_H$, we get the desired conclusion.

"If part: " By Proposition 4.2, we know that $k<x,y>/I$ is a tame algebra. So the sufficiency is gotten from Lemma 4.14. □

REMARK 4.16. (1) In order to apply Lemma 4.14, we need kG to be semisimple and thus the hypothesis char$k \nmid |G|$, posed at the beginning of this subsection, is needed. Note that at the end of proof of Lemma 4.13, this hypothesis was also used to guarantee the existence of the right integral t satisfying $\varepsilon(t) = 1$.

(2) By a conclusion of Radford or Majid (see [**61**][**44**]), if Λ is a braided Hopf algebra in $^{(kG)^*}_{(kG)^*}\mathcal{YD}$ for some finite group G, then we can form the bosonization $\Lambda \times (kG)^*$ which is a Hopf algebra. For a tame local Frobenius algebra A, above theorem dose not imply the existence of finite group G satisfying A is a braided Hopf algebra in $^{(kG)^*}_{(kG)^*}\mathcal{YD}$.

PROBLEM 4.1. *For a tame local Frobenius algebra A, give an effective method to determine that whether there is a finite group G satisfying A is a braided Hopf algebra in $^{(kG)^*}_{(kG)^*}\mathcal{YD}$. If such a G exists, then find all of them.*

A similar problem was given in [**41**] ([**41**], Problem 5.1). For a tame local radically graded Frobenius algebra, this problem has been solved by the author with his co-workers. The details will appear elsewhere.

EXAMPLE 4.1. (**Tensor products of Taft algebras**) Let $T_{n^2}(q)$, $T_{m^2}(q')$ be two Taft algebras. Direct computation shows that

$$T_{n^2}(q) \otimes_k T_{m^2}(q') \cong k<x,y>/I \# k(\mathbb{Z}_n \times \mathbb{Z}_m)$$

where $I = (x^n, \ y^m, \ xy - yx)$. Thus by Theorem 4.15, $T_{n^2}(q) \otimes_k T_{m^2}(q')$ is tame if and only if $m = n = 2$.

EXAMPLE 4.2. (**Book Algebras**) Let q be a n-th primitive root of unity and m a positive integer satisfying $(m,n) = 1$. Let $H = \mathbf{h}(q,m) = k<y,x,g>/(x^n, y^n, g^n - 1, gx - qxg, gy - q^m yg, xy - yx)$ and with comultiplication, antipode and counit given by

$$\Delta(x) = x \otimes g + 1 \otimes x, \quad \Delta(y) = y \otimes 1 + g^m \otimes y, \quad \Delta(g) = g \otimes g$$
$$S(x) = -xg^{-1}, \quad S(y) = -g^{-m}y, \quad S(g) = g^{-1}, \quad \varepsilon(x) = \varepsilon(y) = 0, \quad \varepsilon(g) = 1.$$

It is a Hopf algebra and called book algebra. As in [**41**], we have

$$h(q, m) \cong k < x, y > /I \# k\mathbb{Z}_n$$

where $I = (x^n,\ y^n,\ xy - q^m yx)$. Thus by Theorem 4.15, $\mathbf{h}(q, m)$ is tame if and only if $n = 2$. In this case, q must equal to -1 and $m = 1$. Thus only $\mathbf{h}(-1, 1)$ is tame and the others are all wild.

EXAMPLE 4.3. (**The dual of Frobenius-Lusztig kernel**) Let p be an odd number and q a p-th primitive root of unity. By definition, the Frobenius-Lusztig kernel $\mathbf{u}_q(\mathfrak{sl}_2)$ is an associative algebra generated by E, F, K with relations

$$K^p = 1,\ E^p = 0,\ F^p = 0,\ KE = q^2 EK,\ KF = q^{-2} FK,\ EF - FE = \frac{K - K^{p-1}}{q - q^{-1}}.$$

Its comultiplication, counit and antipode are defined by

$$\Delta(E) = 1 \otimes E + E \otimes K,\ \ \Delta(F) = K^{-1} \otimes F + F \otimes 1,\ \ \Delta(K) = K \otimes K;$$

$$\varepsilon(E) = \varepsilon(F) = 0,\ \ \varepsilon(K) = 1;$$

$$S(E) = -q^2 K^{-1} E,\ \ S(F) = -KF,\ \ S(K) = K^{-1}.$$

It is a pointed Hopf algebra and thus $\mathbf{u}_q(\mathfrak{sl}_2)^*$ is a basic Hopf algebra. We now give the Hopf structure of $\mathbf{u}_q(\mathfrak{sl}_2)^*$ explicitly.

It is known that $\mathbf{u}_q(\mathfrak{sl}_2)$ has a basis $\{K^l E^i F^j | 0 \le l, i, j \le p - 1\}$ and thus $dim_k \mathbf{u}_q(\mathfrak{sl}_2) = p^3$. We denote by $(K^l E^i F^j)^*$ the element of $\mathbf{u}_q(\mathfrak{sl}_2)^*$ which sent $K^l E^i F^j$ to 1 and the other element in the above basis to 0.

Let

$$a = \sum_{i=0}^{p-1} q^i (K^i)^* + \sum_{i=0}^{p-1} q^i (K^i EF)^*,\ \ b = \sum_{i=0}^{p-1} q^i (K^i E)^*,$$

$$c = \sum_{i=0}^{p-1} q^{-i} (K^i F)^*,\ \ \ \ d = \sum_{i=0}^{p-1} q^{-i} (K^i)^*.$$

By direct computations, the following relations hold.

$$ba = qab,\ \ db = qbd,\ \ ca = qac.\ dc = qcd,\ \ bc = cb,$$

$$ad - da = (q^{-1} - q)bc,\ \ da - qbc = 1,\ \ d^p = 1,\ \ c^p = b^p = 0.$$

For example, let us check the relation $bc = cb$ and the other relations can be checked similarly. By definition, $bc = \sum_{i,j} q^{i-j} (K^i E)^* (K^j F)^*$. In order to make $(K^i E)^* (K^j F)^* (K^l E^m F^n) \ne 0$, we must have $m = n = 1$. But

$$\Delta(K^l EF) = K^{l-1} \otimes K^l EF + q^2 K^{l-1} E \otimes K^{l+1} F + K^l F \otimes K^l E + K^l EF \otimes K^l.$$

This implies if $(K^i E)^* (K^j F)^* \ne 0$ then $j = i + 2$. Thus

$$bc = \sum_{i,j} q^{i-j} (K^i E)^* (K^j F)^* = \sum_{l=0}^{p-1} q^{-2} q^2 (K^l EF)^* = \sum_{l=0}^{p-1} (K^l EF)^*.$$

Similarly, we can show $cb = \sum_{l=0}^{p-1} (K^l EF)^*$ also.

By $da - qbc = 1$ and $d^p = 1$, we have $a = d^{-1}(1 + qbc)$. It is straightforward to show that the algebra, which is generated by a, b, c, d with above relations, has dimension p^3. Thus algebra is just $\mathbf{u}_q(\mathfrak{sl}_2)^*$. The comultiplication, counit and the antipode are given as follows.

$$\Delta(a) = a \otimes a + b \otimes c,\ \ \ \Delta(b) = a \otimes b + b \otimes d;$$

$$\Delta(c) = c \otimes a + d \otimes c, \quad \Delta(d) = c \otimes b + d \otimes d;$$

$$\varepsilon(a) = \varepsilon(d) = 1, \quad \varepsilon(b) = \varepsilon(c) = 0;$$

$$S(a) = d, \ S(b) = -qb, \ S(c) = -q^{-1}c, \ S(d) = a.$$

Clearly, $J_{\mathbf{u}_q(\mathfrak{sl}_2)^*} = (b, c)$ and $\mathbf{u}_q(\mathfrak{sl}_2)^*/(b, c) \cong k\mathbb{Z}_p$. Thus

$$\mathbf{u}_q(\mathfrak{sl}_2)^* \cong R_{\mathbf{u}_q(\mathfrak{sl}_2)^*} \#_\sigma k\mathbb{Z}_p$$

where by definition $R_{\mathbf{u}_q(\mathfrak{sl}_2)^*} = \{x \in \mathbf{u}_q(\mathfrak{sl}_2)^* | (id \otimes \pi)\Delta(x) = x \otimes 1\}$. Here $\pi : \mathbf{u}_q(\mathfrak{sl}_2)^* \to \mathbf{u}_q(\mathfrak{sl}_2)^*/(b, c)$ is the canonical map. Thus it is easy to see that $dc, \ d^{-1}b \in R_{\mathbf{u}_q(\mathfrak{sl}_2)^*}$ which generate $R_{\mathbf{u}_q(\mathfrak{sl}_2)^*}$ and satisfy the following relations

$$(dc)^p = 0, \quad (d^{-1}b)^p = 0, \quad dc \cdot d^{-1}b = q^2(d^{-1}b) \cdot dc.$$

Denote dc by x and $d^{-1}b$ by y, we have

$$R_{\mathbf{u}_q(\mathfrak{sl}_2)^*} \cong k < x, y > /I$$

where $I = (x^p, \ y^p, \ xy - q^2yx)$ which is not an algebra in Theorem 4.1. Thus, by Theorem 4.15, $\mathbf{u}_q(\mathfrak{sl}_2)^*$ is wild.

At last, I want to take this chance to give an addendum to [41]. In Section 3 of [41], we give the following conclusion (see Theorem 4.1 in [41]).

"Let Λ be a tame local graded Frobenius algebra. If char$k \neq 2$, then $\Lambda \cong k < x, y > /I$ where I is one of forms:
(1): $I = (x^2 - y^2, \ yx - ax^2, \ xy)$ for $0 \neq a \in k$;
(2): $I = (x^2, \ y^2, \ (xy)^m - a(yx)^m)$ for $0 \neq a \in k$ and $m \geq 1$;
(3): $I = (x^n - y^n, \ xy, \ yx)$ for $n \geq 2$;
(4): $I = (x^2, \ y^2, \ (xy)^m x - (yx)^m y)$ for $m \geq 1$;
(5): $I = (yx - x^2, \ y^2)$."

By the Lemma 4.11, we know that the case (5) of above conclusion will not appear. Thus the better form of Theorem 4.1 of [41] is the following.

Theorem *Let Λ be a tame local graded Frobenius algebra. If char$k \neq 2$, then $\Lambda \cong k < x, y > /I$ where I is one of forms:*
(1): $I = (x^2 - y^2, \ yx - ax^2, \ xy)$ for $0 \neq a \in k$;
(2): $I = (x^2, \ y^2, \ (xy)^m - a(yx)^m)$ for $0 \neq a \in k$ and $m \geq 1$;
(3): $I = (x^n - y^n, \ xy, \ yx)$ for $n \geq 2$;
(4): $I = (x^2, \ y^2, \ (xy)^m x - (yx)^m y)$ for $m \geq 1$.
Of course, Theorem 5.4 of [41] should be changed accordingly. That is, delete the case (5) of Theorem 5.4 in [41].

Acknowledgement

I would like to express my hearty thanks to Professor Nanhua Xi for his encouragements. The work in Section 4 was finished when the author visit the Oxford University under the financial support from the Leverhulme Trust through the Academic Interchange Network Algebras, Representations and Applications. I am grateful to Professors K. Erdmann and A. Henke for stimulating conversations. I thank their hospitality from the university and the financial support from the Leverhulme Trust. I also want to thank my supervisor, Professor Fang Li, for his helpful comments. I am very grateful to the organizers for their excellent work. I also would like to thank the referee for very valuable comments.

References

[1] N. Andruskiewitsch, About finite-dimensional Hopf Algebras, Contemp.Math 294 (2002), 1-57.

[2] N. Andruskiewitsch, I. Angiono, On Nichols Algebras with Generic Braiding, to appear in "Modules and Comodules", Proceedings of a conference dedicated to Robert Wisbauer, arXiv:math/070392v2.

[3] N. Andruskiewitsch, P. Etingof, S. Gelaki, Triangular Hopf Algebras with the Chevalley Property, Michigan Math. J. 49 (2001), no. 2, 277–298.

[4] N. Andruskiewitsch, S. Natale, Counting Arguments for Hopf Algebras of Low Dimension, Tsukuba Math J. 25 (2001), no. 1, 187-201.

[5] N. Andruskiewitsch, H. -J. Schneider, A Characterization of Quantum Groups, J. Reine Angew. Math 577 (2004), 81-104.

[6] N. Andruskiewitsch, H. -J. Schneider, Pointed Hopf Algebras, in "New direction in Hopf algebras", 1-68, Math. Sci. Res. Inst. Publ. 43, Cambridge Univ. Press, Cambridge, 2002.

[7] N. Andruskiewitsch, H. -J. Schneider, Finite Quantum Groups and Cartan Matrices, Adv. Math 154 (2000), 1-45.

[8] N. Andruskiewitsch, H. -J. Schneider, Hopf Algebras of Order p^2 and Braided Hopf Algebras of Order p, J. Algebra 199 (1998), 430-454.

[9] N. Andruskiewitsch, H. -J. Schneider, Lifting of Quantum Linear Spaces and Pointed Hopf Algebras of Order p^3, J. Algebra 209 (1998), 659-691.

[10] N. Andruskiewitsch, H. -J. Schneider, On the Classification of Finite-dimensional Pointed Hopf Algebras, to appear in Ann. Math.

[11] M. Auslander, I. Reiten, S. Smalφ, Representation Theory of Artin Algebras, Cambridge University Press, 1995.

[12] M. Beattie, S. Dascalescu, Hopf Algebras of Dimension 14, J. LMS. II 69 (2001), no. 1, 65-78.

[13] Xiao-Wu Chen, Hua-Lin Huang, Yu Ye, Pu Zhang, Monomial Hopf Algebras, J. Algebra 275 (2004), 212-232.

[14] C. Cibils, M. Rosso, Hopf Quiver, J. Algebra 254 (2002), 241-251.

[15] M. Cohen, D. Fishman, Hopf Algebra Actions, J. Algebra 100 (1986), 363-379.

[16] S. Dascalescu, Pointed Hopf Algebras with Large Coradical, Comm. Alg 27 (1999), no. 10, 4821-4826.

[17] Yu. A. Drozd, Tame and Wild Matrix Problems, Representations and Quadratic Forms, Inst.Math., Acad.Sciences.Ukrainian SSR, Kiev 1979, 39-74. Amer.Math.Soc. Transl. 128 (1986), 31-55.

[18] K. Erdmann, Blocks of Tame Representation Type and Related Algebras, LNM 1428, Springer-Verlag, 1990.

[19] K. Erdmann, E. L. Green, N. Snashall, R. Taillefer, Representation Theory of the Drinfeld Double of a Family of Hopf Algebras, J. Pure Appl. Algebra 204 (2006), no. 2, 413-454.

[20] P. Etingof, S. Gelaki, On Finite-dimensional Semisimple and Cosemisimple Hopf Algebras in Positive Characteristic, International Mathematics Research Notices 16 (1998), 851-864.

[21] P. Etingof, S. Gelaki, The Classification of Triangular Semisimple and Cosemisimple Hopf Algebras over an Algebraically Closed Field, International Mathematics Research Notices 5 (1997), 223-234.

[22] P. Etingof, S. Gelaki, Semisimple Hopf Algebras of Dimension pq Are Trivial, J. Algebra 210 (1998), no. 2, 664-669.

[23] P. Etingof, S. Gelaki, On Hopf Algebras of Dimension pq, J. Algebra 276 (2004), no. 1, 399-406.

[24] P. Etingof, S. Gelaki, On Cotriangular Hopf Algebras, Amer. J. Math 123 (2001), no 4, 699-713.

[25] P.Etingof, D. Nikshych, V. Ostrik, On Fusion Categories, Ann. Math 162 (2005), 581-642.

[26] D. Fischman, S. Montgomery, H. -J. Schneider, Frobenius Extensions of Subalgebras of Hopf Algebras, Trans. AMS 349 (1997), 4857-4895.

[27] C. Galindo, S. Natale, Normal Hopf Subalgebras in Cocycle Deformations of Finite Groups, to appear in Manuscripta Mathematica. arXiv: 0708.3407v3.

[28] C. Galindo, S. Natale, Simple Hopf Algebras and Deformations of Finite Groups, Math. Res. Lett 14 (2007), 943-954.

[29] G. Garcia, C. Vay, Hopf Algebras of Dimension 16, arXiv: 0712.0405v1.

[30] S. Gelaki, S. Westreich, On Semisimple Hopf Algebras of dimension pq, Proc. AMS 128 (2000), no. 1, 39-47.

[31] M. Graña, On Nichols Algebras of Low Dimension, in "New Trends in Hopf Algebra Theory", Contemp. Math 267 (2000), 111-134.

[32] E. Green, Ø. Solberg, Basic Hopf Algebras and Quantum Groups, Math. Z 229 (1998), 45-76.

[33] I. Heckenberger, Examples of Finite Dimensional Rank 2 Nichols Algebras of Diagonal Type, Compos. Math 143 (2007), no. 1, 165-190.

[34] I. Heckenberger, Rank 2 Nichols Algebras with Finite Arithmetic Root System, published in Algebras and Representation Theory, DOI 10.1007/s10468-007-9060-7.

[35] I. Heckenberger, The Wyle Groupoid of a Nichols Algebra of Diagonal Type, Invent. Math 164 (2006), 175-188.

[36] I. Heckenberger, Classification of Arithmetic Root Systems, arXiv: math/0605795v1.

[37] H. L. Huang, H. X. Chen, P. Zhang, Generalized Taft Algebras, Alg. Colloq. 11 (2004), no. 3, 313-320.

[38] H. Krause, Stable Equivalence Preserves Representation Type, Comment. Math. Helv 72 (1997), 266-284.

[39] R. G. Larson, D. E. Radford, Finite Dimensional Cosemisimple Hopf Algebras in Characteristic 0 Are Semisimple. J. Algebra 117 (1988), 267-289.

[40] F. Li, M. Zhang, Invariant Properties of Representations under Cleft Extensions, Sciences in China 50 (2007), no. 1, 121-131.

[41] G. Liu, On the Structure of Tame Graded Basic Hopf Algebras, J. Algebra 299 (2006), 841-853.

[42] G. Liu, F. Li, Pointed Hopf Algebras of Finite Corepresentation Type and Their Classification, Proc. AMS 135 (2007), no. 3, 649-657.

[43] D. -M. Lu, Q. -S. Wu and J. J. Zhang, Homological Integral of Hopf Algebras, Trans. AMS, 359 (2007), no. 10, 4945-4975.

[44] S. Majid, Crossed Products by Braided Groups and Bosonization, J. Algebra 163 (1994), 165-190.

[45] A. Masuoka, Self Dual Hopf Algebras of Dimension p^3 Obtained by Extension, J. Algerba 178 (1995), 791-806.

[46] A. Masuoka, Semisimple Hopf Algebras of Dimension $2p$, Comm. Algebra 23 (1995), 1931-1940.

[47] A. Masuoka, Calculations of Some Groups of Hopf Extensions, J. Algebra, 191 (1997), 568-588.

[48] A. Masouka, The p^n Theorem for Semisimple Hopf Algebras, Proc. AMS 124(1996), no. 3, 735-737.

[49] S. Montgomery, Hopf Algebras and Their Actions on Rings. **CBMS**, Lecture in Math.; Providence, RI, (1993); Vol. 82.

[50] S. Montgomery, Classifying Finite Dimensional Semisimple Hopf Algebras, Contemp. Math 229 (1998), 265-279.

[51] T. Nakayama, On Frobenusean Algebras I and II, Ann. of Math 40 (1939), 611-633 and Ann. of Math 42 (1941), 1-21.

[52] S. Natale, On Semisimple Hopf Algerbas of Dimension pq^2, J. Algebra 221 (1999), 242-278.

[53] S. Natale, On Semisimple Hopf Algebras of Dimension pq^2 II, Algebra Representation Theory, 2003.

[54] S. Natale, Semisolvability of Semismple Hopf Algebras of Low Dimension, Mem. AMS 186 (2007).

[55] S-H. Ng, Non-semisimple Hopf Algebras of Dimension p^2, J. Algebra 255 (2002), no. 1, 182-197.

[56] S-H. Ng, Hopf Algebras of Dimension pq, J. Algebra 276 (2004), no. 1, 399-406.

[57] S-H. Ng, Hopf Algebras of Dimension pq II, J. Algebra, In Press. arXiv: 0704.2428v1

[58] S-H. Ng, Hopf Algebras of Dimension $2p$, Proc. AMS 133 (2005), no. 8, 2237-2242.

[59] D. Nikshych, Non Group-theoretical Semisimple Hopf Algebras from Group Actions on Fusion Categories, arXiv: 0712.0585v1.

[60] Fred van Oystaeyen, Pu Zhang, Quiver Hopf Algebras, J. Algebra 280 (2004), 577-589.

[61] D. Radford, The Structure of Hopf Algebras with a Projection, J. Algebra 92 (1985), 322-347.
[62] C. M. Ringel, The Representation Type of Local Algebras, In Representation of Algebras, LNM 488, Springer-Verlag (1975), 282-305.
[63] C. M. Ringel, Tame Algebras and Integral Quadratic Forms, LNM 1099, Springer-Verlag, Berlin, Heidelberg, New York, Tokyo, 1984.
[64] H. -J. Schneider, Normal Basis and Transitivity of Crossed Products for Hopf Algebras, J. Algebra 152 (1992), 289-312.
[65] Y. Zhu, Hopf Algebras of Prime Dimension, International Mathematical Research Notices 1 (1994), 53-59.

DEPARTMENT OF MATHEMATICS, NANJING UNIVERSITY, NANJING 210093, CHINA
E-mail address: gxliu@nju.edu.cn

Contemporary Mathematics
Volume **478**, 2009

Twelve bridges from a reductive group to its Langlands dual

G. Lusztig

Introduction. These notes are based on a series of lectures given by the author at the Central China Normal University in Wuhan (July 2007). The aim of the lectures was to provide an introduction to the Langlands philosophy.

According to a theorem of Chevalley, connected reductive split algebraic groups over a fixed field A are in bijection with certain combinatorial objects called root data. Now there is a natural involution on the collection of root data in which roots and coroots are interchanged. This corresponds by Chevalley's theorem to an involution on the collection of connected reductive split algebraic groups over A. The image of a group under this involution is called the Langlands dual of that group.

In the 1960's Langlands made the remarkable discovery that some features about the representations of a reductive group (such as classification) should be recorded in terms of data in the Langlands dual group. He thus formulated two conjectures: one involving groups over a local field and one involving automorphic representations with respect to a group over a global field.

In these notes we try to give several examples of "bridges" which connect some aspect of the collection (G_A) of Chevalley groups attached to a root datum $\mathcal{R}$ and to various commutative rings A and some aspect of the analogous collection (G_A^*) of Chevalley groups attached to the dual root datum $\mathcal{R}^*$ and to various commutative rings A. By "aspect" we mean something about the structure of one of the groups G_A or of its representations or of an associated object such as the affine Hecke algebra $\mathcal{H}$. In each case the existence of the bridge is surprising due to the fact that (G_A) and (G_A^*) are related only through a very weak connection (via their root data); in particular there is no direct, elementary construction which produces the Langlands dual group from a given group.

In fact we describe twelve such bridges (some conjectural) the first three of which are very famous and were found by Langlands himself.

(I) A (conjectural) bridge from irreducible admissible representations of G_K (where K is a finite extension of $\mathbf{Q}_p$) to certain conjugacy classes of homomorphisms of the Weil group W_K to $G_{\mathbf{C}}^*$. (See §10.) This bridge contains almost as a special

Supported in part by the National Science Foundation

case a bridge from irreducible representations of $\mathcal{H}$ specialized at a non-root of 1 and conjugacy classes of certain pairs of elements in $G_\mathbf{C}^*$. (See §9.)

(II) A bridge from irreducible admissible representations of $G_\mathbf{R}$ to certain conjugacy classes of homomorphisms of the Weil group $W_\mathbf{R}$ to $G_\mathbf{C}^*$. (See §11.)

(III) A (conjectural) bridge connecting certain automorphic representations attached to G_k (k a function field over $\mathbf{F}_p$) and certain homomorphisms of $\mathrm{Gal}(\bar{k}/k)$ into $G_\mathbf{C}^*$. (See §12.)

(IV) A bridge from cells in the affine Weyl group constructed from $\mathcal{H}$ to unipotent classes in $G_\mathbf{C}^*$. (See §13.)

(V) A bridge from "special unipotent pieces" in $G_{\overline{\mathbf{F}}_p}$ to "special unipotent pieces" in $G_{\overline{\mathbf{F}}_p}^*$. (See §14.)

(VI) A bridge from irreducible representations of $G_{\mathbf{F}_q}$ to certain "special" conjugacy classes in $G_\mathbf{C}^*$. (See §16.)

(VII) A bridge from character sheaves on $G_\mathbf{C}$ to "special" conjugacy classes in $G_\mathbf{C}^*$. (See §17.)

(VIII) A bridge constructed by Vogan connecting certain intersection cohomology spaces associated to symmetric spaces of $G_\mathbf{C}$ and similar objects for $G_\mathbf{C}^*$. (See §18.)

(IX) A (partly conjectural) bridge connecting multiplicities in standard modules of G_K (as in (I)) or $G_\mathbf{R}$ with intersection cohomology spaces arising from the geometry of $G_\mathbf{C}^*$. (See §19.)

(X) A bridge connecting the tensor product of two irreducible finite dimensional representations of $G_\mathbf{C}$ with the convolution of certain perverse sheaves on the affine Grassmannian attached to $G_{\mathbf{C}((\epsilon))}^*$. (See §20.)

(XI) A bridge connecting the canonical basis of the plus part of the enveloping algebra attached to $G_\mathbf{Q}$ with certain subsets of the totally positive part of the upper triangular subgroup of G_A^* where $A = \mathbf{R}[[\epsilon]]$. (See §21.)

(XII) A (partly conjectural) bridge connecting the characters of irreducible modular representations of $G_{\overline{\mathbf{F}}_p}$ with certain intersection cohomology spaces associated with the geometry of $G_{\mathbf{C}((\epsilon))}^*$. (See §22.)

Note that in some cases (such as the very important Case III) our treatment is only a very brief sketch. Moreover to simplify the exposition we restrict ourselves to the case of split groups.

We now describe the contents of these notes. In §1 we introduce root data. In §2 we use an idea of McKay (extended by Slodowy) to construct the irreducible simply connected root data. In §3 we introduce the affine Weyl group. In §4 we introduce the affine Hecke algebra and its asymptotic version [**L9**]. In §5 we define the **Z**-form of the coordinate ring of a Chevalley group. We do not follow the original approach of [**C**] but rather the approach of Kostant [**Ko**]. In §6 we define the Chevalley groups. In §7 we define the Weyl modules. In §8 we define the Langlands dual group. In §9-§22 we discuss the various bridges mentioned above.

I wish to thank David Vogan for some useful comments on a first version of these notes.

1. Root data. A *root datum* is a collection
$$\mathcal{R} = (Y, X, \langle,\rangle, \check{\alpha}_i, \alpha_i (i \in I))$$
where Y, X are finitely generated free abelian groups, $\langle,\rangle : Y \times X \to \mathbf{Z}$ is a perfect bilinear pairing, I is a finite set, $\check{\alpha}_i (i \in I)$ are elements of Y and $\alpha_i (i \in I)$ are

elements of X such that $\langle \check{\alpha}_i, \alpha_i \rangle = 2$ for all i, $\langle \check{\alpha}_i, \alpha_j \rangle \in -\mathbf{N}$ for all $i \neq j$; it is assumed that the following (equivalent) conditions are satisfied:

(i) there exist $c_i \in \mathbf{Z}_{>0} (i \in I)$ such that the matrix $(c_i \langle \check{\alpha}_i, \alpha_j \rangle)_{i,j \in I}$ is symmetric, positive definite;

(ii) there exist $c_i' \in \mathbf{Z}_{>0} (i \in I)$ such that the matrix $(\langle \check{\alpha}_i, \alpha_j \rangle c_j')_{i,j \in I}$ is symmetric, positive definite.

By the equivalence of (i),(ii),

$$\mathcal{R}^* = (X, Y, \langle, \rangle', \alpha_i, \check{\alpha}_i (i \in I))$$

where $\langle x, y \rangle' = \langle y, x \rangle$ for $x \in X, y \in Y$, is again a root datum, said to be the dual of $\mathcal{R}$. Note that $\mathcal{R}^{**} = \mathcal{R}$ in an obvious way.

For $\mathcal{R}$ as above let W be the (finite) subgroup of $\mathrm{Aut}(Y)$ generated by the automorphisms $s_i : y \mapsto y - \langle y, \alpha_i \rangle \check{\alpha}_i$ $(i \in I)$ or equivalently the subgroup of $\mathrm{Aut}(X)$ generated by the automorphisms $s_i : x \mapsto x - \langle \check{\alpha}_i, x \rangle \alpha_i$ $(i \in I)$; these two subgroups may be identified by taking contragredients. We say that W is the *Weyl group* of $\mathcal{R}$; it is also the Weyl group of $\mathcal{R}^*$. Let

$$X^+ = \{\lambda \in X; \langle \check{\alpha}_i, \lambda \rangle \in \mathbf{N} \text{ for all } i \in I\},$$
$$Y^+ = \{y \in Y; \langle y, \alpha_i \rangle \in \mathbf{N} \text{ for all } i \in I\}.$$

For $\lambda, \lambda' \in X$ write $\lambda' \geq \lambda$ if $\lambda' - \lambda \in \sum_i \mathbf{N}\alpha_i$ and $\lambda' > \lambda$ if $\lambda' \geq \lambda, \lambda' \neq \lambda$.

We say that $\mathcal{R}$ is simply connected if $Y = \sum_i \mathbf{Z}\check{\alpha}_i$. We say that $\mathcal{R}$ is adjoint if $X = \sum_i \mathbf{Z}\alpha_i$. We say that $\mathcal{R}$ is semisimple if $X/\sum_i \mathbf{Z}\alpha_i$ is finite or equivalently $Y/\sum_i \mathbf{Z}\check{\alpha}_i$ is finite. We say that $\mathcal{R}$ is irreducible if $I \neq \emptyset$ and there is no partition $I = I' \cup I''$ of I such that I', I'' are $\neq \emptyset$ and $\langle \check{\alpha}_i, \alpha_j \rangle = 0$ for all $i \in I', j \in I''$.

2. Subgroups of $SL_2(\mathbf{C})$ and root data. In [**MK**] McKay discovered a remarkable direct connection between finite subgroups of $SL_2(\mathbf{C})$ and "simply laced affine Dynkin diagrams". Slodowy [**SL**] extended this to a connection between certain pairs of subgroups of $SL_2(\mathbf{C})$ (one contained in the other) and "affine Dynkin diagrams".

Let Γ, Γ' be two finite subgroups of $SL_2(\mathbf{C})$ such that Γ is a normal subgroup of Γ' with Γ'/Γ cyclic. We show how to attach to (Γ, Γ') a root datum (we use an argument generalizing one in [**L12, 1.2**]). Let $\mathcal{X}$ be the category of finite dimensional complex representations of Γ which can be extended to representations of Γ'. Let $\rho_i (i \in \tilde{I})$ be the indecomposable objects of $\mathcal{X}$ up to isomorphism, that is, the representations of Γ which are restrictions of irreducible representations of Γ'. Let $i_0 \in \tilde{I}$ be such that ρ_{i_0} is the trivial representation of Γ on $\mathbf{C}$. Let σ be the obvious representation of Γ on $\mathbf{C}^2$; we have $\sigma \in \mathcal{X}$. Let V be the $\mathbf{R}$-vector space with basis $\{i; i \in \tilde{I}\}$. Any object ρ of $\mathcal{X}$ gives rise to a vector $\underline{\rho} = \sum_i n_i i \in V$ where $\rho \cong \oplus_i \rho_i^{\oplus n_i}$; here $n_i \in \mathbf{N}$. For $i \in \tilde{I}$ we have $\rho_i \otimes \sigma \in \mathcal{X}$ and $\underline{\rho_i \otimes \sigma} = \sum_{j \in \tilde{I}} c_{ij} j$ with $c_{ij} \in \mathbf{N}$.

For $i \in \tilde{I}$, ρ_i is the direct sum of m_i irreducible representations of Γ (each with multiplicity 1). Let $[,]$ be the bilinear form on V with values in $\mathbf{R}$ given by $[i, j] = (2\delta_{ij} - c_{ij})m_j$ for $i, j \in \tilde{I}$. Let $x = \sum_i x_i i \in V$, $x' = \sum_i x_i' i \in V$ where $x_i, x_i' \in \mathbf{R}$. For $\gamma \in \Gamma$ let λ_g be an eigenvalue of γ on $\mathbf{C}^2$. We have

$$[x, x'] = |\Gamma|^{-1} \sum_{i,j;\gamma \in \Gamma} x_i x_j' \mathrm{tr}(\gamma, \rho_i)\overline{\mathrm{tr}(\gamma, \rho_j)}(2 - \lambda_g - \overline{\lambda_g})$$
$$= |\Gamma|^{-1} \sum_{\gamma \in \Gamma} (\sum_i x_i \mathrm{tr}(\gamma, \rho_i)|1 - \lambda_\gamma|)(\sum_j x_j' \overline{\mathrm{tr}(\gamma, \rho_j)}|1 - \lambda_\gamma|).$$

In particular,

$$[x, x] = |\Gamma|^{-1} \sum_{\gamma \in \Gamma} |\sum_i x_i \mathrm{tr}(\gamma, \rho_i)| 1 - \lambda_\gamma||^2 \geq 0.$$

If $[x, x] = 0$ then for any $\gamma \in \Gamma$ we have $\sum_i x_i \mathrm{tr}(\gamma, \rho_i)|1 - \lambda_\gamma| = 0$, that is, for any $\gamma \in \Gamma - \{1\}$ we have $\sum_i x_i \mathrm{tr}(\gamma, \rho_i) = 0$, that is, there exists $c \in \mathbf{R}$ such that $x = c\mathfrak{r}$ where $\mathfrak{r} \in \mathcal{X}$ is the regular representation of Γ; if in addition we have $x_{i_0} = 0$ then we see that $c = 0$ hence $x = 0$.

Let $I = \tilde{I} - \{i_0\}$. Let Y be the subgroup of V generated by $\{i; i \in I\}$. For $i \in I$ let $\check{\alpha}_i = i \in Y$. Let $X = \mathrm{Hom}(Y, \mathbf{Z})$. Let $\langle, \rangle : Y \times X \to \mathbf{Z}$ be the obvious pairing. For $j \in I$ define $\alpha_j \in X$ by $\langle \check{\alpha}_i, \alpha_j \rangle = (i, j)$. We have $\langle \check{\alpha}_i, \alpha_i \rangle = 2$ for all $i \in I$ and $\langle \check{\alpha}_i, \alpha_j \rangle = -c_{ij} \in -\mathbf{N}$ for $i \neq j$ in I. By the argument above the matrix $(\langle \check{\alpha}_i, \alpha_j \rangle m_j)_{i,j \in I}$ is symmetric and positive definite. Hence $\mathcal{R} = (Y, X, \langle, \rangle, \check{\alpha}_i, \alpha_i (i \in I))$ is a (simply connected) root datum.

Note that all simply connected irreducible root data are obtained by this construction exactly once (up to isomorphism) from pairs (Γ, Γ') as above (up to conjugacy) with $\Gamma \neq \{1\}$ and with the property that any element of Γ' which commutes with any element of Γ is contained in Γ. Such pairs are classified as follows:

(a) $\Gamma = \Gamma'$ is a cyclic group $\mathbf{Z}_n$ of order $n \geq 2$;
(b) $\Gamma = \Gamma'$ is a binary dihedral group $\mathcal{D}_{4n}$ of order $4n \geq 8$;
(c) $\Gamma = \Gamma'$ is a binary tetrahedral group G_{24} of order 24;
(d) $\Gamma = \Gamma'$ is a binary octahedral group G_{48} of order 48;
(e) $\Gamma = \Gamma'$ is a binary icosahedral group G_{120} of order 120;
(f) $\Gamma' = \mathcal{D}_{4n}, \Gamma = \mathbf{Z}_{2n}$ with $n \geq 2$;
(g) $\Gamma' = \mathcal{D}_{8n}, \Gamma = \mathcal{D}_{4n}$ with $n \geq 2$;
(h) $\Gamma' = G_{48}, \Gamma = G_{24}$;
(i) $\Gamma' = G_{24}, \Gamma = \mathcal{D}_8$.

3. Affine Weyl group. Let $\mathcal{R} = (Y, X, \langle, \rangle, \check{\alpha}_i, \alpha_i (i \in I))$ be a root datum. Let W, $s_i (i \in I)$ be as in §1. Let $\widetilde{W}$ be the semidirect product $W \cdot Y$. We have $\widetilde{W} = \{wa^y; w \in W, y \in Y\}$ where a is a symbol; the multiplication is given by $(wa^y)(w'a^{y'}) = ww'a^{w'^{-1}(y)+y'}$ for $w, w' \in W$, $y, y' \in Y$. We identify Y with its image under the homomorphism $y \mapsto 1a^y$ (a normal subgroup of $\widetilde{W}$) and W with its image under the homomorphism $w \mapsto wa^0$.

Let R be the set of elements of X of the form $w(\alpha_i)$ for some $w \in W, i \in I$. Let $\check{R}$ be the set of elements of Y of the form $w(\check{\alpha}_i)$ for some $w \in W, i \in I$. There is a unique W-equivariant bijection $\alpha \leftrightarrow \check{\alpha}$ between R and $\check{R}$ such that $\alpha_i \leftrightarrow \check{\alpha}_i$ for any $i \in I$. For $\alpha \in R$ we set $s_\alpha = w s_i w^{-1} \in W$ where $\alpha = w(\alpha_i), w \in W, i \in I$. Note that s_α is well defined. Let R_{min} be the set of all $\alpha \in R$ such that the following holds: if $\alpha' \in R, \alpha' \leq \alpha$ then $\alpha' = \alpha$.

Let $R^+ = R \cap (\sum_i \mathbf{N}\alpha_i)$, $R^- = -R^+$. We have $R = R^+ \cup R^-$. Following Iwahori and Matsumoto [**IM**], we define a function $l : \widetilde{W} \to \mathbf{N}$ by

$$l(wa^y) = \sum_{\alpha \in R^+; w(\alpha) \in R^-} |\langle y, \alpha \rangle + 1| + \sum_{\alpha \in R^+; w(\alpha) \in R^+} |\langle y, \alpha \rangle|.$$

Let S be the subset of W consisting of the involutions $s_i (i \in I)$ and the involutions $s_\alpha a^{\check{\alpha}}$ with $\alpha \in R_{min}$. Note that $l|_S = 1$.

If $y \in Y^+$ we have $l(a^y) = \sum_{\alpha \in R^+} \langle y, \alpha \rangle$. Hence for $y, y' \in Y^+$ we have $l(a^y \cdot a^{y'}) = l(a^{y+y'}) = l(a^y) + l(a^{y'})$.

4. Affine Hecke algebra. We preserve the notation of §3. Let $\mathcal{A} = \mathbf{Z}[v, v^{-1}]$ where v is an indeterminate. Let $\mathcal{H}$ be the associative $\mathcal{A}$-algebra with 1 with generators $T_w (w \in \widetilde{W})$ and relations

$(T_s - v)(T_s + v^{-1}) = 0$ for $s \in S$,

$T_w T_{w'} = T_{ww'}$ for $w, w' \in \widetilde{W}$ such that $l(w) + l(w') = l(ww')$.

We have $T_1 = 1$ and $\{T_w; w \in \widetilde{W}\}$ is an $\mathcal{A}$-basis of $\mathcal{H}$. Let $h \mapsto h^\dagger$ be the $\mathcal{A}$-algebra involution of $\mathcal{H}$ such that $T_w^\dagger = (-1)^{l(w)} T_w^{-1}$. Let $h \mapsto \bar{h}$ be the ring involution of $\mathcal{H}$ such that $\overline{T_w} = T_{w^{-1}}^{-1}$ for $w \in \widetilde{W}$ and $\overline{v^n} = v^{-n}$ for $n \in \mathbf{Z}$. Let $z \in \widetilde{W}$. According to [**KL1**] there is a unique element $c_z \in \mathcal{H}$ such that $\overline{c_z} = c_z$ and $c_z = \sum_{w \in \widetilde{W}} p_{w,z} T_w$ where $p_{w,z} \in v^{-1}\mathbf{Z}[v^{-1}]$ for all $w \neq z$, $p_{z,z} = 1$ and $p_{w,z} = 0$ for all but finitely many w. Note that $\{c_w; w \in \widetilde{W}\}$ is an $\mathcal{A}$-basis of $\mathcal{H}$. For $x, y \in \widetilde{W}$ we write $c_x c_y = \sum_{z \in \widetilde{W}} h_{x,y,z} c_z$ where $h_{x,y,z} \in \mathcal{A}$ is 0 for all but finitely many z. Let $z \in \widetilde{W}$. According to [**L9**] there is a unique $a(z) \in \mathbf{N}$ such that $h_{x,y,z} \in v^{a(z)}\mathbf{Z}[v^{-1}]$ for any x, y and $h_{x,y,z} \notin v^{a(z)-1}\mathbf{Z}[v^{-1}]$ for some x, y. We have $h_{x,y,z} = \gamma_{x,y,z^{-1}} v^{a(z)} \mod v^{a(z)-1}\mathbf{Z}[v^{-1}]$ where $\gamma_{x,y,z^{-1}} \in \mathbf{Z}$. Let J be the free abelian group with basis $\{t_w; w \in \mathbf{W}\}$. Consider the $\mathbf{Z}$-algebra structure on J such that $t_x t_y = \sum_{z \in \widetilde{W}} \gamma_{x,y,z^{-1}} t_z$ for all x, y in $\widetilde{W}$. This structure is associative and the ring J has a unit element 1 of the form $\sum_{d \in \mathcal{D}} t_d$ where $\mathcal{D}$ is a finite subset of $\widetilde{W}$ consisting of involutions [**L9**]. For x, y in $\widetilde{W}$ we write $x \sim y$ when $t_x t_u t_y \neq 0$ for some $u \in \widetilde{W}$. This is an equivalence relation on $\widetilde{W}$; the equivalence classes are called two-sided cells. For any two-sided cell c let J_c be the subgroup of J spanned by $\{t_z; z \in c\}$. This is a subring of J with unit $\sum_{d \in \mathcal{D} \cap c} t_d$. We have $J = \oplus_c J_c$ as rings. Consider the $\mathcal{A}$-linear map $\phi : \mathcal{H} \to \mathcal{A} \otimes J$ given by

$$\phi(c_x^\dagger) = \sum_{z \in \widetilde{W}, d \in \mathcal{D}; a(d) = a(z)} h_{x,d,z} t_z$$

for any $x \in \widetilde{W}$. This is an $\mathcal{A}$-algebra homomorphism [**L9**].

5. Coordinate ring. Let $\mathcal{R} = (Y, X, \langle , \rangle, \check{\alpha}_i, \alpha_i (i \in I))$ be a root datum. Let $\underline{f}$ be the associative $\mathbf{Q}$-algebra with 1 defined by the generators $\theta_i (i \in I)$ and the Serre relations

$$\sum_{a,b \in \mathbf{N}; a+b=1-\langle \check{\alpha}_i, \alpha_j \rangle} (-1)^a (\theta_i^a / a!) \theta_j (\theta_i^b / b!)$$

for $i \neq j$ in I. Let $^0\underline{U}$ be the symmetric algebra of $\mathbf{Q} \otimes Y$. Let $\underline{U}$ be the $\mathbf{Q}$-algebra with 1 defined by the generators $x^+, x^- (x \in \underline{f})$, $\underline{a} \in {}^0\underline{U}$ and the relations:

$x \mapsto x^+$ is an algebra homomorphism $\underline{f} \to \underline{U}$ respecting 1;

$x \mapsto x^-$ is an algebra homomorphism $\underline{f} \to \underline{U}$ respecting 1;

$\underline{a} \mapsto \underline{a}$ is an algebra homomorphism $^0\underline{U} \to \underline{U}$ respecting 1;

$y\theta_i^+ - \theta_i^+ y = \langle y, \alpha_i \rangle \theta_i^+$ for $y \in Y, i \in I$;

$y\theta_i^- - \theta_i^- y = -\langle y, \alpha_i \rangle \theta_i^-$ for $y \in Y, i \in I$;

$\theta_i^+ \theta_j^- - \theta_j^- \theta_i^+ = \delta_{ij} \check{\alpha}_i$ for i, j in I.

Define a $\mathbf{Q}$-algebra homomorphism $\Delta : \underline{U} \to \underline{U} \otimes \underline{U}$ by $\Delta(\theta_i^+) = \theta_i^+ \otimes 1 + 1 \otimes \theta_i^+$, $\Delta(\theta_i^-) = \theta_i^- \otimes 1 + 1 \otimes \theta_i^-$ for $i \in I$, $\Delta(y) = y \otimes 1 + 1 \otimes y$ for $y \in Y$. Define a $\mathbf{Q}$-algebra isomorphism $\Sigma : \underline{U} \to \underline{U}^{opp}$ by $\Sigma(\theta_i^+) = -\theta_i^+$, $\Sigma(\theta_i^-) = -\theta_i^-$ for $i \in I$, $\Sigma(y) = -y$ for $y \in Y$.

Let $\underline{f}_{\mathbf{Z}}$ be the subring of $\underline{f}$ generated by the elements $\theta_i^{(n)} := \theta_i^n/n!$, ($i \in I, n \in \mathbf{N}$). Note that $\underline{f}_{\mathbf{Z}}$ is a lattice in the $\mathbf{Q}$-vector space $\underline{f}$. Following [**Ko**] we define $\underline{U}_{\mathbf{Z}}$ to be the subring of $\underline{U}$ generated by the elements x^+, x^- ($x \in \underline{f}_{\mathbf{Z}}$) and $\binom{y}{n} := \frac{y(y-1)\dots(y-n+1)}{n!}$, ($y \in Y, n \in \mathbf{N}$). Note that $\underline{U}_{\mathbf{Z}}$ is a lattice in the $\mathbf{Q}$-vector space $\underline{U}$. Hence $\underline{U}_{\mathbf{Z}} \otimes_{\mathbf{Z}} \underline{U}_{\mathbf{Z}}$ is a lattice in $\underline{U} \otimes_{\mathbf{Q}} \underline{U}$. Note that Δ restricts to a ring homomorphism $\underline{U}_{\mathbf{Z}} \to \underline{U}_{\mathbf{Z}} \otimes_{\mathbf{Z}} \underline{U}_{\mathbf{Z}}$ denoted again by Δ. Also Σ restricts to a ring isomorphism $\underline{U}_{\mathbf{Z}} \to \underline{U}_{\mathbf{Z}}^{opp}$ denoted again by Σ.

For any $\underline{U}$-module M and any $x \in X$ we set
$$M^x = \{m \in M; ym = \langle y, x \rangle m \text{ for any } y \in Y\}.$$
Let $\mathcal{C}$ be the category whose objects are $\underline{U}$-modules M with $\dim_{\mathbf{Q}} M < \infty$ such that $M = \oplus_{x \in X} M^x$. For any $\mathbf{Q}$-vector space V we set $V^\dagger = \operatorname{Hom}_{\mathbf{Q}}(V, \mathbf{Q})$. For $M \in \mathcal{C}$ we define $c_M : M \otimes M^\dagger \to \underline{U}^\dagger$ by $m \otimes \xi \mapsto [u \mapsto \xi(um)]$. Let

$$\mathcal{O} = \sum_{M \in \mathcal{C}} c_M(M \otimes M^\dagger),$$

a $\mathbf{Q}$-subspace of $\underline{U}^\dagger$. (This agrees with the definition in [**Ko**] when $\mathcal{R}$ is semisimple.) For $f, f' \in \mathcal{O}$ we define $ff' : \underline{U} \to \mathbf{Q}$ by $u \mapsto \sum_s f(u_s)f'(u_s')$ where $\Delta(u) = \sum_s u_s \otimes u_s'$, $u_s, u_s' \in \underline{U}$. We have $ff' \in \mathcal{O}$. This defines a structure of associative, commutative algebra on $\mathcal{O}$. This algebra has a unit element: the algebra homomorphism $\underline{U} \to \mathbf{Q}$ such that $\theta_i^+ \mapsto 0$, $\theta_i^- \mapsto 0$ for $i \in I$, $y \mapsto 0$ for $y \in Y$ and $1 \mapsto 1$. For $f \in \mathcal{O}$ we define a linear function $\delta(f) : \underline{U} \otimes \underline{U} \to \mathbf{Q}$ by $u_1 \otimes u_2 \mapsto f(u_1 u_2)$. Note that $\mathcal{O} \otimes \mathcal{O}$ is naturally a subspace of $(\underline{U} \otimes \underline{U})^\dagger$ and that the image of $\delta : \mathcal{O} \to (\underline{U} \otimes \underline{U})^\dagger$ is contained in the subspace $\mathcal{O} \otimes \mathcal{O}$ so that δ defines a linear map $\mathcal{O} \to \mathcal{O} \otimes \mathcal{O}$ denoted again by δ. This is an algebra homomorphism. For $f \in \mathcal{O}$ we define a linear function $\sigma(f) : \underline{U} \to \mathbf{Q}$ by $u \mapsto f(\Sigma(u))$. We have $\sigma(f) \in \mathcal{O}$; thus σ defines a linear map $\mathcal{O} \to \mathcal{O}$ denoted again by σ. This is an algebra homomorphism. Define $\epsilon : \mathcal{O} \to \mathbf{Q}$ by $f \mapsto f(1)$. Note that the commutative algebra $\mathcal{O}$ with the comultiplication δ, the antipode σ and the counit ϵ is a Hopf algebra over $\mathbf{Q}$.

Let $f \in \mathcal{O}$. There is a unique collection $(f^x)_{x \in X}$ of numbers in $\mathbf{Q}$ such that $f^x = 0$ for all but finitely many $x \in X$ and

$$f(y_1^{n_1} y_2^{n_2} \dots y_r^{n_r}) = \sum_{x \in X} \langle y_1, x \rangle^{n_1} \langle y_2, x \rangle^{n_2} \dots \langle y_r, x \rangle^{n_r} f^x$$

for any $y_1, y_2, \dots, y_r$ in Y and $n_1, n_2, \dots, n_r$ in $\mathbf{N}$. For example if $f = c_M(m \otimes \xi)$ where $M \in \mathcal{C}, m \in M, \xi \in M^\dagger$, we have $f^x = \xi(m_x)$ where $m_x \in M^x$ are defined by $m = \sum_{x \in X} m_x$. Note that for any $x \in X$, $f \mapsto f^x$ is a linear function $\mathcal{O} \to \mathbf{Q}$.

Let $\mathcal{O}_{\mathbf{Z}} = \{f \in \mathcal{O}; f(\underline{U}_{\mathbf{Z}}) \subset \mathbf{Z}\}$. (This agrees with the definition in [**Ko**] when $\mathcal{R}$ is semisimple.) Note that $\mathcal{O}_{\mathbf{Z}}$ is a subring of $\mathcal{O}$. One can show that $\mathcal{O}_{\mathbf{Z}}$ is a lattice in the $\mathbf{Q}$-vector space $\mathcal{O}$. Hence $\mathcal{O}_{\mathbf{Z}} \otimes_{\mathbf{Z}} \mathcal{O}_{\mathbf{Z}}$ is a lattice in the $\mathbf{Q}$-vector space $\mathcal{O} \otimes \mathcal{O}$. Note that $\delta : \mathcal{O} \to \mathcal{O} \otimes \mathcal{O}$ restricts to a ring homomorphism $\delta_{\mathbf{Z}} : \mathcal{O}_{\mathbf{Z}} \to \mathcal{O}_{\mathbf{Z}} \otimes_{\mathbf{Z}} \mathcal{O}_{\mathbf{Z}}$; $\sigma : \mathcal{O} \to \mathcal{O}$ restricts to a ring isomorphism $\sigma_{\mathbf{Z}} : \mathcal{O}_{\mathbf{Z}} \to \mathcal{O}_{\mathbf{Z}}$; $\epsilon : \mathcal{O} \to \mathbf{Q}$ restricts to a ring homomorphism $\epsilon_{\mathbf{Z}} : \mathcal{O}_{\mathbf{Z}} \to \mathbf{Z}$. The commutative ring $\mathcal{O}_{\mathbf{Z}}$ together with the comultiplication $\delta_{\mathbf{Z}}$, the antipode $\sigma_{\mathbf{Z}}$ and the counit $\epsilon_{\mathbf{Z}}$ is a Hopf ring. For any $x \in X$ and $f \in \mathcal{O}_{\mathbf{Z}}$ we have $f^x \in \mathbf{Z}$.

For any commutative ring A with 1 we set $\mathcal{O}_A = A \otimes \mathcal{O}_{\mathbf{Z}}$. By extension of scalars, from $\delta_{\mathbf{Z}}, \sigma_{\mathbf{Z}}, \epsilon_{\mathbf{Z}}$ we get A-algebra homomorphisms $\delta_A : \mathcal{O}_A \to \mathcal{O}_A \otimes_A \mathcal{O}_A$, $\sigma_A : \mathcal{O}_A \to \mathcal{O}_A$, $\epsilon_A : \mathcal{O}_A \to A$. The commutative A-algebra $\mathcal{O}_A$ together with the comultiplication δ_A, the antipode σ_A and the counit ϵ_A is a Hopf algebra over A. For any $x \in X$ the homomorphism $\mathcal{O}_{\mathbf{Z}} \to \mathbf{Z}$ gives rise by extension of scalars to an A-linear map $\mathcal{O}_A \to A$ denoted again by f^x.

The following two properties are proved in [**L19**]:

(i) the A-algebra $\mathcal{O}_A$ is finitely generated;

(ii) if A is an integral domain then $\mathcal{O}_A$ is an integral domain.

6. Chevalley groups. We preserve the notation of §5. Let W, s_i be as in §1. Let A be a commutative ring with 1. As in [**Ko**] we define G_A to be the set of A-algebra homomorphisms $\mathcal{O}_A \to A$ respecting 1. For $g, g' \in G_A$ we define $gg' : \mathcal{O}_A \to A$ by $f \mapsto \sum_s g(f_s)g'(f'_s)$ where $\delta_A(f) = \sum_s f_s \otimes f'_s$ with f_s, f'_s in $\mathcal{O}_A$. Then $gg' \in G_A$ and $(g, g') \mapsto gg'$ is a group structure on G_A with unit element ϵ_A. We say that G_A is the *Chevalley group* attached to the root datum $\mathcal{R}$ and to the commutative ring A. We write also $G_A^{\mathcal{R}}$ instead of G_A when we want to emphasize the dependence on $\mathcal{R}$.

If $\kappa : A \to A'$ is a homomorphism of commutative rings with 1 and $g : \mathcal{O}_A \to A$ is in G_A then applying to g the functor $A' \otimes_A ?$ (where A' is regarded as an A-algebra via κ) we obtain an A'-algebra homomorphism $\mathcal{O}_{A'} \to A'$ respecting 1 which is denoted by $\tilde{g}$. Now $g \mapsto \tilde{g}$ is a group homomorphism $G_A \to G_{A'}$ said to be induced by κ.

For any $i \in I, b \in A$ we define an A-linear map $x_i(b) : \mathcal{O}_A \to A$ by

$$\sum_s a_s \otimes f_s \mapsto \sum_s \sum_{n \in \mathbf{N}} a_s b^n f_s((\theta_i^{(n)})^+).$$

Here $a_s \in A, f_s \in \mathcal{O}_{\mathbf{Z}}$. Since $f_s \in \mathcal{O}$ we have $f_s((\theta_i^{(n)})^+) = 0$ for large enough n so that the last sum is finite. From the definitions we see that $x_i(b) \in G_A$ and that $b \mapsto x_i(b)$ is an (injective) group homomorphism $A \to G_A$.

Similarly, for any $i \in I, b \in A$ we define an A-linear map $y_i(b) : \mathcal{O}_A \to A$ by

$$\sum_s a_s \otimes f_s \mapsto \sum_s \sum_{n \in \mathbf{N}} a_s b^n f_s((\theta_i^{(n)})^-).$$

Here $a_s \in A, f_s \in \mathcal{O}_{\mathbf{Z}}$. Again the last sum is finite. From the definitions we see that $y_i(b) \in G_A$ and that $b \mapsto y_i(b)$ is an (injective) group homomorphism $A \to G_A$.

Let A^* be the group of units of A. Let $t = \sum_r a_r \otimes y_r \in A^* \otimes Y$ with $a_r \in A^*, y_r \in Y$. We define $\underline{t} : \mathcal{O}_A \to A$ by

$$f \mapsto \sum_r \sum_{x \in X} a_r^{\langle y_r, x \rangle} f^x.$$

(Since $a_r \in A^*$, a_r^n is defined for any $n \in \mathbf{Z}$.) From the definitions we see that $\underline{t} \in G_A$ and that $t \mapsto \underline{t}$ is an injective group homomorphism $A^* \otimes Y \to G_A$ with image denoted by T_A (a commutative subgroup of G_A). We identify $A^* \otimes Y$ with T_A via this homomorphism. We write also $T_A^{\mathcal{R}}$ instead of T_A when we want to emphasize the dependence on $\mathcal{R}$.

For $i \in I$ we define an element $\dot{s}_i \in G_A$ by $\dot{s}_i = x_i(1)y_i(-1)x_i(1)$. We have $\dot{s}_i^2 = (-1) \otimes \check{\alpha}_i \in A^* \otimes Y = T_A$. Moreover, for $i \neq j$ we have $\dot{s}_i \dot{s}_j \dot{s}_i \cdots = \dot{s}_j \dot{s}_i \dot{s}_j \ldots$ (both sides have n factors where n is the order of $s_i s_j$ in W). It follows that for any $w \in W$ there is a well defined element $\dot{w} \in G_A$ such that $\dot{w} = \dot{s}_{i_1} \dot{s}_{i_2} \ldots \dot{s}_{i_r}$ whenever $w = s_{i_1} s_{i_2} \ldots s_{i_r}$ with $r = l(w)$. Note that $\dot{w} T_A \dot{w}^{-1} = T_A$. More precisely, for

$t \in T_A$ we have $\dot{w}t\dot{w}^{-1} = w(t)$ where $w : t \mapsto w(t)$ is the W-action on T_A given by $a \otimes y \mapsto a \otimes w(y)$ for $a \in A^*$, $y \in Y$.

For any sequence $i_1, i_2, \ldots, i_r$ in I such that $l(s_{i_1}s_{i_2} \ldots s_{i_r}) = r = |R^+|$, the map $A^r \to G_A$ given by

$$(a_1, a_2, \ldots, a_r) \mapsto$$
$$x_{i_1}(a_1)\dot{s}_{i_1}x_{i_2}(a_2)\dot{s}_{i_1}^{-1} \ldots \dot{s}_{i_1}\dot{s}_{i_2} \ldots \dot{s}_{i_{r-1}}x_{i_r}(a_r)\dot{s}_{i_{r-1}}^{-1} \ldots \dot{s}_{i_2}^{-1}\dot{s}_{i_1}^{-1}$$

is injective and its image is a subgroup U_A^+ of G_A independent of the choice of $i_1, i_2, \ldots, i_r$. (See [**L19**].)

Similarly, for any sequence $i_1, i_2, \ldots, i_r$ in I such that $l(s_{i_1}s_{i_2} \ldots s_{i_r}) = r = |R^+|$, the map $A^r \to G_A$ given by

$$(a_1, a_2, \ldots, a_r) \mapsto$$
$$y_{i_1}(a_1)\dot{s}_{i_1}y_{i_2}(a_2)\dot{s}_{i_1}^{-1} \ldots \dot{s}_{i_1}\dot{s}_{i_2} \ldots \dot{s}_{i_{r-1}}y_{i_r}(a_r)\dot{s}_{i_{r-1}}^{-1} \ldots \dot{s}_{i_2}^{-1}\dot{s}_{i_1}^{-1}$$

is injective and its image is a subgroup U_A^- of G_A independent of the choice of $i_1, i_2, \ldots, i_r$.

The subgroups U_A^+, U_A^- are normalized by T_A. We set $B_A^+ = U_A^+T_A = T_AU_A^+$, $B_A^- = U_A^-T_A = T_AU_A^-$.

If A is a field, we have a partition $G_A = \cup_{w \in W} B_A^+\dot{w}B_A^+$.

7. Weyl modules. We preserve the notation of §6. For any $\lambda \in X^+$ let $\mathcal{T}_\lambda = \sum_i \underline{f}\theta_i^{\langle \check{\alpha}_i, \lambda \rangle + 1}$, a left ideal of $\underline{f}$; let $\Lambda_\lambda = \underline{f}/\mathcal{T}_\lambda$, a finite dimensional **Q**-vector space. Let η be the image of $1 \in \underline{f}$ in Λ_λ. We regard Λ_λ as a $\underline{U}$-module in which x^- acts as left multiplication by $x(x \in \underline{f})$; $\theta_i^+\eta = 0$ for $i \in I$; $y\eta = \langle y, \lambda \rangle\eta$ for $y \in Y$. We say that Λ_λ is a *Weyl module*. We have $\Lambda_\lambda \in \mathcal{C}$.

For $\lambda \in X^+$ let

$$\mathcal{T}_{\lambda, \mathbf{Z}} = \sum_{i, n; n \geq \langle \check{\alpha}_i, \lambda \rangle + 1} \underline{f}\theta_i^{(n)} = \mathcal{T}_\lambda \cap \underline{f}_{\mathbf{Z}},$$

a left ideal of $\underline{f}_{\mathbf{Z}}$. Let $\Lambda_{\lambda, \mathbf{Z}} = \underline{f}_{\mathbf{Z}}/\mathcal{T}_{\lambda, \mathbf{Z}}$. Then $\Lambda_{\lambda, \mathbf{Z}}$ is a lattice in the **Q**-vector space Λ_λ and a $\underline{U}_{\mathbf{Z}}$-submodule of Λ_λ. For a commutative ring A with 1 we set $\Lambda_{\lambda, A} = \Lambda_{\lambda, \mathbf{Z}} \otimes A$.

We write $\mathcal{O}^{opp}, \mathcal{O}_A^{opp}$ for $\mathcal{O}, \mathcal{O}_A$ with the opposite comultiplication. Define $\Xi : \Lambda_\lambda \to \mathcal{O} \otimes \Lambda_\lambda$ by $e \mapsto \sum_j c_{\Lambda_l}(e \otimes \xi_j') \otimes e_j$ where (e_j) is a **Q**-basis of Λ_λ and (ξ_j) is the dual basis of $\Lambda_\lambda^\dagger$. This makes Λ_λ into a $\mathcal{O}^{opp}$-comodule. Now $\mathcal{O}_{\mathbf{Z}} \otimes_{\mathbf{Z}} \Lambda_{\lambda, \mathbf{Z}}$ is a lattice in $\mathcal{O} \otimes \Lambda_\lambda$ and Ξ restricts to $\Xi_{\mathbf{Z}} : \Lambda_{\lambda, \mathbf{Z}} \to \mathcal{O}_{\mathbf{Z}} \otimes_{\mathbf{Z}} \Lambda_{\lambda, \mathbf{Z}}$. By extension of scalars we obtain an A-linear map $\Xi_A : \Lambda_{\lambda, A} \to \mathcal{O}_A \otimes_A \Lambda_{\lambda, A}$ making $\Lambda_{\lambda, A}$ into a $\mathcal{O}_A^{opp}$-comodule. For any $g : \mathcal{O}_A \to A$ which is in G_A, we define $\rho_g : \Lambda_{\lambda, A} \to \Lambda_{\lambda, A}$ by $e \mapsto \sum_h g(f_h) \otimes e_h$ where $\Xi_A(e) = \sum_h f_h \otimes e_h$, $f_h \in \mathcal{O}_A$, $e_h \in \Lambda_{\lambda, A}$. Note that $g \mapsto \rho_g$ is a group action. Thus $\Lambda_{\lambda, A}$ is a G_A-module.

8. The Langlands dual group. If A is an algebraically closed field then $G_A = G_A^{\mathcal{R}}$ is a reductive connected algebraic group over A with coordinate ring $\mathcal{O}_A$. Moreover, according to Chevalley, $\mathcal{R} \mapsto G_A^{\mathcal{R}}$ is a bijection

$$\{\text{root data up to isom.}\} \xrightarrow{\sim}$$

$$\{\text{reductive connected algebraic groups over } A \text{ up to isom.}\}.$$

An element of G_A is said to be semisimple if it is conjugate to an element in T_A. An element of G_A is said to be unipotent if it is conjugate to an element in U_A^+.

(a) *In the remainder of these notes (except in §15) we fix a root datum* $\mathcal{R} = (Y, X, \langle, \rangle, \check{\alpha}_i, \alpha_i (i \in I))$. Define $W, \widetilde{W}, \mathcal{H}$ in terms of $\mathcal{R}$ as in §1,§3,§4. Define $\widetilde{W}^*, \mathcal{H}^*$ like $\widetilde{W}, \widetilde{W}^*$ but in terms of $\mathcal{R}^*$ instead of $\mathcal{R}$.

For a commutative ring A with 1 we write G_A, T_A instead of $G_A^{\mathcal{R}}, T_A^{\mathcal{R}}$ and G_A^*, T_A^* instead of $G_A^{\mathcal{R}^*}, T_A^{\mathcal{R}^*}$. We say that G_A^* is the Langlands dual group to G_A. Let $\mathcal{B}_A$ be the set of subgroups of G_A that are conjugate to B_A^+ (or equivalently to B_A^-). Let $\mathcal{B}_A^*$ be the analogous set defined in terms of $\mathcal{R}^*$ instead of $\mathcal{R}$. For $g \in G_A^*$ we set $\mathcal{B}_{A,g}^* = \{B \in \mathcal{B}_A^*; g \in B\}$.

9. Representations of affine Hecke algebras. Assume that $X / \sum_i \mathbf{Z}\alpha_i$ has no torsion. We fix $\underline{v} \in \mathbf{C}^*$. Let $\mathcal{H}_{\underline{v}} = \mathbf{C} \otimes_{\mathcal{A}} \mathcal{H}$ where $\mathbf{C}$ is regarded as an $\mathcal{A}$-algebra via $v \mapsto \underline{v}$. Let $\Psi_{\underline{v}}$ be the set of all pairs (s, u) up to $G_{\mathbf{C}}^*$-conjugation where $s \in G_{\mathbf{C}}^*$ is semisimple, $u \in G_{\mathbf{C}}^*$ is unipotent and $sus^{-1} = u^{\underline{v}^2}$. (There is a unique morphism of algebraic groups $\mathbf{C} \to G_{\mathbf{C}}^*$, $z \mapsto u^z$ such that for any $n \in \mathbf{N}$, u^n is the n-th power of u.)

Assume that $\underline{v}$ is not a root of 1. The Deligne-Langlands conjecture states that there is a canonical finite to one surjective map

(a) $\qquad\qquad\qquad$ {irred. $\mathcal{H}_{\underline{v}}$-modules up to isom.} $\to \Psi_{\underline{v}}$.

A refinement of this conjecture was stated in [**L5**] namely that the fibre of (a) at (s, u) should be in natural bijection with the set of irreducible representations (up to isomorphism) of the group of connected components of the centralizer of (s, u) in $G_{\mathbf{C}}^*$ which appear in the natural representation of this group on the cohomology of $\{B \in \mathcal{B}_{\mathbf{C}}^*; s \in B, u \in B\}$. This came from a study of examples connected with "subregular" unipotent elements in $G_{\mathbf{C}}^*$. In [**L7**] it was shown that $\mathcal{H}_{\underline{v}}$ acts naturally on the equivariant K-theory of $\mathcal{B}_{\mathbf{C}}^*$ where the parameter of the Hecke algebra comes from equivariance with respect to a $\mathbf{C}^*$-action. In [**L7**] it was also suggested that one should construct representations of $\mathcal{H}_{\underline{v}}$ using the equivariant K-theory of the varieties $\mathcal{B}_{\mathbf{C},u}^*$ for $u \in G_{\mathbf{C}}^*$ unipotent. This was established in [**KL2**] which gave a proof of (a) (in the refined form).

In [**Xi**] it is shown (by a reduction to [**KL2**]) that a statement analogous to (a) (in the refined form) holds also when $\underline{v}$ is allowed to be a root of 1 in the complement of a specific finite set of roots of 1 depending on $\mathcal{R}$.

10. p-adic groups. Let K be a finite extension of the field of p-adic numbers (p a prime number). Let $\mathfrak{a}$ be the integral closure in K of the ring of p-adic integers and let $\mathfrak{m}$ be the unique maximal ideal of $\mathfrak{a}$ so that $\mathfrak{a}/\mathfrak{m}$ is a finite field with q elements. Let $\bar{K}$ be an algebraic closure of K. Let $\bar{\mathfrak{a}}$ be the integral closure of $\mathfrak{a}$ in $\bar{K}$ and let $\bar{\mathfrak{m}}$ be the unique maximal ideal of $\bar{\mathfrak{a}}$ so that $\bar{\mathfrak{a}}/\bar{\mathfrak{m}}$ is an algebraic closure of $\mathfrak{a}/\mathfrak{m}$. Let W_K be the Weil group of K, that is, the inverse image under the natural homomorphism $\pi : \mathrm{Gal}(\bar{K}/K) \to \mathrm{Gal}(\bar{\mathfrak{a}}/\bar{\mathfrak{m}}, \mathfrak{a}/\mathfrak{m})$ of the subgroup $\mathbf{Z}$ of $\mathrm{Gal}(\bar{\mathfrak{a}}/\bar{\mathfrak{m}}, \mathfrak{a}/\mathfrak{m})$ consisting of the integer powers of the automorphism $x \mapsto x^q$. Let $\mathcal{I}$ be the kernel of π. We have an exact sequence $1 \to \mathcal{I} \to W_K \xrightarrow{\omega} \mathbf{Z} \to 1$ where ω is the restriction of π. A homomorphism $\rho : W_K \to G_{\mathbf{C}}^*$ is said to be admissible if $\rho(\mathcal{I})$ is finite and $\rho(\gamma)$ is semisimple in $G_{\mathbf{C}}^*$ for some/any $\gamma \in \omega^{-1}(1)$. Let $\Phi^K(G_{\mathbf{C}}^*)$ be the set of all pairs (ρ, u) (up to $G_{\mathbf{C}}^*$-conjugacy) where $\rho : W_K \to G_{\mathbf{C}}^*$ is an admissible homomorphism and $u \in G_{\mathbf{C}}^*$ is a unipotent element such that

$\rho(w)u\rho(w)^{-1} = u^{q^{\omega(w)}}$ for any $w \in W_K$. We regard G_K as a topological group with the p-adic topology. An irreducible representation $G_K \to GL(E)$ (where E is a $\mathbf{C}$-vector space) is said to be admissible if the stabilizer of any vector of E is open in G_K and if for any open subgroup H of G_K the space of H-invariant vectors in E has finite dimension. According to the local Langlands conjecture there is a canonical finite to one surjective map

(a)   {irred. admissible representations of G_K up to isom.} $\to \Phi^K(G_{\mathbf{C}}^*)$.

This is known to be true in the case where $G_K = GL_n(K)$, see [**HT**]. In the general case but assuming that $\mathcal{R}$ is adjoint, a class of irreducible admissible representations (called "unipotent") has been described in [**L15**] where a canonical finite to one surjective map

(b)   {unipotent representations of G_K up to isom.} $\to \Phi_1^K(G_{\mathbf{C}}^*)$

was constructed; here $\Phi_1^K(G_{\mathbf{C}}^*) = \{(\rho, u) \in \Phi(G_{\mathbf{C}}^*); \rho(\mathcal{I}) = \{1\}\}$. Note that $\Phi_1^K(G_{\mathbf{C}}^*)$ may be identified with the set $\Psi_{\sqrt{q}}$ in §9 and that (b) constitutes a verification of (a) in a special case. Note that some of the unipotent representations can be understood by the method described in §9; to understand the remaining ones one needs the theory of character sheaves and a geometric construction of certain affine Hecke algebras with unequal parameters in terms of equivariant homology.

For any $(\rho, u) \in \Phi^K(G_{\mathbf{C}}^*)$ we denote by $\Xi_{\rho,u}$ the set of irreducible representations (up to isomorphism) of the group of connected components of the simultaneous centralizer of (ρ, u) in $G_{\mathbf{C}}^*$ on which the action of the centre of $G_{\mathbf{C}}^*$ is trivial.

According to [**L15**], for any $(\rho, u) \in \Phi_1^K(G_{\mathbf{C}}^*)$ the fibre of the map (b) at (ρ, u) is in bijection with $\Xi_{\rho,u}$. This suggests that more generally for any $(\rho, u) \in \Phi^K(G_{\mathbf{C}}^*)$, the fibre of the (conjectural) map (a) at (ρ, u) is in bijection with $\Xi_{\rho,u}$.

Note that in general neither side of (a) is well understood. But recent results of J.-L.Kim [**Ki**] give a classification of "supercuspidal representations" of G_K (assuming that p is sufficiently large) which gives some hope that the left hand side of (a) can be understood for such p.

11. Real groups. The Weil group of $\mathbf{R}$ is by definition $W_{\mathbf{R}} = \mathbf{C}^* \times \text{Gal}(\mathbf{C}/\mathbf{R})$ with the group structure $(z_1, \tau_1)(z_2, \tau_2) = ((-1)^{\epsilon(\tau_1)\epsilon(\tau_2)} z_1\tau_1(z_2), \tau_1\tau_2)$ where $\epsilon(\tau) = 0$ if $\tau = 1$ and $\epsilon(\tau) = 1$ if $\tau \neq 1$. We can identify $W_{\mathbf{R}}$ with the subgroup of the group of nonzero quaternions $a + bi + cj + dk$ generated by $\{a + bi; (a, b) \in \mathbf{R}^2 - \{0\}\}$ and by j. We regard $W_{\mathbf{R}}$ as a Lie group with two connected components. Let $\Phi^{\mathbf{R}}(G_{\mathbf{C}}^*)$ be the set of all continuous homomorphisms $W_{\mathbf{R}} \to G_{\mathbf{C}}^*$ whose image consists of semisimple elements, up to conjugation by $G_{\mathbf{C}}^*$. Let $\mathcal{K}$ be a maximal compact subgroup of the Lie group $G_{\mathbf{R}}$. An "irreducible admissible" representation of $G_{\mathbf{R}}$ is by definition a $\mathbf{C}$-vector space E with an action of $\mathcal{K}$ and one of Lie $(G_{\mathbf{R}})$ such that any vector in E is contained in a finite dimensional $\mathcal{K}$-stable subspace of E; the two actions induce the same action on Lie $(\mathcal{K})$; the action of Lie $(G_{\mathbf{R}})$ is compatible with the $\mathcal{K}$-action on Lie $(G_{\mathbf{R}})$ and the $\mathcal{K}$-action on E. Moreover, this should be irreducible in the obvious sense.

According to Langlands [**La2**] there is a canonical finite to one surjective map

(a)   {irred. admissible representations of $G_{\mathbf{R}}$ up to isom.} $\to \Phi^{\mathbf{R}}(G_{\mathbf{C}}^*)$.

The fibres of (a) have been described by Knapp and Zuckerman.

We now give some examples of elements of $\Phi^{\mathbf{R}}(G_{\mathbf{C}}^*)$ in the case where $G_{\mathbf{C}}^* = GL(V)$ with V a finite dimensional $\mathbf{C}$-vector space. Assume that we are given a direct sum decomposition $V = \oplus_{p,q} V^{p,q}$ with $(p,q) \in \mathbf{C} \times \mathbf{C}, p - q \in \mathbf{Z}$; assume also that we are given a $\mathbf{C}$-linear isomorphism $\varpi : V \to V$ such that $\varpi^2 = 1$ and $\varpi(V^{p,q}) = V^{q,p}$ for all p, q. We define an action of $W_{\mathbf{R}}$ on V by specifying the action of $(z, 1)$ with $z \in \mathbf{C}^*$ and the action of $(1, \tau)$ with $\tau \in \mathrm{Gal}(\mathbf{C}/\mathbf{R}) - \{1\}$. If (p, q) runs only in $\mathbf{Z} \times \mathbf{Z}$ we have $(z, 1) \cdot x = z^p \bar{z}^q x$ for $z \in \mathbf{C}^*, x \in V^{p,q}$. The same formula holds in the general case: we interpret $z^p \bar{z}^q$ as $(z\bar{z})^{(p+q)/2}(\frac{z}{\sqrt{z\bar{z}}})^{p-q}$. (The strictly positive real number $z\bar{z}$ can be raised to any complex power.) We have $(1, \tau) \cdot x = \sqrt{(-1)^{p-q}} \varpi(x)$ for $x \in V^{p,q}$. This is an object of $\Phi^{\mathbf{R}}(G_{\mathbf{C}}^*)$.

12. Global fields. Let k be a field which is a finite algebraic extension of the field of rational functions in one variable over the finite field $\mathbf{F}_p$. Let A be the ring of adeles of k and let $k \to A$ be the canonical imbedding. Let $\bar{k}$ be an algebraic closure of k. Let l be a prime number $\neq p$.

The global Langlands conjecture [**La1**] predicts a connection between the set consisting of irreducible "cuspidal" representations of G_A with nonzero vectors fixed by G_k on the one hand and a certain set of homomorphisms of $\mathrm{Gal}(\bar{k}/k)$ into $G_{\bar{\mathbf{Q}}_l}^*$ which are irreducible in a suitable sense, on the other hand.

This conjecture has been proved in the case where $G_k = GL_n(k)$. (See [**Dr**] for $n = 2$ and [**Lf**] for any n.)

There is an analogous conjecture in which k is replaced by a finite extension of $\mathbf{Q}$ and also a geometric analogue of the conjecture in which the curve over a finite field represented by k is replaced by a smooth projective curve over $\mathbf{C}$. (See [**KW**].)

13. Cells in affine Weyl groups and unipotent classes. Define $\phi : \mathcal{H} \to \mathcal{A} \otimes J$ in terms of $\mathcal{R}$ as in §4. Let K be an algebraic closure of the field $\mathbf{C}(v)$ of rational functions with coefficients in $\mathbf{C}$ in an indeterminate v. Let c be a two-sided cell of $\widetilde{W}$. Let J_c be the corresponding direct summand of the ring J. We can find some simple module E of the $\mathbf{C}$-algebra $\mathbf{C} \otimes J_c$. It is necessarily of finite dimension over $\mathbf{C}$. We can regard $K \otimes_{\mathbf{C}} E$ as a $K \otimes_{\mathbf{Q}} J$-module in which the summands $K \otimes_{\mathbf{Q}} J_{c'}$ act as zero for $c' \neq c$. For y, y' in Y^+ we have $T_{a^y} T_{a^{y'}} = T_{a^{y'}} T_{a^y} = T_{a^{y+y'}}$ (see §3). Hence the operators $\phi(T_{a^y}) : K \otimes_{\mathbf{C}} E \to K \otimes_{\mathbf{C}} E$ (with $y \in Y^+$) commute. We can find $e \in K \otimes_{\mathbf{C}} E - \{0\}$ which is a simultaneous eigenvector for these operators. Thus we have $\phi(T_{a^y})e = b(y)e$ for all $y \in Y^+$ where $b(y) \in K^*$ satisfy $b(y)b(y') = b(y + y')$ for any y, y' in Y^+. There is a unique element $t \in K^* \otimes X$ such that, if $t = \sum_s k_s \otimes x_s$ with $k_s \in K^*, x_s \in X$ then $b(y) = \prod_s k_s^{\langle y, x \rangle}$ for any $y \in Y^+$. One can show [**L9**] that t is a very special element of $K^* \otimes T$: we can write uniquely $t = t't''$ where $t'' \in C^* \otimes X$ and $t' \in \{v^n; n \in \mathbf{Z}\} \otimes X \subset T_K^* \subset G_K^*$ is equal to $\vartheta \begin{pmatrix} v & 0 \\ 0 & v^{-1} \end{pmatrix}$ for some homomorphism of algebraic groups $\vartheta : SL_2(K) \to G_K^*$.

Let C be the conjugacy class in $G_{\mathbf{C}}^*$ such that $\vartheta \begin{pmatrix} 1 & 1 \\ 0 & 1 \end{pmatrix}$ is conjugate in G_K^* to some element of C. One can show [**L9**] that C is well defined by c (it is independent of the choice of E, e, ϑ) and that $c \mapsto C$ is a bijection

(a)   {two-sided cells of $\widetilde{W}$} $\xrightarrow{\sim}$ {unipotent conjugacy classes in $G_{\mathbf{C}}^*$}.

14. Special unipotent classes. We preserve the setup in §13. The intersection of W with a two-sided cell of $\widetilde{W}$ is said to be a two-sided cell of W if it is nonempty. Note that the two-sided cells of W form a partition of W. A unipotent conjugacy class in $G_{\mathbf{C}}^*$ is said to be *special* if it corresponds under the bijection §13(a) to a two-sided cell of $\widetilde{W}$ which has a nonempty intersection with W.

The special unipotent classes of $G_{\mathbf{C}}^*$ were introduced in a different (but equivalent) way in [**L1**] as the unipotent classes such that the corresponding irreducible representation of W (under the Springer correspondence) is in the class $\mathcal{S}_W$ defined in [**L1**]. This definition makes sense when $\mathbf{C}^*$ is replaced by any algebraically closed field A. For $\rho \in \mathcal{S}_W$ we denote by $C_{\rho,A}$ the corresponding special unipotent element of G_A and by $C_{\rho,A}^*$ the corresponding special unipotent element of G_A^*. (The sets $\mathcal{S}_W$ for $\mathcal{R}, \mathcal{R}^*$ coincide.) Let $\hat{C}_{\rho,A} = \bar{C}_{\rho,A} - \cup_{C'} \bar{C}'$ where $\bar{C}_{\rho,A}$ is the closure of $C_{\rho,A}$ and C' runs over the special unipotent classes contained in $\bar{C}_{\rho,A} - C_{\rho,A}$. It is known that the subsets $\hat{C}_{\rho,A}$ form a partition of the unipotent variety of G_A into locally closed subvarieties which are rational homology manifolds. We define similarly the subvarieties $\hat{C}_{\rho,A}^*$ of the unipotent variety of G_A^*. We have the following result (see [**L16**],[**L17**]):

For any $\rho \in \mathcal{S}_W$ there exists a polynomial P_ρ with integer coefficients such that for any q (a power of a prime number) we have

$$|\hat{C}_{\rho,\bar{\mathbf{F}}_q} \cap G_{\mathbf{F}_q}| = |\hat{C}_{\rho,\bar{\mathbf{F}}_q}^* \cap G_{\mathbf{F}_q}^*| = P_\rho(q).$$

15. Preparatory results. Let Y, X be two free abelian groups of finite rank and let $\langle , \rangle : Y \times X \to \mathbf{Z}$ be a perfect pairing. Let $\mathfrak{A} : Y \to Y$ be a homomorphism such that $\det(\mathfrak{A}) \neq 0$, that is, such that $|Y/\mathfrak{A}Y| < \infty$. We then have $|Y/\mathfrak{A}Y| = \pm \det(\mathfrak{A})$. Define a homomorphism $\mathfrak{A}' : X \to X$ by $\langle y, \mathfrak{A}'(x) \rangle = \langle \mathfrak{A}(y), x \rangle$ for all $y \in Y, x \in X$. Then $\det(\mathfrak{A}') = \det(\mathfrak{A})$ hence $|X/\mathfrak{A}'(X)| < \infty$. Now $\mathfrak{A}$ (resp. $\mathfrak{A}'$) induces endomorphisms of $\mathbf{Q} \otimes Y$ and of $\mathbf{Q}/\mathbf{Z} \otimes Y$ (resp. $\mathbf{Q} \otimes X$ and $\mathbf{Q}/\mathbf{Z} \otimes X$) denoted again by $\mathfrak{A}$ (resp. $\mathfrak{A}'$). Also, $\langle , \rangle$ induces a $\mathbf{Q}$-linear pairing $(\mathbf{Q} \otimes Y) \times (\mathbf{Q} \otimes X) \to \mathbf{Q}$ denoted again by $\langle , \rangle$. We define a pairing $(,) : Y/\mathfrak{A}(Y) \times X/\mathfrak{A}'(X) \to \mathbf{Q}/\mathbf{Z}$ by

$$(y, x) = \langle \mathfrak{A}^{-1}(y), x \rangle \mod \mathbf{Z} = \langle y, \mathfrak{A}'^{-1}(x) \rangle \mod \mathbf{Z}$$

where $y \in Y, \mathfrak{A}^{-1}(y) \in \mathbf{Q} \otimes Y, x \in X, \mathfrak{A}'^{-1}(x) \in \mathbf{Q} \otimes X$. Now $x \mapsto [y \mapsto (y, x)]$ is an isomorphism
  (a) $X/\mathfrak{A}'(X) \xrightarrow{\sim} \operatorname{Hom}(Y/\mathfrak{A}(Y), \mathbf{Q}/\mathbf{Z})$.
We define an isomorphism $Y/\mathfrak{A}(Y) \xrightarrow{\sim} (\mathbf{Q}/\mathbf{Z} \otimes Y)^{\mathfrak{A}+1}$ (fixed point set of $\mathfrak{A}+1$) by

$$y \mapsto \text{image of } \mathfrak{A}^{-1}(y) \text{ under } \mathbf{Q} \otimes Y \to \mathbf{Q}/\mathbf{Z} \otimes Y.$$

Similarly we have an isomorphism $X/\mathfrak{A}'(X) \xrightarrow{\sim} (\mathbf{Q}/\mathbf{Z} \otimes X)^{\mathfrak{A}'+1}$. Via the last two isomorphisms, (a) becomes an isomorphism

$$(\mathbf{Q}/\mathbf{Z} \otimes X)^{\mathfrak{A}'+1} \xrightarrow{\sim} \operatorname{Hom}((\mathbf{Q}/\mathbf{Z} \otimes Y)^{\mathfrak{A}+1}, \mathbf{Q}/\mathbf{Z}).$$

This is induced by $\xi \mapsto [\eta \mapsto \langle \mathfrak{A}(\eta), \xi \rangle \mod \mathbf{Z}$ where $\xi \in \mathbf{Q} \otimes X, \mathfrak{A}'(\xi) \in X$, $\eta \in \mathbf{Q} \otimes Y, \mathfrak{A}(\eta) \in Y$. Let p be a prime number and let $(\mathbf{Q}/\mathbf{Z})'$ be the subgroup of $\mathbf{Q}/\mathbf{Z}$ consisting of elements of order not divisible by p. Assume now that p does not divide $\det(\mathfrak{A})$. Then $(\mathbf{Q}/\mathbf{Z} \otimes Y)^{\mathfrak{A}+1} = ((\mathbf{Q}/\mathbf{Z})' \otimes Y)^{\mathfrak{A}+1}$ and we get an isomorphism

  (b) $(\mathbf{Q}/\mathbf{Z} \otimes X)^{\mathfrak{A}'+1} \xrightarrow{\sim} \operatorname{Hom}(((\mathbf{Q}/\mathbf{Z})' \otimes Y)^{\mathfrak{A}+1}, (\mathbf{Q}/\mathbf{Z})').$

Now let k be an algebraic closure of the finite field $\mathbf{F}_p$. Let $(k^* \otimes Y)^{\mathfrak{A}+1}$ be the fixed point set of the endomorphism $z \otimes y \mapsto z \otimes (\mathfrak{A}+1)y$ of $k^* \otimes Y$. We define a canonical isomorphism

(c) $\qquad \operatorname{Hom}(((\mathbf{Q}/\mathbf{Z})' \otimes Y)^{\mathfrak{A}+1}, (\mathbf{Q}/\mathbf{Z})') \xrightarrow{\sim} \operatorname{Hom}((k^* \otimes Y)^{\mathfrak{A}+1}, k^*)$

as follows. We choose an isomorphism $\zeta : (\mathbf{Q}/\mathbf{Z})' \xrightarrow{\sim} k^*$. Then $\zeta \otimes 1 : (\mathbf{Q}/\mathbf{Z})' \otimes Y) \xrightarrow{\sim} k^* \otimes Y$ restricts to an isomorphism $\zeta_1 : (\mathbf{Q}/\mathbf{Z})' \otimes Y)^{\mathfrak{A}+1} \xrightarrow{\sim} (k^* \otimes Y)^{\mathfrak{A}+1}$ and (c) carries a homomorphism $\phi : ((\mathbf{Q}/\mathbf{Z})' \otimes Y)^{\mathfrak{A}+1} \to (\mathbf{Q}/\mathbf{Z})'$ to $\zeta\phi\zeta_1^{-1}$. We must show that the map (c) is independent of the choice of ζ. Let $\kappa : (\mathbf{Q}/\mathbf{Z})' \xrightarrow{\sim} (\mathbf{Q}/\mathbf{Z})'$ be an isomorphism. Then $\kappa \otimes 1 : (\mathbf{Q}/\mathbf{Z})' \otimes Y) \xrightarrow{\sim} (\mathbf{Q}/\mathbf{Z})' \otimes Y$ restricts to an isomorphism $\kappa_1 : (\mathbf{Q}/\mathbf{Z})' \otimes Y)^{\mathfrak{A}+1} \xrightarrow{\sim} ((\mathbf{Q}/\mathbf{Z})' \otimes Y)^{\mathfrak{A}+1}$ and it is enough to show that for any homomorphism $\phi : ((\mathbf{Q}/\mathbf{Z})' \otimes Y)^{\mathfrak{A}+1} \to (\mathbf{Q}/\mathbf{Z})'$ we have $\kappa\phi\kappa_1^{-1} = \phi$. Since $(\mathbf{Q}/\mathbf{Z})'$ is an injective $\mathbf{Z}$-module, there exists a homomorphism $\tilde{\phi} : (\mathbf{Q}/\mathbf{Z})' \otimes Y \to (\mathbf{Q}/\mathbf{Z})'$ whose restriction to $((\mathbf{Q}/\mathbf{Z})' \otimes Y)^{\mathfrak{A}+1}$ is ϕ. It is enough to show that $\kappa\tilde{\phi}(\kappa \otimes 1)^{-1} = \tilde{\phi}$. By choosing a basis of Y we see that it is enough to show that for any homomorphism $\psi : (\mathbf{Q}/\mathbf{Z})' \to (\mathbf{Q}/\mathbf{Z})'$ we have $\kappa\psi\kappa^{-1} = \psi$. This follows from the fact that the ring of endomorphisms of the group $(\mathbf{Q}/\mathbf{Z})'$ is commutative (it is a product of rings of l-adic integers for various primes $l \neq p$).

Let μ be the group of roots of 1 in $\mathbf{C}$. We note that the isomorphism $\zeta' : \mathbf{Q}/\mathbf{Z} \to \mu$ given by $r \mapsto \exp(2\pi i r)$ induces an isomorphism $\zeta' \otimes 1 : \mathbf{Q}/\mathbf{Z} \otimes X \to \mu \otimes X$ and this restricts to an isomorphism $\zeta_1' : (\mathbf{Q}/\mathbf{Z} \otimes X)^{\mathfrak{A}'+1} \xrightarrow{\sim} (\mu \otimes X)^{\mathfrak{A}'+1}$ where $(\mu \otimes X)^{\mathfrak{A}'+1}$ is the fixed point set of the endomorphism $z \otimes x \mapsto z \otimes (\mathfrak{A}'+1)x$ of $\mu \otimes X$. Via ζ_1' and (c), the isomorphism (b) becomes a canonical isomorphism

(d) $\qquad\qquad (\mu \otimes X)^{\mathfrak{A}'+1} \xrightarrow{\sim} \operatorname{Hom}((k^* \otimes Y)^{\mathfrak{A}+1}, k^*).$

16. Groups over $\mathbf{F}_q$. We return to the setup in §8(a). Let k be an algebraic closure of the finite field $\mathbf{F}_p$. Let K be an algebraically closed field of characteristic 0 with a fixed imbedding of groups $\iota : k^* \to K^*$.

We have $T_k = k^* \otimes Y$, $T_{\mathbf{C}}^* = \mathbf{C}^* \otimes X$. Let $\mathbf{F}_q$ be the subfield of k such that $|\mathbf{F}_q| = q$. The ring homomorphism $k \to k, c \mapsto c^q$ induces (as in §6) a group homomorphism $F : G_k \to G_k$ (Frobenius map) whose fixed point set is the finite group $G_{\mathbf{F}_q}$. Following [**DL**] we consider for any $w \in W$ the set $\dot{X}_w = \{g \in G_k; g^{-1}F(g) \in \dot{w}U_k^+\}$, an algebraic variety over k. Let $T_k^w = \{t \in T_k; t^q = w^{-1}(t)\}$, a finite subgroup of T_k. The finite group $G_{\mathbf{F}_q} \times T_k^w$ acts on $\dot{X}_w$ by $(g_1, t) : g \mapsto g_1 g t^{-1}$. Let $\chi_w : G_{\mathbf{F}_q} \times T_k^w \to \mathbf{Z}$ be the class function which to any $(g_1, t) \in G_{\mathbf{F}_q} \times T_k^w$ associates the alternating sum of traces of $(g_1, t)^*$ on the l-adic cohomology with compact support of $\dot{X}_w$. (Here l is any prime number $\neq p$ but the resulting class function is known to be independent of l; see [**DL**].) For any irreducible $G_{\mathbf{F}_q}$-module ρ over K let $\mathcal{E}_\rho$ be the set of all pairs (w, θ) where $w \in W$, $\theta \in \operatorname{Hom}(T_k^w, K^*)$ and $\sum_{(g_1,t) \in G_{\mathbf{F}_q} \times T_k^w} \theta(t)\operatorname{tr}(g_1, \rho)\chi_w(g_1, t) \neq 0$. According to [**DL**] we have $\mathcal{E}_\rho \neq \emptyset$ for any ρ.

To any $(w, \theta) \in \mathcal{E}_\rho$ we associate an element $\hat{\theta} \in T_{\mathbf{C}}^{*w^{-1}} := \{t \in T_{\mathbf{C}}^*; t^q = w(t)\}$ as follows. Define $\mathfrak{A} : Y \to Y$ by $y \mapsto qw(y) - y$ and $\mathfrak{A}' : X \to X$ by $x \mapsto qw^{-1}(x) - x$. Then $T_k^w = (k^* \otimes Y)^{\mathfrak{A}+1}$ is a finite group of order prime to p. Hence $\theta : T_k^w \to K^*$ has values in the group of roots of 1 of order prime to p in K^* which can be identified with k^* via ι. Thus θ can be viewed as an element of $\operatorname{Hom}((k^* \otimes Y)^{\mathfrak{A}+1}, k^*)$ so that it corresponds under §15(d) to an element $\hat{\theta}$ of $(\mu \otimes X)^{\mathfrak{A}'+1}$. This last group is a subgroup of $(\mathbf{C}^* \otimes X)^{\mathfrak{A}'+1}$ (the fixed point set of the endomorphism $z \otimes x \mapsto z \otimes (\mathfrak{A}' + 1)x$ of $\mathbf{C}^* \otimes X$) which is the same as $T_{\mathbf{C}}^{*w^{-1}}$. From the results in [**DL**] we see that the W-orbit of $\hat{\theta}$ in $T_{\mathbf{C}}^*$ depends only on ρ and not on the choice of (w, θ) in $\mathcal{E}_\rho$. We thus have a well defined map

$$\{\text{irred. } G_{\mathbf{F}_q}\text{-modules over } K \text{ up to isom.}\} \rightarrow$$

(a) $\{\text{semisimple conjugacy classes in } G_{\mathbf{C}}^* \text{ stable under } g \mapsto g^q\};$

it is given by $\rho \mapsto G_{\mathbf{C}}^*$-conjugacy class of $\hat{\theta}$ (as above). This map appears in [**DL**] in a somewhat different form. In [**DL**] $G_{\mathbf{C}}^*$ is replaced by G_k^*. But the method of [**DL**] is less canonical: it is based on two choices (see [**DL**, (5.0.1), (5,0.2)]) while the present method is based on only one choice, that of ι; the choice of ι can be also eliminated as we will see below).

An element $g \in G_{\mathbf{C}}^*$ is said to be special if the unipotent part g_u of g is a special unipotent element (see §14) of the connected centralizer of the semisimple part g_s of g (a reductive connected group). A conjugacy class in $G_{\mathbf{C}}^*$ is said to be special if some/any element of it is special. The map (a) can be refined to a canonical map

$$\{\text{irred. } G_{\mathbf{F}_q}\text{-modules over } K \text{ up to isom.}\} \rightarrow$$

(b) $\{\text{special conjugacy classes in } G_{\mathbf{C}}^* \text{ stable under } g \mapsto g^q\}.$

(See [**L6**].) Note that (a) is the composition of (b) with the map which to the $G_{\mathbf{C}}^*$-conjugacy class of a special element g associates the $G_{\mathbf{C}}^*$-conjugacy class of g_s. The map (b) is surjective and its fibres are described explicitly in [**L6**], [**L10**].

Note that the maps (a),(b) depend on the choice of the imbedding $\iota : k^* \to K^*$. However if we take K to be an algebraic closure of the quotient field of the ring of Witt vectors of k then there is a canonical choice of ι and the maps (a),(b) become completely canonical.

17. Character sheaves. Define $B_{\mathbf{C}}^+, U_{\mathbf{C}}^+, \dot{w}$ in terms of $\mathbf{C}, \mathcal{R}$ as in §6. Let $\mathcal{E}$ be a $\mathbf{C}$-local system of rank 1 on $T_{\mathbf{C}}$ with finite monodromy. The monodromy of $\mathcal{E}$ is a homomorphism $f : Y \to \mathbf{C}^*$ with finite image which can be viewed as an element of finite order $\chi_{\mathcal{E}} \in \mathbf{C}^* \otimes X$ given by $\chi_{\mathcal{E}} = \sum_j f(y_j) \otimes x_j$ where (y_j) is a basis of Y and (x_j) is the dual basis of X. Moreover $\mathcal{E} \mapsto \chi_{\mathcal{E}}$ is a bijection

$$\{\mathbf{C} - \text{local systems of rank 1 on } T_{\mathbf{C}} \text{ with finite monodromy up to isom.}\} \xrightarrow{\sim}$$

(a)
$$\{\text{elements of finite order of } T_{\mathbf{C}}^*\}.$$

Let $c : G_{\mathbf{C}} \to G_{\mathbf{C}}/U_{\mathbf{C}}^+$ be the obvious map. An irreducible intersection cohomology complex K on $G_{\mathbf{C}}$ is said to be a *character sheaf* on $G_{\mathbf{C}}$ if it is $G_{\mathbf{C}}$-equivariant and if for any $w \in W$ and any $j \in \mathbf{Z}$ the j-th cohomology sheaf of $c_! K$ restricted to $B_{\mathbf{C}}^+ \dot{w} B_{\mathbf{C}}^+ / U_{\mathbf{C}}^+$ is a local system $\mathcal{L}_{K,w,j}$ with finite monodromy. We can find w, j as above and a local system $\mathcal{E}$ of rank 1 on $T_{\mathbf{C}}$ with finite monodromy such that $\mathcal{E}$ is a direct summand of the inverse image of $\mathcal{L}_{K,w,j}$ under the map $T_{\mathbf{C}} \to B_{\mathbf{C}}^+ \dot{w} B_{\mathbf{C}}^+ / U_{\mathbf{C}}^+$, $t \mapsto \dot{w} t U_{\mathbf{C}}^+$. One can show that the corresponding element $\chi_{\mathcal{E}} \in T_{\mathbf{C}}^*$ is well defined (up to the action of W) that is, it does not depend on the choice of $w, j, \mathcal{E}$. Thus we have a well defined map

$$\{\text{character sheaves on } G_{\mathbf{C}} \text{ up to isom.}\} \rightarrow$$

(a) $\{\text{conjugacy classes of elements of finite order in } G_{\mathbf{C}}^*\};$

it is given by $K \mapsto G_{\mathbf{C}}^*$-conjugacy class of $\chi_{\mathcal{E}}$ (as above). The map (a) can be refined to a canonical map

{character sheaves on $G_{\mathbf{C}}$ up to isom.} $\rightarrow$

(b)

{special conjugacy classes in $G_{\mathbf{C}}^*$ of elements g with g_s of finite order}.

(See [**L8**].) Note that (a) is the composition of (b) with the map which to the $G_{\mathbf{C}}^*$-conjugacy class of a special element g associates the $G_{\mathbf{C}}^*$-conjugacy class of g_s. The map (b) is surjective and its fibres are described explicitly in [**L8**].

18. Vogan duality. To $G_{\mathbf{C}}$ we associate a finite collection of polynomials: those recording the restrictions of the cohomology sheaves of the simple perverse sheaves on $\mathcal{B}_{\mathbf{C}}$, equivariant under the conjugation action of the centralizers of the various involutions of $G_{\mathbf{C}}$, to the various orbits of these centralizers. (These polynomials were studied in [**LV**].) We consider also the analogous collection of polynomials associated to $G_{\mathbf{C}}^*$. *Vogan duality* [**V1**] states that these two collections of polynomials are related to each other by a simple algebraic rule: essentially the inversion of a matrix. This is a generalization of the inversion formula in [**KL1**].

19. Multiplicities in standard modules. Let K be either as in §10 or $K = \mathbf{R}$. The Grothendieck group whose basis consists of admissible irreducible representations of G_K has another natural basis consisting of "standard representations" in natural bijection with the first basis. The representations in the second basis are easier to describe and understand. Hence the (upper triangular) matrix $\mathcal{M}$ expressing the second basis in terms of the first basis ("multiplicity matrix") is of interest. In every known case the entries of $\mathcal{M}$ can be expressed in terms of intersection cohomology coming from the geometry of $G_{\mathbf{C}}^*$. For the case where $K = \mathbf{R}$ we refer the reader to [**ABV**]; in this case some of the polynomials in §18 (attached to $G_{\mathbf{C}}^*$) evaluated at 1 appear as entries of $\mathcal{M}$. In the remainder of this subsection we assume that K is as in §10. For simplicity we assume that $\mathcal{R}$ is of adjoint type. We use the notation in §10.

We fix an element $w^0 \in W_K$ such that $\omega(w^0) = 1$. Let $\tilde{\Phi}^K$ be the set of all triples (ρ, u, E) (up to $G_{\mathbf{C}}^*$-conjugacy) where $\rho : W_K \rightarrow G_{\mathbf{C}}^*$ is an admissible homomorphism, u is a unipotent element of $G_{\mathbf{C}}^*$ such that $\rho(w)u\rho(w)^{-1} = u^{q^{\omega(w)}}$ for all $w \in W_K$ and E is an irreducible representation of the group of connected components of $G_{\rho,u}^* := \{g \in G_{\mathbf{C}}^*; \rho(w)g\rho(w)^{-1} = g \text{ for all } w \in W_K, gu = ug\}$ on which the image of the centre of $G_{\mathbf{C}}^*$ acts trivially.

By §10 it is expected that $\tilde{\Phi}^K$ is an index set for both the rows and the columns of $\mathcal{M}$. We shall describe a matrix $\mathcal{M}'$ indexed by $\tilde{\Phi}^K$ which is defined in terms of geometry of $G_{\mathbf{C}}^*$.

Let Ψ be the set of homomorphisms $\psi : \mathcal{I} \rightarrow G_{\mathbf{C}}^*$ such that $\psi(\mathcal{I})$ is finite and such that $\Gamma_\psi := \{g \in G_{\mathbf{C}}^*; g\psi(w)g^{-1} = \psi(w^0ww^{0-1})\}$ for all $w \in \mathcal{I}\}$ is non-empty. Let $\bar{\Psi}$ be the set of $G_{\mathbf{C}}^*$-orbits (by conjugacy) on Ψ.

Define $\kappa : \tilde{\Phi}^K \rightarrow \bar{\Psi}$ by $(\rho, u, E) \mapsto \rho|_{\mathcal{I}}$. The entries $m_{(\rho,u,E),(\rho',u',E')}$ of $\mathcal{M}'$ can be described as follows. If $(\rho, u, E), (\rho', u', E')$ are not in the same fibre of κ then $m_{(\rho,u,E),(\rho',u',E')} = 0$.

We now fix $\psi \in \Psi$. Let $G_\psi^* = \{g \in G_{\mathbf{C}}^*; g\psi(w)g^{-1} = \psi(w) \text{ for all } w \in \mathcal{I}\}$. This is the centralizer of a finite subgroup of $G_{\mathbf{C}}^*$ hence is a (possibly disconnected) reductive subgroup of $G_{\mathbf{C}}^*$. Let $G_\psi'^*$ be the normalizer of $\psi(\mathcal{I})$ in $G_{\mathbf{C}}^*$. Note that

G_ψ^* is a normal subgroup of finite index of $G_\psi'^*$ and that Γ_ψ is a single G_ψ^*-coset in $G_\psi'^*$. The fibre of κ at ψ can be identified with the set of all triples (s, N, E) (up to G_ψ^*-conjugacy) where $s \in G_\psi'^*$ is a semisimple element such that $s \in \Gamma_\psi$, N is an element of $X_s := \{N_1 \in \mathrm{Lie}\,(G_\psi^*); \mathrm{Ad}(s)N_1 = qN\}$ (necessarily nilpotent) and E is an irreducible representation of the group of connected components of $G_{\psi,s,N}^* := \{g \in G_\psi^*; gs = sg, \mathrm{Ad}(g)N = N\}$ on which the image of the centre of $G_\mathbf{C}^*$ acts trivially. (The identification is given by $(\rho, u, E) \mapsto (\rho(w^0), \log(u), E)$ where $\rho|_\mathcal{I} = \psi$.) We now consider two elements $(s, N, E), (s', N', E')$ in $\kappa^{-1}(\psi)$. If s, s' are not in the same G_ψ^*-orbit then $m_{(s,N,E),(s',N',E')} = 0$. Now assume that $(s, N, E), (s', N', E')$ are such that s, s' are in the same G_ψ^*-orbit. We can assume that $s = s'$. Let $G_{\psi,s}^* = \{g \in G_\psi^*; gs = sg\}$. This is an algebraic group which acts on X_s by conjugation with finitely many orbits. Let C be the $G_{\psi,s}^*$-orbit of N and let C' be the $G_{\psi,s}^*$-orbit of N'. Note that E (resp. E') determines a local system $\underline{E}$ (resp. $\underline{E}'$) on C (resp. C') which is $G_{\psi,s}^*$-equivariant and is irreducible as a $G_{\psi,s}^*$-equivariant local system. If C is not contained in the closure of C' then $m_{(s,N,E),(s',N',E')} = 0$. Now assume that C is contained in the closure of C'. Let $\underline{E}'^\sharp$ be the intersection cohomology complex on the closure of C' determined by $\underline{E}'$. For every integer j we consider the j-th cohomology sheaf of $\underline{E}'^\sharp$ restricted to C; this is a $G_{\psi,s}^*$-equivariant local system on C in which $\underline{E}$ appears say n_j times. We set $m_{(s,N,E),(s',N',E')} = \sum_j (-1)^j n_j$.

We see that the intersection cohomology complexes on X_s considered above are essentially of the type considered in [**L18**]. We also see that the objects in $\kappa^{-1}(\psi)$ behave like the parameters for the unipotent representations for a collection of not necessarily split and not necessarily connected p-adic groups smaller than G_K.

It is known that $\mathcal{M}$, $\mathcal{M}'$ coincide as far as the entries with both indices contained in $\kappa^{-1}(1)$ are concerned; these correspond to unipotent representations. (This was conjectured by the author and independently, in a special case connected with GL_n, in [**Ze**]; the proof was given by Ginzburg in a special case connected with the affine Hecke algebra $\mathcal{H}$ and by the author in the general case.) We expect that $\mathcal{M} = \mathcal{M}'$. (See also [**V2**].)

20. Multiplicities in tensor products.

Assume that $\mathcal{R}$ is simply connected. For $\lambda, \lambda', \lambda''$ in X^+ let $m_{\lambda,\lambda',\lambda''}$ be the multiplicity of $\Lambda_{\lambda''}$ in the tensor product $\Lambda_\lambda \otimes \Lambda_{\lambda'}$ (an object of $\mathcal{C}$, see §5). On the other hand let $l^* : \widetilde{W}^* \to \mathbf{N}, T_w, c_w, p_{w,z}$ be defined like $l : \widetilde{W} \to \mathbf{N}, T_w, c_w, p_{w,z}$ in §3,§4 but with respect to $\mathcal{R}^*$ instead of $\mathcal{R}$. We have $\widetilde{W}^* = \{wa^x; w \in W, x \in X\}$.

For any $\lambda \in X^+$ there is a unique element M_λ in the double coset $Wa^\lambda W$ on which $l^* : Wa^\lambda W \to \mathbf{N}$ achieves its maximum value. For any λ, λ' in X^+ we have in $\mathcal{H}^*$:

$$(P^{-1}c_{M_\lambda})(P^{-1}c_{M_{\lambda'}}) = \sum_{\lambda'' \in X^+} \tilde{m}_{\lambda,\lambda',\lambda''}(P^{-1}c_{M_{\lambda''}})$$

where $P \in \mathcal{A}$ is given by $c_{M_0}c_{M_0} = Pc_{M_0}$ and $\tilde{m}_{\lambda,\lambda',\lambda''} \in \mathcal{A}$. In [**L4**] it is shown that

(a) $$m_{\lambda,\lambda',\lambda''} = \tilde{m}_{\lambda,\lambda',\lambda''};$$

in particular,

(b) $\tilde{m}_{\lambda,\lambda',\lambda''}$ is a constant.

(In the special case where $G_{\mathbf{C}}$ is a general linear group, this was proved earlier in [**L3**] using the theory of Hall-Littlewood functions.) In [**L4**] it is also shown that for λ, λ' in X^+,

(c) $\dim \Lambda_\lambda^{\lambda'}$ (that is, the multiplicity of the weight λ' in the $\underline{U}$-module Λ_λ) is equal to $p_{M_{l'},M_\lambda}(1)$.

Note that at the time when [**L4**] was written it was known that the product of elements of the form $P^{-1}c_{M_\lambda}$ corresponds to the convolution of $G^*_{\mathbf{C}[[\epsilon]]}$-equivariant simple perverse sheaves on the "affine Grassmannian" $G^*_{\mathbf{C}((\epsilon))}/G^*_{\mathbf{C}[[\epsilon]]}$ so that (b) is equivalent to the statement that such a convolution is a direct sum of simple perverse sheaves of the same type (without shift). Thus it was clear that the category whose objects are finite direct sums of $G^*_{\mathbf{C}[[\epsilon]]}$-equivariant simple perverse sheaves on the "affine Grassmannian" has a natural monoidal structure given by convolution; moreover (b) showed that this monoidal category was very similar to that of representations of $G_{\mathbf{C}}$ (identical at the level of Grothendieck groups). But it was not clear how to construct the commutativity isomorphism for the convolution product. This was accomplished around 1989 by V.Ginzburg [**Gi**] and later in a more elegant form by V.Drinfeld. As a result, $G_{\mathbf{C}}$ can be reconstructed from the tensor category of $G^*_{\mathbf{C}[[\epsilon]]}$-equivariant perverse sheaves on $G^*_{\mathbf{C}((\epsilon))}/G^*_{\mathbf{C}[[\epsilon]]}$, see [**Gi**].

21. Canonical bases. Define f as in §5 in terms of $\mathcal{R}$. Let $A = \mathbf{R}[[\epsilon]]$ where ϵ is an indeterminate. Define U_A^{*+} in terms of $\mathcal{R}^*$ in the same way as U_A^+ was defined in §6 in terms of $\mathcal{R}$. By [**L14**,§10] there is a canonical bijection between the *canonical basis* of f (defined as in [**L11**], [**L13**]) and a certain collection of subsets of U_A^{*+} which form a partition of the totally positive part of U_A^{*+}. The bijection is not defined directly; instead it is shown that both sets are parametrized by the same combinatorial objects.

22. Modular representations. Let k be an algebraic closure of the finite field $\mathbf{F}_p$. Assume that $\mathcal{R}$ is simply connected. For $\lambda \in X^+$ the G_k-module $\Lambda_{\lambda,k}$ (see §7) is not necessarily irreducible but has a unique irreducible quotient $\Lambda_{\lambda,k}^\sharp$. For λ, λ' in X^+ let $m_{\lambda,\lambda'}$ be the number of times that $\Lambda_{\lambda',k}^\sharp$ appears in a composition series of the G_k-module $\Lambda_{\lambda,k}$. Note that the knowledge of the multiplicities $m_{\lambda,\lambda'}$ implies the knowledge of the character of the G_k-modules $\Lambda_{\lambda,k}^\sharp$ since the character of $\Lambda_{\lambda,k}$ is known by Weyl's character formula. Conjecturally (see [**L2**]) if p is sufficiently large with respect to $\mathcal{R}$, the multiplicities $m_{\lambda,\lambda'}$ can be expressed in terms of polynomials $p_{w,z}$ (as in §4) where w, z are elements in $\widetilde{W}^*$ which have maximal length in their left W-coset; they can be also expressed in terms of certain intersection cohomology spaces associated with the geometry of $G^*_{\mathbf{C}((\epsilon))}$, where ϵ is an indeterminate. A proof of the conjecture (without an explicit bound for p) is provided by combining [**AJS**], [**KT**], [**KL3**] or alternatively by combining [**AJS**], [**ABG**].

References

[ABV] J. Adams, D. Barbasch and D. A. Vogan, Jr., *The Langlands classification of irreducible characters of real reductive groups*, Progress in Math (1992), Birkhauser.

[AJS] H. H. Andersen, W. Soergel and J. C. Jantzen, *Representations of quantum groups at a p-th root of unity and of semisimple groups in characteristic p: independence of p*, Astérisque **220** (1994).

[ABG]   S. Arkhipov, R. Bezrukavnikov and V. Ginzburg, *Quantum groups, the loop grassmannian and the Springer resolution*, Jour.Amer.Math.Soc. **17** (2004), 595-678.

[C]   C. Chevalley, *Certains schémas de groupes semi-simples*, Sém. Bourbaki 1960/61, Soc. Math. France, 1995.

[DL]   P. Deligne and G. Lusztig, *Representations of reductive groups over finite fields*, Ann. Math. **103** (1976), 103-161.

[Dr]   V. Drinfeld, *Two dimensional l-adic representations of the fundamental group of a curve over a finite field and automorphic forms on GL(2)*, Amer.Jour.Math. **105** (1983), 85-114.

[Gi]   V. Ginzburg, *Perverse sheaves on a loop group and Langlands duality*, math.AG/9511007.

[HT]   M. Harris and R. Taylor, *On the geometry and cohomology of some simple Shimura varieties*, Ann.Math.Studies, vol. 151, Princeton Univ.Press, 2001.

[IM]   N. Iwahori and H. Matsumoto, *On some Bruhat decomposition and the structure of the Hecke ring of p-adic Chevalley groups*, Publ.Math.IHES **25** (1965), 5-48.

[KW]   A. Kapustin and E. Witten, *Electric-magnetic duality and the geometric Langlands program*, arXiv:hep-th/0604151.

[KT]   M. Kashiwara and T. Tanisaki, *The Kazhdan-Lusztig conjecture for affine Lie algebras with negative level*, Duke Math.J. **77** (1995), 21-62.

[KL1]   D. Kazhdan and G. Lusztig, *Representations of Coxeter groups and Hecke algebras*, Invent.Math. **53** (1979), 165-184.

[KL2]   D. Kazhdan and G. Lusztig, *Proof of the Deligne-Langlands conjecture for Hecke algebras*, Invent.Math. **87** (1987), 153-215.

[KL3]   D. Kazhdan and G. Lusztig, *Tensor structures arising from affine Lie algebras, IV*, Jour. Amer. Math. Soc. **7** (1994), 383-453.

[Ki]   J.-L. Kim, *Supercuspidal representations: an exhaustion theorem*, Jour.Amer.Math.Soc. **20** (2007), 273-320.

[Ko]   B. Kostant, *Groups over **Z***, Algebraic Groups and Their Discontinuous Subgroups, Proc. Symp. Pure Math., vol. 8 publ. Amer. Math. Soc., 1966, pp. 90-98.

[Lf]   L. Lafforgue, *Chtoukas de Drinfeld et correspondance de Langlands*, Invent.Math. **147** (2002), 1-242.

[La1]   R. P. Langlands, *Problems in the theory of automorphic forms*, Lectures in Modern Analysis and Applications, Lecture Notes in Math, vol. 170, Springer Verlag, 1970, pp. 18-61.

[La2]   R. P. Langlands, *On the classification of irreducible representations of real algebraic groups*, Representation theory and harmonic analysis on semisimple Lie groups, Math. Surveys Monogr., vol. 31 Amer. Math. Soc., Providence, RI, 1989, pp. 101-170.

[L1]   G. Lusztig, *A class of irreducible representations of a Weyl group*, Proc. Kon. Nederl. Akad. (A) **82** (1979), 323-335.

[L2]   G. Lusztig, *Some problems in the representation theory of finite Chevalley groups*, Proc. Symp. Pure Math. Amer. Math. Soc. **37** (1980), 313-317.

[L3]   G. Lusztig, *Green polynomials and singularities of unipotent classes*, Adv. Math. **42** (1981), 169-178.

[L4]   G. Lusztig, *Singularities, character formulas and a q-analog of weight multiplicities*, Astérisque **101-102** (1983), 208-229.

[L5]   G. Lusztig, *Some examples of square integrable representations of semisimple p-adic groups*, Trans.Amer.Math.Soc. **227** (1983), 623-653.

[L6]   G. Lusztig, *Characters of reductive groups over a finite field*, Ann.Math.Studies, vol. 107, Princeton Univ.Press, 1984.

[L7]   G. Lusztig, *Equivariant K-theory and representations of Hecke algebras*, Proc. Amer. Math. Soc. **94** (1985), 337-342.

[L8]   G. Lusztig, *Character sheaves, V*, Adv.Math. **61** (1986), 103-155.

[L9]   G. Lusztig, *Cells in affine Weyl groups, II*, J.Alg. **109** (1987), 536-548; IV, J. Fac. Sci. Tokyo U.(IA) **36** (1989), 297-328.

[L10]   G. Lusztig, *On representations of reductive groups with disconnected center*, Astérisque **168** (1988), 157-166.

[L11]   G. Lusztig, *Canonical bases arising from quantized enveloping algebras*, Jour. Amer. Math. Soc. **3** (1990), 447-498.

[L12]   G. Lusztig, *Affine quivers and canonical bases*, Publ.Math.IHES **76** (1992), 111-163.

[L13]   G. Lusztig, *Introduction to quantum groups*, Progress in Math., vol. 110, Birkhauser, 1993.

[L14]   G. Lusztig, *Total positivity in reductive groups*, Lie theory and geometry, Progr.in Math., vol. 123, Birkhäuser Boston, 1994, pp. 531-568.

[L15]   G. Lusztig, *Classification of unipotent representations of simple p-adic groups*, Int. Math. Res. Notices (1995), 517-589; II, Represent.Th. **6** (2002), 243-289.

[L16]   G. Lusztig, *Notes on unipotent classes*, Asian J.Math. **1** (1997), 194-207.

[L17]   G. Lusztig, *Unipotent elements in small characteristic*, Transform. Groups **10** (2005), 449-487; II,arXiv:RT/0612320.

[L18]   G. Lusztig, *Graded Lie algebras and intersection cohomology, arXiv:RT/0604535*.

[L19]   G. Lusztig, *Study of a $\mathbf{Z}$-form of the coordinate ring of a reductive group*, arxiv:0709.1286.

[LV]   G. Lusztig and D.A.Vogan, Jr., *Singularities of closures of K-orbits on a flag manifold*, Invent.Math. **71** (1983), 365-379.

[MK]   J. McKay, *Graphs, singularities and finite groups*, Proc. Symp. Pure Math. Amer. Math. Soc. **37** (1980), 183-186.

[SL]   P. Slodowy, *Simple singularities and simple algebraic groups*, Lecture Notes in Math., vol. 815, Springer Verlag, 1980.

[V1]   D. A. Vogan, Jr., *Irreducible characters of semisimple Lie groups, IV: character multiplicity duality*, Duke Math.J. **4** (1982), 943-1073.

[V2]   D. A. Vogan, Jr., *The local Langlands conjecture*, Representation theory of groups and algebras, Contemp. Math., vol. 145 Amer. Math. Soc., Providence, RI, 1993, pp. 305-379.

[Xi]   N. Xi, *Representations of affine Hecke algebras and based rings of affine Weyl groups*, Jour.Amer.Math.Soc. **20** (2007), 211-217.

[Ze]   A. Zelevinsky, *A p-adic analogue of the Kazhdan-Lusztig conjecture*, Funkt.Anal.Pril. **15** (1981), 9-21.

DEPARTMENT OF MATHEMATICS, M.I.T., CAMBRIDGE, MA 02139
E-mail address: gyuri@math.mit.edu

Contemporary Mathematics
Volume **478**, 2009

Some New Highest Weight Categories

Brian J. Parshall and Leonard L. Scott

ABSTRACT. Let G be a semisimple, simply connected algebraic group defined over an algebraically closed field k of positive characteristic p. Assume that p is larger than the Coxeter number of G and that the Lusztig character formula holds for regular restricted dominant weights. In this paper, we introduce two new highest weight categories $\mathcal{C}_{\mathrm{even}}^{\mathrm{reg}}$ and $\mathcal{C}_{\mathrm{odd}}^{\mathrm{reg}}$, both full subcategories of the category $\mathcal{C}$ of rational G-modules. The standard and costandard modules for these categories arise from "reduction mod p" from the quantum enveloping algebra associated to G.

1. Introduction

Let U_ζ be a quantum enveloping algebra (Lusztig form) associated to a finite root system Φ and a pth root of unity ζ. Here, p is a positive prime integer. For each dominant weight λ, let $L_\zeta(\lambda)$ be the finite dimensional, irreducible (type 1 and integrable) representation of U_ζ of high weight λ. Let k be an algebraically closed field of characteristic p, and let G be the semisimple, simply connected algebraic group over k having root system Φ. By a process of "reduction mod p," the module $L_\zeta(\lambda)$ defines rational modules for G. The rational module obtained from a minimal (resp., maximal) lattice (with respect to an appropriate integral form of U_ζ over a PID $\mathcal{O}$) is denoted $\Delta^{\mathrm{red}}(\lambda)$ (resp., $\nabla_{\mathrm{red}}(\lambda)$). These modules were (with a different notation) first defined by Lusztig [**10**]. The homological properties of the modules $\Delta^{\mathrm{red}}(\lambda)$ and $\nabla_{\mathrm{red}}(\lambda)$ were investigated in [**6**] (see also [**12**]). In that work, the modules played an important role in the authors' efforts (with E. Cline) to attack growth issues on 1-cohomology arising from Guralnick's universal bound conjecture [**7**] for finite groups.

Let $\mathcal{C}$ be the category of rational G-modules; it is a highest weight category in a precise sense [**3**]. Assume that $p > h$ (the Coxeter number of G). For each regular dominant weight λ, write $\lambda = w \cdot \lambda^-$, where λ^- belongs to the anti-dominant p-alcove (the alcove containing -2ρ, ρ being the half-sum of the positive roots) and where $w \in W_p$, the affine p-Weyl group. Define the length $\ell(\lambda)$ of λ to be the length $\ell(w)$ of w in the Coxeter group W_p. Then let $\mathcal{C}_{\mathrm{even}}^{\mathrm{reg}}$ (resp., $\mathcal{C}_{\mathrm{odd}}^{\mathrm{reg}}$) be the full subcategory of $\mathcal{C}$ generated by the irreducible rational G-modules $L(\lambda)$ of high weight λ for λ regular and $\ell(\lambda)$ an even integer (resp., odd integer). The main result, proved in Theorem 3.2, shows that, if the Lusztig character formula holds

Research supported in part by the National Science Foundation.

for all restricted, regular dominant weights, then $\mathcal{C}^{\mathrm{reg}}_{\mathrm{even}}$ and $\mathcal{C}^{\mathrm{reg}}_{\mathrm{odd}}$ are highest weight categories with standard objects $\Delta^{\mathrm{red}}(\lambda)$ and costandard objects $\nabla_{\mathrm{red}}(\lambda)$.

Given a dominant weight λ, write $\lambda = \lambda_0 + p\lambda_1$, where λ_0 is restricted dominant and λ_1 is dominant. Then $\Delta^{\mathrm{red}}(\lambda)$ has an elegant alternative description, due to Z. Lin [9]. Explicitly, $\Delta^{\mathrm{red}}(\lambda) \cong \Delta^{\mathrm{red}}(\lambda_0) \otimes \Delta(\lambda_1)^{(1)}$, where $\Delta(\lambda_1)^{(1)}$ denotes the twist of $\Delta(\lambda_1)$ through the Frobenius map on G. Also, define $\Delta^{\mathrm{red}}(\lambda)' = L(\lambda_0) \otimes \Delta(\lambda_1)^{(1)}$. When p is sufficiently large so that the Lusztig character formula holds for the irreducible modules $L(\lambda_0)$, we have $\Delta^{\mathrm{red}}(\lambda) \cong \Delta^{\mathrm{red}}(\lambda)'$. Generally, one naturally asks if the standard modules $\Delta(\lambda)$ have Δ^{red}-filtrations, i. e., filtrations by G-submodules with sections of the form $\Delta^{\mathrm{red}}(\mu)$. The same question, but using the modules $\Delta^{\mathrm{red}}(\lambda)'$, goes back at least to J. Jantzen [8, §§3.8–12, 5.9], who answers it positively subject to restrictions on λ; see [1] for some history of this problem.[1] This question remains open, but it is conjectured in [6] to have a positive answer for all dominant weights λ, so long as $p > h$, and [6, Thm. 6.9] does present a partial result in this direction. We revisit this result in Theorem 4.3 below, providing another proof based on the methods of this paper; see Remark 4.4. Finally, Theorem 4.5 uses the main result again, giving a new description of the modules $\Delta^{\mathrm{red}}(\lambda)$ and $\nabla_{\mathrm{red}}(\lambda)$ for λ regular.

2. Review of some preliminary notation and results

Let G be a simply connected, semisimple algebraic group over k, defined and split over $\mathbb{F}_p$. Fix a maximal split torus T and a Borel subgroup $B \supset T$. Let $X = X(T)$ (resp., $X^\vee = X^\vee(T)$) be the character (resp., cocharacter) group on T. There is a natural pairing $X \times X^\vee \to \mathbb{Z}$, $\gamma, \sigma \mapsto (\gamma, \sigma) \in \mathbb{Z}$.

We let $\Phi \subset X$ be the set of roots of T in the Lie algebra $\mathfrak{g}$ of G, ordered so that the set Φ^+ of positive roots coincides with the Borel subgroup $B^+ \supset T$ opposite to $B = B^-$. Let $\Pi = \{\alpha_1, \ldots, \alpha_n\}$ be the simple roots in Φ^+, listed as in [2, Appendix]. For convenience, we will assume that Φ is irreducible (so that G is a simple algebraic group).

Let $X^+ = X^+(T)$ be the set of dominant weights on T; thus, X is a free $\mathbb{Z}$-module with basis the fundamental dominant weights $\varpi_1, \ldots, \varpi_n$ which are defined by $(\varpi_i, \alpha_j^\vee) = \delta_{i,j}$. Here, for a root $\alpha \in \Phi$, $\alpha^\vee \in X^\vee$ is the corresponding coroot. Let $\rho = \varpi_1 + \cdots + \varpi_n$ be the Weyl weight, and let α_0 be the maximal short root in Φ^+. Let $h = (\rho, \alpha_0^\vee) + 1$ be the Coxeter number of Φ.

For a positive integer r, X_r^+ is the set of dominant weights λ which satisfy $(\lambda, \alpha^\vee) < p^r$ for all $\alpha \in \Pi$. The elements of X_1^+ are the restricted dominant weights. Let X_{reg} be the set of regular weights, i. e., $\lambda \in X_{\mathrm{reg}}$ if and only if $(\lambda + \rho, \alpha^\vee) \not\equiv 0 \bmod p$ for all roots α. Then $X_{\mathrm{reg}}^+ := X^+ \cap X_{\mathrm{reg}}$ is the set of regular dominant weights. We have $X_{\mathrm{reg}}^+ \neq \infty$ if and only if $p \geq h$.

The set X^+ is a poset, putting $\lambda \leq \mu$ if and only if $\mu - \lambda \in \mathbb{N}\Phi^+$.

Let $\mathcal{C}$ be the category of rational G-modules. For $\lambda \in X^+$, let $L(\lambda)$ (resp., $\Delta(\lambda)$, $\nabla(\lambda)$) be the irreducible (standard, costandard) rational G-module of high weight λ. Thus, $\Delta(\lambda)$ (resp., $\nabla(\lambda)$) has head (resp., socle) isomorphic to $L(\lambda)$. The modules $\Delta(\lambda)$ (resp., $\nabla(\lambda)$) are obtained by reduction mod p using a minimal (resp., maximal) lattice in the complex irreducible module of high weight λ for

---

[1] In particular, [1, Cor. 3.7] would imply that if $p \geq 2h - 2$, then every $\Delta(\lambda)$ has a $\Delta^{\mathrm{red}'}$-filtration, for each $\lambda \in X^+$. However, as pointed out to us by Andersen (in a private communication), the proof of [1, Lemma 3.3] does not hold, and so [1, Cor. 3.7] remains unproved.

the complex simple Lie algebra associated to G. Thus, at the character level, ch $\Delta(\lambda)$ = ch $\nabla(\lambda)$, and these characters are given by Weyl's character formula.

The category $\mathcal{C}$ is a highest weight category in the sense of [**3**]. This means the following conditions hold:

(HWC1) For $\lambda \in X^+$, $\nabla(\lambda)$ has socle $L(\lambda)$, and if $L(\mu)$ is a composition factor of $\nabla(\lambda)$ with $\mu \neq \lambda$, then $\mu < \lambda$.

(HWC2) For $\lambda \in X^+$, let $I(\lambda)$ be the injective envelope in $\mathcal{C}$ of $L(\lambda)$. Then $I(\lambda)$ has an increasing filtration

$$F^\bullet = F^\bullet(\lambda) : \ 0 = F^0 \subset F^1 \subset F^2 \cdots ,$$

such that $\bigcup F^i = I(\lambda)$, $F^1/F^0 \cong \nabla(\lambda)$, and, for $i > 1$, $F^i/F^{i-1} \cong \nabla(\mu_i)$ for some $\mu_i \in X^+$ satisfying $\mu_i > \lambda$.

The category $\mathcal{C}$ admits a strong duality D (in the sense of [**4**]) and we have $D\nabla(\lambda) = \Delta(\lambda)$. However, $\mathcal{C}$ does not have enough projective objects.

Let $F : G \to G$ be the Frobenius morphism defined by the $\mathbb{F}_p$-structure on G. If $V \in \mathcal{C}$ and r is a positive integer, $V^{(r)}$ denotes the rational G-module obtained from V by making $g \in G$ act on V by $F^r(g)$. In this paper, we only consider the case of $r = 1$.

Let $\mathbb{E} := \mathbb{R} \otimes_{\mathbb{Z}} X$ be endowed with a positive definite, symmetric bilinear form $(\ ,\ )$, invariant under the Weyl group W of Φ. We identify $X^\vee$ as a subgroup of $\mathbb{E}$, so that $\alpha^\vee = \frac{2}{(\alpha,\alpha)}\alpha$ and the pairing $X \times X^\vee \to \mathbb{Z}$ is compatible with the inner product. The affine Weyl group $W_p = p\mathbb{Z}\Phi \rtimes W$ is the group of affine transformations on $\mathbb{E}$ generated by W and the normal subgroup consisting of translations by elements in $p\mathbb{Z}\Phi$. If $\alpha \in \Phi$ and $r \in \mathbb{Z}$, define $s_{\alpha,r} : \mathbb{E} \to \mathbb{E}$ by $s_{\alpha,r}(x) = x - ((x, \alpha^\vee) - rp)\alpha$. Then $s_{\alpha,r} \in W_p$. Also, (W_p, S_p) is a Coxeter system, putting $S_p = \{s_{\alpha_1}, \cdots, s_{\alpha_n}, s_{\alpha_0,-1}\}$.

In this paper, we use the "dot" action of W_p on $\mathbb{E}$, given by setting $w \cdot x = w(x + \rho) - \rho$. Let $C^+ \subset \mathbb{E}$ be the dominant fundamental alcove; it consists of all $x \in \mathbb{E}$ satisfying the inequalities $0 < (x + \rho, \alpha_i^\vee)$, $i = 1, \cdots, n$, and $(x + \rho, \alpha_0^\vee) < p$. Let $w_0 \in W$ be the longest word, and put $C^- = w_0 \cdot C^+$, the anti-dominant alcove. The closures $\overline{C^+}$ and $\overline{C^-}$ are fundamental domains for the action of W_p on $\mathbb{E}$. The subsets $w \cdot C^+ \subset \mathbb{E}$, $w \in W_p$, are the alcoves for W_p. If $C = w \cdot C^+$ is an alcove, put $C_{\mathbb{Z}} = C \cap X$ and $\overline{C}_{\mathbb{Z}} = \overline{C} \cap X$.

A regular dominant weight λ has the form $\lambda = w \cdot \lambda^-$, where $\lambda^- \in C^-_{\mathbb{Z}}$. We say that λ satisfies the Lusztig character formula (LCF) provided that the formal character of $L(\lambda)$ is given by

$$\text{ch } L(\lambda) = \chi_{\text{KL}}(\lambda) := \sum_{y \in W_p, y\cdot\lambda^- \in X^+} (-1)^{l(w)-l(y)} P_{y,w}(-1) \text{ ch } \Delta(y \cdot \lambda^-),$$

where $P_{y,w}$ is the Kazhdan-Lusztig polynomial associated to the pair (y, w).

Let U_ζ be the quantum enveloping algebra (Lusztig form) associated to G over a pth root of unity. We will assume that the prime p is odd, and, if G has type G_2, then $p > 3$.

We fix a p-modular system $(K, \mathcal{O}, k)$. Thus, $\mathcal{O}$ is a discrete valuation ring with maximal ideal $\mathfrak{m} = (\pi)$, fraction field K of characteristic 0, and residue field $k = \mathcal{O}/\mathfrak{m}$. We can assume that $\mathcal{O}$ has a primitive pth root of unity ζ. (See [**6**, Rem. 1.4] and the discussion there.) Choose an $\mathcal{O}$-form $\widetilde{U}_\zeta$ of U_ζ. Put $\overline{U}_\zeta = \widetilde{U}_\zeta/\pi\widetilde{U}_\zeta$, and let I be the ideal in $\overline{U}_\zeta$ generated by the images of the elements $K_i - 1$, $1 \leq i \leq n$, in the usual notation. By [10, (8.15)],

(2.1) $\overline{U}_\zeta / I \cong \mathrm{hy}(G)$,

the distribution algebra of G over k.

The category of finite dimensional integrable, type 1 U_ζ-modules will be denoted by $\mathcal{C}_\zeta$. It is a highest weight category with irreducible (resp. standard, costandard) modules $L_\zeta(\lambda)$ (resp., $\Delta_\zeta(\lambda)$, $\nabla_\zeta(\lambda)$), $\lambda \in X^+$. For $\mu \in X^+$, $\mathrm{ch}\,\Delta_\zeta(\mu) = \mathrm{ch}\,\nabla_\zeta(\mu) = \mathrm{ch}\,\Delta(\mu)$. We will always assume that given $\lambda \in X_{\mathrm{reg}}^+$, $\mathrm{ch}\,L_\zeta(\lambda) = \chi_{KL}(\lambda)$. This assumption is always valid if $p > h$, though in many cases this restriction is too severe; see [13] for the precise result.

As discussed in [6, §1.5], for $\lambda \in X^+$, the module $L_\zeta(\lambda)$ has a minimal (resp., maximal) admissible lattice $\widetilde{L}^{\min}(\lambda)$ (resp., $L^{\max}(\lambda)$ for $\widetilde{U}_\zeta$ which, upon reduction to k, defines a rational G-module $\Delta^{\mathrm{red}}(\lambda)$ (resp., $\nabla_{\mathrm{red}}(\lambda)$). We recall that these modules have, according to Lin [9], an alternative description as $\Delta^{\mathrm{red}}(\lambda) \cong \Delta^{\mathrm{red}}(\lambda_0) \otimes \Delta(\lambda_1)^{(1)}$ and $\nabla_{\mathrm{red}}(\lambda) \cong \nabla_{\mathrm{red}}(\lambda_0) \otimes \nabla_{\mathrm{red}}(\lambda_1)^{(1)}$ if $\lambda = \lambda_0 + p\lambda_1$, with $\lambda_0 \in X_1^+$, $\lambda_1 \in X^+$; see also [6, Prop. 1.6].[2] By [6, Thm. 5.4, Thm.6.7] if $p > h$ and if the LCF holds for all regular weights in X_1^+, then

(2.2) $\dim \mathrm{Ext}_G^n(\Delta^{\mathrm{red}}(\lambda), \nabla_{\mathrm{red}}(\mu)) = \dim \mathrm{Ext}_{\mathcal{C}_\zeta}^n(L_\zeta(\lambda), L_\zeta(\mu))$,

for all $\lambda, \mu \in X_{\mathrm{reg}}^+$. The groups $\mathrm{Ext}_{\mathcal{C}_\zeta}^n(L_\zeta(\lambda), L_\zeta(\mu))$ can be explicitly calculated; see, e. g., [6, (1.4.2)]. In particular, we will often use without mention the fact that, given $\lambda, \mu \in X_{\mathrm{reg}}^+$,

$\mathrm{Ext}_{\mathcal{C}_\zeta}^n(L_\zeta(\lambda), L_\zeta(\mu)) \neq 0 \implies n \equiv \ell(\lambda) - \ell(\mu) \bmod 2$.

We record the following result, which we have largely already discussed. See [9] and [6, §1.5].

LEMMA 2.1. *Let $\lambda \in X^+$.*

(1) $\Delta^{\mathrm{red}}(\lambda)$ (resp., $\nabla_{\mathrm{red}}(\lambda)$) has irreducible head (resp., socle) $L(\lambda)$. All other composition factors $L(\mu)$ of $\Delta^{\mathrm{red}}(\lambda)$ (resp., $\nabla_{\mathrm{red}}(\lambda)$) satisfy $\mu < \lambda$.

(2) Assume that $p > h$ and that the LCF holds for all regular weights in X_1^+. Then $\Delta^{\mathrm{red}}(\lambda) \cong L(\lambda_0) \otimes \Delta(\lambda_1)^{(1)}$ and $\nabla_{\mathrm{red}}(\lambda) \cong L(\lambda_0) \otimes \nabla_{\mathrm{red}}(\lambda_1)^{(1)}$. Here $\lambda = \lambda_0 + p\lambda_1$ as above.

For a regular dominant weight λ, write $\lambda = w \cdot \lambda^-$, where $\lambda^- \in C_{\mathbb{Z}}^-$. As in the introduction, define $\ell(\lambda) = \ell(w)$. We say that $L(\lambda)$ has even (resp., odd) parity provided that $\ell(\lambda) \equiv 0$ (resp., $\ell(\lambda) \not\equiv 0$) mod 2. Let $X_{\mathrm{reg, even}}^+$ (resp., $X_{\mathrm{reg,odd}}^+$) be set of regular dominant weights λ such that $\ell(\lambda) \equiv 0$ (resp., $\ell(\lambda) \not\equiv 0$) mod 2.

3. The highest weight categories

Let $\mathcal{C}_{\mathrm{even}}^{\mathrm{reg}}$ (resp., $\mathcal{C}_{\mathrm{odd}}^{\mathrm{reg}}$) be the full subcategory of $\mathcal{C}$ generated by the irreducible modules $L(\lambda)$ having even (resp., odd) parity. For example, a rational G-module belongs to $\mathcal{C}_{\mathrm{even}}^{\mathrm{reg}}$ if the composition factors of any finite dimensional submodule have even parity.

---

[2]More generally, [9] describes, in the same spirit, the rational G-modules obtained from the irreducible modules for quantum enveloping algebras at a p^rth root of unity by a reduction mod p process. The result we have quoted is merely the $r = 1$ case of this result (which is sufficient for the applications in [6]). We do not investigate here how the results of this paper might generalize to the $r > 1$ case.

LEMMA 3.1. *Assume that $p > h$ and that the LCF holds for all regular weights in X_1^+. Let $\lambda, \mu \in X_{\mathrm{reg,\,even}}^+$, and let $\tau \in X^+$. Then:*

(1) If $[\Delta^{\mathrm{red}}(\lambda) : L(\tau)] \neq 0$ or $[\nabla_{\mathrm{red}}(\lambda) : L(\tau)] \neq 0$, then $\tau \in X_{\mathrm{reg,\,even}}^+$. In particular, $\Delta^{\mathrm{red}}(\lambda)$ and $\nabla_{\mathrm{red}}(\lambda)$ both belong to $\mathcal{C}_{\mathrm{even}}^{\mathrm{reg}}$.

(2) If $\mathrm{Ext}_G^1(\Delta^{\mathrm{red}}(\lambda), L(\mu)) \neq 0$ or $\mathrm{Ext}_G^1(L(\mu), \nabla_{\mathrm{red}}(\lambda))$, then $\mu > \lambda$.

(3) If $\mathrm{Ext}_G^1(\Delta^{\mathrm{red}}(\lambda), \Delta^{\mathrm{red}}(\mu)) \neq 0$ or $\mathrm{Ext}_G^1(\nabla_{\mathrm{red}}(\mu), \nabla_{\mathrm{red}}(\lambda))$, then $\mu > \lambda$.

Similar statements holds with $X_{\mathrm{reg,\,even}}^+$ replaced throughout by $X_{\mathrm{reg,odd}}^+$.

PROOF. We only prove the Δ^{red}-statements for the poset $X_{\mathrm{reg,\,even}}^+$.

Writing $\lambda = \lambda_0 + p\lambda_1$ with $\lambda_0 \in X_1^+$ and $\lambda_1 \in X^+$, we have that $\Delta^{\mathrm{red}}(\lambda) \cong L(\lambda_0) \otimes \Delta(\lambda_1)^{(1)}$. Thus, if $L(\tau)$ is a composition factor of $\Delta^{\mathrm{red}}(\lambda)$, then $\tau = \lambda_0 + p\sigma$, where $L(\sigma)$ is a composition factor of $\Delta(\lambda_1)$. Hence, $\sigma = \lambda_1 - \delta$, where $\delta \in \mathbb{Z}\Phi$. Therefore, $\tau = \lambda - p\delta$, so τ and λ have the same parity. This proves (1).

It is easy to see that (3) follows from (2). So suppose $\mathrm{Ext}_G^1(\Delta^{\mathrm{red}}(\lambda), L(\mu)) \neq 0$. Since λ and μ have the same parity, (2.2) implies that

$$\dim \mathrm{Ext}_G^1(\Delta^{\mathrm{red}}(\lambda), \nabla_{\mathrm{red}}(\lambda)) = \dim \mathrm{Ext}_{U_\zeta}^1(L_\zeta(\lambda), L_\zeta(\mu)) = 0.$$

Therefore, forming the exact sequence $0 \to L(\mu) \to \nabla_{\mathrm{red}}(\mu) \to Q(\mu) \to 0$, we see that $\mathrm{Ext}_G^1(\Delta^{\mathrm{red}}(\lambda), L(\mu))$ is a homomorphic image of $\mathrm{Hom}_G(\Delta^{\mathrm{red}}(\lambda), Q(\mu))$. Since $\Delta^{\mathrm{red}}(\lambda)$ has head $L(\lambda)$, it follows that if this Hom-space is non-zero, then $\lambda < \mu$, as required. $\qquad\square$

Now we can prove the main result of this section.

THEOREM 3.2. *Assume that $p > h$ and that the LCF holds for all regular restricted dominant weights. Then $\mathcal{C}_{\mathrm{even}}^{\mathrm{reg}}$ is a highest weight category with weight poset $X_{\mathrm{reg,\,even}}^+$. For $\lambda \in X_{\mathrm{reg,\,even}}^+$, the corresponding standard (resp., costandard) object is $\Delta^{\mathrm{red}}(\lambda)$ (resp., $\nabla_{\mathrm{red}}(\lambda)$).*

Similarly, $\mathcal{C}_{\mathrm{odd}}^{\mathrm{reg}}$ is a highest weight category with weight poset $X_{\mathrm{reg,odd}}^+$. For $\lambda \in X_{\mathrm{reg,odd}}^+$, the corresponding standard (resp., costandard) object is $\Delta^{\mathrm{red}}(\lambda)$ (resp., $\nabla_{\mathrm{red}}(\lambda)$).

PROOF. We will prove only the assertion for $\mathcal{C}_{\mathrm{even}}^{\mathrm{reg}}$. We follow the definition of highest weight categories as given in [**3**]. By Lemma 3.1(1), each $\nabla_{\mathrm{red}}(\lambda)$, $\lambda \in X_{\mathrm{reg,\,even}}^+$, belongs to $\mathcal{C}_{\mathrm{even}}^{\mathrm{reg}}$. In addition, by Lemma 2.1(2), $\nabla_{\mathrm{red}}(\lambda)$ has socle isomorphic to $L(\lambda)$, while the other composition factors $L(\mu)$ satisfy $\mu < \lambda$ (and, of course, $\mu \in X_{\mathrm{reg,\,even}}^+$).

Therefore, it remains to show that, given $\lambda \in X_{\mathrm{reg,\,even}}^+$, $L(\lambda)$ has injective envelope in $\mathcal{C}_{\mathrm{even}}^{\mathrm{reg}}$ which has an ∇_{red}-filtration with bottom section $\nabla_{\mathrm{red}}(\lambda)$ and higher sections $\nabla_{\mathrm{red}}(\mu)$ for $\mu > \lambda$. Consider the set of dominant weights $\mu \in X_{\mathrm{reg,\,even}}^+$ which satisfy $\mu \geq \lambda$. Enumerate this set as $\lambda_0 = \lambda, \lambda_1, \lambda_2, \ldots$ so that $\lambda \leq \mu < \tau$, then $\mu = \lambda_i$ and $\tau = \lambda_j$ for $i < j$. Let $\Gamma_i = \{\lambda_0, \ldots, \lambda_i\}$ for each $i \geq 0$.

We construct a sequence $I_0(\lambda) \subseteq I(\lambda_1) \subseteq \cdots$ of submodules of $\mathcal{C}_{\mathrm{even}}^{\mathrm{reg}}$ with the following properties:

(i) $I_j(\lambda)$ has a filtration with sections $\nabla_{\mathrm{red}}(\lambda_i)$, $i \leq j$;

(ii) soc $I_j(\lambda) = L(\lambda)$;

(iii) $\mathrm{Ext}_G^1(L(\mu), I_j(\lambda)) \neq 0$ implies that $\mu = \lambda_i$ for some $i > j$.

(iv) $I_j(\lambda) \cong \widetilde{I}_j(\lambda)/\pi\widetilde{I}_j(\lambda)$ for some $\widetilde{U}_\zeta$-lattice $\widetilde{I}_j(\lambda)$.

Then

$$I(\lambda) := \bigcup_j I_j(\lambda)$$

is the required injective envelope of $L(\lambda)$ in $\mathcal{C}^{\mathrm{reg}}_{\mathrm{even}}$.

To begin, let $I_0(\lambda) = \nabla_{\mathrm{red}}(\lambda)$. Then Lemma 3.1(1),(2) implies that (i)–(iii) hold, while we can take $\widetilde{I}_0(\lambda) = \widetilde{L}^{\min}(\lambda)$ to satisfy (iv). Suppose then that $I_j(\lambda)$ has been constructed satisfying properties (i)–(iv). Choose a basis $\xi_1, \ldots, \xi_n$ of $\mathrm{Ext}^1_G(\nabla_{\mathrm{red}}(\lambda_{j+1}), I_j(\lambda))$. Of course, $n < \infty$. Define $I_{j+1}(\lambda)$ by means of an extension

$$(3.1) \qquad \xi: \quad 0 \to I_j(\lambda) \to I_{j+1}(\lambda) \to \nabla_{\mathrm{red}}(\lambda_{j+1})^{\oplus n} \to 0,$$

with the following property: if $\iota_i : \nabla_{\mathrm{red}}(\lambda_{j+1}) \hookrightarrow \nabla_{\mathrm{red}}(\lambda_{j+1})^{\oplus n}$ maps $\nabla_{\mathrm{red}}(\lambda_{j+1})$ isomorphically onto the ith coordinate of $\nabla_{\mathrm{red}}(\lambda_{j+1})^{\oplus n}$, then the pull-back $\iota_i^* \xi$ defines ξ_i.

The map

$$(3.2) \qquad \mathrm{Ext}^1_G(\nabla_{\mathrm{red}}(\lambda_{j+1}), I_j(\lambda)) \to \mathrm{Ext}^1_G(L(\lambda_{j+1}), I_j(\lambda))$$

is injective, because $\nabla_{\mathrm{red}}(\lambda_{j+1})/L(\lambda_{j+1})$ has only composition factors with high weights smaller than λ_{j+1}. The exact sequence $0 \to \widetilde{I}_j(\lambda) \overset{\pi}{\to} \widetilde{I}_j(\lambda) \to I_j(\lambda) \to 0$ gives an exact sequence

$$0 \to \mathrm{Hom}_G(\nabla_{\mathrm{red}}(\lambda_{j+1}), I_j(\lambda))$$

$$\overset{\alpha}{\to} \mathrm{Ext}^1_{\widetilde{U}_\zeta}(\widetilde{L}^{\max}(\lambda_{j+1}), \widetilde{I}_j(\lambda))$$

$$\overset{\pi}{\to} \mathrm{Ext}^1_{\widetilde{U}_\zeta}(\widetilde{L}^{\max}(\lambda_{j+1}), \widetilde{I}_j(\lambda)) \overset{\beta}{\to} \mathrm{Ext}^1_G(\nabla_{\mathrm{red}}(\lambda_{j+1}), I_j(\lambda)).$$

(See [**6**, (1.4.5)]. Notice that $\mathrm{Ext}^1_{\widetilde{U}_\zeta}(\widetilde{L}^{\max}(\lambda_{j+1}), \widetilde{I}_j(\lambda))$ is a torsion module by [**6**, (1.4.4)] and the fact that λ_{j+1} and the high weights of the composition factors of $\widetilde{I}_j(\lambda)_K$ have even parity. The image of α is the submodule of all elements killed by π, which has the same dimension as $\mathrm{Im}\,\beta$. So, we have

$$\dim \mathrm{Hom}_G(\nabla_{\mathrm{red}}(\lambda_{j+1}), I_j(\lambda)) \quad = \quad \dim \mathrm{Im}\,\beta$$

$$\leq \quad \dim \mathrm{Ext}^1_G(\nabla_{\mathrm{red}}(\lambda_{j+1}), I_j(\lambda))$$

$$\leq \quad \dim \mathrm{Ext}^1_G(L(\lambda_{j+1}), I_j(\lambda))$$

by the injectivity of (3.2). Using (2.2), we obtain by the long exact sequence of cohomology that $\mathrm{Ext}^1_G(L(\lambda_{j+1}), I_j(\lambda)) \cong \mathrm{Hom}_G(\mathrm{rad}\,\Delta^{\mathrm{red}}(\lambda_{j+1}), I_j(\lambda))$, which has the same dimension as $\mathrm{Hom}_G(\nabla_{\mathrm{red}}(\lambda_{j+1})/L(\lambda_{j+1}), I_j(\lambda)) \cong \mathrm{Hom}_G(\nabla_{\mathrm{red}}(\lambda_{j+1}), I_j(\lambda))$. (To see the equality of dimension, first observe that $\mathrm{rad}\,\Delta^{\mathrm{red}}(\lambda_{j+1})$ and $\nabla_{\mathrm{red}}(\lambda_{j+1})/L(\lambda_{j+1})$ have the same composition factors. Then use condition (iii), which implies that the functor $\mathrm{Hom}_G(-, I_j(\lambda))$ is exact on the subcategory of G-modules having composition factors $L(\lambda_i)$, $i = 0, \ldots, j$.)

Thus, the inequalities in (3.3) are all equalities. In particular, we see that the map (3.2) is an isomorphism. Now diagram chasing easily implies that $I_{j+1}(\lambda)$ satisfies conditions (i)–(iii). For example, the long exact sequence of cohomology applied to the exact sequence (3.1), together with Lemma 3.1, gives an exact sequence

$$0 \to \operatorname{Hom}_G(L(\lambda_{j+1}), \nabla_{\mathrm{red}}(\lambda_{j+1})^{\oplus n}) \xrightarrow{\epsilon} \operatorname{Ext}^1_G(L(\lambda_{j+1}), I_j)$$

$$\to \operatorname{Ext}^1_G(L(\lambda_{j+1}), I_{j+1})) \to 0.$$

But

$$\dim \operatorname{Hom}_G(L(\lambda_j), \nabla_{\mathrm{red}}(\lambda_{j+1})^{\oplus n}) = n,$$

which also equals

$$\dim \operatorname{Ext}^1_G(\nabla_{\mathrm{red}}(\lambda_{j+1}), I_j(\lambda)).$$

Now the isomorphism (3.2) implies that ϵ is an isomorphism. Hence, we conclude that $\operatorname{Ext}^1_G(L(\lambda_{j+1}), I_{j+1}(\lambda)) = 0$.

Finally, we have also shown that β is surjective, so that $I_{j+1}(\lambda)$ lifts to a lattice $\widetilde{I}_{j+1}(\lambda)$ for $\widetilde{U}_\zeta$. Thus, condition (iv) also holds for $I_{j+1}(\lambda)$. □

REMARK 3.3. It is interesting to compare the above result to the discussion in [**11**, §5]. Let Γ be a finite ideal in $X^+_{\mathrm{reg, even}}$. (A similar discussion would work for $X^+_{\mathrm{reg,odd}}$.) Form the derived category $\mathcal{D} = D^b(\mathcal{C})$, regarding $\mathcal{C}$ as fully embedding in $\mathcal{D}$ as complexes concentrated in degree 0. Using Lemma 2.1, we see that the modules $\Delta^{\mathrm{red}}(\lambda)$, $\lambda \in \Gamma$, satisfy the conditions of [**11**, Th. 5.9]. Therefore, there exists a strict full subcategory $\mathcal{C}_\Gamma$ of $\mathcal{D}$ and a highest weight category $\overline{\mathcal{C}}_\Gamma$ (having weight poset Γ) with the following properties: $\mathcal{C}_\Gamma$ is the exact full subcategory of $\overline{\mathcal{C}}_\Gamma$ consisting of objects with a Δ-filtration. Each $\Delta^{\mathrm{red}}(\lambda) \in \mathcal{C}_\Gamma$ identifies in $\overline{\mathcal{C}}_\Gamma$ with the standard module indexed by $\lambda \in \Gamma$. For $\lambda \in \Gamma$, the projective indecomposable cover $P_\Gamma(\lambda)$ of $L(\lambda)$ in $\overline{\mathcal{C}}_\Gamma$ is obtained by a recursive construction exactly dual to the construction of the $I_j(\lambda)$ in the proof of the above theorem. Letting $T = \oplus_{\lambda \in \Gamma} P_\Gamma(\lambda)$, we can take $\overline{\mathcal{C}}_\Gamma = (\operatorname{End}_\mathcal{C}(T)\text{-mod})^{\mathrm{op}}$. However, in the abstract setting of [**11**], there is no guarantee that the constructed highest weight category $\overline{\mathcal{C}}_\Gamma$ will fully embed into the original category $\mathcal{D}$ (or, in the present case, $\mathcal{C}$). The above theorem shows that this embedding does happen in the case of $\mathcal{C}^{\mathrm{reg}}_{\mathrm{even}}$.

4. Applications

Maintain the notation of the previous section. Let $i^+_* : \mathcal{C}^{\mathrm{reg}}_{\mathrm{even}} \to \mathcal{C}$ (resp., $i^-_* : \mathcal{C}^{\mathrm{reg}}_{\mathrm{odd}} \to \mathcal{C}$) denote the canonical full embedding of categories. The functor i^+_* is exact, and admits left (resp., right) exact right (resp., left) adjoint $i^!_+ : \mathcal{C} \to \mathcal{C}^{\mathrm{reg}}_{\mathrm{even}}$ (resp., $i^*_+ : \mathcal{C} \to \mathcal{C}^{\mathrm{reg}}_{\mathrm{even}}$). Explicitly, for $M \in \mathcal{C}$, $i^!_+ M$ (resp., $i^*_+ M$) is the largest submodule (resp., quotient module) of M having composition factors $L(\lambda)$ with $\lambda \in X^+_{\mathrm{reg, even}}$. Similarly, i^-_* admits a right (resp., left) adjoint $i^!_-$ (resp., i^*_-). Suppose that $M, N \in \mathcal{C}^{\mathrm{reg}}_{\mathrm{even}}$ (resp., $M, N \in \mathcal{C}^{\mathrm{reg}}_{\mathrm{odd}}$), then there is an induced homomorphism $R^n i^+_*(M, N) : \operatorname{Ext}^n_{\mathcal{C}^{\mathrm{reg}}_{\mathrm{even}}}(M, N) \to \operatorname{Ext}^n_G(M, N)$ (resp., $R^n i^-_*(M, N) : \operatorname{Ext}^n_{\mathcal{C}^{\mathrm{reg}}_{\mathrm{odd}}}(M, N) \to \operatorname{Ext}^n_G(M, N)$). Clearly,

LEMMA 4.1. *For* $M, N \in \mathcal{C}^{\mathrm{reg}}_{\mathrm{even}}$,

$$R^n i^+_*(M, N) : \operatorname{Ext}^n_{\mathcal{C}^{\mathrm{reg}}_{\mathrm{even}}}(M, N) \to \operatorname{Ext}^n_G(M, N)$$

is an isomorphism for $n = 0, 1$. *The analogous statement holds for* $\mathcal{C}^{\mathrm{reg}}_{\mathrm{odd}}$.

For $n > 1$, $R^n i_*^+(M, N)$ is generally not an isomorphism, i. e., i_*^+ does not induce a full embedding at the level of derived categories. (And, of course, a similar statement holds for $R^n i_*^-(M, N)$.) In fact, take $\lambda, \mu \in X_{\mathrm{reg, even}}^+$ so that $\mathrm{Ext}_{\mathcal{C}_\zeta}^2(L_\zeta(\lambda), L_\zeta(\mu)) \neq 0$. Then $\mathrm{Ext}_G^2(\Delta^{\mathrm{red}}(\lambda), \nabla_{\mathrm{red}}(\mu)) \neq 0$ by (2.2). On the other hand, $\mathrm{Ext}_{\mathcal{C}_{\mathrm{even}}^{\mathrm{reg}}}^2(\Delta^{\mathrm{red}}(\lambda), \nabla_{\mathrm{red}}(\lambda)) = 0$ by Theorem 3.2 and standard properties of highest weight categories (e. g., use Lemma 4.2 below).

We will make use of the following well-known criterion, first proved independently in the context of rational G-modules by S. Donkin and L. Scott.

LEMMA 4.2. *Let $\mathcal{C}'$ be a highest weight category with finite poset Λ. Let $M \in \mathcal{C}'$. Then M has a Δ-filtration (resp., ∇-filtration) if and only if $\mathrm{Ext}_{\mathcal{C}'}^1(M, \nabla(\omega)) = 0$ (resp., $\mathrm{Ext}_{\mathcal{C}'}^1(\Delta(\omega), M) = 0$) for all $\omega \in \Lambda$.*

For $\lambda \in X_{\mathrm{reg}}^+$, define $\mathrm{rad}^{\mathrm{red}}(\lambda)$ by means of the following exact sequence

$$(4.1) \qquad 0 \to \mathrm{rad}^{\mathrm{red}}(\lambda) \to \Delta(\lambda) \to \Delta^{\mathrm{red}}(\lambda) \to 0.$$

Also, define $\mathrm{soc}_{\mathrm{red}}(\lambda)$ by the following exact sequence

$$(4.2) \qquad 0 \to \nabla_{\mathrm{red}}(\lambda) \to \nabla(\lambda) \to \mathrm{soc}_{\mathrm{red}}(\lambda) \to 0.$$

We state the following theorem for $\mathcal{C}_{\mathrm{even}}^{\mathrm{reg}}$. A similar result holds for $\mathcal{C}_{\mathrm{odd}}^{\mathrm{reg}}$.

THEOREM 4.3. *Assume that $p > h$ and that the LCF holds for all regular restricted dominant weights. Let $\lambda \in X_{\mathrm{reg, even}}^+$. Then $i_-^* \mathrm{rad}^{\mathrm{red}}(\lambda)$ (resp., $i_-^! \mathrm{soc}_{\mathrm{red}}(\lambda)$) has a Δ^{red}-filtration (resp., ∇_{red}-filtration) in $\mathcal{C}_{\mathrm{odd}}^{\mathrm{reg}}$.*

PROOF. We prove the assertion for $i_-^* \mathrm{rad}^{\mathrm{red}}(\lambda)$, leaving the dual case for $i_-^! \mathrm{soc}_{\mathrm{red}} \nabla(\lambda)$ to the reader. First, write $i_-^* \mathrm{rad}^{\mathrm{red}}(\lambda)$ as a quotient $\mathrm{rad}^{\mathrm{red}}(\lambda)/M(\lambda)$, and let $E(\lambda) := \Delta(\lambda)/M(\lambda)$. Thus, we have an exact sequence

$$0 \to i_-^* \mathrm{rad}^{\mathrm{red}}(\lambda) \to E(\lambda) \to \Delta^{\mathrm{red}}(\lambda) \to 0.$$

Claim 1: Let $\omega \in X_{\mathrm{reg, odd}}^+$ and assume that $\omega < \lambda$. Then $\mathrm{Ext}_G^1(E(\lambda), L(\omega)) = 0$.

In fact, suppose that $0 \to L(\omega) \to F \overset{\iota}{\to} E(\lambda) \to 0$ is an extension. The module $E(\lambda)$ is a cyclic G-module generated by a high weight vector v_λ of weight λ. Thus, the universal mapping property of $\Delta(\lambda)$ implies there is a morphism $\sigma : \Delta(\lambda) \to F$ such that $\iota \circ \sigma$ is surjective. Because $\omega \in X_{\mathrm{reg, odd}}^+$, $\sigma(M(\lambda)) = 0$, so that (up to a nonzero scalar), the morphism ι is split by the induced morphism $\bar\sigma : E(\lambda) = \Delta(\lambda)/M(\lambda) \to F$. This proves Claim 1.

Claim 2: Let $\omega \in X_{\mathrm{reg, odd}}^+$ and assume that $\omega < \lambda$. Then $\mathrm{Ext}_G^1(E(\lambda), \nabla_{\mathrm{red}}(\omega)) = 0$.

This claim follows immediately from Claim 1, since all the composition factors $L(\tau)$ of $\nabla_{\mathrm{red}}(\omega)$ satisfy $\tau \in X_{\mathrm{reg, odd}}^+$ and $\tau < \lambda$.

Claim 3: Assume that $\omega \in X_{\mathrm{reg, odd}}^+$ and $\omega < \lambda$. Then $\mathrm{Ext}_G^1(i_-^* \mathrm{rad}^{\mathrm{red}}(\lambda), \nabla_{\mathrm{red}}(\omega)) = 0$.

For convenience, denote $i_-^* \mathrm{rad}^{\mathrm{red}}(\lambda)$ by $R(\lambda)$. The long exact sequence of $\mathrm{Ext}_G^\bullet$ applied to the short exact sequence above Claim 1 give the following exact sequence

$$0 \to \mathrm{Ext}_G^1(R(\lambda), \nabla_{\mathrm{red}}(\omega)) \to \mathrm{Ext}_G^2(\Delta^{\mathrm{red}}(\lambda), \nabla_{\mathrm{red}}(\omega))$$

in view of Claim 2. Since ω and λ have opposite parity, Claim 3 follows from (2.2).

Now consider the highest weight category $\mathcal{C}_{\mathrm{odd}}^{\mathrm{reg}}[\Gamma]$, where $\Gamma = (-\infty, \lambda)$, the ideal in $X_{\mathrm{reg,odd}}^+$ consisting of dominant weights $< \lambda$. It is the full subcategory of $\mathcal{C}_{\mathrm{odd}}^{\mathrm{reg}}$ whose objects have composition factors $L(\gamma)$, $\gamma \in \Gamma$. Then $i_-^* \, \mathrm{rad}^{\mathrm{red}}(\lambda) \in \mathcal{C}_{\mathrm{odd}}^{\mathrm{reg}}[\Gamma]$. Since

$$\mathrm{Ext}_G^1(i_-^* \, \mathrm{rad}^{\mathrm{red}}(\lambda), \nabla_{\mathrm{red}}(\omega)) = 0$$

for all $\omega \in \Gamma$, Lemma 4.1 and Lemma 4.2 imply that $i_-^* \, \mathrm{rad}^{\mathrm{red}}(\lambda)$ has a Δ^{red}-filtration, as required. □

REMARK 4.4. It is interesting to compare the above result with that given in [**6**, Thm. 6.9], which proves a similar filtration result. We will show in this remark that the results are the same, though we require the original argument for [**6**, Thm. 6.9] to see the equivalence. We still assume that $p > h$ and that the LCF holds for all regular weights in X_1^+. For $\lambda \in X_{\mathrm{reg}}^+$, put

$$E_\zeta(\lambda) = \Delta_\zeta(\lambda)/\mathrm{rad}^2 \Delta_\zeta(\lambda).$$

Let $\widetilde{E}(\lambda)$ be the image of $\widetilde{\Delta}_\zeta(\lambda)$ in $E_\zeta(\lambda)$, and set $E'(\lambda)$ be the rational G-module obtained by reducing $\widetilde{E}'(\lambda)$ to the field k. Let $\widetilde{D}(\lambda)$ be the kernel of the natural surjection $\widetilde{E}(\lambda) \twoheadrightarrow \widetilde{L}^{\min}(\lambda)$. Then [**6**, Thm. 6.9] shows that $E'(\lambda)$ has a Δ^{red}-filtration with top section $\Delta^{\mathrm{red}}(\lambda)$.

An essential step in the proof of this result was to show that if $\omega \in X_{\mathrm{reg}}^+$ satisfies $\omega < \lambda$ and if ω has parity opposite to that of λ, then $\mathrm{Ext}_G^1(E'(\lambda), L(\omega)) = 0$. (See, for example, Claim 4 in the proof given in [**6**].) It follows easily from this fact that $E'(\lambda) \cong E(\lambda)$ (as defined in the proof of Theorem 4.3. In particular, this isomorphism means that $D(\lambda) \cong \widetilde{D}(\lambda)/\pi\widetilde{D}(\lambda)$ is isomorphic to $i_+^* \, \mathrm{rad}^{\mathrm{red}}(\lambda)$ if $\lambda \in X_{\mathrm{reg, even}}^+$, and so the conclusion of Theorem 4.3 and [**6**, Thm. 6.9] are really the same, at least over k. (It is then easy to extend Theorem 4.3 to the integral case.)

Finally, we conclude this section with the following alternative characterization of the modules $\Delta^{\mathrm{red}}(\lambda)$ and $\nabla_{\mathrm{red}}(\lambda)$.

THEOREM 4.5. *Assume that $p > h$ and that the LCF holds for all regular restricted dominant weights. Let $\lambda \in X_{\mathrm{reg, even}}^+$. Then*

$$i_+^* \Delta(\lambda) \cong \Delta^{\mathrm{red}}(\lambda) \text{ and } i_+^! \nabla(\lambda) \cong \nabla_{\mathrm{red}}(\lambda).$$

A similar result holds if $\lambda \in X_{\mathrm{reg,odd}}^+$, using i_-^* and $i_-^!$.

PROOF. We only consider the case $\lambda \in X_{\mathrm{reg, even}}^+$, and prove that $i_+^* \Delta(\lambda) \cong \Delta^{\mathrm{red}}(\lambda)$ for $\lambda \in X_{\mathrm{reg, even}}^+$. The dual statement is left to the reader. Apply i_+^* to (4.1) to obtain using Lemma 3.1 the exact sequence

$$i_+^* \, \mathrm{rad}^{\mathrm{red}}(\lambda) \to i_+^* \Delta(\lambda) \to \Delta^{\mathrm{red}}(\lambda) \to 0.$$

Thus, it suffices to prove that $i_+^* \, \mathrm{rad}^{\mathrm{red}}(\lambda) = 0$. If not, there exists $\nu \in X_{\mathrm{reg, even}}^+$ such that $\nu < \lambda$ and $L(\nu)$ lies in the head of $i_+^* \, \mathrm{rad}^{\mathrm{red}}(\lambda)$. Since $\Delta(\lambda)$ has irreducible head $L(\lambda)$, it follows from Lemma 4.1 that

$$\mathrm{Ext}_G^1(\Delta^{\mathrm{red}}(\lambda), L(\nu)) = \mathrm{Ext}_{\mathcal{C}_{\mathrm{even}}^{\mathrm{reg}}}^1(\Delta^{\mathrm{red}}(\lambda), L(\nu)) \neq 0.$$

But this situation is impossible by standard highest weight theory, cf. [**3**, Lemma 3.2(b)] (use the dual version for standard modules). □

References

[1] H. Andersen, p-Filtrations and the Steinberg module, *J. Algebra* **244** (2001), 664–683.

[2] N. Bourbaki, *Groupes et algèbres de Lie*, IV, V, VI, Hermann (1968).

[3] E. Cline, B. Parshall, Finite dimensional algebras and highest weight categories, *J. reine angew. Math.* **391** (1988), 85-99.

[4] E. Cline, B. Parshall, and L. Scott, Duality in highest weight categories, *Comtemp. Math.* **82** (1989), 7-22.

[5] E. Cline, B. Parshall, and L. Scott, Abstract Kazhdan-Lusztig theories, *Tôhoku Math. J.* **45** (1993), 511-534.

[6] E. Cline, B. Parshall, and L. Scott, Reduced standard modules and cohomology, *Transactions of Amer. Math. Soc.*, to appear.

[7] R. Guralnick, The dimension of the first cohomology group, in: *Representation Theory, II*, Ottawa (1984), in: Lectures Notes in Math. **1178**, Springer (1986), 94–97.

[8] J. C. Jantzen, *Darstellungen halbeinfacher Gruppen und ihrer Frobenius-Kerne, J. Reine und Angew. Math.* **317** (1980), 157-199. .

[9] Z. Lin, Highest weight modules for algebraic groups arising from quantum groups, *J. Algebra* **208** (1998), 276–303.

[10] G. Lusztig, Quantum groups at roots of 1, *Geom. Dedicata* **35** (1990), 89–114.

[11] B. Parshall and L. Scott, Derived categories, quasi-hereditary algebras, and algebraic groups, *Carlton Univ. Math. Notes* **3** (1989), 1–111.

[12] B. Parshall and L. Scott, Beyond the Jantzen region, Oberwolfach Report No. 15/2006 (2006), 26–30.

[13] T. Tanisaki, Character formulas of Kazhdan-Lusztig type, *Representations of finite dimensional algebras and related topics in Lie theory and geometry*, Fields Inst. Commun. **40**, Amer. Math. Soc., Providence, RI (2004) 261–276.

DEPARTMENT OF MATHEMATICS, UNIVERSITY OF VIRGINIA, CHARLOTTESVILLE, VA 22903
E-mail address: bjp8w@virginia.edu

DEPARTMENT OF MATHEMATICS, UNIVERSITY OF VIRGINIA, CHARLOTTESVILLE, VA 22903
E-mail address: lls2l@virginia.edu

Contemporary Mathematics
Volume **478**, 2009

Classification of quasi-trigonometric solutions of the classical Yang–Baxter equation

Iulia Pop and Alexander Stolin

ABSTRACT. It was proved by Montaner and Zelmanov that up to classical twisting Lie bialgebra structures on $\mathfrak{g}[u]$ fall into four classes. Here $\mathfrak{g}$ is a simple complex finite-dimensional Lie algebra. It turns out that classical twists within one of these four classes are in a one-to-one correspondence with the so-called quasi-trigonometric solutions of the classical Yang-Baxter equation. In this paper we give a complete list of the quasi-trigonometric solutions in terms of sub-diagrams of the certain Dynkin diagrams related to $\mathfrak{g}$. We also explain how to quantize the corresponding Lie bialgebra structures.

1. Introduction

The present paper constitutes a step towards the classification of quantum groups. We describe an algorithm for the quantization of all Lie bialgebra structures on the polynomial Lie algebra $P = \mathfrak{g}[u]$, where $\mathfrak{g}$ is a simple complex finite-dimensional Lie algebra.

Lie bialgebra structures on P, up to so-called *classical twisting*, have been classified by F. Montaner and E. Zelmanov in [**6**]. We recall that given a Lie co-bracket δ on P, a *classical twist* is an element $s \in P \wedge P$ such that

$$(1.1) \qquad \mathrm{CYB}(s) + \mathrm{Alt}(\delta \otimes \mathrm{id})(s) = 0,$$

where CYB is the l.h.s. of the classical Yang-Baxter equation.

We also note that a classical twist does not change the classical double $D_\delta(P)$ associated to a given Lie bialgebra structure δ. If δ^s is the twisting co-bracket via s, then the Lie bialgebras (P, δ) and (P, δ^s) are in the same class, i.e. there exists a Lie algebra isomorphism between $D_\delta(P)$ and $D_{\delta^s}(P)$, preserving the canonical forms and compatible with the canonical embeddings of P into the doubles.

According to the results of Montaner and Zelmanov, there are four Lie bialgebra structures on P up to classical twisting. Let us present them:

2000 *Mathematics Subject Classification.* Primary 17B37, 17B62; Secondary 17B81.

Key words and phrases. Classical Yang–Baxter equation, r-matrix, Manin triple, parabolic subalgebra, generalized Belavin–Drinfeld data.

Acknowledgment. The first author was supported by European Community Research Training Network LIEGRITS, grant no. MRTN-CT-2003-505078.

Case 1. Consider $\delta_1 = 0$. Consequently, $D_1(P) = P + \varepsilon P^*$, where $\varepsilon^2 = 0$. The symmetric nondegenerate invariant form Q is given by the canonical pairing between P and εP^*.

Lie bialgebra structures which fall in this class are the elements $s \in P \wedge P$ satisfying $\mathrm{CYB}(s) = 0$. Such elements are in a one-to-one correspondence with finite-dimensional quasi-Frobenius Lie subalgebras of P.

Case 2. Let us consider the co-bracket δ_2 given by

$$(1.2) \qquad \delta_2(p(u)) = [r_2(u,v), p(u) \otimes 1 + 1 \otimes p(v)],$$

where $r_2(u,v) = \Omega/(u-v)$. Here Ω denotes the quadratic Casimir element on $\mathfrak{g}$.

It was proved in [**7**] that the associated classical double is $D_2(P) = \mathfrak{g}((u^{-1}))$, together with the canonical invariant form

$$(1.3) \qquad Q(f(u), g(u)) = Res_{u=0} K(f,g),$$

where K denotes the Killing form of the Lie algebra $\mathfrak{g}((u^{-1}))$ over $\mathbb{C}((u^{-1}))$.

Moreover, the Lie bialgebra structures which are obtained by twisting δ_2 are in a one-to-one correspondence with so-called *rational solutions* of the CYBE, according to [**7**].

Case 3. In this case, let us consider the Lie bialgebra structure given by

$$(1.4) \qquad \delta_3(p(u)) = [r_3(u,v), p(u) \otimes 1 + 1 \otimes p(v)],$$

with $r_3(u,v) = v\Omega/(u-v) + \Sigma_\alpha e_\alpha \otimes f_\alpha + \frac{1}{2}\Omega_0$, where e_α, f_α are root vectors of $\mathfrak{g}$ and Ω_0 is the Cartan part of Ω.

It was proved in [**4**] that the associated classical double is $D_3(P) = \mathfrak{g}((u^{-1})) \times \mathfrak{g}$, together with the invariant nondegenerate form Q defined by

$$(1.5) \qquad Q((f(u), a), (g(u), b)) = K(f(u), g(u))_0 - K(a, b),$$

where the index zero means that one takes the free term in the series expansion. According to [**4**], there is a one-to-one correspondence between Lie bialgebra structures which are obtained by twisting δ_3 and so-called *quasi-trigonometric solutions* of the CYBE.

Case 4. We consider the co-bracket on P given by

$$(1.6) \qquad \delta_4(p(u)) = [r_4(u,v), p(u) \otimes 1 + 1 \otimes p(v)],$$

with $r_4(u,v) = uv\Omega/(v-u)$.

It was shown in [**10**] that the classical double associated to the Lie bialgebra structure δ_4 is $D_4(P) = \mathfrak{g}((u^{-1})) \times (\mathfrak{g} \otimes \mathbb{C}[\varepsilon])$, where $\varepsilon^2 = 0$. The form Q is described as follows: if $f(u) = \sum_{-\infty}^{N} a_k u^k$ and $g(u) = \sum_{-\infty}^{N} b_k u^k$, then

$$Q(f(u) + A_0 + A_1\varepsilon, g(u) + B_0 + B_1\varepsilon) = Res_{u=0} u^{-2} K(f,g) - K(A_0, B_1) - K(A_1, B_0).$$

Lie bialgebra structures which are in the same class as δ_4 are in a one-to-one correspondence with *quasi-rational r-matrices*, as it was proved in [**10**].

Regarding the quantization of these Lie bialgebra structures on P, the following conjecture stated in [**4**] and proved by G. Halbout in [**3**] plays a crucial role.

THEOREM 1.1. *Any classical twist can be extended to a quantum twist, i.e., if* (L, δ) *is any Lie bialgebra, s is a classical twist, and* (A, Δ, ε) *is a quantization of* (L, δ), *there exists* $F \in A \otimes A$ *such that*

(1) $F = 1 + O(\hbar)$ *and* $F - F^{21} = \hbar s + O(\hbar^2)$,

(2) $(\Delta \otimes \mathrm{id})(F)F^{12} - (\mathrm{id} \otimes \Delta)(F)F^{23} = 0$,

(3) $(\varepsilon \otimes \mathrm{id})(F) = (\mathrm{id} \otimes \varepsilon)(F) = 1$.

Moreover gauge equivalence classes of quantum twists for A are in bijection with gauge equivalence classes of $\hbar$-*dependent classical twists* $s_\hbar = \hbar s_1 + O(\hbar^2)$ *for* L.

Let us suppose that we have a Lie bialgebra structure δ on P. Then δ is obtained by twisting one of the four structures δ_i from Cases 1–4. This above theorem implies that in order to find a quantization for (P, δ), it is sufficient to determine the quantization of δ_i and then find the quantum twist whose classical limit is s. Let us note that the quantization of (P, δ_3) is well-known. The corresponding quantum algebra was introduced by V. Tolstoy in [11] and it is denoted by $U_q(\mathfrak{g}[u])$.

The quasi-trigonometric solutions of the CYBE were studied in [5], where it was proved that they fall into classes, which are in a one-to-one correspondence with vertices of the extended Dynkin diagram of $\mathfrak{g}$. Let us consider corresponding roots, namely simple roots $\alpha_1, \alpha_2, \cdots \alpha_r$ and $\alpha_0 = -\alpha_{\mathrm{max}}$. In [5] quasi-trigonometric solutions corresponding to the simple roots which have coefficient one in the decomposition of the maximal root were classified. It was also proved there that quasi-trigonometric solutions corresponding to α_0 are in a one-to-one correspondence with constant solutions of the modified CYBE classified in [1] and the polynomial part of these solutions is constant. The aim of our paper is to obtain a complete classification of quasi-trigonometric solutions of the CYBE. In particular, we describe all the quasi-trigonometric solutions with non-trivial polynomial part for $\mathfrak{g} = o(5)$.

2. Lie bialgebra structures associated with quasi-trigonometric solutions

DEFINITION 2.1. A solution X of the CYBE is called *quasi-trigonometric* if it is of the form $X(u, v) = v\Omega/(u - v) + p(u, v)$, where p is a polynomial with coefficients in $\mathfrak{g} \otimes \mathfrak{g}$.

The class of quasi-trigonometric solutions is closed under *gauge transformations*. We first need to introduce the following notation: Let R be a commutative ring and let L be a Lie algebra over R. Let us denote by $\mathrm{Aut}_R(L)$ the group of automorphisms of L over R. In other words we consider such automorphisms of L, which satisfy the condition $f(rl) = rf(l)$, where $r \in R$, $l \in L$.

At this point we note that there exists a natural embedding

$$\mathrm{Aut}_{\mathbb{C}[u]}(\mathfrak{g}[u]) \hookrightarrow \mathrm{Aut}_{\mathbb{C}((u^{-1}))}(\mathfrak{g}((u^{-1}))),$$

defined by the formula

$$\sigma(u^{-k}x) = u^{-k}\sigma(x),$$

for any $\sigma \in \mathrm{Aut}_{\mathbb{C}[u]}(\mathfrak{g}[u])$ and $x \in \mathfrak{g}[u]$.

Now if X is a quasi-trigonometric solution and $\sigma(u) \in \mathrm{Aut}_{\mathbb{C}[u]}(\mathfrak{g}[u])$, one can check that the function $Y(u, v) := (\sigma(u) \otimes \sigma(v))(X(u, v))$ is again a quasi-trigonometric solution. X and Y are said to be *gauge equivalent*.

THEOREM 2.2. *There exists a natural one-to-one correspondence between quasi-trigonometric solutions of CYBE for $\mathfrak{g}$ and linear subspaces W of $\mathfrak{g}((u^{-1})) \times \mathfrak{g}$ which satisfy the following properties:*

(1) W is a Lie subalgebra of $\mathfrak{g}((u^{-1})) \times \mathfrak{g}$ and $W \supseteq u^{-N}\mathfrak{g}[[u^{-1}]]$ for some positive integer N.

(2) $W \oplus \mathfrak{g}[u] = \mathfrak{g}((u^{-1})) \times \mathfrak{g}$.

(3) W is a Lagrangian subspace of $\mathfrak{g}((u^{-1})) \times \mathfrak{g}$ with respect to the invariant bilinear form Q given by (1.5).

Let $\sigma(u) \in \mathrm{Aut}_{\mathbb{C}[u]}(\mathfrak{g}[u])$. Let $\tilde{\sigma}(u) = \sigma(u) \oplus \sigma(0)$ be the induced automorphism of $\mathfrak{g}((u^{-1})) \times \mathfrak{g}$.

DEFINITION 2.3. We will say that W_1 and W_2 are *gauge equivalent* if there exists $\sigma(u) \in \mathrm{Aut}_{\mathbb{C}[u]}(\mathfrak{g}[u])$ such that $W_1 = \tilde{\sigma}(u)W_2$.

It was checked in [4] that two quasi-trigonometric solutions are gauge equivalent if and only if the corresponding subalgebras are gauge equivalent.

Let $\mathfrak{h}$ be a Cartan subalgebra of $\mathfrak{g}$ with the corresponding set of roots R and a choice of simple roots Γ. Denote by $\mathfrak{g}_\alpha$ the root space corresponding to a root α. Let $\mathfrak{h}(\mathbb{R})$ be the set of all $h \in \mathfrak{h}$ such that $\alpha(h) \in \mathbb{R}$ for all $\alpha \in R$. Consider the valuation on $\mathbb{C}((u^{-1}))$ defined by $v(\sum_{k \geq n} a_k u^{-k}) = n$. For any root α and any $h \in \mathfrak{h}(\mathbb{R})$, set $M_\alpha(h) := \{f \in \mathbb{C}((u^{-1})) : v(f) \geq \alpha(h)\}$. Consider

$$(2.1) \qquad \mathbb{O}_h := \mathfrak{h}[[u^{-1}]] \oplus (\oplus_{\alpha \in R} M_\alpha(h) \otimes \mathfrak{g}_\alpha).$$

As it was shown in [5], any maximal order W which corresponds to a quasi-trigonometric solution of the CYBE, can be embedded (up to some gauge equivalence) into $\mathbb{O}_h \times \mathfrak{g}$. Moreover h may be taken as a vertex of the standard simplex $\Delta_{st} = \{h \in \mathfrak{h}(\mathbb{R}) : \alpha(h) \geq 0 \text{ for all } \alpha \in \Gamma \text{ and } \alpha_{\max} \leq 1\}$.

Vertices of the above simplex correspond to vertices of the extended Dynkin diagram of $\mathfrak{g}$, the correspondence being given by the following rule:

$$0 \leftrightarrow \alpha_{\max}$$

$$h_i \leftrightarrow \alpha_i,$$

where $\alpha_i(h_j) = \delta_{ij}/k_j$ and k_j are given by the relation $\sum k_j \alpha_j = \alpha_{\max}$. We will write $\mathbb{O}_\alpha$ instead of $\mathbb{O}_h$ if α is the root which corresponds to the vertex h.

By straightforward computations, one can check the following two results:

LEMMA 2.4. *Let R be the set of all roots and α an arbitrary simple root. Let k be the coefficient of α in the decomposition of $\alpha_{\max}$.*

For each r, $-k \leq r \leq k$, let R_r denote the set of all roots which contain α with coefficient r. Let $\mathfrak{g}_0 = \mathfrak{h} \oplus \sum_{\beta \in R_0} \mathfrak{g}_\beta$ and $\mathfrak{g}_r = \sum_{\beta \in R_r} \mathfrak{g}_\beta$. Then

$$(2.2) \qquad \mathbb{O}_\alpha = \sum_{r=1}^{k} u^{-1}\mathbb{O}\mathfrak{g}_r + \sum_{r=1-k}^{0} \mathbb{O}\mathfrak{g}_r + u\mathbb{O}\mathfrak{g}_{-k},$$

where $\mathbb{O} := \mathbb{C}[[u^{-1}]]$.

LEMMA 2.5. *Let α be a simple root and k its coefficient in the decomposition of $\alpha_{\max}$. Let Δ_α denote the set of all pairs (a, b), $a \in \mathfrak{g}_0 + \mathfrak{g}_{-k}$, $b \in \mathfrak{g}_0 + \mathfrak{g}_{-1} + \ldots + \mathfrak{g}_{-k}$, $a = a_0 + a_{-k}$, $b = b_0 + b_{-1} + \ldots + b_{-k}$ and $a_0 = b_0$. Then*

(i) The orthogonal complement of $\mathbb{O}_\alpha \times \mathfrak{g}$ with respect to Q is given by

$$(2.3) \qquad (\mathbb{O}_\alpha \times \mathfrak{g})^\perp = \sum_{r=-k}^{-1} \mathbb{O}\mathfrak{g}_r + \sum_{r=0}^{k-1} u^{-1}\mathbb{O}\mathfrak{g}_r + u^{-2}\mathbb{O}\mathfrak{g}_k.$$

(ii) There exists an isomorphism σ

$$(2.4) \qquad \frac{\mathbb{O}_\alpha \times \mathfrak{g}}{(\mathbb{O}_\alpha \times \mathfrak{g})^\perp} \cong (\mathfrak{g}_k \oplus \mathfrak{g}_0 \oplus \mathfrak{g}_{-k}) \times \mathfrak{g}$$

given by

$$\sigma((f,a) + (\mathbb{O}_\alpha \times \mathfrak{g})^\perp) = (a_0 + b_0 + c_0, a),$$

where the element $f \in \mathbb{O}_\alpha$ is decomposed according to Lemma 2.4:

$$f = u^{-1}(a_0 + a_1 u^{-1} + ...) + (b_0 + b_1 u^{-1} + ...) + u(c_0 + c_1 u^{-1} + ...) + ...,$$

$a_i \in \mathfrak{g}_k$, $b_i \in \mathfrak{g}_0$, $c_i \in \mathfrak{g}_{-k}$ and $a \in \mathfrak{g}$.
(iii) $(\mathbb{O}_\alpha \times \mathfrak{g}) \cap \mathfrak{g}[u]$ is sent via the isomorphism σ to Δ_α.

Let us make an important remark. The Lie subalgebra $\mathfrak{g}_k + \mathfrak{g}_0 + \mathfrak{g}_{-k}$ of $\mathfrak{g}$ coincides with the semisimple Lie algebra whose Dynkin diagram is obtained from the extended Dynkin diagram of $\mathfrak{g}$ by crossing out α. Let us denote this subalgebra by L_α. The Lie algebra $L_\alpha \times \mathfrak{g}$ is endowed with the following invariant bilinear form:

$$(2.5) \qquad Q'((a,b),(c,d)) = K(a,c) - K(b,d),$$

for any $a, c \in L_\alpha$ and $b, d \in \mathfrak{g}$.

On the other hand, $\mathfrak{g}_0 + \mathfrak{g}_{-k}$ is the parabolic subalgebra $P^+_{-\alpha_{\max}}$ of L_α which corresponds to $-\alpha_{\max}$. The Lie subalgebra $\mathfrak{g}_0 + \mathfrak{g}_{-1} + ... + \mathfrak{g}_{-k}$ is the parabolic subalgebra P^-_α of $\mathfrak{g}$ which corresponds to the root α and contains the negative Borel subalgebra. Let us also note that $\mathfrak{g}_0$ is precisely the reductive part of P^-_α and of $P^+_{-\alpha_{\max}}$. We can conclude that the set Δ_α consists of all pairs $(a,b) \in P^+_{-\alpha_{\max}} \times P^-_\alpha$ whose reductive parts are equal.

THEOREM 2.6. *Let α be a simple root. There is a one-to-one correspondence between Lagrangian subalgebras W of $\mathfrak{g}((u^{-1})) \times \mathfrak{g}$ which are contained in $\mathbb{O}_\alpha \times \mathfrak{g}$ and transversal to $\mathfrak{g}[u]$, and Lagrangian subalgebras $\mathfrak{l}$ of $L_\alpha \times \mathfrak{g}$ transversal to Δ_α (with respect to the bilinear form Q').*

PROOF. Since W is a subspace of $\mathbb{O}_\alpha \times \mathfrak{g}$, let $\mathfrak{l}$ be its image in $L_\alpha \times \mathfrak{g}$. Because W is transversal to $\mathfrak{g}[u]$, one can check that $\mathfrak{l}$ is transversal to the image of $(\mathbb{O}_\alpha \times \mathfrak{g}) \cap \mathfrak{g}[u]$ in $L_\alpha \times \mathfrak{g}$, which is exactly Δ_α. The fact that W is Lagrangian implies that $\mathfrak{l}$ is also Lagrangian.

Conversely, if $\mathfrak{l}$ is a Lagrangian subalgebra of $L_\alpha \times \mathfrak{g}$ transversal to Δ_α, then its preimage W in $\mathbb{O}_\alpha \times \mathfrak{g}$ is transversal to $\mathfrak{g}[u]$ and Lagrangian as well.

$\square$

The Lagrangian subalgebras $\mathfrak{l}$ of $L_\alpha \times \mathfrak{g}$ which are transversal to Δ_α, can be determined using results of P. Delorme [2] on the classification of Manin triples. We are interested in determining Manin triples of the form $(Q', \Delta_\alpha, \mathfrak{l})$.

Let us recall Delorme's construction of so-called *generalized Belavin-Drinfeld data*. Let $\mathfrak{r}$ be a finite-dimensional complex, reductive, Lie algebra and B a symmetric, invariant, nondegenerate bilinear form on $\mathfrak{r}$. The goal in [2] is to classify all

Manin triples of $\mathfrak{r}$ up to conjugacy under the action on $\mathfrak{r}$ of the simply connected Lie group $\mathcal{R}$ whose Lie algebra is $\mathfrak{r}$.

One denotes by $\mathfrak{r}_+$ and $\mathfrak{r}_-$ respectively the sum of the simple ideals of $\mathfrak{r}$ for which the restriction of B is equal to a positive (negative) multiple of the Killing form. Then the derived ideal of $\mathfrak{r}$ is the sum of $\mathfrak{r}_+$ and $\mathfrak{r}_-$.

Let $\mathfrak{j}_0$ be a Cartan subalgebra of $\mathfrak{r}$, $\mathfrak{b}_0$ a Borel subalgebra containing $\mathfrak{j}_0$ and $\mathfrak{b}_0'$ be its opposite. Choose $\mathfrak{b}_0 \cap \mathfrak{r}_+$ as Borel subalgebra of $\mathfrak{r}_+$ and $\mathfrak{b}_0' \cap \mathfrak{r}_-$ as Borel subalgebra of $\mathfrak{r}_-$. Denote by Σ_+ (resp., Σ_-) the set of simple roots of $\mathfrak{r}_+$ (resp., $\mathfrak{r}_-$) with respect to the above Borel subalgebras. Let $\Sigma = \Sigma_+ \cup \Sigma_-$ and denote by $\mathcal{W} = (H_\alpha, X_\alpha, Y_\alpha)_{\alpha \in \Sigma_+}$ a Weyl system of generators of $[\mathfrak{r}, \mathfrak{r}]$.

DEFINITION 2.7 (Delorme, [2]). One calls $(A, A', \mathfrak{i}_\mathfrak{a}, \mathfrak{i}_{\mathfrak{a}'})$ *generalized Belavin-Drinfeld data* with respect to B when the following five conditions are satisfied:

(1) A is a bijection from a subset Γ_+ of Σ_+ on a subset Γ_- of Σ_- such that

$$B(H_{A\alpha}, H_{A\beta}) = -B(H_\alpha, H_\beta), \alpha, \beta \in \Gamma_+.$$

(2) A' is a bijection from a subset Γ_+' of Σ_+ on a subset Γ_-' of Σ_- such that

$$B(H_{A'\alpha}, H_{A'\beta}) = -B(H_\alpha, H_\beta), \alpha, \beta \in \Gamma_+'.$$

(3) If $C = A^{-1}A'$ is the map defined on $\mathrm{dom}(C) = \{\alpha \in \Gamma_+' : A'\alpha \in \Gamma_-\}$ by $C\alpha = A^{-1}A'\alpha$, then C satisfies:

For all $\alpha \in \mathrm{dom}(C)$, there exists a positive integer n such that $\alpha, ..., C^{n-1}\alpha \in \mathrm{dom}(C)$ and $C^n\alpha \notin \mathrm{dom}(C)$.

(4) $\mathfrak{i}_\mathfrak{a}$ (resp., $\mathfrak{i}_{\mathfrak{a}'}$) is a complex vector subspace of $\mathfrak{j}_0$, included and Lagrangian in the orthogonal $\mathfrak{a}$ (resp., $\mathfrak{a}'$) to the subspace generated by H_α, $\alpha \in \Gamma_+ \cup \Gamma_-$ (resp., $\Gamma_+' \cup \Gamma_-'$).

(5) If $\mathfrak{f}$ is the subspace of $\mathfrak{j}_0$ generated by the family $H_\alpha + H_{A\alpha}$, $\alpha \in \Gamma_+$, and $\mathfrak{f}'$ is defined similarly, then

$$(\mathfrak{f} \oplus \mathfrak{i}_\mathfrak{a}) \cap (\mathfrak{f}' \oplus \mathfrak{i}_{\mathfrak{a}'}) = 0.$$

Let R_+ be the set of roots of $\mathfrak{j}_0$ in $\mathfrak{r}$ which are linear combinations of elements of Γ_+. One defines similarly R_-, R_+' and R_-'. The bijections A and A' can then be extended by linearity to bijections from R_+ to R_- (resp., R_+' to R_-'). If A satisfies condition (1), then there exists a unique isomorphism τ between the subalgebra $\mathfrak{m}_+$ of $\mathfrak{r}$ spanned by X_α, H_α and Y_α, $\alpha \in \Gamma_+$, and the subalgebra $\mathfrak{m}_-$ spanned by X_α, H_α and Y_α, $\alpha \in \Gamma_-$, such that $\tau(H_\alpha) = H_{A\alpha}$, $\tau(X_\alpha) = X_{A\alpha}$, $\tau(Y_\alpha) = Y_{A\alpha}$ for all $\alpha \in \Gamma_+$. If A' satisfies (2), then one defines similarly an isomorphism τ' between $\mathfrak{m}_+'$ and $\mathfrak{m}_-'$.

THEOREM 2.8 (Delorme, [2]). *(i) Let $\mathcal{BD} = (A, A', \mathfrak{i}_\mathfrak{a}, \mathfrak{i}_{\mathfrak{a}'})$ be generalized Belavin-Drinfeld data, with respect to B. Let $\mathfrak{n}$ be the sum of the root spaces relative to roots α of $\mathfrak{j}_0$ in $\mathfrak{b}_0$, which are not in $R_+ \cup R_-$. Let $\mathfrak{i} := \mathfrak{k} \oplus \mathfrak{i}_\mathfrak{a} \oplus \mathfrak{n}$, where $\mathfrak{k} := \{X + \tau(X) : X \in \mathfrak{m}_+\}$.*

Let $\mathfrak{n}'$ be the sum of the root spaces relative to roots α of $\mathfrak{j}_0$ in $\mathfrak{b}_0'$, which are not in $R_+' \cup R_-'$. Let $\mathfrak{i}' := \mathfrak{k}' \oplus \mathfrak{i}_{\mathfrak{a}'} \oplus \mathfrak{n}'$, where $\mathfrak{k}' := \{X + \tau'(X) : X \in \mathfrak{m}_+'\}$.

Then $(B, \mathfrak{i}, \mathfrak{i}')$ is a Manin triple.

(ii) Every Manin triple is conjugate by an element of $\mathcal{R}$ to a unique Manin triple of this type.

Let us consider the particular case $\mathfrak{r} = L_\alpha \times \mathfrak{g}$. We set $\Sigma_+ := (\Gamma^{ext} \setminus \{\alpha\}) \times \{0\}$, $\Sigma_- := \{0\} \times \Gamma$ and $\Sigma := \Sigma_+ \cup \Sigma_-$.

Denote by $(X_\gamma, Y_\gamma, H_\gamma)_{\gamma \in \Gamma}$ a Weyl system of generators for $\mathfrak{g}$ with respect to the root system Γ. Denote $-\alpha_{\max}$ by α_0. Let H_{α_0} be the coroot of α_0. We choose $X_{\alpha_0} \in \mathfrak{g}_{\alpha_0}$, $Y_{\alpha_0} \in \mathfrak{g}_{-\alpha_0}$ such that $[X_{\alpha_0}, Y_{\alpha_0}] = H_{\alpha_0}$.

A Weyl system of generators in $L_\alpha \times \mathfrak{g}$ (with respect to the root system Σ) is the following: $X_{(\beta,0)} = (X_\beta, 0)$, $H_{(\beta,0)} = (H_\beta, 0)$, $Y_{(\beta,0)} = (Y_\beta, 0)$, for any $\beta \in \Gamma^{ext} \setminus \{\alpha\}$, and $X_{(0,\gamma)} = (0, X_\gamma)$, $H_{(0,\gamma)} = (0, H_\gamma)$, $Y_{(0,\gamma)} = (0, Y_\gamma)$, for any $\gamma \in \Gamma$.

By applying the general result of Delorme, one can deduce the description of the Manin triples of the form $(Q', \Delta_\alpha, \mathfrak{l})$.

COROLLARY 2.9. *Let* $S := \Gamma \setminus \{\alpha\}$ *and* $\zeta_S := \{h \in \mathfrak{h} : \beta(h) = 0, \forall \beta \in S\}$. *For any Manin triple* $(Q', \Delta_\alpha, \mathfrak{l})$, *there exists a unique generalized Belavin-Drinfeld data* $\mathcal{BD} = (A, A', \mathfrak{i}_\mathfrak{a}, \mathfrak{i}_{\mathfrak{a}'})$ *where* $A : S \times \{0\} \longrightarrow \{0\} \times S$, $A(\gamma, 0) = (0, \gamma)$ *and* $\mathfrak{i}_\mathfrak{a} = \mathrm{diag}(\zeta_S)$, *such that* $(Q', \Delta_\alpha, \mathfrak{l})$ *is conjugate to the Manin triple* $\mathcal{T}_{\mathcal{BD}} = (Q', \mathfrak{i}, \mathfrak{i}')$. *Moreover, up to a conjugation which preserves* Δ_α, *one has* $\mathfrak{l} = \mathfrak{i}'$.

PROOF. Let us suppose that $(Q', \Delta_\alpha, \mathfrak{l})$ is a Manin triple. Then there exists a unique generalized Belavin-Drinfeld data $\mathcal{BD} = (A, A', \mathfrak{i}_\mathfrak{a}, \mathfrak{i}_{\mathfrak{a}'})$ such that the corresponding $\mathcal{T}_{\mathcal{BD}} = (Q', \mathfrak{i}, \mathfrak{i}')$ is conjugate to $(Q', \Delta_\alpha, \mathfrak{l})$. Since $\mathfrak{i}$ and Δ_α are conjugate and Δ_α is "under" the parabolic subalgebra $P_{\alpha_0}^+ \times P_\alpha^-$, it follows that $\mathfrak{i}$ is also "under" this parabolic and thus $\mathfrak{a} = \zeta_S \times \zeta_S$. According to [**2**], p. 136, the map A should be an isometry between $S \times \{0\}$ and $\{0\} \times S$, so it is given by an isometry $\tilde{A}$ of S.

Let $\mathfrak{m}$ be spanned by $X_\beta, H_\beta, Y_\beta$ for all $\beta \in S$. Then $\mathfrak{i}$ contains $\mathfrak{k} := \{(X, \tau(X)) : X \in \mathfrak{m}\}$, where τ has the property that $\tau(X_\beta) = X_{\tilde{A}\beta}$, $\tau(H_\beta) = H_{\tilde{A}\beta}$, $\tau(Y_\beta) = Y_{\tilde{A}\beta}$, for all $\beta \in S$.

Since $\mathfrak{i}$ and Δ_α are conjugate, τ has to be an inner automorphism of $\mathfrak{m}$ and $\mathfrak{i}_\mathfrak{a} = \mathrm{diag}(\zeta_S)$. It follows that $\tilde{A} = \mathrm{id}$. This ends the proof. $\square$

We will consider triples of the form $(\Gamma_1', \Gamma_2', \tilde{A}')$, where $\Gamma_1' \subseteq \Gamma^{ext} \setminus \{\alpha\}$, $\Gamma_2' \subseteq \Gamma$ and $\tilde{A}'$ is an isometry between Γ_1' and Γ_2'.

We say that such a triple is of *type I* if $\alpha \notin \Gamma_2'$ and $(\Gamma_1', \Gamma_2', \tilde{A}')$ is an admissible triple in the sense of [**1**]. A triple is of *type II* if $\alpha \in \Gamma_2'$ and $\tilde{A}'(\beta) = \alpha$, for some $\beta \in \Gamma_1'$ and $(\Gamma_1' \setminus \{\beta\}, \Gamma_2' \setminus \{\alpha\}), \tilde{A}')$ is an admissible triple in the sense of [**1**].

By using the definition of generalized Belavin-Drinfeld data, one can easily check the following:

LEMMA 2.10. *Let* $A : S \times \{0\} \longrightarrow \{0\} \times S$, $A(\gamma, 0) = (0, \gamma)$ *and* $\mathfrak{i}_\mathfrak{a} = \mathrm{diag}(\zeta_S)$. *A quadruple* $(A, A', \mathfrak{i}_\mathfrak{a}, \mathfrak{i}_{\mathfrak{a}'})$ *is generalized Belavin-Drinfeld data if and only if the pair* $(A', \mathfrak{i}_{\mathfrak{a}'})$ *satisfies the following conditions:*

(1) $A' : \Gamma_1' \times \{0\} \longrightarrow \{0\} \times \Gamma_2'$ *is given by* $A'(\gamma, 0) = (0, \tilde{A}'(\gamma))$ *and* $(\Gamma_1', \Gamma_2', \tilde{A}')$ *is of type I or II from above.*

(2) Let $\mathfrak{f}'$ *be the subspace of* $\mathfrak{h} \times \mathfrak{h}$ *spanned by pairs* $(H_\beta, H_{\tilde{A}'(\beta)})$ *for all* $\beta \in \Gamma_1'$. *Let* $\mathfrak{i}_{\mathfrak{a}'}$ *be Lagrangian subspace of* $\mathfrak{a}' := \{(h_1, h_2) \in \mathfrak{h} \times \mathfrak{h} : \beta(h_1) = 0, \gamma(h_2) = 0, \forall \beta \in \Gamma_1', \forall \gamma \in \Gamma_2'\}$. *Then*

$$(2.6) \qquad (\mathfrak{f}' \oplus \mathfrak{i}_{\mathfrak{a}'}) \cap \mathrm{diag}(\mathfrak{h}) = 0.$$

REMARK 2.11. One can always find $\mathfrak{i}_{\mathfrak{a}'}$ which is a Lagrangian subspace of $\mathfrak{a}'$ and satisfies condition (2.6). This is a consequence of [**2**] Remark 2, p. 142. A proof of this elementary fact can also be found in [**9**], Lemma 5.2.

Summing up the previous results we conclude the following:

THEOREM 2.12. *Let α be a simple root. Suppose that $\mathfrak{l}$ is a Lagrangian subalgebra of $L_\alpha \times \mathfrak{g}$ transversal to Δ_α. Then, up to a conjugation which preserves Δ_α, one has $\mathfrak{l} = \mathfrak{i}'$, where $\mathfrak{i}'$ is constructed from a pair formed by a triple $(\Gamma_1', \Gamma_2', \tilde{A}')$ of type I or II and a Lagrangian subspace $\mathfrak{i}_{\mathfrak{a}'}$ of $\mathfrak{a}'$ such that (2.6) is satisfied.*

Let us apply this result to classify solutions in $\mathfrak{g} = o(5)$.

COROLLARY 2.13. *Let α_1, α_2 be the simple roots in $o(5)$ and $\alpha_0 = -2\alpha_1 - \alpha_2$. Up to gauge equivalence, there exist two quasi-trigonometric solutions with nontrivial polynomial part.*

PROOF. The root α_2 has coefficient $k = 1$ in the decomposition of the maximal root. The only possible choice for a triple $(\Gamma_1', \Gamma_2', \tilde{A}')$ with $\Gamma_1' \subseteq \{\alpha_0, \alpha_1\}$, $\Gamma_2' \subseteq \{\alpha_1, \alpha_2\}$ to be of type I or II is $\Gamma_1' = \{\alpha_0,\}$, $\Gamma_2' = \{\alpha_2,\}$, $\tilde{A}'(\alpha_0) = \alpha_2$. One can check that $\mathfrak{i}_{\mathfrak{a}'}$ is a 1-dimensional space spanned by the following pair $(\text{diag}(-2, 1, 0, -1, 2), \text{diag}(0, \sqrt{5}, 0, -\sqrt{5}, 0))$. The Lagrangian subalgebra $\mathfrak{i}_2'$ constructed from this triple is transversal to Δ_{α_2} in $\mathfrak{g} \times \mathfrak{g}$.

The root α_1 has coefficient $k = 2$ in the decomposition of the maximal root. The only possible choice for a triple $(\Gamma_1', \Gamma_2', \tilde{A}')$ with $\Gamma_1' \subseteq \{\alpha_0, \alpha_2\}$, $\Gamma_2' \subseteq \{\alpha_1, \alpha_2\}$ is again $\Gamma_1' = \{\alpha_0,\}$, $\Gamma_2' = \{\alpha_2,\}$, $\tilde{A}'(\alpha_0) = \alpha_2$ and $\mathfrak{i}_{\mathfrak{a}'}$ is as in the previous case. The Lagrangian subalgebra $\mathfrak{i}_1'$ constructed from this triple is transversal to Δ_{α_1} in $L_{\alpha_1} \times \mathfrak{g}$.

$\square$

References

[1] A. Belavin, V. Drinfeld, *Triangle equations and simple Lie algebras.* Math. Phys. Reviews, Vol. **4** (1984), Harwood Academic, 93–165.

[2] P. Delorme, *Classification des triples de Manin pour les algèbres de Lie réductives complexes.* J. Algebra, **246** (2001), 97–174.

[3] G. Halbout, *Formality theorem for Lie bialgebras and quantization of twists and coboundary r-matrices.* Adv. Math. **207** (2006), 617–633.

[4] S. Khoroshkin, I. Pop, A. Stolin, V. Tolstoy, *On some Lie bialgebra structures on polynomial algebras and their quantization.* Preprint no. 21, 2003/2004, Mittag-Leffler Institute, Sweden.

[5] S. Khoroshkin, I. Pop, M. Samsonov, A. Stolin, V. Tolstoy, *On some Lie bialgebra structures on polynomial algebras and their quantization.* ArXiv math. QA/0706.1651v1. To appear in Comm. Math. Phys.

[6] F. Montaner, E. Zelmanov, *Bialgebra structures on current Lie algebras.* Preprint, University of Wisconsin, Madison, 1993.

[7] A. Stolin, *On rational solutions of Yang-Baxter equations. Maximal orders in loop algebra.* Comm. Math. Phys. **141** (1991), 533–548.

[8] A. Stolin, *A geometrical approach to rational solutions of the classical Yang-Baxter equation.* Part I. *Symposia Gaussiana, Conf.A,* Walter de Gruyter, Berlin, New York, 1995, 347–357.

[9] A. Stolin, *Some remarks on Lie bialgebra structures for simple complex Lie algebras.* Comm. Alg. **27** (9) (1999), 4289–4302.

[10] A. Stolin, J. Yermolova–Magnusson, *The 4th structure.* Czech. J. Phys, Vol. **56**, No. 10/11 (2006), 1293–1927.

[11] V. Tolstoy, *From quantum affine Kac–Moody algebras to Drinfeldians and Yangians.* in *Kac–Moody Lie algebras and related topics,* Contemp. Math. **343**, Amer. Math. Soc. 2004, 349–370.

DEPARTMENT OF MATHEMATICAL SCIENCES, UNIVERSITY OF GOTHENBURG, SWEDEN
E-mail address: iulia@math.chalmers.se, astolin@math.chalmers.se

Contemporary Mathematics
Volume **478**, 2009

The Relevance and the Ubiquity of Prüfer Modules

Claus Michael Ringel

ABSTRACT. Let R be a ring. An R-module M is called a Prüfer module provided there exists a locally nilpotent, surjective endomorphism of M with kernel of finite length. We want to outline the relevance, but also the ubiquity of Prüfer modules. The main assertion will be that any Prüfer module which is not of finite type gives rise to a generic module, thus to infinite families of indecomposable modules with fixed endo-length (here we are in the setting of the second Brauer-Thrall conjecture). In addition, we will report on a construction procedure which yields a wealth of Prüfer modules. Unfortunately, we do not know which modules obtained in this way are of finite type.

Introduction.

This is the written account of a lecture given at 4th International Conference on Representation Theory (ICRT-IV), Lhasa, July 16–20, 2007. Section 1 recalls the definition of a Prüfer module as introduced in [R6] and provides some examples, in Section 2 we show that the degeneration theory of modules concerns certain Prüfer modules of finite type, whereas Section 4 provides a construction of Prüfer modules using pairs of monomorphisms $U_0 \to U_1$; these two sections 2 and 4 are reports on some of the results of [R6]. In the last Section 6 we discuss the question how to search for pairs of monomorphisms $U_0 \to U_1$. This is an announcement of results of the forthcoming paper [R9], where "take-off categories" are introduced. The central part is section 3. There, we show that the existence of a Prüfer module which is not of finite type implies the existence of a generic module, thus of infinite families of indecomposable modules with fixed (and arbitrarily large) endolength. This has not yet appeared in print (but see [R8]) and has been announced under the title: *Prüfer modules which are not of finite type*. Also section 5 is new, here we show that given a tame hereditary algebra, a finite length module N can generate a non-finite-type module M only in case N has a preprojective direct summand which is sincere. This gives an indication why it seems to be reasonable to look for module embeddings in take-off categories.

1991 *Mathematics Subject Classification.* Primary 16D10, 16G60. Secondary: 16D70, 16D90 16S50, 16G20, 16P99.

1. Prüfer modules.

(1.1). Let R be any ring. We deal with (left) R-modules. An R-module M is called a *Prüfer module* provided there exists an endomorphism ϕ of M with the following properties: ϕ is locally nilpotent, surjective, and the kernel W of ϕ is non-zero and of finite length. The module W is called the *basis* of M. Let $W[n]$ be the kernel of ϕ^n, then $M = \bigcup_n W[n]$, thus we also may write $M = W[\infty]$.

(1.2) The classical example. Let $R = \mathbb{Z}$ and p a prime number. Let $S = \mathbb{Z}[p^{-1}]$ the subring of $\mathbb{Q}$ generated by p^{-1}. Then

$$S/R = \lim_{\rightarrow} \mathbb{Z}/\mathbb{Z}p^n$$

is the Prüfer group for the prime p. These Prüfer groups are all the indecomposable $\mathbb{Z}$-modules which are Prüfer modules.

(1.3). More generally, let R be a Dedekind ring. Then any indecomposable R-module of finite length is of the from $W[n]$, where W is the basis of an indecomposable Prüfer module and $n \in \mathbb{N}$. In particular, this applies to $R = \mathbb{Z}$, but also say to the polynomial ring $R = k[T]$ in one variable with coefficients in the field k. Note that a $k[T]$-module is just a pair (V, f), where V is a k-space and f is a linear operator on V. If char $k = 0$, then the pair $(k[T], \frac{\mathrm{d}}{\mathrm{d}t})$ is a Prüfer module.

As Atiyah [A] has shown, a corresponding assertion holds for the coherent sheaves over an elliptic curve: any indecomposable coherent sheaf of finite length is of the from $W[n]$, where W is the basis of an indecomposable Prüfer module and $n \in \mathbb{N}$.

(1.4) Example. Consider the Kronecker algebra $\Lambda = kQ$, this is the path algebra of the quiver Q

$$\circ \rightleftarrows \circ$$

The embedding functor

$$\operatorname{mod} k[T] \longrightarrow \operatorname{mod} k\Lambda \qquad\qquad (V, f) \mapsto V \underset{f}{\overset{1}{\rightleftarrows}} V$$

preserves Prüfer modules. Using this functor, we obtain all indecomposable Prüfer modules for the Kronecker algebra with one exception, the remaining one is of the form

$$V \underset{1}{\overset{f}{\rightleftarrows}} V$$

where (V, f) is the indecomposable Prüfer $k[T]$-module such that f has 0 as eigenvalue.

(1.5). We also should mention a famous theorem of Crawley-Boevey [CB1]: Let Λ be a finite-dimensional k-algebra and k an algebraically closed field. If Λ is **tame,** and $d \in \mathbb{N}$, then almost all indecomposable Λ-modules of length d are of the from $W[n]$, where $W[\infty]$ is an indecomposable Prüfer module.

(1.6) Warning. The Prüfer modules as defined above do not have to be indecomposable: for example the countable direct sum $W^{(\mathbb{N})}$ of copies of W with the shift endomorphism $(w_1, w_2, \dots) \mapsto (w_2, w_3, \dots)$ is a Prüfer module: the *trivial* Prüfer module with basis W. Less trivial examples will be seen in the next section.

(1.7) LEMMA. *Let M be a Prüfer module with basis W. If the endomorphism ring of W is a division ring, then either $M = W^{(\mathbb{N})}$ or else M is indecomposable.*

PROOF. This is an immediate consequence of the process of simplification [R1]. □

A module M is of *finite type* provided it is the direct sum of finitely generated modules and such that there are only finitely many isomorphism classes of indecomposable direct summands. Our main concern will be Prüfer modules which are not of finite type. But first we consider some Prüfer modules of finite type.

2. Degenerations of modules

In this and the next section we consider artin algebras (these are rings which are module finite over the center, the center being artinian).

(2.1). In order to motivate the notion of a degeneration, let us consider first the case of Λ being a finite-dimensional k-algebra where k is an algebraically closed field.

Assume that Λ is generated as a k-algebra by $a_1, a_2, \ldots, a_t$ subject to relations ρ_i. For $d \in \mathbb{N}$, we consider the variety

$$\mathcal{M}(d) = \{(A_1, \ldots, A_t) \in M(d \times d, k)^t \mid \rho_i(A_1, \ldots, A_t) = 0 \text{ for all } i\},$$

of d-dimensional Λ-modules: its elements are the d-dimensional Λ-modules with underlying vector space k^d (thus, up to isomorphism, all d-dimensional Λ-modules). The group $\mathrm{GL}(d, k)$ operates on $\mathcal{M}(d)$ by simultaneous conjugation. Elements of $\mathcal{M}(d)$ belong to the same orbit if and only if they are isomorphic.

THEOREM (ZWARA). *Let X, Y be d-dimensional Λ-modules. Then Y is in the orbit closure of X if and only if there exists a finitely generated Λ-module U and an exact sequence*

$$0 \to U \to X \oplus U \to Y \to 0.$$

Such a sequence should be called a *Riedtmann-Zwara sequence*.

(2.2). Now let Λ be an arbitrary artin algebra and X, Y Λ-modules of finite length. We call Y a *degeneration* of X provided there exists a finitely generated Λ-module U and an exact sequence

$$0 \to U \to X \oplus U \to Y \to 0$$

PROPOSITION (ZWARA). *Y is a degeneration of X if and only if there is a Prüfer module M with basis Y such that $Y[t + 1] \simeq Y[t] \oplus X$ for some t, or, equivalently, for almost all t.*

PROOF. See [Z2] and [R6]. □

Note that an isomorphism $Y[t + 1] \simeq Y[t] \oplus X$ yields directly a Riedtmann-Zwara sequence as well as a co-Riedtmann-Zwara sequence, using the canonical exact sequences

$$0 \to Y[t] \to Y[t + 1] \to Y \to 0 \,,$$
$$0 \to Y \to Y[t + 1] \to Y[t] \to 0$$

and replacing the middle term by $Y[t] \oplus X$.

(2.3). Note that the Prüfer modules $Y[\infty]$ obtained in (2.2) satisfy $Y[\infty] \simeq Y[t] \oplus X^{(\mathbb{N})}$ for some t. In particular, these are modules of finite type.

3. Prüfer modules and the second Brauer-Thrall conjecture

(3.1). Let Λ be an artin algebra. The Krull-Remak-Schmidt Theorem asserts that any finitely generated Λ-module can be written as a direct sum of indecomposable modules, and such a decomposition is unique up to isomorphism.

The artin algebra Λ is called *representation-infinite* provided there are infinitely many isomorphism classes of indecomposable Λ-modules, otherwise *representation-finite*.

(3.2). The first Brauer-Thrall conjecture was solved by Roiter in 1968:

THEOREM (ROITER [RO]). *If Λ is a representation-infinite artin algebra, then there are indecomposable modules of arbitrarily large finite length.*

(3.3). The second Brauer-Thrall conjecture has been solved only for finite dimensional k-algebras, where k is an algebraically closed (or at least perfect) field:

THEOREM (BAUTISTA [BA], BONGARTZ [BO]). *If Λ is a representation-infinite k-algebra, where k is an infinite perfect field, then there are infinitely many natural numbers d such that there are infinitely many indecomposable Λ-modules of length d.*

It has been conjectured by Brauer-Thrall that the assertion holds for any infinite field. For finite fields, or, more generally, for an arbitrary artin algebra Λ, one may conjecture the following: if Λ is representation-infinite, then there are infinitely many natural numbers d such that there are infinitely many indecomposable Λ-modules of endo-length d. (The *endo-length* of a module M is the length of M when M is considered as a module over its endomorphism ring).

(3.3). Let Λ be an artin algebra. A Λ-module M is said to be *generic* provided M is indecomposable, of infinite length, but of finite endo-length.

THEOREM (CRAWLEY-BOEVEY [CB2]). *Let Λ be a finite-dimensional k-algebra (k a field). Let M be a generic Λ-module. Then there are infinitely many natural numbers d such that there are infinitely many indecomposable Λ-modules of endo-length d.*

(3.4) Prüfer modules yield generic modules.

THEOREM. *Let M be a Prüfer module. The following conditions are equivalent:*
(i) *M is not of finite type.*
(ii) *There is an infinite index I set such that the product module M^I has a generic direct summand.*
(iii) *For every infinite index I set, the product module M^I has a generic direct summand.*

PROOF. The implication (iii) $\implies$ (ii) is trivial. Also (ii) $\implies$ (i) is obvious: If M is of finite type, then all product modules M^I are of finite type. We only have to show (i) $\implies$ (iii). (It is sufficient to consider $I = \mathbb{N}$ in (iii), since any infinite index set I can be written as the disjoint union of $\mathbb{N}$ and some other index set I', and then $M^I = M^{\mathbb{N}} \oplus M^{I'}$, however, there is no problem to work in general.)

Now assume that I is an infinite index set and that M^I has no indecomposable direct summand which is endo-finite and of infinite length. Since M is a Prüfer module, there is a surjective, locally nilpotent endomorphism ϕ with kernel $W = W[1]$ non-zero and of finite length. Let $W[n]$ be the kernel of ϕ^n. Thus

$$W[1] \subset W[2] \subset \cdots \subset \bigcup_n W[n] = M$$

is a filtration of M with finite length modules $W[n]$. We obtain a corresponding chain of inclusions

$$W[1]^I \subset W[2]^I \subset \cdots \subset \bigcup_n W[n]^I = M'.$$

It has been shown in [R3] (see also [K]) that M' is isomorphic to a direct sum of copies of M and itself a direct summand of M^I; there is an endo-finite submodule E of M^I such that
$$M^I = M' \oplus E.$$

Any endo-finite module E can be written as a direct sum of copies of finitely many indecomposable endo-finite modules, say $E_1, \ldots, E_t$. By assumption, all these modules E_i are of finite length. A well-known lemma of Auslander asserts that any indecomposable direct summand of M^I of finite length is a direct summand of M itself, thus the modules $E_1, \ldots, E_t$ occur as direct summands of M.

Since M is artinian as a module over its endomorphism ring, M is Σ-algebraic compact, thus it is a direct sum of indecomposable modules with local endomorphism ring. Write $M = A \oplus B$, where A is a direct sum of copies of the various E_i and B has no direct summand of the form E_i, for any i. We want to show that B is of finite length. This then shows that M is of finite type.

The modules A, B are also filtered, with $A_n = A \cap W[n]$, $B_n = B \cap W[n]$ (it is obvious that $A = \bigcup_n A_n$, $B = \bigcup_n B_n$). For any n there is some n' with $W[n] \subseteq A_{n'} \oplus B_{n'}$. (Namely, let $x \in W[n]$, write $x = a + b$ with $a \in A$, $b \in B$. Then there is some n' with $a, b \in M_{n'}$, thus $a \in A_{n'}$, $b \in B_{n'}$.) We write $A' = \bigcup_i A_i^I$ and $B' = \bigcup_i B_i^I$. Then
$$M' = A' \oplus B'$$

(the inclusion $\supseteq$ is obvious, the other follows from $W[n]^I \subseteq (A_{n'} \oplus B_{n'})^I = A_{n'}^I \oplus B_{n'}^I \subseteq A' \oplus B'$.). We see that

$$(A^I/A') \oplus (B^I/B') = M^I/M' = E,$$

thus $A^I/A' = E_A$ and $B^I/B' = E_B$ with $E = E_A \oplus E_B$. In particular, E_A and E_B are direct sums of copies of $E_1, \ldots, E_t$. Since the direct sum of the inclusion maps

$$A' \to A^I \quad \text{and} \quad B' \to B^I$$

is a split monomorphism, the maps themselves are split monomorphisms, thus

$$A^I \simeq A' \oplus E_A \quad \text{and} \quad B^I \simeq B' \oplus E_B.$$

Consider the last isomorphism. If E_i is a direct summand of E_B, then it is a direct summand of B (Auslander Lemma), impossible. This shows that $E_B = 0$. But then $B' = B^I$ implies that $B = B_n$ for some n, thus $B \subseteq W[n]$. This shows that B is of finite length.   □

May-be one should record: *Assume that M^I/M' is the direct sum of copies of indecomposable modules $E_1, \ldots, E_t$ of finite length, then M is the direct sum of a finite length module B and of copies of the modules E_i.*

(3.5). As we have mentioned, the second Brauer-Thrall conjecture claims the following: if Λ is a representation-infinite algebra, then there are infinitely many natural numbers d such that there are infinitely many indecomposable Λ-modules of endo-length d.

Note that this is an assertion which concerns only modules of finite length. But it seems that it may be reasonable to look for a solution using modules of infinite length. As we have seen, it will be sufficient to show that a representation-infinite algebra has a Prüfer module which is not of finite type, since this implies the existence of a generic module and thus the existence of infinitely many indecomposable Λ-modules of endo-length d.

4. The ladder construction of Prüfer modules

We return to rings and modules in general.

(4.1). This construction was exhibited in [R6], let us recall here the essential steps: We start with a proper inclusion $U_0 \subset U_1$ (say with cokernel W) and a map $v_0 \colon U_0 \to U_1$, and we form the pushout of w_0 and v_0

$$
\begin{array}{ccccccccc}
0 & \longrightarrow & U_0 & \xrightarrow{w_0} & U_1 & \longrightarrow & W & \longrightarrow & 0 \\
& & \downarrow{v_0} & & \downarrow{v_1} & & \| & & \\
0 & \longrightarrow & U_1 & \xrightarrow{w_1} & U_2 & \longrightarrow & W & \longrightarrow & 0
\end{array}
$$

we obtain a module U_2, as well as a monomorphism $w_1 \colon U_1 \subset U_2$ (again with cokernel W) and a map $v_1 \colon U_1 \to U_2$.

Using induction, we obtain in this way modules U_i, monomorphisms $w_i \colon U_i \subset U_{i+1}$ (all with cokernel W) as well as maps $v_i \colon U_i \to U_{i+1}$ such that $v_{i+1}w_i = w_{i+1}v_i$ for all $i \geq 0$. This means that we obtain the following ladder of commutative squares:

$$
\begin{array}{ccccccccc}
U_0 & \xrightarrow{w_0} & U_1 & \xrightarrow{w_1} & U_2 & \xrightarrow{w_2} & U_3 & \xrightarrow{w_3} & \cdots \\
\downarrow{v_0} & & \downarrow{v_1} & & \downarrow{v_2} & & \downarrow{v_3} & & \\
U_1 & \xrightarrow{w_1} & U_2 & \xrightarrow{w_2} & U_3 & \xrightarrow{w_3} & U_4 & \xrightarrow{w_4} & \cdots
\end{array}
$$

We form the inductive limit $U_\infty = \bigcup_i U_i$ (along the maps w_i). Since all the squares commute, the maps v_i induce a map $U_\infty \to U_\infty$ which we denote by v_∞:

$$
\begin{array}{ccccccccc}
U_0 & \xrightarrow{w_0} & U_1 & \xrightarrow{w_1} & U_2 & \xrightarrow{w_2} & U_3 & \xrightarrow{w_3} & \cdots \qquad \bigcup_i U_i = U_\infty \\
\downarrow{\scriptstyle v_0} & & \downarrow{\scriptstyle v_1} & & \downarrow{\scriptstyle v_2} & & \downarrow{\scriptstyle v_3} & & \qquad\qquad\quad \downarrow{\scriptstyle v_\infty} \\
U_1 & \xrightarrow{w_1} & U_2 & \xrightarrow{w_2} & U_3 & \xrightarrow{w_3} & U_4 & \xrightarrow{w_4} & \cdots \qquad \bigcup_i U_i = U_\infty
\end{array}
$$

We also may consider the factor modules U_∞/U_0 and U_∞/U_1. The map $v_\infty \colon U_\infty \to U_\infty$ maps U_0 into U_1, thus it induces a map

$$\bar{v} \colon U_\infty/U_0 \longrightarrow U_\infty/U_1.$$

and *this map $\bar{v}$ is an isomorphism.* Namely, there are the commutative diagrams with exact rows:

$$
\begin{array}{ccccccccc}
0 & \longrightarrow & U_{i-1} & \xrightarrow{w_{i-1}} & U_i & \longrightarrow & W & \longrightarrow & 0 \\
& & \downarrow{\scriptstyle v_{i-1}} & & \downarrow{\scriptstyle v_i} & & \| & & \\
0 & \longrightarrow & U_i & \xrightarrow{w_i} & U_{i+1} & \longrightarrow & W & \longrightarrow & 0
\end{array}
$$

which means that the cokernel $U_i/U_{i-1} = W$ of w_{i-1} is mapped under the restriction $\bar{v}_i$ of $\bar{v}$ isomorphically onto the cokernel $U_{i+1}/U_i = W$ of w_i. Thus, we see that the map $\bar{v}$ is a map from a filtered module with factors U_i/U_{i-1} (where $i \geq 1$) to a filtered module with factors U_{i+1}/U_i (again with $i \geq 1$), and the maps $\bar{v}_i$ are just those induced on the factors. Since all the maps $\bar{v}_i$ are isomorphisms, also $\bar{v}$ itself is an isomorphism.

It follows: The composition of maps

$$U_\infty/U_0 \xrightarrow{p} U_\infty/U_1 \xrightarrow{\bar{v}^{-1}} U_\infty/U_0$$

(p the projection map) is an epimorphism ϕ with kernel U_1/U_0. It is easy to see that ϕ is locally nilpotent.

PROPOSITION. *The module U_∞/U_0 is a Prüfer module with respect to the endomorphism $\phi = \bar{v}^{-1} \circ p$, its basis is $W = U_1/U_0$.*

This shows that starting with a proper inclusion $U_0 \subset U_1$ and a map $v_0 \colon U_0 \to U_1$, the ladder construction yields a Prüfer module U_∞/U_0 with bases U_1/U_0.

(4.2). In case also v_0 is injective, we obtain a second Prüfer module. Namely, there is the following chessboard:

$$
\begin{array}{ccccccccc}
U_0 & \xrightarrow{\ w_0\ } & U_1 & \xrightarrow{\ w_1\ } & U_2 & \xrightarrow{\ w_2\ } & U_3 & \xrightarrow{\ w_3\ } & \cdots \\
\ \downarrow{\scriptstyle v_0} & & \ \downarrow{\scriptstyle v_1} & & \ \downarrow{\scriptstyle v_2} & & \ \downarrow{\scriptstyle v_3} & & \\
U_1 & \xrightarrow{\ w_1\ } & U_2 & \xrightarrow{\ w_2\ } & U_3 & \xrightarrow{\ w_3\ } & \cdots & & \\
\ \downarrow{\scriptstyle v_1} & & \ \downarrow{\scriptstyle v_2} & & \ \downarrow{\scriptstyle v_3} & & & & \\
U_2 & \xrightarrow{\ w_2\ } & U_3 & \xrightarrow{\ w_3\ } & \cdots & & & & \\
\ \downarrow{\scriptstyle v_2} & & \ \downarrow{\scriptstyle v_3} & & & & & & \\
U_4 & \xrightarrow{\ w_3\ } & \cdots & & & & & & \\
\ \downarrow{\scriptstyle v_3} & & & & & & & & \\
\cdots & & & & & & & &
\end{array}
$$

We see both horizontally as well as vertically ladders: the horizontal ladders yield U_∞ and its endomorphism v_∞; the vertical ladders yield U'_∞ with an endomorphism w_∞.

(4.3) Examples. First, let us show that the ordinary Prüfer groups (as considered in abelian group theory) are obtained in this way. Let $R = \mathbb{Z}$ be the ring of integers. Module homomorphisms $\mathbb{Z} \to \mathbb{Z}$ are given by the multiplication with some integer n, thus we denote such a map just by n. Let $U_0 = U_1 = \mathbb{Z}$ and $w_0 = 2$, $v_0 = n$. *If n is odd, then the Prüfer module U_∞/U_0 is just the Prüfer group for the prime 2* (and $U_\infty(2, n) = \mathbb{Z}[\frac{1}{2}]$ is the subring of $\mathbb{Q}$ generated by $\frac{1}{2}$). Note that if n is even, then the Prüfer module U_∞/U_0 is an elementary abelian 2-group.

Second, let $R = K(2)$ be the Kronecker algebra over some field k. Let U_0 be simple projective, U_1 indecomposable projective of length 3 and $w_0 \colon U_0 \to U_1$ a non-zero map with cokernel H (one of the indecomposable modules of length 2). For any map $v_0 \colon U_0 \to U_1$, we obtain a Prüfer module $M = U_\infty/U_0$. In case $v_0 \notin kw_0$, this module M is indecomposable (and it is the Prüfer module $H[\infty]$ as considered in [R2]), otherwise M it is a direct sum of copies of H.

(4.4) LEMMA. *The modules U_∞ as well as U_∞/U_0 are generated by U_1.*

PROOF. We only have to consider $U_\infty = \bigoplus_i U_i$. The pushout construction shows that for $i \geq 2$, the module U_i is a factor module of $U_{i-1} \oplus U_{i-1}$, thus by induction U_i is generated by U_1. □

(4.5). A self-extension $0 \to W \to W[2] \to W \to 0$ is called a *ladder extension* provided there is a commutative diagram with exact rows

$$
\begin{array}{ccccccccc}
0 & \longrightarrow & U_0 & \xrightarrow{\ w_0\ } & U_1 & \xrightarrow{\ q\ } & W & \longrightarrow & 0 \\
& & \ \downarrow{\scriptstyle \alpha} & & \downarrow & & \| & & \\
0 & \longrightarrow & W & \longrightarrow & W[2] & \longrightarrow & W & \longrightarrow & 0
\end{array}
$$

such that $\alpha = qv_0$ for some $v_0 \colon U_0 \to U_1$. In this case, the given self-extension is the $W[2]$ part of the Prüfer module $W[\infty]$ which is obtained from the maps w_0, v_0 using the ladder construction.

One should note that not every self-extension of a module is a ladder extension (for example, if S is a simple R-module, where R is artinian, then no non-trivial self-extension of S is a ladder extension). On the other hand, for R hereditary, every self-extension is a ladder extension [R6].

(4.5). Assume that Λ is a finite-dimensional hereditary k-algebra. (For example, Λ may be the path algebra kQ of a finite quiver Q without oriented cycles.)

Recall that the Euler characteristic

$$\sum_{i \geq 0} (-1)^i \dim \mathrm{Ext}^i(M, M')$$

yields a quadratic form q on the Grothendieck group $K_0(\Lambda)$ and q is positive definite if and only if Λ is representation-finite. (In the quiver case, this means that Q is the disjoint union of quivers of Dynkin type A_n, D_n, E_6, E_7, E_8.)

We see: *Any Λ-module M with $\mathrm{End}(M)$ a division ring and $q([M]) \leq 0$ is the basis of an indecomposable Prüfer module. The Prüfer module is unique if and only if $q([M]) = 0$.*

5. The search for pairs of embeddings

Let Λ be a representation-infinite artin algebra. The aim is to find pairs of embeddings $w_0, v_0 \colon U_0 \to U_1$ such that the corresponding Prüfer module U_∞ / U_0 is not of finite type.

(5.1). For dealing with Prüfer modules obtained using the ladder construction, it seems to be of interest to relate the finite type properties of U_∞ and U_∞ / U_0.

LEMMA. *U_∞ / U_0 is of finite type iff U_∞ is of finite type.*

PROOF. First, assume that U_∞ is of finite type, say $U_\infty = \bigoplus_{i \in I} M_i$ with all M_i indecomposable of finite length, and with only finitely many isomorphism classes of modules involved. Now $U_0 \subseteq \bigoplus_{i \in I'} M_i = M'$ with I' a finite subset of I. Then

$$U_\infty / U_0 = M' / U_0 \oplus \bigoplus_{i \in I \setminus I'} M_i,$$

is a direct sum of indecomposable modules of finite length (one has to decompose M'/U_0) and only finitely many isomorphism classes are involved.

The converse follows from Roiter's extension argument, see for example [R5]. $\square$

(5.2) PROPOSITION. *Assume that Λ is tame hereditary (or tame concealed). Let M, N be Λ-modules such that N is of finite length and generates M. If M is not of finite type, then N has a direct summand which is sincere and preprojective.*

Note that we cannot claim that N has an indecomposable direct summand which is sincere and preprojective. A typical example will be $N = {}_\Lambda\Lambda$, this module generates all the Λ-modules, but usually has no indecomposable sincere direct summand.

PROOF OF PROPOSITION. Let k be the center of Λ, this is a field and Λ is a finite-dimensional k-algebra. We will assume that Λ is hereditary (the case when Λ is concealed requires only few modifications). Let $N = P \oplus R \oplus Q$ with P preprojective, R regular and Q preinjective. Assume P is not sincere, thus there is a simple Λ-module S which does not occur as a composition factor of P.

We can order the indecomposable preprojective modules $P_1, P_2, \ldots$ and the indecomposable preinjective modules $\ldots, Q_2, Q_1$ such that $\mathrm{Hom}(P_i, P_j) = 0$ for $i > j$ and $\mathrm{Hom}(Q_i, Q_j) = 0$ for $i < j$. There is an index n such that all the modules P_i with $i > n$ are sincere.

Note that there is a bound b such that $\dim_k \mathrm{Hom}(R, Q_i) \le b$ for all i. Namely, $Q(i) = \tau^t I(y)$ for some $t \ge 0$ and some vertex y in the quiver of Λ. Since R is regular, it is τ-periodic with period at most 6, thus $\dim_k \mathrm{Hom}(R, Q_i)$ is bounded by the maximum of the numbers $\dim_k \mathrm{Hom}(R, \tau^t I(y))$ with $0 \le t \le 5$ and y a vertex of the quiver.

This implies the following: If Q_i is generated by R, then there is a surjective map $R^b \to Q_i$, and therefore the multiplicity $[Q_i : S]$ of S as a composition factor of Q_i is bounded by $b[R : S]$. There are only finitely many Q_i with $[Q_i : S] \le b[R : S]$, thus there is some m such that $[Q_i : S] > b[R : S]$ for all $i > m$. But this implies that a module Q_i with $i > m$ cannot be generated by $P \oplus R$ (the trace of R in Q_i is a submodule with at most $b[R : S]$ composition factors S and the trace of P in Q_i does not provide any such composition factor).

We can assume in addition that m is chosen in such a way that all the indecomposable direct summands of Q are of the form Q_i with $i \le m$. Then we see that the modules Q_i with $i > m$ cannot be generated by $N = P \oplus R \oplus Q$.

Now let M be a (not necessarily finitely generated) module which is generated by N. We want to show that M is of finite type. According to [R2], we can write $M = M_1 \oplus M_2 \oplus M_3$ where M_1 is a direct sum of modules of the form P_i with $1 \le i \le n$, where M_3 is a direct sum of modules of the form Q_i with $1 \le i \le m$, and where M_2 has no direct summand of the form P_i with $1 \le i \le n$, or Q_i with $1 \le i \le m$. With M also M_2 is generated by N, and we want to see that M_2 is of finite type (then also M is of finite type).

Thus we see that we can assume that $M = M_2$, this means that we consider a module generated by N which has no direct summand of the form P_i with $1 \le i \le n$, or Q_i with $1 \le i \le m$. First of all, M cannot have any indecomposable preprojective direct summand M'. Namely, $\mathrm{Hom}(R \oplus Q, M') = 0$, thus M' would be generated by P, but P is not sincere, whereas M' is sincere. Second, we note that M cannot have any indecomposable preinjective direct summand. Namely, it would be generated by N, but an indecomposable preinjective module Q_i which is generated by N satisfies $i \le m$.

This means that M is regular (as defined in [R2]). Also, we see that $\mathrm{Hom}(Q, M) = 0$, thus M is generated by $P \oplus R$. Let M' be the trace of R in M. Since R is a regular module of finite length, it follows that M' is regular and of finite type (it is a direct sum of copies of the regular factor modules of R). Now M/M' is generated by P, thus it is not sincere and therefore of finite type. Write M/M' as a direct sum of indecomposable modules, and collect these modules according to the property of being preprojective, regular or preinjective. Thus $M/M' = P' \oplus R' \oplus Q'$, where P' is a direct sum of modules of finite length modules which are preprojective, R' is a direct sum of modules of finite length which are regular, and Q' is a direct sum of modules of finite length which are preinjective.

Now $P' = 0$, since otherwise M would have a proper factor module which is preprojective (and therefore a direct summand). Also, $Q' = 0$, since otherwise M cannot be regular.

This shows that M is an extension of M' by $M/M' = R'$. As we have seen, M' is a direct sum of copies of the regular factor modules of R, whereas R' is a direct sum of finite length modules which are regular and do not contain the composition factor S. But this implies that M is regular and is of finite type.   □

6. Take-off subcategories

Let Λ be a representation-infinite artin algebra and $\operatorname{mod}\Lambda$ the category of finitely generated Λ-modules.

A full subcategory $\mathcal{C}$ of $\operatorname{mod}\Lambda$ is said to be a *take-off* subcategory, provided the following conditions are satisfied:
(1) $\mathcal{C}$ is closed under direct sums and under submodules.
(2) $\mathcal{C}$ contains infinitely many isomorphism classes of indecomposable modules.
(3) No proper subcategory of $\mathcal{C}$ satisfies (1) and (2).

(6.1) THEOREM. *Any subcategory satisfying* (1) *and* (2) *contains a take-off subcategory.*

In particular, this means that the module category of any representation-infinite artin algebra has at least one take-off subcategory: take-off subcategories always do exist!

(6.2) Examples. If Λ is a connected hereditary algebra which is representation-infinite, then the preprojective modules form a take-off subcategory.

In general, there may be several take-off subcategories: For example, if Λ has several minimal representation-infinite factor algebras, then any such factor algebra yields a take-off subcategory of $\operatorname{mod}\Lambda$.

Remark. Observe that the existence of take-off subcategories is in sharp contrast to the usual characterization of "infinity" (a set is infinite iff it contains proper subsets of the same cardinality)!

(6.3) Properties of a take-off subcategory $\mathcal{C}$.

Let $\mathcal{C}$ be a take-off subcategory of $\operatorname{mod}\Lambda$.

(1) *For any d, there are only finitely many isomorphism classes of modules of length d which belong to $\mathcal{C}$.*

Thus: *$\mathcal{C}$ contains indecomposable modules of arbitrarily large finite length.*

Let $\overline{\mathcal{C}}$ be the class of all Λ-modules M such that any finitely generated submodule of M belongs to $\mathcal{C}$.

(2) *There are indecomposable modules M in $\overline{\mathcal{C}}$ of infinite length.*

(3) *If M is an indecomposable module M in $\overline{\mathcal{C}}$ of infinite length, then any indecomposable module N in $\mathcal{C}$ embeds into M — even a countable direct sum $N^{(\mathbb{N})}$ embeds into M.*

(6.4). We have seen that $\operatorname{mod}\Lambda$ contains take-off subcategories $\mathcal{C}$, such a subcategory $\mathcal{C}$ contains indecomposable modules M of arbitrarily large finite length, and thus indecomposable modules with arbitrarily large socle.

CONJECTURE. *Let U be an indecomposable Λ-module belonging to a take-off subcategory. If there is a simple module S such that S^7 embeds into U, then there are two embeddings $w_0, v_0 \colon S \to U$ such that the corresponding Prüfer module is not of finite type.*

Remark. The bound 7 cannot be lowered, as the path algebra kQ of the $\tilde{E}_8$-quiver Q with subspace orientation shows. Note that kQ is a tame hereditary algebra, and for a tame hereditary algebra, the preprojective modules form the unique take-off subcategory. Now the indecomposable representation U with dimension vector

$$
\begin{array}{ccccc}
5 & 4 & 3 & 2 & 0 \\
& 6 & 4 & 2 & \\
& & 3 & &
\end{array}
$$

is preprojective and S^6 embeds into U, where S is the simple projective kQ-module. Note that U is not faithful, thus the Prüfer modules constructed by pairs $w_0, v_0 \colon S \to U$ are also not faithful. But all non-faithful kQ-modules are of finite type.

References

[A]   Atiyah, M., *Vector bundles over an elliptic curve*, Proceedings of the London Mathematical Society **7** (1957), 414–452.

[Ba]   Bautista, R., *On algebras of strongly unbounded representation type*, Comment. Math. Helv. **60** (1985), 392–399.

[Bo]   Bongartz, K., *Indecomposables are standard*, Comment. Math. Helv. **60** (1985), 400–410.

[CB1] Crawley-Boevey, W.W., *On tame algebras and bocses*, Proc. London Math. Soc.(3) **56** (1988), 451–483.

[CB2] Crawley-Boevey, W.W., *Modules of finite length over their endomorphism ring*, Representations of algebras and related topics (ed: Brenner, Tachikawa) London Math. Soc. Lec. Note Series, vol. 168, 1992, pp. 127–184.

[K]   Krause, H., *Generic modules over artin algebras*, Proc. London Math. Soc. **76** (1998), 276–306.

[R1]   Ringel, C.M., *Representations of k-species and bimodules*, J.Algebra **41** (1976), 269–302.

[R2]   Ringel, C.M., *Infinite dimensional representations of finite dimensional hereditary algebras*, Symposia Math. **23**, 321–412.

[R3]   Ringel, C.M., *A construction of endofinite modules*, Advances in Algebra and Model Theory, (ed. M. Droste, R. Göbel), Gordon-Breach, London, 1997, pp. 387–399.

[R4]   Ringel, C.M., *The Gabriel-Roiter measure*, Bull. Sci. math. **129** (2005), 726–748.

[R5]   Ringel, C.M., *Foundation of the Representation Theory of Artin Algebras, Using the Gabriel-Roiter Measure.*, Proceedings ICRA 11. Queretaro 2004. Contemporary Math., vol. 406., Amer.Math.Soc., 2006, pp. 105–135.

[R6]   Ringel, C.M., *The ladder construction of Prüfer modules*, Revista de la Union Matematica Argentina **48-2** (2007), 47–65.

[R7]   Ringel, C.M., *The first Brauer-Thrall conjecture*, Models, Modules and Abelian Groups, In Memory of A.L.S. Corner. Walter de Gruyter, Berlin 2008 (ed: R. Göbel, B. Goldsmith) (To appear).

[R8]   Ringel, C.M., *Prüfer modules of finite type.*, Selected Topics in Representation Theory. Lecture Notes. Bielefeld, 2006.

[R9]   Ringel, C.M., *Take-off subcategories.*, (In preparation). A preliminary version is in: Selected Topics in Representation Theory. Lecture Notes. Bielefeld, 2006.

[Ro]   Roiter, A.V., *Unboundedness of the dimension of the indecomposable representations of an algebra which has infinitely many indecomposable representations*, Izv. Akad. Nauk SSSR. Ser. Mat. **32** (1968), 1275–1282.

[Z1]   Zwara, G., *A degeneration-like order for modules*, Arch. Math. **71** (1998), 437–444.
[Z2]   Zwara, G., *Degenerations of finite-dimensional modules are given by extensions*, Compositio Mathematica **121** (2001), 205–218.

FAKULTÄT MATHEMATIK, UNIVERSITÄT BIELEFELD, POBOX 100131, D33501, BIELEFELD, GERMANY
  E-mail address: `ringel@math.uni-bielefeld.de`

Contemporary Mathematics
Volume **478**, 2009

Quivers and the Euclidean group

Alistair Savage

ABSTRACT. We show that the category of representations of the Euclidean group of orientation-preserving isometries of two-dimensional Euclidean space is equivalent to the category of representations of the preprojective algebra of type A_∞. We also consider the moduli space of representations of the Euclidean group along with a set of generators. We show that these moduli spaces are quiver varieties of the type considered by Nakajima. Using these identifications, we prove various results about the representation theory of the Euclidean group. In particular, we prove it is of wild representation type but that if we impose certain restrictions on weight decompositions, we obtain only a finite number of indecomposable representations.

1. Introduction

The Euclidean group $E(n) = \mathbb{R}^n \rtimes SO(n)$ is the group of orientation-preserving isometries of n-dimensional Euclidean space. The study of these objects, at least for $n = 2, 3$, predates even the concept of a group. In this paper we will focus on the Euclidean group $E(2)$. Even in this case, much is still unknown about the representation theory.

Since $E(2)$ is solvable, all its finite-dimensional irreducible representations are one-dimensional. The finite-dimensional unitary representations, which are of interest in quantum mechanics, are completely reducible and thus isomorphic to direct sums of such one-dimensional representations. The infinite-dimensional unitary irreducible representations have received considerable attention (see [**1, 3, 4**]). There also exist finite-dimensional nonunitary indecomposable representations (which are not irreducible) and much less is known about these. However, they play an important role in mathematical physics and the representation theory of the Poincaré group. The Poincaré group is the group of isometries of Minkowski spacetime. It is the semidirect product of the translations of $\mathbb{R}^3$ and the Lorentz transformations. In 1939, Wigner [**25**] studied the subgroups of the Lorentz group leaving invariant the four-momentum of a given free particle. The maximal such subgroup is called the *little group*. The little group governs the internal space-time symmetries of the relativistic particle in question. The little groups of massive particles are locally

2000 *Mathematics Subject Classification.* Primary: 17B10, 22E47; Secondary: 22E43.

This research was supported in part by the Natural Sciences and Engineering Research Council (NSERC) of Canada.

isomorphic to the group $O(3)$ while the little groups of massless particles are locally isomorphic to $E(2)$. That is, their Lie algebras are isomorphic to those of $O(3)$ and $E(2)$ respectively. We refer the reader to [2, 5, 16, 20] for further details.

The group $E(2)$ also appears in the Chern-Simons formulation of Einstein gravity in $2+1$ dimensions. In the case when the space-time has Euclidean signature and the cosmological constant vanishes, the phase space of gravity is the moduli space of flat $E(2)$-connections.

In the current paper, we relate the representation theory of the Euclidean group $E(2)$ to the representation theory of preprojective algebras of quivers of type A_∞. In fact, we show that the categories of representations of the two are equivalent. To prove this, we introduce a modified enveloping algebra of the Lie algebra of $E(2)$ and show that it is isomorphic to the preprojective algebra of type A_∞. Furthermore, we consider the moduli space of representations of $E(2)$ along with a set of generators. We show that these moduli spaces are quiver varieties of the type considered by Nakajima in [21, 22]. These identifications allow us to draw on known results about preprojective algebras and quiver varieties to prove various statements about representations of $E(2)$. In particular, we show that $E(2)$ is of wild representation type but that if we impose certain restrictions on the weight decomposition of a representation, we obtain only a finite number of indecomposable representations. We conclude with some potential directions for future investigation.

2. The Euclidean algebra

Let $E(2) = \mathbb{R}^2 \rtimes SO(2)$ be the Euclidean group of motions in the plane and let $\mathfrak{e}(2)$ be the complexification of its Lie algebra. We call $\mathfrak{e}(2)$ the (three-dimensional) Euclidean algebra. It has basis $\{p_+, p_-, l\}$ and commutation relations

$$(2.1) \qquad [p_+, p_-] = 0, \quad [l, p_\pm] = \pm p_\pm.$$

Since $SO(2)$ is compact, the category of finite-dimensional $E(2)$-modules is equivalent to the category of finite-dimensional $\mathfrak{e}(2)$-modules in which l acts semisimply with integer eigenvalues. Will will use the term $\mathfrak{e}(2)$-module to refer only to such modules. For $k \in \mathbb{Z}$, we shall write V_k to indicate the eigenspace of l with eigenvalue k (the k-weight space). Thus, for an $\mathfrak{e}(2)$-module V, we have the weight space decomposition

$$V = \bigoplus_k V_k, \quad V_k = \{v \in V \mid l \cdot v = kv\}, \quad k \in \mathbb{Z},$$

and

$$p_+ V_k \subseteq V_{k+1}, \quad p_- V_k \subseteq V_{k-1}.$$

We may form the tensor product of any representation V with the character χ_n for $n \in \mathbb{Z}$. Here χ_n is the one-dimensional module $\mathbb{C}$ on which $p_\pm$ act by zero and l acts by multiplication by n. Then a weight space V_k of weight k becomes a weight space $V_k \otimes \chi_n$ of weight $k + n$. In this way, we may "shift weights" as we please.

For $k \in \mathbb{Z}$, let $\mathbf{e}^k$ be the element of $(\mathbb{Z}_{\geq 0})^{\mathbb{Z}}$ with kth component equal to one and all others equal to zero. For an $\mathfrak{e}(2)$-module V we define

$$\mathbf{dim}\, V = \sum_{k \in \mathbb{Z}} (\dim V_k) \mathbf{e}^k.$$

Let U be the universal enveloping algebra of $\mathfrak{e}(2)$ and let U^+, U^- and U^0 be the subalgebras generated by p_+, p_- and l respectively. Then we have the triangular decomposition

$$U \cong U^+ \otimes U^0 \otimes U^- \quad \text{(as vector spaces)}.$$

Note that the category of representations of U is equivalent to the category of representations of $\mathfrak{e}(2)$. In [**18**, Chapter 23], Lusztig introduced the modified quantized enveloping algebra of a Kac-Moody algebra. Following this idea, we introduce the *modified enveloping algebra* $\tilde{U}$ by replacing U^0 with a sum of 1-dimensional algebras

$$\tilde{U} = U^+ \otimes \left(\bigoplus_{k \in \mathbb{Z}} \mathbb{C}a_k \right) \otimes U^-.$$

Multiplication is given by

$$a_k a_l = \delta_{kl} a_k,$$

$$p_+ a_k = a_{k+1} p_+, \quad p_- a_k = a_{k-1} p_-,$$

$$p_+ p_- a_k = p_- p_+ a_k.$$

One can think of a_k as projection onto the kth weight space. Note that $\tilde{U}$ is an algebra without unit. We say a $\tilde{U}$-module V is *unital* if

(1) for any $v \in V$, we have $a_k v = 0$ for all but finitely many $k \in Z$, and
(2) for any $v \in V$, we have $\sum_{k \in \mathbb{Z}} a_k v = v$.

A unital $\tilde{U}$-module can be thought of as a U-module with weight decomposition. Thus we have the following proposition.

PROPOSITION 2.1. *The category of unital $\tilde{U}$-modules is equivalent to the category of U-modules and hence the category of $\mathfrak{e}(2)$-modules.*

3. Preprojective algebras

In this section, we review some basic results about preprojective algebras. The reader is referred to [**12**] for further details.

A *quiver* is a 4-tuple $(I, H, \text{out}, \text{in})$ where I and H are disjoint sets and out and in are functions from H to I. The sets I and H are called the *vertex set* and *arrow set* respectively. We think of an element $h \in H$ as an arrow from the vertex $\text{out}(h)$ to the vertex $\text{in}(h)$.

An arrow $h \in H$ is called a *loop* if $\text{out}(h) = \text{in}(h)$. A quiver is said to be *finite* if both its vertex and arrow sets are finite.

We shall be especially concerned with the following quivers. For $a, b \in \mathbb{Z}$ with $a \leq b$, let $Q_{a,b}$ be the quiver with vertex set $I = \{k \in \mathbb{Z} \mid a \leq k \leq b\}$ and arrows $H = \{h_i \mid a \leq i \leq b-1\}$ with $\text{out}(h_i) = i$ and $\text{in}(h_i) = i+1$. We say that $Q_{a,b}$ is a quiver of type A_{b-a+1} since this is the type of its underlying graph. The quiver Q_∞ has vertex set $I = \mathbb{Z}$ and arrows $H = \{h_i \mid i \in \mathbb{Z}\}$ with $\text{out}(h_i) = i$ and $\text{in}(h_i) = i+1$. We say that the quiver Q_∞ is of type A_∞. Note that the quivers $Q_{a,b}$ are finite while the quiver Q_∞ is not.

Let $Q = (I, H, \text{out}, \text{in})$ be a quiver without loops and let $Q^* = (I, H^*, \text{out}^*, \text{in}^*)$ be the *double quiver* of Q. By definition,

$$H^* = \{h \mid h \in H\} \cup \{\bar{h} \mid h \in H\},$$

$$\text{out}^*(h) = \text{out}(h), \quad \text{in}^*(h) = \text{in}(h), \quad \text{out}^*(\bar{h}) = \text{in}(h), \quad \text{in}^*(\bar{h}) = \text{out}(h).$$

From now on, we will write in and out for in^* and out^* respectively. Since $\text{in}^* |_H = \text{in}$ and $\text{out}^* |_H = \text{out}$, this should cause no confusion.

A *path* in a quiver Q is a sequence $p = h_n h_{n-1} \cdots h_1$ of arrows such that $\text{in}(h_i) = \text{out}(h_{i+1})$ for $1 \le i \le n - 1$. We call the integer n the *length* of p and define $\text{out}(p) = \text{out}(h_1)$ and $\text{in}(p) = \text{in}(h_n)$. The *path algebra* $\mathbb{C}Q$ is the algebra spanned by the paths in Q with multiplication given by

$$p \cdot p' = \begin{cases} pp' & \text{if } \text{in}(p') = \text{out}(p) \\ 0 & \text{otherwise} \end{cases}$$

and extended by linearity. We note that there is a trivial path ϵ_i starting and ending at i for each $i \in I$. The path algebra $\mathbb{C}Q$ has a unit (namely $\sum_{i \in I} \epsilon_i$) if and only if the quiver Q is finite.

A *relation* in a quiver Q is a sum of the form $\sum_{j=1}^{k} a_j p_j$, $a_j \in \mathbb{C}$, p_j a path for $1 \le j \le k$. For $i \in I$ let

$$r_i = \sum_{h \in H, \, \text{out}(h)=i} \bar{h}h - \sum_{h \in H, \, \text{in}(h)=i} h\bar{h}$$

be the *Gelfand-Ponomarev relation* in Q^* associated to i. The *preprojective algebra* $P(Q)$ corresponding to Q is defined to be

$$P(Q) = \mathbb{C}Q^* / J$$

where J is the two-sided ideal generated by the relations r_i for $i \in I$.

Let $\mathcal{V}(I)$ denote the category of finite-dimensional I-graded vector spaces with morphisms being linear maps respecting the grading. For $V \in \mathcal{V}(I)$, we let $\mathbf{dim}\, V = (\dim V_i)_{i \in I}$ be the I-graded dimension of V. A *representation* of the quiver Q^* is an element $V \in \mathcal{V}(I)$ along with a linear map $x_h : V_{\text{out}(h)} \to V_{\text{in}(h)}$ for each $h \in H^*$. We let

$$\text{rep}(Q^*, V) = \bigoplus_{h \in H^*} \text{Hom}_{\mathbb{C}}(V_{\text{out}(h)}, V_{\text{in}(h)})$$

be the affine variety consisting of representations of Q^* with underlying graded vector space V. A representation of a quiver can be naturally interpreted as a $\mathbb{C}Q^*$-module structure on V. For a path $p = h_n h_{n-1} \ldots h_1$ in Q^*, we let

$$x_p = x_{h_n} x_{h_{n-1}} \cdots x_{h_1}.$$

We say a representation $x \in \text{rep}(Q^*, V)$ *satisfies the relation* $\sum_{j=1}^{k} a_j p_j$, if

$$\sum_{j=1}^{k} a_j x_{p_j} = 0.$$

If R is a set of relations, we denote by $\text{rep}(Q^*, R, V)$ the set of all representations in $\text{rep}(Q^*, V)$ satisfying all relations in R. This is a closed subvariety of $\text{rep}(Q^*, V)$. Every element of $\text{rep}(Q^*, J, V)$ can be naturally interpreted as a $P(Q)$-module structure on V and so we also write

$$\text{mod}(P(Q), V) = \text{rep}(Q^*, J, V)$$

for the affine variety of $P(Q)$-modules with underlying vector space V.

The algebraic group $G_V = \prod_{i \in I} GL(V_i)$ acts on $\mathrm{mod}(P(Q), V)$ by

$$g \cdot x = (g_i)_{i \in I} \cdot (x_h)_{h \in H^*} = (g_{\mathrm{in}(h)} x_h g_{\mathrm{out}(h)}^{-1})_{h \in H^*}.$$

Two $P(Q)$-modules are isomorphic if and only if they lie in the same orbit. For a dimension vector $\mathbf{v} \in (\mathbb{Z}_{\geq 0})^I$, let

$$V^{\mathbf{v}} = \bigoplus_{i \in I} \mathbb{C}^{\mathbf{v}_i}, \quad \mathrm{mod}(P(Q), \mathbf{v}) = \mathrm{mod}(P(Q), V^{\mathbf{v}}), \quad G_{\mathbf{v}} = G_{V^{\mathbf{v}}}.$$

Then we have that $\mathrm{mod}(P(Q), V) \cong \mathrm{mod}(P(Q), \mathbf{dim}\, V)$ for all $V \in \mathcal{V}(I)$. Therefore, we will blur the distinction between $\mathrm{mod}(P(Q), V)$ and $\mathrm{mod}(P(Q), \mathbf{dim}\, V)$.

We say an element $x \in \mathrm{mod}(P(Q), V)$ is *nilpotent* if there exists an $N \in \mathbb{Z}_{>0}$ such that for any path p of length greater than N, we have $x_p = 0$. Denote the closed subset of nilpotent elements of $\mathrm{mod}(P(Q), V)$ by $\Lambda_{V,Q}$ and let $\Lambda_{\mathbf{v},Q} = \Lambda_{V^{\mathbf{v}},Q}$. The varieties $\Lambda_{V,Q}$ are called *nilpotent varieties* or *Lusztig quiver varieties*. Lusztig [17, Theorem 12.3] has shown that the $\Lambda_{V,Q}$ have pure dimesion $\dim(\mathrm{rep}(Q, V))$.

PROPOSITION 3.1. *For a quiver Q, the following are equivalent:*

(1) $P(Q)$ *is finite-dimensional,*
(2) $\Lambda_{V,Q} = \mathrm{mod}(P(Q), V)$ *for all $V \in \mathcal{V}(I)$,*
(3) Q *is a Dynkin quiver (i.e. its underlying graph is of ADE type).*

PROOF. The equivalence of (1) and (3) is well-known (see for example [23]). That (2) implies (3) was proven by Crawley-Boevey [6] and the converse was proven by Lusztig [17, Proposition 14.2]. $\qquad\square$

Thus, for a Dynkin quiver Q, nilpotency holds automatically and $\Lambda_{V,Q}$ is just the variety of representations of the preprojective algebra $P(Q)$ with underlying vector space V.

The representation type of the preprojective algebras is known.

PROPOSITION 3.2 ([7, 13]). *Let Q be a finite quiver. Then the following hold:*

(1) $P(Q)$ *is of finite representation type if and only if Q is of Dynkin type A_n, $n \leq 4$, and*
(2) $P(Q)$ *is of tame representation type if and only if Q is of Dynkin type A_5 or D_4.*

Thus $P(Q)$ is of wild representation type if Q is not of Dynkin type A_n, $n \leq 5$, or D_4.

In the sequel, we will refer to the preprojective algebra $P(Q_\infty)$. While Q_∞ is not a finite quiver, any finite-dimensional representation is supported on finitely many vertices and thus is a representation of a quiver of type A_n for sufficiently large n. Thus we deduce the following.

COROLLARY 3.3. *All finite-dimensional representations of $P(Q_\infty)$ are nilpotent and $P(Q_\infty)$ is of wild representation type.*

For a finite quiver Q, let $\mathfrak{g}_Q$ denote the Kac-Moody algebra whose Dynkin graph is the underlying graph of Q and let $U(\mathfrak{g}_Q)^-$ denote the lower half of its universal enveloping algebra. It turns out that Lusztig quiver varieties are intimately related to $U(\mathfrak{g}_Q)^-$. Namely, Lusztig [17] has shown that there is a space of constructible functions on the varieties $\Lambda_{\mathbf{v},Q}$, $\mathbf{v} \in (\mathbb{Z}_{\geq 0})^I$, and a natural convolution product such that this space of functions is isomorphic as an algebra to $U(\mathfrak{g}_Q)^-$. The functions on an individual $\Lambda_{\mathbf{v},Q}$ correspond to the weight space of weight $-\sum_{i \in I} \mathbf{v}_i \alpha_i$, where the α_i are the simple roots of $\mathfrak{g}_Q$. Furthermore, the irreducible components of $\Lambda_{\mathbf{v},Q}$ are in one-to-one correspondence with a basis of this weight space. Under this correspondence, each irreducible component is associated to the unique function equal to one on an open dense subset of that component and equal to zero on an open dense subset of all other components. The set of these functions yields a basis of $U(\mathfrak{g}_Q)^-$, called the *semicanonical basis*, with very nice integrality and positivity properties (see [**19**]). If instead of constructible functions one works with the Grothendieck group of a certain class of perverse sheaves, a similar construction yields a realization of (the lower half of) the quantum group $U_q(\mathfrak{g}_Q)^-$ and the *canonical basis* (see [**17**]).

4. Representations of the Euclidean algebra and preprojective algebras

In this section we examine the close relationship between representations of the Euclidean algebra $\mathfrak{e}(2)$ and the preprojective algebras of types A_n and A_∞.

THEOREM 4.1. *The modified universal enveloping algebra $\tilde{U}$ is isomorphic to the preprojective algebra $P(Q_\infty)$.*

PROOF. Define a map $\psi : \mathbb{C}Q_\infty^* \to \tilde{U}$ by

$$\psi(\epsilon_i) = a_i, \quad \psi(h_i) = p_+ a_i = a_{i+1} p_+, \quad \psi(\bar{h}_i) = a_i p_- = p_- a_{i+1}, \quad i \in I.$$

It is easily verified that this extends to a surjective map of algebras with kernel J and thus the result follows.   □

Let $\mathbf{Mod}\, \mathfrak{e}(2)$ be the category of $\mathfrak{e}(2)$-modules. For $a \leq b$, let $\mathbf{Mod}_{a,b}\, \mathfrak{e}(2)$ be the full subcategory consisting of representations V such that $V_k = 0$ for $k < a$ or $k > b$. For $\mathbf{v} \in (\mathbb{Z}_{\geq 0})^{\mathbb{Z}}$, we also define $\mathbf{Mod}_{a,b}^{\mathbf{v}}\, \mathfrak{e}(2)$ and $\mathbf{Mod}^{\mathbf{v}}\, \mathfrak{e}(2)$ to be the full subcategories of $\mathbf{Mod}_{a,b}\, \mathfrak{e}(2)$ and $\mathbf{Mod}\, \mathfrak{e}(2)$ consisting of representations V such that $\dim V = \mathbf{v}$.

Let $\mathbf{Mod}\, P(Q)$ be the category of finite-dimensional $P(Q)$-modules and for $\mathbf{v} \in (\mathbb{Z}_{\geq 0})^I$, let $\mathbf{Mod}^{\mathbf{v}}\, P(Q)$ be the full subcategory consisting of modules of graded dimension $\mathbf{v}$.

COROLLARY 4.2. *We have the following equivalences of categories.*

(1) $\mathbf{Mod}^{\mathbf{v}}\, \mathfrak{e}(2) \cong \mathbf{Mod}^{\mathbf{v}}\, P(Q_\infty)$, $\mathbf{Mod}\, \mathfrak{e}(2) \cong \mathbf{Mod}\, P(Q_\infty)$,

(2) $\mathbf{Mod}_{a,b}^{\mathbf{v}}\, \mathfrak{e}(2) \cong \mathbf{Mod}^{\mathbf{v}}\, P(Q_{a,b})$, $\mathbf{Mod}_{a,b}\, \mathfrak{e}(2) \cong \mathbf{Mod}\, P(Q_{a,b})$.

PROOF. Statement (1) follows from Theorem 4.1 and Proposition 2.1. Statement (2) is obtained by restricting weights to lie between a and b.   □

THEOREM 4.3. *The following statements hold.*

(1) *The Euclidean algebra $\mathfrak{e}(2)$, and hence the Euclidean group $E(2)$, have wild representation type, and*

(2) *for $a, b \in \mathbb{Z}$ with $0 \leq b - a \leq 3$, there are a finite number of isomorphism classes of indecomposable $\mathfrak{e}(2)$-modules V whose weights lie between a and b; that is, such that $V_k = 0$ for $k < a$ or $k > b$.*

PROOF. These statements follow immediately from Corollary 4.2, Proposition 3.2 and Corollary 3.3.   □

COROLLARY 4.4. *Let A be a finite subset of $\mathbb{Z}$ with the property that A does not contain any five consecutive integers. Then there are a finite number of isomorphism classes of indecomposable $\mathfrak{e}(2)$-modules V with the property that $V_k = 0$ if $k \notin A$.*

PROOF. Partition A into subsets $A_1, \ldots, A_n$ such that $A_j = \{a_j, a_j+1, \ldots, a_j+m_j\}$ and $|a - b| > 1$ for $a \in A_i$, $b \in A_j$ with $i \neq j$. By hypothesis, we have $m_j \leq 3$ for $1 \leq j \leq n$. Let V be an $\mathfrak{e}(2)$-module such that $V_k = 0$ if $k \notin A$. Then V decomposes as a direct sum of modules $V = \bigoplus_{j=1}^n V^j$ where $V_k^j = 0$ if $k < a_j$ or $k > a_j + m_k$. Thus, if V is indecomposable, we must have $V = V^j$ for some j. But there are a finite number of such V^j, up to isomorphism, by Theorem 4.3. The result follows.   □

For $a \in \mathbb{Z}$, we say an $\mathfrak{e}(2)$-module V has *lowest weight* a if $V_a \neq 0$ and $V_k = 0$ for $k < a$.

COROLLARY 4.5. *For all $a \in \mathbb{Z}$, there are a finite number of isomorphism classes of indecomposable $\mathfrak{e}(2)$-modules with lowest weight a and dimension less than or equal to five.*

PROOF. By tensoring with the character χ_{-a} we may assume that $a = 0$. In order for an $\mathfrak{e}(2)$-module to be indecomposable, its set of weights must be a set of consecutive integers. By Corollary 4.4, it suffices to consider the modules of dimension 5. Again, by Corollary 4.4, we need only consider the case when $\dim V_k = 1$ for $0 \leq k \leq 4$. We consider the equivalent problem of classifying the G_V-orbits of indecomposable elements $x \in \Lambda_{V,Q_{0,4}}$ where $V_k = \mathbb{C}$ for $0 \leq k \leq 4$. Fixing the standard basis in each V_k, we can view the maps x_h, $h \in H^*$, as complex numbers. Considering the Gelfand-Ponomarev relation r_0, we see that $x_{\bar{h}_0} x_{h_0} = 0$. Then the relation r_1 implies $x_{\bar{h}_1} x_{h_1} = 0$. Continuing in this manner, we see that $x_{\bar{h}_i} x_{h_i} = 0$ for $0 \leq i \leq 3$. Thus $x_{h_i} = 0$ or $x_{\bar{h}_i} = 0$ for $0 \leq i \leq 3$. Since x is indecomposable, we cannot have both $x_{h_i} = 0$ and $x_{\bar{h}_i} = 0$ for any i. Thus, there are precisely $2^4 = 16$ G_V-orbits in $\Lambda_{V,Q_{0,4}}$. Representatives for these orbits correspond to setting one of x_{h_i} or $x_{\bar{h}_i}$ equal to one and the other to zero for each $0 \leq i \leq 3$.   □

We note that Douglas [9] has shown that there are finitely many indecomposable $\mathfrak{e}(2)$-modules (up to isomorphism) of dimensions five and six. The proof of Corollary 4.5 shows how Corollary 4.4 can simply such proofs. We also point out that the graphs appearing in [9] roughly correspond, under the equivalence of categories in Corollary 4.2, to the diagrams appearing in the enumeration of irreducible components of quiver varieties given in [11].

REMARK 4.6. *As noted at the end of Section 3, the Lusztig quiver varieties $\Lambda_{\mathbf{v},Q}$ are closely related to the Kac-Moody algebra $\mathfrak{g}_Q$. Thus, the results of this section show that there is a relationship between the representation theory of the Euclidean group $E(2)$ and the Lie algebra $\mathfrak{sl}_\infty$ (or the Lie groups $SL(n)$).*

5. Nakajima quiver varieties

In this section we briefly review the quiver varieties introduced by Nakajima [21, 22]. We restrict our attention to the case when the quiver involved is of type A.

Let Q be the quiver Q_∞ or $Q_{a,b}$ for some $a \leq b$. For $V, W \in \mathcal{V}(I)$ define

$$L_Q(V,W) = \Lambda_{V,Q} \oplus \bigoplus_{i \in I} \mathrm{Hom}_{\mathbb{C}}(W_i, V_i).$$

We denote points of $L_Q(V,W)$ by (x,s) where $x = (x_h)_{h \in H^*} \in \Lambda_{V,Q}$ and $s = (s_i)_{i \in I} \in \bigoplus_{i \in I} \mathrm{Hom}_{\mathbb{C}}(W_i, V_i)$. We say an I-graded subspace S of V is x-invariant if $x_h(S_{\mathrm{out}(h)}) \subseteq S_{\mathrm{in}(h)}$ for all $h \in H^*$. We say a point $(x,s) \in L_Q(V,W)$ is stable if the following property holds: If S is an I-graded x-invariant subspace of V containing $\mathrm{im}\, s$, then $S = V$. We denote by $L_Q(V,W)^{\mathrm{st}}$ the set of stable points.

The group G_V acts on $L_Q(V,W)$ by

$$g \cdot (x,s) = (g_i)_{i \in I} \cdot ((x_h)_{h \in H^*}, (s_i)_{i \in I}) = ((g_{\mathrm{in}(h)} x_h g_{\mathrm{out}(h)}^{-1})_{h \in H^*}, (g_i s_i)_{i \in I}).$$

The action of G_V preserves the stability condition and the stabilizer in G_V of a stable point is trivial. We form the quotient

$$\mathcal{L}_Q(V,W) = L_Q(V,W)^{\mathrm{st}}/G_V.$$

The $\mathcal{L}_Q(V,W)$ are called *Nakajima quiver varieties*. For $\mathbf{v}, \mathbf{w} \in (\mathbb{Z}_{\geq 0})^I$, we set

$$L_Q(\mathbf{v},\mathbf{w}) = L_Q(V^{\mathbf{v}}, V^{\mathbf{w}}), \quad L_Q(\mathbf{v},\mathbf{w})^{\mathrm{st}} = L_Q(V^{\mathbf{v}}, V^{\mathbf{w}})^{\mathrm{st}}, \quad \mathcal{L}_Q(\mathbf{v},\mathbf{w}) = \mathcal{L}_Q(V^{\mathbf{v}}, V^{\mathbf{w}}).$$

We then have

$$L_Q(V,W) \cong L_Q(\dim V, \dim W), \quad L_Q(V,W)^{\mathrm{st}} \cong L_Q(\dim V, \dim W)^{\mathrm{st}},$$
$$\mathcal{L}_Q(V,W) \cong \mathcal{L}_Q(\dim V, \dim W),$$

and so we often blur the distinction between these pairs of isomorphic varieties.

Let $\mathrm{Irr}\, \Lambda_{V,Q}$ (resp. $\mathrm{Irr}\, \mathcal{L}_Q(V,W)$) denote the set of irreducible components of $\Lambda_{V,Q}$ (resp. $\mathcal{L}_Q(V,W)$). Then $\mathrm{Irr}\, \mathcal{L}_Q(V,W)$ can be identified with

$$\left\{ Y \in \mathrm{Irr}\, \Lambda_{V,Q} \,\middle|\, \left(Y \oplus \bigoplus_{i \in I} \mathrm{Hom}_{\mathbb{C}}(W_i, V_i) \right)^{\mathrm{st}} \neq \emptyset \right\}.$$

Specifically, the irreducible components of $\mathcal{L}_Q(V,W)$ are precisely those

$$\left(\left(Y \oplus \bigoplus_{i \in I} \mathrm{Hom}_{\mathbb{C}}(W_i, V_i) \right)^{\mathrm{st}} \right)/G_V$$

which are nonempty.

PROPOSITION 5.1 ([22, Corollary 3.12]). *The dimension of the Nakajima quiver varieties associated to the quiver Q_∞ are given by*

$$\dim_{\mathbb{C}} \mathcal{L}_{Q_\infty}(\mathbf{v},\mathbf{w}) = \sum_{i \in \mathbb{Z}} (\mathbf{v}_i \mathbf{w}_i - \mathbf{v}_i^2 + \mathbf{v}_i \mathbf{v}_{i+1}).$$

In a manner analogous to the way in which Lusztig quiver varieties are related to $U(\mathfrak{g}_Q)^-$ (see Section 3), Nakajima quiver varieties are closely related to the representation theory of $\mathfrak{g}_Q$. In particular, Nakajima [22] has shown that $\bigoplus_{\mathbf{v}} H_{\mathrm{top}}(\mathcal{L}_Q(\mathbf{v},\mathbf{w}))$ is isomorphic to the irreducible integrable highest-weight representation of $\mathfrak{g}_Q$ of highest weight $\sum_{i \in I} \mathbf{w}_i \omega_i$ where the ω_i are the fundamental weights of $\mathfrak{g}_Q$. Here H_{top} is top-dimensional Borel-Moore homology. The action of the Chevalley generators of $\mathfrak{g}_Q$ are given by certain convolution operations. The vector space $H_{\mathrm{top}}(\mathcal{L}_Q(\mathbf{v},\mathbf{w}))$ corresponds to the weight space of weight

$\sum_{i \in I} (\mathbf{w}_i \omega_i - \mathbf{v}_i \alpha_i)$. In [21], Nakajima gave a similar realization of these representations using a space of constructible functions on the quiver varieties rather than their homology. The irreducible components of Nakajima quiver varieties enumerate a natural basis in the representations of $\mathfrak{g}_Q$. These bases are given by the fundamental classes of the irreducible components in the Borel-Moore homology construction and by functions equal to one on an open dense subset of an irreducible component (and equal to zero on an open dense subset of all other irreducible components) in the constructible function realization.

6. Moduli spaces of representations of the Euclidean algebra

Given that $\mathfrak{e}(2)$ has wild representation type, it is prudent to restrict one's attention to certain subclasses of modules and to attempt a classification of the modules belonging to these classes. One possible approach is to impose a restriction on the number of generators of a representation (see [8, 9] for some results in this direction and [10] for other classes). In this section we will examine the relationship between moduli spaces of representations of the Euclidean algebra along with a set of generating vectors and Nakajima quiver varieties.

Let V be a finite-dimensional $\mathfrak{e}(2)$-module. For $u_1, u_2, \ldots, u_n \in V$, we denote by $\langle u_1, \ldots, u_n \rangle$ the submodule of V generated by $\{u_1, \ldots, u_n\}$. It is defined to be the smallest submodule of V containing all the u_i. A element $u \in V$ is called a *weight vector* if it lies in some weight space V_k of V. For a weight vector u, we let wt $u = k$ where $u \in V_k$. We say that $\{u_1, \ldots, u_n\}$ is a set of *generators* of V if each u_i is a weight vector and $\langle u_1, \ldots, u_n \rangle = V$. For $\mathbf{v} \in (\mathbb{Z}_{\geq 0})^{\mathbb{Z}}$, we let $|\mathbf{v}| = \sum_{k \in \mathbb{Z}} v_k$.

DEFINITION 6.1. *For* $\mathbf{v}, \mathbf{w} \in (\mathbb{Z}_{\geq 0})^{\mathbb{Z}}$, *let* $E(\mathbf{v}, \mathbf{w})$ *be the set of all*

$$(V, (u_k^j)_{k \in \mathbb{Z},\, 1 \leq j \leq \mathbf{w}_k})$$

where V *is a finite-dimensional* $\mathfrak{e}(2)$-*module with* $\dim V = \mathbf{v}$ *and* $\{u_k^j\}_{k \in \mathbb{Z},\, 1 \leq j \leq \mathbf{w}_k}$ *is a set of generators of* V *such that* wt $u_k^j = k$. *We say that two elements* $(V, (u_k^j))$ *and* $(\tilde{V}, (\tilde{u}_k^j))$ *of* $E(\mathbf{v}, \mathbf{w})$ *are* equivalent *if there exists a* $\mathfrak{e}(2)$-*module isomorphism* $\phi : V \to \tilde{V}$ *such that* $\phi(u_k^j) = \tilde{u}_k^j$. *We denote the set of equivalence classes by* $\mathcal{E}(\mathbf{v}, \mathbf{w})$.

THEOREM 6.2. *There is a natural one-to-one correspondence between* $\mathcal{E}(\mathbf{v}, \mathbf{w})$ *and* $\mathcal{L}_{Q_\infty}(\mathbf{v}, \mathbf{w})$.

PROOF. Let $(V, (u_k^j)_{k \in \mathbb{Z},\, 1 \leq j \leq \mathbf{w}_k}) \in E(\mathbf{v}, \mathbf{w})$ and let $V = \bigoplus V_k$ be the weight space decomposition of V. Thus V_k is isomorphic to $\mathbb{C}^{\mathbf{v}_k}$ and we identity the two via this isomorphism. We then define a point $\varphi(V, (u_k^j)) = (x, s) \in L_{Q_\infty}(\mathbf{v}, \mathbf{w})$ by setting

$$x_{h_i} = p_+|_{V_i}, \quad x_{\bar{h}_i} = p_-|_{V_{i+1}}, \quad i \in \mathbb{Z},$$
$$s(w_k^j) = u_k^j, \quad k \in \mathbb{Z}, \quad 1 \leq j \leq \mathbf{w}_k,$$

where $\{w_k^j\}_{1 \leq j \leq \mathbf{w}_k}$ is the standard basis of $\mathbb{C}^{\mathbf{w}_k}$ and the map s is extended by linearity. It follows from the results of Section 4 that $x \in \Lambda_{V^{\mathbf{v}}, Q}$ and so $(x, s) \in L_{Q_\infty}(\mathbf{v}, \mathbf{w})$. Furthermore, it follows from the fact that (u_k^j) is a set of generators, that (x, s) is a stable point. Thus $\varphi : E(\mathbf{v}, \mathbf{w}) \to L_{Q_\infty}(\mathbf{v}, \mathbf{w})^{\mathrm{st}}$. It is easily verified that two elements $(V, (u_k^j))$ and $(\tilde{V}, (\tilde{u}_k^j))$ are equivalent if and only if $\varphi(V, (u_k^j))$ and $\varphi(\tilde{V}, (\tilde{u}_k^j))$ lie in the same $G_{\mathbf{v}}$-orbit. Thus φ induces a map

$\varphi' : \mathcal{E}(\mathbf{v}, \mathbf{w}) \to \mathcal{L}_{Q_\infty}(\mathbf{v}, \mathbf{w})$ which is independent of the isomorphism $V \cong \mathbb{C}^{\mathbf{v}_k}$ chosen in our construction. It is easily seen that φ' is a bijection.   □

As noted in Section 5, the irreducible components of Nakajima quiver varieties can be identified with the irreducible components of Lusztig quiver varieties that are not killed by the stability condition. In the language of $\mathfrak{e}(2)$-modules, passing from Lusztig quiver varieties to Nakajima quiver varieties amounts to imposing the condition that the module be generated by a set of $|\mathbf{w}|$ weight vectors with weights prescribed by $\mathbf{w}$.

A *partition* is a sequence of non-increasing natural numbers $\lambda = (\lambda_1, \lambda_2, \dots, \lambda_l)$. The corresponding *Young diagram* is a collection of rows of square boxes which are left justified, with λ_i boxes in the ith row, $1 \le i \le l$. We will identify a partition and its Young diagram and we denote by $\mathcal{Y}$ the set of all partitions (or Young diagrams). If b is a box in a Young diagram λ, we write $x \in \lambda$ and we denote the box in the ith column and jth row of λ by $x_{i,j}$ (if such a box exists). The *residue* of $x_{i,j} \in \lambda$ is defined to be $\operatorname{res} x_{i,j} = i - j$. For $\lambda \in \mathcal{Y}$ and $a \in \mathbb{Z}$, define $\mathbf{v}^{\lambda,a} \in (\mathbb{Z}_{\ge 0})^{\mathbb{Z}}$ by setting $\mathbf{v}^{\lambda,a}_{i+a}$ to be the number of boxes in λ of residue i.

PROPOSITION 6.3. *For $\lambda \in \mathcal{Y}$, there exists a unique $\mathfrak{e}(2)$-module V (up to isomorphism) with a single generator of weight $a \in \mathbb{Z}$ and $\dim V = \mathbf{v}^{\lambda,a}$. It is given by*

$$V = \operatorname{Span}_{\mathbb{C}}\{x \mid x \in \lambda\}$$
$$l(x_{i,j}) = (a + \operatorname{res} x_{i,j})x_{i,j} = (a + i - j)x_{i,j}$$
$$p_+(x_{i,j}) = x_{i+1,j}$$
$$p_-(x_{i,j}) = x_{i,j+1},$$

where we set $x_{i,j} = 0$ if there is no box of λ in the ith column and jth row.

For $\mathbf{v} \in (\mathbb{Z}_{\ge 0})^{\mathbb{Z}}$ such that $\mathbf{v} \ne \mathbf{v}^{\lambda,a}$ for all $\lambda \in \mathcal{Y}$ and $a \in \mathbb{Z}$, there are no $\mathfrak{e}(2)$-modules V with a single generator and $\dim V = \mathbf{v}$

PROOF. By tensoring with an appropriate χ_n, we may assume that the generator of our module has weight zero. It is shown in [**11**, §5.1] that

$$\dim_{\mathbb{C}} \mathcal{L}_{Q_\infty}(\mathbf{v}, \mathbf{w}^0) = \begin{cases} 1 & \text{if } \mathbf{v} = \mathbf{v}^{\lambda,0}, \ \lambda \in \mathcal{Y}, \\ 0 & \text{otherwise} \end{cases},$$

where $\mathbf{w}^0_0 = 1$ and $\mathbf{w}^0_i = 0$ for $i \ne 0$ (the first case can be deduced from the dimension formula in Proposition 5.1). It then follows from Theorem 6.2 that if V is an $\mathfrak{e}(2)$-module with a single generator v of weight zero, we must have $\dim V = \mathbf{v}^{\lambda,0}$. Furthermore, up to isomorphism, there is only one such pair (V, v) and thus only one such module V.   □

Thus $\mathfrak{e}(2)$-modules with a single generator of a fixed weight are determined completely by the dimensions of their weight spaces. This was proven directly by Gruber and Henneberger in [**14**]. As in the proof of Corollary 4.5, we see that our knowledge of the precise relationship between quivers and the Euclidean algebra allows us to use known results about quivers and quiver varieties to simplify such proofs.

REMARK 6.4. *As explained at the end of Section 5, the Nakajima quiver varieties $\mathcal{L}_Q(\mathbf{v}, \mathbf{w})$ are closely connected to the representation theory of $\mathfrak{g}_Q$. Therefore,* *the relationship noted in Remark 4.6 between the representation theory of the Euclidean group and the Lie algebra $\mathfrak{sl}_\infty$ (or the Lie groups $SL(n)$) is emphasized further by the above results. Namely, the moduli space of representations of the Euclidean group along with a set of generators is closely related to the representation theory of $\mathfrak{sl}_\infty$ and the Lie groups $SL(n)$.*

REMARK 6.5. *Although Theorem 4.3 tells us that the Euclidean group has wild representation type, the results of this section produce a method of approaching the unwieldy problem of classifying its representations. Namely, if we fix the cardinality and weights of a generating set, the resulting moduli space of representations (along with a set of generators) is enumerated by a countable number of finite-dimensional varieties, one variety for the representations of each graded dimension.*

7. Further directions

The ideas presented in this paper open up some possible avenues of further investigation. We present here two of these.

Consider the Euclidean algebra over a field k of characteristic p instead of over the complex numbers. This algebra is still spanned by $\{p_+, p_-, l\}$ with commutation relations (2.1) but the weights of representations are elements of $\mathbb{Z}/p\mathbb{Z}$ (if we restrict our attention to "integral" weights as usual) instead of $\mathbb{Z}$. One can then show that this category of representations is equivalent to the category of representations of the preprojective algebra of the quiver of affine type $\hat{A}_{p-1}$. In this case, the representations with one generator are, in general, more complicated than in the complex case. We refer the reader to [**11**] for an analysis of the corresponding quiver varieties. There a graphical depiction of the irreducible components of these varieties is developed. These quiver varieties are related to moduli spaces of solutions of anti-self-dual Yang-Mills equations and Hilbert schemes of points in $\mathbb{C}^2$ and it would be interesting to further examine the relationship between these spaces and the Euclidean algebra.

In [**15**] and [**24**], Kashiwara and Saito defined a crystal structure on the sets of irreducible components of Lusztig and Nakajima quiver varieties. Using this structure, each irreducible component can be identified with a sequence of crystal operators acting on the highest weight element of the crystal. Under the identification of quiver varieties with (moduli spaces of) $\mathfrak{e}(2)$-modules, these sequences correspond to the Jordan-Hölder decomposition of $\mathfrak{e}(2)$-modules. It could be fruitful to further examine the implications of this correspondence.

References

[1] H. Ahmedov and I. H. Duru. Unitary representations of the two-dimensional Euclidean group in the Heisenberg algebra. *J. Phys. A*, 33(23):4277–4281, 2000.

[2] A. O. Barut and R. Raczka. *Theory of group representations and applications.* World Scientific Publishing Co., Singapore, second edition, 1986.

[3] K. Baumann. Vector and ray representations of the Euclidean group $E(2)$. *Rep. Math. Phys.*, 34(2):171–180, 1994.

[4] A. M. Boyarskiĭ and T. V. Skrypnik. Singular orbits of a coadjoint representation of Euclidean groups. *Uspekhi Mat. Nauk*, 55(3(333)):169–170, 2000.

[5] G. Cassinelli, G. Olivieri, P. Truini, and V. S. Varadarajan. On some nonunitary representations of the Poincaré group and their use for the construction of free quantum fields. *J. Math. Phys.*, 30(11):2692–2707, 1989.

[6] W. Crawley-Boevey. Geometry of the moment map for representations of quivers. *Compositio Math.*, 126(3):257–293, 2001.

 [7] V. Dlab and C. M. Ringel. The module theoretical approach to quasi-hereditary algebras. In
     Representations of algebras and related topics (Kyoto, 1990), volume 168 of *London Math.
     Soc. Lecture Note Ser.*, pages 200–224. Cambridge Univ. Press, Cambridge, 1992.
 [8] A. Douglas. *A classification of the finite dimensional, indecomposable representations of the
     Euclidean algebra $\mathfrak{e}(2)$ having two generators*. PhD thesis, University of Toronto, 2006.
 [9] A. Douglas. Finite dimensional representations of the Euclidean algebra $\mathfrak{e}(2)$ having two
     generators. *J. Math. Phys.*, 47(5):053506, 14, 2006.
[10] A. Douglas and A. Premat. A class of nonunitary, finite dimensional representations of the
     Euclidean algebra $e(2)$. *Comm. Algebra*, 35(5):1433–1448, 2007.
[11] I. B. Frenkel and A. Savage. Bases of representations of type A affine Lie algebras via quiver
     varieties and statistical mechanics. *Int. Math. Res. Not.*, (28):1521–1547, 2003.
[12] C. Geiss, B. Leclerc, and J. Schröer. Semicanonical bases and preprojective algebras. *Ann.
     Sci. École Norm. Sup. (4)*, 38(2):193–253, 2005.
[13] C. Geiss and J. Schröer. Varieties of modules over tubular algebras. *Colloq. Math.*, 95(2):163–
     183, 2003.
[14] B. Gruber and W. C. Henneberger. Representations of the Euclidean group in the plane.
     Nuovo Cimento B (11), 77(2):203–233, 1983.
[15] M. Kashiwara and Y. Saito. Geometric construction of crystal bases. *Duke Math. J.*, 89(1):9–
     36, 1997.
[16] Y. S. Kim and M. E. Noz. *Theory and applications of the Poincaré group*. Fundamental
     Theories of Physics. D. Reidel Publishing Co., Dordrecht, 1986.
[17] G. Lusztig. Quivers, perverse sheaves, and quantized enveloping algebras. *J. Amer. Math.
     Soc.*, 4(2):365–421, 1991.
[18] G. Lusztig. *Introduction to quantum groups*, volume 110 of *Progress in Mathematics*.
     Birkhäuser Boston Inc., Boston, MA, 1993.
[19] G. Lusztig. Semicanonical bases arising from enveloping algebras. *Adv. Math.*, 151(2):129–
     139, 2000.
[20] R. Mirman. Poincaré zero-mass representations. *Internat. J. Modern Phys. A*, 9(1):127–156,
     1994.
[21] H. Nakajima. Instantons on ALE spaces, quiver varieties, and Kac-Moody algebras. *Duke
     Math. J.*, 76(2):365–416, 1994.
[22] H. Nakajima. Quiver varieties and Kac-Moody algebras. *Duke Math. J.*, 91(3):515–560, 1998.
[23] I. Reiten. Dynkin diagrams and the representation theory of algebras. *Notices Amer. Math.
     Soc.*, 44(5):546–556, 1997.
[24] Y. Saito. Crystal bases and quiver varieties. *Math. Ann.*, 324(4):675–688, 2002.
[25] E. Wigner. On unitary representations of the inhomogeneous Lorentz group. *Ann. of Math.
     (2)*, 40(1):149–204, 1939.

UNIVERSITY OF OTTAWA, OTTAWA, ONTARIO, CANADA
E-mail address: `alistair.savage@uottawa.ca`

Contemporary Mathematics
Volume **478**, 2009

$\mathfrak{eu}_2$-Lie admissible algebras and Steinberg unitary Lie algebras

Shikui Shang and Yun Gao

ABSTRACT. In this paper, we will give a necessary and sufficient condition such that a k-algebra R becomes $\mathfrak{eu}_2$-Lie admissible. Then, we will work out the second homology group of the Lie algebra $\mathfrak{eu}_2(R,-,\boldsymbol{\gamma})$.

Introduction

In this paper, we will study the $\mathfrak{eu}_2$-Lie admissible algebras and Steinberg unitary Lie algebras $\mathfrak{stu}_2(R,-,\boldsymbol{\gamma})$.

Steinberg unitary Lie algebras were introduced by Allison and Faulkner [AF] as a generalization of Steinberg Lie algebras (see [KL] and [F]). They were further studied in [G1] for the case R is associative and in [AG] for the case R is structurable. More precisely, let k be a field and R be an associative k-algebras with identity, equipped with an involution $-$(an anti-automorphism which squares to the identity). The elementary unitary Lie algebra $\mathfrak{eu}_n(R,-,\boldsymbol{\gamma})$ is the subalgebra of the matrice Lie algebra $gl_n(R)$ generated by $\xi_{ij}(a) = e_{ij}(a) - \gamma_i\gamma_j^{-1}e_{ji}(\bar{a})$, for $1 \leq i \neq j \leq n$, $a \in R$, where e_{ij} is the standard matrix unit and $\boldsymbol{\gamma}$ is an n-tuple of nonzero scalars $(\gamma_1,\cdots,\gamma_n)$, $\gamma_i \in k^\times$, $1 \leq i \leq n$. The elements $\xi_{ij}(a)$ satisfy certain canonical relations. When $n \geq 3$, the Steinberg unitary Lie algebra $\mathfrak{stu}_n(R,-,\boldsymbol{\gamma})$ is defined by generators $u_{ij}(a)$ corresponding to $\xi_{ij}(a)$ and those same canonical relations. So one has a Lie algebra homomorphism ϕ from $\mathfrak{stu}_n(R,-,\boldsymbol{\gamma})$ to $\mathfrak{eu}_n(R,-,\boldsymbol{\gamma})$. It is known that the above homomorphism ϕ yields a central extension(see [AF]). It is easy to see that both $\mathfrak{stu}_n(R,-,\boldsymbol{\gamma})$ and $\mathfrak{eu}_n(R,-,\boldsymbol{\gamma})$ are perfect since the generators are contained in their derived algebras. With some assumptions on R, ϕ becomes the universal covering of $\mathfrak{eu}_n(R,-,\boldsymbol{\gamma})$ with $\ker\phi \cong {}_{-1}HD_1(R)$, where ${}_{-1}HD_1(R)$ is the first skew-dihedral homology group of R (see [G1]).

For $n \geq 3$ and an arbitrary nonassociative algebra $(R,-)$ with involution, the canonical relations might cause a nontrivial relation $u_{ij}(a) = 0$ with $a \neq 0$. Denote the kernel of u_{ij} by $I_n = I_n(R,-,\boldsymbol{\gamma})$, which is independent of the choice of γ and

Key words and phrases. Steinberg unitary Lie algebra, $\mathfrak{eu}_2$-Lie admissible algebra, Lie triple system.

2000 Mathematics Subject Classification: 17B60, 17B55, 17D25, 17A30.

Research of the second author (the corresponding author) was partially supported by NSERC of Canada and Chinese Academy of Science.

$i \neq j$. It is easy to show that I_n is an ideal of R and $\mathfrak{stu}_n(\bar{R}, -, \gamma) \cong \mathfrak{stu}_n(R, -, \gamma)$ where $\bar{R} = R/I_n$. If $I_n(R, -, \gamma) = 0$, we say that $(R, -)$ is n-faithful. In fact, if $n \geq 4$, $(R, -)$ is n-faithful if and only if R is associative. Allison and Faulkner proved that $(R, -)$ is a 3-faithful nonassociative algebra if and only if $(R, -)$ is structurable in [AF].

On the other hand, when $n = 2$, $sl_2(R)$ and $\mathfrak{st}_2(R)$ were studied in [G2], which demonstrated some intrigue phenomena. In this paper, we will deal with $\mathfrak{eu}_2(R, -, \gamma)$ and introduce the Steinberg unitary Lie algebra $\mathfrak{stu}_2(R, -, \gamma)$. We will further study their relations. Our approach is to use the Lie triple system and some technics developed in [H] and [S]. This not only simplified a lot of verifications as was done in [G2] but also enabled us to obtain a necessary and sufficient condition for $I_2 = 0$ which was not able to do in [G2]. This revised approach (using the Lie triple system) is due to the generous referee who provided many critical and instructive suggestions with great details.

The organization of this paper is as follows. In Section 1, we study $\mathfrak{eu}_2(R, -, \gamma)$ and give a necessary and sufficient condition for $(R, -)$ such that $\mathfrak{eu}_2(R, -, \gamma)$ becomes a Lie algebra. Then in Section 2, we define the Steinberg Lie algebra $\mathfrak{stu}_2(R, -, \gamma)$ and give a necessary and sufficient condition for $I_2(R, -, \gamma) = 0$ using some knowledge of Lie triple systems. Moreover, we prove that $\mathfrak{stu}_2(R, -, \gamma)$ is the universal covering of $\mathfrak{eu}_2(R, -, \gamma)$ when R is an $\mathfrak{eu}_2$-Lie admissible algebra satisfying some assumptions given in Section 1. We investigate the structure of $\ker \phi$ in Section 3. In the last section, we apply the results of Section 2 and 3 to study $sl_2(S)$ and $\mathfrak{st}_2(S)$, and recover the main theorems in [G2].

We would like to thank the referee for his lengthy comments by showing us the adoption of Lie triple system which significantly improved this new version of our paper.

1. Some basics for the algebra $\mathfrak{eu}_2(R, -, \gamma)$

Let k be a field and R be a nonassociative k-algebra with identity, equipped with an involution $-$. We always assume that char $k \neq 2$. Then, we have $R = R_+ \oplus R_-$ where $R_+ = \{a \in R | \ \bar{a} = a\}$ and $R_- = \{a \in R | \ \bar{a} = -a\}$. Any $a \in R$ has a decomposition $a = \frac{1}{2}(a + \bar{a}) + \frac{1}{2}(a - \bar{a})$ such that $\frac{1}{2}(a + \bar{a}) \in R_+$ and $\frac{1}{2}(a - \bar{a}) \in R_-$.

For each positive integer $n \geq 2$, the algebra $M_n(R)$ of $n \times n$ matrices with coefficients in R forms a nonassociative algebra $gl_n(R) = M_n(R)^-$ over k under the commutator product. Assume that $n \geq 2$, Let γ be an n-tuple of nonzero scalars $(\gamma_1, \cdots, \gamma_n)$, $\gamma_i \in k^\times$, $1 \leq i \leq n$. We denote $\xi_{ij}(a) = e_{ij}(a) - \gamma_i \gamma_j^{-1} e_{ji}(\bar{a})$ for $1 \leq i \neq j \leq n, a \in R$, where e_{ij} is the standard matrix unit, and $\mathfrak{eu}_n(R, -, \gamma)$ is the subalgebra of $gl_n(R)$ generated by all $\xi_{ij}(a)$. An algebra $\mathfrak{g}$ with product $[\ ,\ ]$ is called perfect if $[\mathfrak{g}, \mathfrak{g}] = \mathfrak{g}$. When $n \geq 3$, $\mathfrak{eu}_n(R, -, \gamma)$ is perfect. But we will see the case $n = 2$ is a bit more complicated.

In this paper, we use the following notations.

NOTATION . For $a, b, c \in R$, we denote

$$[a, b] = ab - ba; \quad a * b = \frac{1}{2}(ab + ba); \quad (a, b, c) = (ab)c - a(bc);$$

$$\langle a, b \rangle = ab - \overline{ab}; \quad \langle a, b, c \rangle = \langle a, b \rangle c + c \langle b, a \rangle$$

$$J(a, b, c) = [[a, b], c] + [[b, c], a] + [[c, a], b]$$

and $[A, B]$, $A * B$, (A, B, C), $\langle A, B \rangle$, $\langle A, B, C \rangle$ are the linear spans of $[a, b]$, $a *$ b, (a, b, c), $\langle a, b \rangle$, $\langle a, b, c \rangle$, for $a \in A, b \in B$ and $c \in C$. People often call $[a, b]$ (or (a, b, c)) a commutator (or associator) of R.

LEMMA 1.1 . Let R be a nonassociative algebra with involution $-$ and decomposition $R = R_+ \oplus R_-$. For any $a, b, c \in R$,

(i) $\overline{\langle a, b \rangle} = -\langle a, b \rangle$, i.e. $\langle R, R \rangle \subset R_-$,

(ii) $\langle a, b \rangle = \langle ab, 1 \rangle = \langle 1, ab \rangle$, $\langle a, 1 \rangle = \langle 1, a \rangle = a - \bar{a}$,

In particular, $\langle a, 1 \rangle = \langle 1, a \rangle = 2a$, if $a \in R_-$,

(iii) $\overline{[a, b]} = [\bar{b}, \bar{a}] = -[\bar{a}, \bar{b}]$, we have $\overline{[R, R]} = [R, R]$,

(iv) $\langle R, R \rangle = R_-$,

(v) $\mathrm{span}_k \{ \langle a, b \rangle - \langle b, a \rangle \,|\, a, b \in R \} = R_- \cap [R, R]$,

(vi) $[R_-, R_-] \subset R_-$, i.e. R_- is closed under the Lie product,

(vii) $\overline{\langle a, b, c \rangle} = -\langle b, a, \bar{c} \rangle$, we have $\overline{\langle R, R, R \rangle} = \langle R, R, R \rangle$.

PROOF. (i),(ii) and (iii) are obvious. (iv) follows from (i) and (ii).

For (v), since (iii), (iv) and $\langle a, b \rangle - \langle b, a \rangle = ab - \overline{ab} - ba + \overline{ba} = [a, b] - \overline{[a, b]}$, we have

$$\mathrm{span}_k \{ \langle a, b \rangle - \langle b, a \rangle \,|\, a, b \in R \} \subset R_- \cap [R, R].$$

Conversely, if we take $t = \Sigma_i [a_i, b_i] \in R_- \cap [R, R]$, then

$$t = \frac{1}{2}(\Sigma_i [a_i, b_i] - \overline{\Sigma_i [a_i, b_i]}) = \frac{1}{2}\Sigma_i([a_i, b_i] - \overline{[a_i, b_i]}) = \frac{1}{2}\Sigma_i(\langle a_i, b_i \rangle - \langle b_i, a_i \rangle)$$

which shows $\mathrm{span}_k \{ \langle a, b \rangle - \langle b, a \rangle \,|\, a, b \in R \} \supset R_- \cap [R, R]$.

(vi) and (vii) are easily verified. □

Since $\xi_{12}(a) = -\gamma_1 \gamma_2^{-1} \xi_{21}(\bar{a})$, we have $\xi_{12}(R) = \xi_{21}(R)$ and $\mathfrak{eu}_2(R, -, \gamma)$ is generated by $\xi_{12}(R)$. Set $H(a, b) = [\xi_{12}(a), \xi_{21}(b)]$ for $a, b \in R$, we have

$$H(a, b) = [\xi_{12}(a), \xi_{21}(b)]$$
$$= [e_{12}(a) - \gamma_1 \gamma_2^{-1} e_{21}(\bar{a}), e_{21}(b) - \gamma_2 \gamma_1^{-1} e_{12}(\bar{b})]$$
$$= e_{11}(ab - \overline{ab}) - e_{22}(ba - \overline{ba})$$
$$= e_{11}(\langle a, b \rangle) - e_{22}(\langle b, a \rangle)$$

Let $H(R, R) = [\xi_{12}(R), \xi_{21}(R)]$ be the linear span of $H(a, b)$, $a, b \in R$, then

LEMMA 1.2 .

$$H(R, R) = \{A = e_{11}(x) + e_{22}(y)\,|\, x, y \in R_- \text{ and } \mathrm{Tr}(A) \in R_- \cap [R, R]\}.$$

PROOF. "$\subset$" can be obtained from Lemma 1.1 (i) and (v).

For "$\supset$", Let A be a diagonal matrix with $\mathrm{Tr}(A) \in R_- \cap [R, R]$. Since $\mathrm{Tr}(H(a, b)) = \langle a, b \rangle - \langle b, a \rangle$, Lemma 1.1(v) shows $\mathrm{Tr}(H(R, R)) = R_- \cap [R, R]$. Thus, we can adjust A by an element of $H(R, R)$ to assume that $\mathrm{Tr}(A) = 0$; i.e.

$$A = e_{11}(y) - e_{22}(y) = \frac{1}{2}(e_{11}(\langle y, 1 \rangle) - e_{22}(\langle 1, y \rangle))$$

$$= \frac{1}{2}H(y, 1) \in H(R, R),$$

for some $y \in R$. □

LEMMA 1.3 . The nonassociative algebra $\mathfrak{eu}_2(R, -, \boldsymbol{\gamma})$ has a vector space decomposition

$$\mathfrak{eu}_2(R, -, \boldsymbol{\gamma}) = \xi_{12}(R) \oplus H(R, R).$$

PROOF. By Lemma 1.1 (vi), for any $a, b, c \in R$,

$$[H(a, b), \xi_{12}(c)] = [e_{11}(\langle a, b\rangle) - e_{22}(\langle b, a\rangle), e_{12}(c) - \gamma_1 \gamma_2^{-1} e_{21}(\bar{c})]$$

$$= e_{12}(\langle a, b\rangle c + c\langle b, a\rangle) - \gamma_1 \gamma_2^{-1} e_{21}(-(\langle b, a\rangle \bar{c} + \bar{c}\langle a, b\rangle))$$

$$= e_{12}(\langle a, b, c\rangle) - \gamma_1 \gamma_2^{-1} e_{21}(\overline{\langle a, b, c\rangle})$$

$$= \xi_{12}(\langle a, b, c\rangle)$$

We only need to show that $H(R, R)$ is closed under $[\ ,\ ]$. Indeed, we have

$$[H(a, b), H(c, d)] = [e_{11}(\langle a, b\rangle) - e_{22}(\langle b, a\rangle), e_{11}(\langle c, d\rangle) - e_{22}(\langle d, c\rangle)]$$

$$= e_{11}([\langle a, b\rangle, \langle c, d\rangle]) + e_{22}([\langle b, a\rangle, \langle d, c\rangle]).$$

By Lemma 1.1 (i) and (vi), $[\langle a, b\rangle, \langle c, d\rangle] + [\langle b, a\rangle, \langle d, c\rangle] \in R_- \cap [R, R]$, for any $a, b, c, d \in R$.

It is contained in $H(R, R)$ because of Lemma 1.2. $\qquad\square$

Now we give the necessary and sufficient condition for $\mathfrak{eu}_2(R, -, \boldsymbol{\gamma})$ is perfect.

THEOREM 1.1 . $\mathfrak{eu}_2(R, -, \boldsymbol{\gamma})$ is perfect if and only if $R = \langle R, R, R\rangle$.

PROOF. By Lemma 1.3, $\mathfrak{eu}_2(R, -, \boldsymbol{\gamma}) = \xi_{12}(R) \oplus H(R, R)$, then

$$[\mathfrak{eu}_2(R, -, \boldsymbol{\gamma}), \mathfrak{eu}_2(R, -, \boldsymbol{\gamma})] = [H(R, R), \xi_{12}(R)] \oplus H(R, R).$$

We also have $[H(a, b), \xi_{12}(c)] = \xi_{12}(\langle a, b, c\rangle)$. Thus, the theorem is proved. $\qquad\square$

Recall the definition of the center $Z(R)$ of a nonassociative algebra R is the set of all elements in R which commute and associate with all elements. Particularly, $Z(R)$ is contained in the nucleus $N(R)$ of R, where

$$N(R) = \{n \in R | (n, R, R) = (R, n, R) = (R, R, n) = 0\}.$$

We show some assumptions on R, from which imply $R = \langle R, R, R\rangle$.

ASSUMPTION 1.1 . Suppose that there exists an element $e \in R$ such that
(i) $e \in R_-$, namely, $\bar{e} = -e$,
(ii) e lies in the center $Z(R)$ of R,
(iii) e is a unit.

ASSUMPTION 1.2 . Suppose that $R_- * R = R$, where $*$ is the Jordan product $a * b = \frac{1}{2}(ab + ba)$ and $R_- * R = \{\sum_i a_i * b_i | a_i \in R_-, b_i \in R\}$.

Obviously, the Assumption 1.1 implies 1.2. These assumptions are not so restrictive. For example, if -1 has no square root in k, and S is an associative k-algebra with identity, equipped an involution $^-$, then consider the algebra $R = S \otimes_k k(\sqrt{-1}) = S \oplus \sqrt{-1}S$, extending $^-$ on R by $\overline{a + \sqrt{-1}b} = \bar{a} - \sqrt{-1}\,\bar{b}$. One can check that R is an associative k-algebra equipped an involution. In this case, we may choose $e = \sqrt{-1}$ and Assumption 1.1 holds. Also, one may check that Assumption 1.2 holds for any associative composition algebra with the natural involution $-$, but Assumption 1.1 may fail.

For $a \in R_-, b \in R$, since $\langle a, 1 \rangle = \langle 1, a \rangle = a - \bar{a} = 2a$, we see

$$a * b = \frac{1}{2}(ab + ba) = \frac{1}{4}((a - \bar{a})b + b(a - \bar{a})) = \frac{1}{4}\langle a, 1, b \rangle$$

Therefore, $R_- * R \subset \langle R, R, R \rangle$. So, assumption 1.2 implies $\mathfrak{eu}_2(R, -, \gamma)$ is perfect.

Next, we shall investigate when $\mathfrak{eu}_2(R, -, \gamma)$ is a Lie algebra. A nonassociative algebra R with an involution $-$ over k is called $\mathfrak{eu}_2$-Lie admissible if $\mathfrak{eu}_2(R, -, \gamma)$ is a Lie algebra over k. (We will see that it is independent of γ.) Certainly, associative algebras are $\mathfrak{eu}_2$-Lie admissible.

Let T be a nonassociative k-algebra. For $x_1, x_2, x_3 \in T$, we define

$$J(x_1, x_2, x_3) = [[x_1, x_2], x_3] + [[x_2, x_3], x_1] + [[x_3, x_1], x_2].$$

Then, T is a Lie algebra over k if and only if it is anticommutative and $J(x_1, x_2, x_3) = 0$ for any $x_1, x_2, x_3 \in T$.

LEMMA 1.4 . Let T be an anticommutative k-algebra,

(i) $J(x_1, x_2, x_3) = (-1)^{\sigma} J(x_{\sigma(1)}, x_{\sigma(2)}, x_{\sigma(3)})$ for $x_i \in T$ and $\sigma \in S_3$, the permutation group of $\{1, 2, 3\}$.

(ii) If $S \subset T$ generates T and $J(S, T, T) = 0$, then T is a Lie algebra over k.

PROOF. Clearly, (i) holds for $\sigma = (12)$ and $\sigma = (123)$, and hence for all $\sigma \in S_3$.

For $x \in T$, we note that $J(x, T, T) = 0$ if and only if $ad(x) : y \mapsto [x, y]$ is a derivation of T. If $J(x, T, T) = J(y, T, T) = 0$, $ad(x)$ and $ad(y)$ are derivations. Then, $J(x, y, T) = 0$ shows that $ad([x, y]) = [ad(x), ad(y)]$, which is also a derivation of T. Thus, $J([x, y], T, T) = 0$. This shows (ii). $\square$

Now, we prove the main theorem of this section.

THEOREM 1.2 . $(R, -)$ is $\mathfrak{eu}_2$-Lie admissible if and only if the following identities hold

(i) $(a, \bar{b}, c) + (b, \bar{c}, a) + (c, \bar{a}, b) = (a, \bar{c}, b) + (c, \bar{b}, a) + (b, \bar{a}, c)$

(ii) $(a, b, \langle c, d \rangle) + (\langle c, d \rangle, a, b) + (a, \langle d, c \rangle, b) = (\bar{b}, \bar{a}, \langle c, d \rangle) + (\langle c, d \rangle, \bar{b}, \bar{a}) + (\bar{b}, \langle d, c \rangle, \bar{a})$

(iii) $(\langle a, b \rangle, \langle c, d \rangle, e) + (e, \langle b, a \rangle, \langle d, c \rangle) + (\langle a, b \rangle, e, \langle d, c \rangle)$
$\quad = (\langle c, d \rangle, \langle a, b \rangle, e) + (e, \langle d, c \rangle, \langle b, a \rangle) + (\langle c, d \rangle, e, \langle b, a \rangle)$

for any $a, b, c, d, e \in R$.

PROOF. Obviously, $\mathfrak{eu}_2(R, -, \gamma)$ is anticommutative. Thus, it is a Lie algebra if and only if $J(x, y, z) = 0$ for any $x, y, z \in \mathfrak{eu}_2(R, -, \gamma)$.

Since $\mathfrak{eu}_2(R, -, \gamma)$ is generated by $\xi_{12}(a), a \in R$, by Lemma 1.3 and Lemma 1.4, we need to consider the following three cases.

Case 1. $x, y, z \in \xi_{12}(R)$.

We can assume $x = \xi_{12}(a), y = \xi_{12}(b)$ and $z = \xi_{12}(c)$.

$$[[x, y], z] = [[\xi_{12}(a), \xi_{12}(b)], \xi_{12}(c)] = -\gamma_1 \gamma_2^{-1}[[\xi_{12}(a), \xi_{21}(\bar{b})], \xi_{12}(c)]$$
$$= -\gamma_1 \gamma_2^{-1}[H(a, \bar{b}), \xi_{12}(c)] = -\gamma_1 \gamma_2^{-1}\xi_{12}(\langle a, \bar{b}, c \rangle)$$

Thus, $J(x, y, z) = -\gamma_1 \gamma_2^{-1}\xi_{12}(\langle a, \bar{b}, c \rangle + \langle b, \bar{c}, a \rangle + \langle c, \bar{a}, b \rangle)$.

From direct calculation,

$$\langle a, \bar{b}, c \rangle + \langle b, \bar{c}, a \rangle + \langle c, \bar{a}, b \rangle$$
$$= (a\bar{b})c - (b\bar{a})c + c(\bar{b}a) - c(\bar{a}b)$$
$$+ (b\bar{c})a - (c\bar{b})a + a(\bar{c}b) - a(\bar{b}c)$$
$$+ (c\bar{a})b - (a\bar{c})b + b(\bar{a}c) - b(\bar{c}a)$$
$$= ((a, \bar{b}, c) + (b, \bar{c}, a) + (c, \bar{a}, b)) - ((a, \bar{c}, b) + (c, \bar{b}, a) + (b, \bar{a}, c)),$$

which yields the identity (i).

Case 2. $x, y \in \xi_{12}(R)$ and $z \in H(R, R)$.
We assume $x = \xi_{12}(a), y = \xi_{21}(b)$ and $z = H(c, d)$.

$$[[x, y], z] = [[\xi_{12}(a), \xi_{21}(b)], H(c, d)] = [H(a, b), H(c, d)]$$
$$= e_{11}([\langle a, b \rangle, \langle c, d \rangle]) + e_{22}([\langle b, a \rangle, \langle d, c \rangle])$$

$$[[y, z], x] = [[\xi_{21}(b), H(c, d)], \xi_{12}(a)] = [\xi_{21}(b\langle c, d \rangle + \langle d, c \rangle b), \xi_{12}(a)]$$
$$= [\xi_{21}(\langle d, c, b \rangle), \xi_{12}(a)] = -H(a, \langle d, c, b \rangle)$$

$$[[z, x], y] = [[H(c, d), \xi_{12}(a)], \xi_{21}(b)] = [\xi_{12}(\langle c, d, a \rangle), \xi_{21}(b)]$$
$$= H(\langle c, d, a \rangle, b)$$

Then,

$$J(x, y, x) = e_{11}([\langle a, b \rangle, \langle c, d \rangle] - \langle a, \langle d, c, b \rangle \rangle + \langle \langle c, d, a \rangle, b \rangle)$$
$$+ e_{22}([\langle b, a \rangle, \langle d, c \rangle] + \langle \langle d, c, b \rangle, a \rangle - \langle b, \langle c, d, a \rangle \rangle)$$

Furthermore, we have,

$$[\langle a, b \rangle, \langle c, d \rangle] - \langle a, \langle d, c, b \rangle \rangle + \langle \langle c, d, a \rangle, b \rangle$$
$$= \langle a, b \rangle \langle c, d \rangle - \langle c, d \rangle \langle a, b \rangle - \langle a, \langle d, c \rangle b + b \langle c, d \rangle \rangle + \langle \langle c, d \rangle a + a \langle d, c \rangle, b \rangle$$
$$= (ab)\langle c, d \rangle - (\bar{b}\bar{a})\langle c, d \rangle - \langle c, d \rangle(ab) + \langle c, d \rangle(\bar{b}\bar{a}) - a(\langle d, c \rangle b) - a(b \langle c, d \rangle)$$
$$- (\bar{b}\langle d, c \rangle)\bar{a} - (\langle c, d \rangle \bar{b})\bar{a} + (\langle c, d \rangle a)b + (a \langle d, c \rangle)b + \bar{b}(\bar{a}\langle c, d \rangle) + \bar{b}(\langle d, c \rangle \bar{a})$$
$$= (a, b, \langle c, d \rangle) - (\bar{b}, \bar{a}, \langle c, d \rangle) + (\langle c, d \rangle, a, b)$$
$$- (\langle c, d \rangle, \bar{b}, \bar{a}) + (a, \langle d, c \rangle, b) - (\bar{b}, \langle d, c \rangle, \bar{a})$$

Exchanging a, c with b, d, we can obtain that $J(x, y, z) = 0 \iff$ (ii) holds.

Case 3. $x, y \in H(R, R)$ and $z \in \xi_{12}(R)$.
Assume that $x = H(a, b), y = H(c, d)$ and $z = \xi_{12}(e)$,

$$[[x, y], z] = [[H(a, b), H(c, d)], \xi_{12}(e)]$$
$$= [e_{11}([\langle a, b \rangle, \langle c, d \rangle]) + e_{22}([\langle b, a \rangle, \langle d, c \rangle]), \xi_{12}(e)]$$
$$= \xi_{12}([\langle a, b \rangle, \langle c, d \rangle]e - e[\langle b, a \rangle, \langle d, c \rangle])$$

$$[[y, z], x] = [[H(c, d), \xi_{12}(e)], H(a, b)] = [\xi_{12}(\langle c, d, e \rangle), H(a, b)]$$
$$= -\xi_{12}(\langle a, b, \langle c, d, e \rangle \rangle)$$

$$[[z, x], y] = [[\xi_{12}(e), H(a, b)], H(c, d)] = [H(c, d), \xi_{12}(\langle a, b, e \rangle)]$$
$$= \xi_{12}(\langle c, d, \langle a, b, e \rangle \rangle)$$

So in this case,

$$J(x, y, x) =$$
$$\xi_{12}([\langle a, b\rangle, \langle c, d\rangle]e - e[\langle b, a\rangle, \langle d, c\rangle] - \langle a, b, \langle c, d, e\rangle\rangle + \langle c, d, \langle a, b, e\rangle\rangle)$$

and

$$[\langle a, b\rangle, \langle c, d\rangle]e - e[\langle b, a\rangle, \langle d, c\rangle] - \langle a, b, \langle c, d, e\rangle\rangle + \langle c, d, \langle a, b, e\rangle\rangle$$
$$=(\langle a, b\rangle\langle c, d\rangle - \langle c, d\rangle\langle a, b\rangle)e - e(\langle b, a\rangle\langle d, c\rangle - \langle d, c\rangle\langle b, a\rangle)$$
$$- (\langle a, b\rangle\langle c, d, e\rangle + \langle c, d, e\rangle\langle b, a\rangle) + (\langle c, d\rangle\langle a, b, e\rangle + \langle a, b, e\rangle\langle d, c\rangle)$$
$$=(\langle a, b\rangle\langle c, d\rangle)e - (\langle c, d\rangle\langle a, b\rangle)e - e(\langle b, a\rangle\langle d, c\rangle) + e(\langle d, c\rangle\langle b, a\rangle)$$
$$- \langle a, b\rangle(\langle c, d\rangle e) - \langle a, b\rangle(e\langle d, c\rangle) - (\langle c, d\rangle e)\langle b, a\rangle - (e\langle d, c\rangle)\langle b, a\rangle$$
$$+ \langle c, d\rangle(\langle a, b\rangle e) + \langle c, d\rangle(e\langle b, a\rangle) + (\langle a, b\rangle e)\langle d, c\rangle + (e\langle b, a\rangle)\langle d, c\rangle$$
$$=(\langle a, b\rangle, \langle c, d\rangle, e) - (\langle c, d\rangle, \langle a, b\rangle, e) + (e, \langle b, a\rangle, \langle d, c\rangle)$$
$$- (e, \langle d, c\rangle, \langle b, a\rangle) + (\langle a, b\rangle, e, \langle d, c\rangle) - (\langle c, d\rangle, e, \langle b, a\rangle)$$

which shows that $J(x, y, x) = 0 \iff$ (iii).

So, $(R, -)$ is $\mathfrak{eu}_2$-Lie admissible $\iff$ (i)-(iii) hold.   □

COROLLARY 1.1 . Let R be a nonassociative k-algebra with an involution $-$. If R is $\mathfrak{eu}_2$-Lie admissible, then R_- is Lie admissible, i.e R_- is a Lie algebra with $[\,,\,]$ over k.

PROOF. Let $x = H(a, b), y = H(c, d), z = H(e, f) \in \mathfrak{eu}_2(R, -, \boldsymbol{\gamma})$, we can easily see

$$J(x, y, z) = e_{11}(J(\langle a, b\rangle, \langle c, d\rangle, \langle e, f\rangle)) - e_{22}(J(\langle b, a\rangle, \langle d, c\rangle, \langle f, e\rangle)).$$

By Lemma 1.1 (iv) and (vi), $J(x, y, z) = 0$ implies $J(R_-, R_-, R_-) = 0$. We have this corollary.   □

An associative k-algebra is $\mathfrak{eu}_2$-Lie admissible. Finally, we give a class of $\mathfrak{eu}_2$-Lie admissible algebras which are not associative.

Example 1.1. Let $V = X \oplus Y$ be a k-vector space with a symplectic form $f : V \times V \to k$ satisfying $f(X, X) = f(Y, Y) = 0$. Let $W = k1 \oplus V$ be the k-algebra with identity 1 and $uv = f(u, v)1$ for $u, v \in V$ and involution given by $\bar{1} = 1$ and $\overline{x + y} = x - y$ for $x \in X, y \in Y$.

With this involution, we have $W_+ = k1 \oplus X$ and $W_- = Y$. If $u, v, w \in V$, we have

$$(u, v, w) = f(u, v)w - f(v, w)u = f(u, v)w + f(w, v)u.$$

By the non-degeneracy of f, if $f(u, v) \neq 0$, then $(v, v, u) = f(u, v)v \neq 0$. Thus, W is not associative. But we will show that W is $\mathfrak{eu}_2$-Lie admissible, i.e. it satisfies the identities $(i) - (iii)$ in Theorem 1.2.

For (i), it suffices to check $a, b, c \in V$.

$$LHS = (a, \bar{b}, c) + (b, \bar{c}, a) + (c, \bar{a}, b)$$
$$=f(a, \bar{b})c + f(c, \bar{b})a + f(b, \bar{c})a + f(a, \bar{c})b + f(c, \bar{a})b + f(b, \bar{a})c$$
$$=(f(c, \bar{b}) + f(b, \bar{c}))a + (f(a, \bar{c}) + f(c, \bar{a}))b + (f(a, \bar{b}) + f(b, \bar{a}))c,$$

and

$$RHS = (a, \bar{c}, b) + (c, \bar{b}, a) + (b, \bar{a}, c)$$
$$= f(a, \bar{c})b + f(b, \bar{c})a + f(c, \bar{b})a + f(a, \bar{b})c + f(b, \bar{a})c + f(c, \bar{a})b$$
$$= (f(c, \bar{b}) + f(b, \bar{c}))a + (f(a, \bar{c}) + f(c, \bar{a}))b + (f(a, \bar{b}) + f(b, \bar{a}))c.$$

So, (i) holds.

Furthermore, since $\langle k1, k1 \rangle = \langle V, V \rangle = 0$ and $\langle 1, v \rangle = \langle v, 1 \rangle = v - \bar{v}$ for $v \in V$. we see that

$$\langle a, b \rangle = \langle b, a \rangle \in Y$$

for all $a, b \in W$. Also, we have $(u, v, w) = f(u, v)w + f(w, v)u = (w, v, u)$ and

$$(u, v, w) + (v, w, u) + (w, u, v)$$
$$= f(u, v)w + f(w, v)u + f(v, w)u + f(u, w)v + f(w, u)v + f(v, u)w = 0$$

for $u, v, w \in V$.

Therefore, both sides of (ii) and (iii) are 0. We have W is $\mathfrak{eu}_2$-Lie admissible. Particularly, if $\dim_k X = \dim_k Y = n < \infty$, then $\dim_k W = 2n + 1$.

In Section 4, we will give another class of $\mathfrak{eu}_2$-Lie admissible algebras which may be not associative (See Example 4.1).

2. Steinberg unitary Lie algebras $\mathfrak{stu}_2(R, -, \gamma)$

For an nonassociative k-algebra R with identity, equipped with an involution $-$, the Steinberg unitary Lie algebra $\mathfrak{stu}_2(R, -, \gamma)$ is defined by the generators $u_{12}(a), u_{21}(a), h(a, b), a, b \in R$, subject to the relations:

(2.1) $a \mapsto u_{12}(a)$ is a k-linear mapping,

(2.2) $u_{21}(a) = u_{12}(-\gamma_2\gamma_1^{-1}\bar{a})$

(2.3) $h(a, b) = [u_{12}(a), u_{21}(b)]$

(2.4) $[h(a, b), u_{12}(c)] = u_{12}(\langle a, b, c \rangle)$

for all $a, b, c \in R$.

By the result of Allison and Faulkner in [AF], if $n \geq 3$, $I_n(R) = \{a \in R | u_{ij}(a) = 0\}$ are ideals of R. When $n = 2$, the subspace $I_2(R) = \{a \in R | u_{12}(a) = 0\}$ of R may not be an ideal. We will give a necessary and sufficient condition for $I_2(R) = 0$ using Lie triple systems.

Recall that a Lie triple system T over k is a k-vector space with a triple product $[\ ,\ ,\ ] : T \times T \times T \to T$(a trilinear map) satisfying

(LTS1) $[x, x, y] = 0$,

(LTS2) $[x, y, z] + [y, z, x] + [z, x, y] = 0$,

(LTS3) $[x, y, [u, v, w]] = [[x, y, u], v, w] + [u, [x, y, v], w] + [u, v, [x, y, w]]$,

for any $x, y, z, u, v, w \in T$. (More details can be found in [H].)

For any Lie algebra L, we define a triple product on L by

$$[x, y, z] = [[x, y], z]$$

for $x, y, z \in L$. Then L becomes a Lie triple system. On the other hand, for a Lie triple system T, we define $L(x, y)(z) = [x, y, z]$ for $x, y, z \in T$. The standard embedding of T is the k-vector space $S(T) = L(T,T) \oplus T$ with the bracket product

$$[A + x, b + y] = (AB - BA + L(x,y)) + (A(y) - B(x))$$

for $A, B \in L(T,T), x, y \in T$. Then, $S(T)$ is closed under the bracket product $[\,,]$ and becomes a Lie algebra.

Let T be a vector space with any triple product $[\,,\,,]$. The universal embedding $U(T)$ of T is the Lie algebra over k generated by $u(x), x \in T$, subject to the relations:

(2.5) $x \mapsto u(x)$ is a k-linear mapping,

(2.6) $u([x,y,z]) = [[u(x), u(y)], u(z)]$

for all $x, y, z \in T$. From (2.6), we see that $u(T)$ is a Lie triple system, which is a subsystem of the Lie triple system $U(T)$.

LEMMA 2.1 . T is a Lie triple system over k if and only if $\ker u = 0$.

PROOF. Clearly, $u : T \to U(T)$ preserves the triple product. If $\ker u = 0$, then $T \cong u(T)$ is a Lie triple system.

Conversely, if T is a Lie triple system, by the universal property of $U(T)$, there exists a Lie homomorphism $U(T) \to S(T)$ such that $u(x) \mapsto x$. So, $\ker u = 0$.   □

Take $T = T(R, -, \gamma)$ to be R with triple bracket product

$$[a,b,c] = -\gamma_1 \gamma_2^{-1} \langle a, \bar{b}, c \rangle,$$

for $a, b, c \in R$. By comparing the relations (2.5)-(2.6) to (2.1)-(2.4), we have $U(T(R, -, \gamma)) = \mathfrak{stu}_2(R, -, \gamma)$ and $\ker u = I_2(R)$. Thus,

THEOREM 2.1 . If R is a nonassociative k-algebra with identity, equipped with an involution $-$, then $I_2(R) = 0 \iff$ the following equations hold

(i) $(a, \bar{b}, c) + (b, \bar{c}, a) + (c, \bar{a}, b) = (a, \bar{c}, b) + (c, \bar{b}, a) + (b, \bar{a}, c),$

(ii) $\langle a, b, \langle c, d, e \rangle \rangle = \langle \langle a, b, c \rangle, d, e \rangle - \langle c, \langle b, a, d \rangle, e \rangle + \langle c, d, \langle a, b, e \rangle \rangle,$

for any $a, b, c, d, e \in R$.

PROOF. We have seen that $I_2(R) = 0 \iff T(R, -, \gamma)$ is a Lie triple system.

First, if we define $[x,y,z]' = \lambda[x,y,z]$ for some $\lambda \in k^\times$, whether (LTS1)-(LTS3) hold is independent of λ. So, we can assume that $[a,b,c] = \langle a, \bar{b}, c \rangle$.

Since $\langle a, \bar{a} \rangle = a\bar{a} - a\bar{a} = 0$,

$$[a,a,c] = \langle a, \bar{a}, c \rangle = \langle a, \bar{a} \rangle c + c \langle \bar{a}, a \rangle = 0.$$

(LTS1) is automatic in $T(R, -, \gamma)$.

(LTS2) becomes

$$\langle a, \bar{b}, c \rangle + \langle b, \bar{c}, a \rangle + \langle c, \bar{a}, b \rangle = 0,$$

which is just the equation (i) in Theorem 1.2.

Using Lemma 1.1(vii), we see that (LTS3) with $x = a, y = \bar{b}, u = c, v = \bar{d}, w = e$ becomes

$$\langle a, b, \langle c, d, e \rangle \rangle = \langle \langle a, b, c \rangle, d, e \rangle - \langle c, \langle b, a, d \rangle, e \rangle + \langle c, d, \langle a, b, e \rangle \rangle.$$

Thus, $I_2(R) = 0 \iff$ the equations (i) and (ii) hold.   □

If $(R, -)$ is $\mathfrak{eu}_2$-Lie admissible, we have that $\mathfrak{eu}_2(R, -, \gamma)$ is a Lie algebra over k. Since $\xi_{12}(a), \xi_{21}(a)$ and $H(a, b)$ satisfy certain canonical relations (2.1)-(2.4) for $a, b \in R$, by the universal property of $\mathfrak{stu}_2(R, -, \gamma)$, there exits a Lie homomorphism

$$\phi : \mathfrak{stu}_2(R, -, \gamma) \to \mathfrak{eu}_2(R, -, \gamma)$$

such that

$$\phi(\xi_{ij}(a)) = e_{ij}(a), \quad \phi(h(a, b)) = H(a, b),$$

for $a, b \in R, 1 \le i \ne j \le 2$. Since $\xi_{12}(R)$ generates $\mathfrak{eu}_2(R, -, \gamma)$, ϕ is a Lie epimorphism. The restriction of $\phi|_{u_{12}(R)} : u_{12}(R) \to \xi_{12}(R)$ is a linear isomorphism. Thus, $I_2(R) = 0$.

Furthermore, we have

LEMMA 2.2 . If $(R, -)$ is $\mathfrak{eu}_2$-Lie admissible,

$$0 \to \ker \phi \to \mathfrak{stu}_2(R, -, \gamma) \overset{\phi}{\to} \mathfrak{eu}_2(R, -, \gamma) \to 0$$

is a central extension of the Lie algebra $\mathfrak{eu}_2(R, -, \gamma)$.

PROOF. By Jacobi identity, we have

$$[h(a, b), h(c, d)] = [h(a, b), [u_{12}(c), u_{21}(d)]]$$
$$= [u_{12}(\langle a, b, c \rangle), u_{21}(d)] + [u_{12}(c), u_{21}(\overline{\langle a, b, \overline{d} \rangle})]$$
$$= h(\langle a, b, c \rangle, d) - h(c, \langle b, a, d \rangle).$$

Thus,

$$[h(R, R), h(R, R)] \subset h(R, R).$$

We obtain $\mathfrak{stu}_2(R, -, \gamma) = h(R, R) + u_{12}(R)$. Moreover, since

$$\phi(h(R, R) \cap u_{12}(R)) \subset H(R, R) \cap \xi_{12}(R) = 0$$

and $\phi|_{u_{12}(R)}$ is injective, we have $h(R, R) \cap u_{12}(R) = 0$. This shows that

$$\mathfrak{stu}_2(R, -, \gamma) = h(R, R) \oplus u_{12}(R)$$

is a vector space decomposition of $\mathfrak{stu}_2(R, -, \gamma)$.

Because $\phi|_{u_{12}(R)}$ is a linear isomorphism, $\ker \phi \subset h(R, R)$ and

$$[\ker \phi, u_{12}(R)] \subset u_{12}(R).$$

On the other hand, $\ker \phi$ is an ideal of $\mathfrak{stu}_2(R, -, \gamma)$, which yields

$$[\ker \phi, u_{12}(R)] \subset \ker \phi \subset h(R, R).$$

So, $[\ker \phi, u_{12}(R)] = 0$. But $u_{12}(R)$ generates the whole Lie algebra $\mathfrak{stu}_2(R, -, \gamma)$, so $\ker \phi$ is contained in center of $\mathfrak{stu}_2(R, -, \gamma)$.   □

If we assume that R has the element e satisfying the condition in Assumption 1.1, then $\langle R, R, R \rangle = R$, $\mathfrak{stu}_2(R, -, \gamma)$ and $\mathfrak{eu}_2(R, -, \gamma)$ are perfect. For this case, we can prove that $\mathfrak{stu}_2(R, -, \gamma)$ is the universal covering of $\mathfrak{eu}_2(R, -, \gamma)$.

THEOREM 2.2 . If R is $\mathfrak{eu}_2$-Lie admissible and satisfies Assumption 1.1, then $\mathfrak{stu}_2(R, -, \gamma)$ is the universal covering of $\mathfrak{eu}_2(R, -, \gamma)$, i.e. it is centrally closed.

PROOF. With the assumption of the existence of the element e, we have

$$\langle 1, e, e^{-1}a \rangle = 2e \cdot e^{-1}a + e^{-1}a \cdot 2e = 4a$$

for any $a \in R$.

Suppose that

$$0 \longrightarrow V \longrightarrow L \overset{\pi}{\longrightarrow} \mathfrak{eu}_2(R, -, \gamma) \longrightarrow 0$$

is a central extension of $\mathfrak{eu}_2(R, -, \gamma)$. We will show that there exists a Lie homomorphism $\tau : \mathfrak{stu}_2(R, -, \gamma) \to L$ such that $\phi = \pi \circ \tau$.

Let $\{r_\lambda\}_{\lambda \in \Lambda}$ be a basis of R containing $1(\Lambda$ is an index set). Let $\tilde{u}_{12}(r_\lambda)$ be a preimage in L of $\xi_{12}(r_\lambda)$ under π. For general $a \in R$, define the preimage $\tilde{u}_{12}(a)$ of $\xi_{12}(a)$ by linearity and define $\tilde{u}_{21}(a) = \tilde{u}_{12}(-\gamma_2\gamma_1^{-1}\bar{a})$, $\tilde{h}(a, b) = [\tilde{u}_{12}(a), \tilde{u}_{21}(b)]$.

We know that $[H(a, b), \xi_{12}(c)] = \xi_{12}(\langle a, b, c \rangle)$, so

$$[\tilde{h}(1, e), \tilde{u}_{12}(e^{-1}a)] = \tilde{u}_{12}(\langle 1, e, e^{-1}a \rangle) + v(a) = 4\tilde{u}_{12}(a) + v(a),$$

where $v : R \to V$ is a k-linear map. Replace $\tilde{u}_{12}(a)$ by $\tilde{u}_{12}(a) + \frac{1}{4}v(a)$, and set $\tilde{u}_{21}(a) = \tilde{u}_{12}(-\gamma_2\gamma_1^{-1}\bar{a})$. Note that $\tilde{h}(a, b)$ is unchanged since it dose not depend on the choice of preimages of $\tilde{u}_{12}(a)$ and $\tilde{u}_{21}(b)$. So we have

$$(2.7) \qquad\qquad [\tilde{h}(1, e), \tilde{u}_{12}(e^{-1}a)] = 4\tilde{u}_{12}(a).$$

Since e is contained in the center of R, in $\mathfrak{eu}_2(R, -, \gamma)$

$$[H(a, b), H(1, e)]$$
$$= e_{11}([\langle a, b \rangle, \langle 1, e \rangle]) + e_{22}([\langle b, a \rangle, \langle e, 1 \rangle])$$
$$= 2e_{11}([\langle a, b \rangle, e]) + 2e_{22}([\langle b, a \rangle, e]) = 0.$$

Therefore, $[\tilde{h}(a, b), \tilde{h}(1, e)] \in V$. By (2.7), we have

$$[\tilde{h}(a, b), \tilde{u}_{12}(c)] = \frac{1}{4}[\tilde{h}(a, b), [\tilde{h}(1, e), \tilde{u}_{12}(e^{-1}c)]]$$

$$= \frac{1}{4}([[\tilde{h}(a, b), \tilde{h}(1, e)], \tilde{u}_{12}(e^{-1}c)] + [\tilde{h}(1, e), [\tilde{h}(a, b), \tilde{u}_{12}(e^{-1}c)]])$$

$$= \frac{1}{4}(0 + [\tilde{h}(1, e), \tilde{u}_{12}(\langle a, b, e^{-1}c \rangle)])$$

$$= \frac{1}{4}[\tilde{h}(1, e), \tilde{u}_{12}(e^{-1}\langle a, b, c \rangle)] = \tilde{u}_{12}(\langle a, b, c \rangle).$$

Then, $\tilde{u}_{12}(a), \tilde{u}_{21}(a), \tilde{h}(a, b)$ satisfy the relations (2.1)-(2.4). By the universal property of $\mathfrak{stu}_2(R, -, \gamma)$, there exists a unique homomorphism τ from $\mathfrak{stu}_2(R, -, \gamma)$ to L so that $\tau(u_{12}(a)) = \tilde{u}_{12}(a)$, $\tau(u_{21}(a)) = \tilde{u}_{12}(a)$, $\tau(h(a, b)) = \tilde{h}(a, b)$. Evidently, $\pi \circ \tau = \phi$. So $\mathfrak{stu}_2(R, -, \gamma)$ is the universal covering of $\mathfrak{eu}_2(R, -, \gamma)$ and it is centrally closed. $\qquad\qquad\square$

Remark 2.1 Note that $\ker\phi$ is also independent of the choice of $\gamma = (\gamma_1, \gamma_2)$.

3. The structure of $\ker\phi$

Throughout this section we assume that $(R, -)$ is an $\mathfrak{eu}_2$-Lie admissible k-algebra, where $-$ is an involution of R. By Lemma 2.2, $\phi : \mathfrak{stu}_2(R, -, \gamma) \to \mathfrak{eu}_2(R, -, \gamma)$ is a central extension of $\mathfrak{eu}_2(R, -, \gamma)$. We will give the exact and explicit structure of $\ker\phi$. In Section 2, we have seen that $\mathfrak{stu}_2(R, -, \gamma) \simeq U(T)$, where $T = T(R, -, \gamma)$ is R with triple product $[a, b, c] = -\gamma_1\gamma_2^{-1}\langle a, \bar{b}, c \rangle$ and $U(T)$

is the universal embedding of it. When R is $\mathfrak{eu}_2$-Lie admissible, we have $I_2(R) = 0$ and $T(R, -, \gamma)$ is a Lie triple system.

We will use the following theorem for a Lie triple system T(See [S, Theorem 3.5]).

THEOREM 3.1 . Let T be a Lie triple system over k, let I be the subspace of $T^{\otimes 2}$ spanned by all

$$x \otimes y + y \otimes x,$$
$$[x, y, z] \otimes w + z \otimes [x, y, w] + [z, w, x] \otimes y + x \otimes [z, w, y]$$

for $x, y, z, w \in T$, and set $l(x, y) = x \otimes y + I \in T^{\otimes 2}/I$. Then $T^{\otimes 2}/I \oplus T$ with bracket product given by

$$[x, y] = l(x, y)$$
$$[l(x, y), z] = -[z, l(x, y)] = [x, y, z]$$
$$[l(x, y), l(z, w)] = l([x, y, z], w) + l(z, [x, y, w])$$

is a Lie algebra isomorphic to the universal embedding $U(T)$. $\square$

Using Theorem 3.1 for the Lie triple system $T((R, -\gamma))$, we get

COROLLARY 3.1 . Suppose that $(R, -)$ is $\mathfrak{eu}_2$-Lie admissible. Let Δ be the subspace of $R^{\otimes 2}$ spanned by all

(3.1) $a \otimes b + \bar{b} \otimes \bar{a}$

(3.2) $\langle a, b, c \rangle \otimes d - c \otimes \langle b, a, d \rangle + \langle c, d, a \rangle \otimes b - a \otimes \langle d, c, b \rangle$

and set

$$h'(a, b) = a \otimes b + \Delta \in R^{\otimes 2}/\Delta.$$

Let $u'_{12}(R)$ be a copy of R and set $u'_{21}(a) = u'_{12}(-\gamma_2\gamma_1^{-1}\bar{a})$. Then, $R^{\otimes 2}/\Delta \oplus u'_{12}(R)$ with product given by

$$[u'_{12}(a), u'_{21}(b)] = h'(a, b),$$
$$[h'(a, b), u'_{12}(c)] = -[u'_{12}(c), h'(a, b)] = u'_{12}(\langle a, b, c \rangle),$$
$$[h'(a, b), h'(c, d)] = h'(\langle a, b, c \rangle, d) - h'(c, \langle b, a, d \rangle)$$

is a Lie algebra isomorphic to the Steinberg unitary Lie algebra $\mathfrak{stu}_2(R, -, \boldsymbol{\gamma})$.

PROOF. We know that $R = T(R, -, \gamma)$ with $[a, b, c] = -\gamma_1\gamma_2^{-1}\langle a, \bar{b}, c \rangle$ is a Lie triple system. It suffices to show that $R^{\otimes 2}/\Delta \oplus u'_{12}(R)$ is isomorphic to the universal embedding $U(T(R, -\gamma))$ of $T(R, -, \gamma)$.

Let $\tilde{\theta} : R^{\otimes 2} \to R^{\otimes 2}$ be given by

$$\tilde{\theta}(a \otimes b) = -\gamma_1\gamma_2^{-1}a \otimes \bar{b}.$$

Applying $\tilde{\theta}$ to the generators of I, we get

$$\tilde{\theta}(a \otimes b + b \otimes a) = -\gamma_1\gamma_2^{-1}(a \otimes \bar{b} + b \otimes \bar{a}),$$
$$\tilde{\theta}([a, b, c] \otimes d + c \otimes [a, b, d] + [c, d, a] \otimes b + a \otimes [c, d, b])$$
$$= -\gamma_1\gamma_2^{-1}(\langle a, \bar{b}, c \rangle \otimes \bar{d} - c \otimes \langle \bar{b}, a, \bar{d} \rangle + \langle c, \bar{d}, a \rangle \otimes \bar{b} - a \otimes \langle \bar{d}, c, \bar{b} \rangle),$$

so $\tilde{\theta}(I) = \Delta$, and $\tilde{\theta}$ induces a linear isomorphism $\theta : R^{\otimes 2}/I \to R^{\otimes 2}/\Delta$ with $\theta(l(a, b)) = -\gamma_1\gamma_2^{-1}h'(a, \bar{b})$. Extending θ on $R^{\otimes 2}/I \oplus R \to R^{\otimes 2}/\Delta \oplus u'_{12}(R)$ by

$$u + a \mapsto \theta(u) + u'_{12}(a)$$

for $u \in R^{\otimes 2}/I$ and $a \in R$, it is easy to check that θ is an isomorphism of Lie algebra over k. By Theorem 3.1, we have $R^{\otimes 2}/\Delta \oplus u'_{12}(R)$ is isomorphic to the Steinberg unitary Lie algebra $\mathfrak{stu}_2(R, -, \gamma)$.   □

We get the main theorem of this section

THEOREM 3.2 . Suppose that $(R, -)$ is an $\mathfrak{eu}_2$-Lie admissible algebra and $\phi : \mathfrak{stu}_2(R, -, \gamma) \to \mathfrak{eu}_2(R, -, \gamma)$ is the central extension. Let $\partial : R^{\otimes 2} \to R_-$ with

$$\partial(a \otimes b) = \langle a, b \rangle - \langle b, a \rangle.$$

Then,

$$\ker \phi \simeq \ker \partial / \tilde{\Delta}$$

where $\tilde{\Delta} = \Delta \oplus (1 \otimes R_-)$.

PROOF. Define $\partial_i : R^{\otimes 2} \to R_-$, $i = 1, 2$ with $\partial_1(a \otimes b) = \langle a, b \rangle$ and $\partial_2(a \otimes b) = \langle b, a \rangle$. Thus, $\partial = \partial_1 - \partial_2$ and $\partial_i(1 \otimes s) = 2s$ for $s \in R_-$. By Corollary 3.1, We can take $\mathfrak{stu}_2(R, -, \gamma) = R^{\otimes 2}/\Delta \oplus u'_{12}(R)$ with $\phi(u'_{12}(a)) = \xi_{12}(a)$ and

$$\phi(a \otimes b + \Delta) = H(a, b) = e_{11}(\partial_1(a \otimes b)) - e_{22}(\partial_2(a \otimes b)).$$

Thus, $\ker \phi = K/\Delta$ where $K = \ker \partial_1 \cap \ker \partial_2$. We see that $K, 1 \otimes R_- \subset \ker \partial$ and

$$K \cap (1 \otimes R_-) = 0.$$

Also, if $u \in \ker \partial$ and $v = u - 1 \otimes \frac{1}{2}\partial_2(u)$, then

$$\partial_1(v) = \partial_1(u) - \frac{1}{2}\partial_1(1 \otimes \partial_2(u)) = \partial_1(u) - \partial_2(u) = \partial(u) = 0$$

and

$$\partial_2(v) = \partial_2(u) - \frac{1}{2}\partial_2(1 \otimes \partial_2(u)) = \partial_2(u) - \partial_2(u) = 0.$$

So, $v \in K$ and $u = v + 1 \otimes \frac{1}{2}\partial_2(u)$. Thus, $\ker \partial = K \oplus (1 \otimes R_-)$, and the canonical projection $\pi : \ker \partial \to K/\Delta$, $(u, v) \mapsto u + \Delta$ for $u \in K, v \in 1 \otimes R_-$, gives us $K/\Delta \simeq \ker \partial / \ker \pi$, where $\ker \pi = \Delta + (1 \otimes R_-)$ and the sum is direct. Finally, we have

$$\ker \phi \simeq K/\Delta \simeq \ker \partial / \tilde{\Delta}$$

for $\tilde{\Delta} = \Delta \oplus (1 \otimes R_-)$.   □

By Theorem 2.2, if we assume that $(R, -)$ is $\mathfrak{eu}_2$-Lie admissible with Assumption 1.1, then $(\mathfrak{stu}_2(R, -, \gamma), \phi)$ is the universal covering of $\mathfrak{eu}_2(R, -, \gamma)$. Associating with Theorem 3.2, we have

COROLLARY 3.2 . $H_2(\mathfrak{eu}_2(R, -, \gamma)) \cong \ker \partial / \tilde{\Delta}$. □

4. Applications on $sl_2(S)$ and $\mathfrak{st}_2(S)$

In the last section, we use the above theorems for dealing with $sl_2(S)$ and $\mathfrak{st}_2(S)$ for the k-algebra S, and obtain the corresponding results in [G2].

Suppose that S is a nonassociative k-algebra with identity 1. Let $(R, -) = (S \oplus S^{\mathrm{op}}, \mathrm{ex})$ be the k-algebra with identity $(1, 1)$, satisfying $(s_1, s_2)(t_1, t_2) = (s_1 t_1, t_2 s_2)$ and $\overline{(s_1, s_2)} = (s_2, s_1)$ for $s_1, s_2, t_1, t_2 \in S$. Then, we have $R_+ = \{(s, s) \in S \oplus S^{\mathrm{op}} |$ for $s \in S\}$ and $R_- = \{(s, -s) \in S \oplus S^{\mathrm{op}} |$ for $s \in S\}$. It is clear that R is associative if and only if so is S.

As usual, we consider the k-algebra $sl_2(S)$ as the subalgebra of $gl_2(S)$ generated by $e_{12}(s)$ and $e_{21}(s)$ for $s \in S$. Taking $\gamma = (1, 1) \in (k^\times)^2$, there is an isomorphism $\varphi : \mathfrak{eu}_2(R, -, \gamma) \to sl_2(S)$ given by $\varphi(\xi_{12}((s_1, s_2))) = e_{12}(s_1) - e_{21}(s_2)$ and $\varphi(H(((s_1, s_2), (t_1, t_2)))) = e_{11}(s_1 t_1 - t_2 s_2) + e_{22}(s_2 t_2 - t_1 s_1)$ (see Lemma 1.3). Especially, $\varphi(H(((s, 0), (t, 0)))) = e_{11}(st) - e_{22}(ts)$.

In [G2], S is called sl_2-Lie admissible if $sl_2(S)$ is a Lie algebra. Since $sl_2(S) \cong \mathfrak{eu}_2(R, -, \gamma)$, we have that S is sl_2-Lie admissible if and only if R is $\mathfrak{eu}_2$-Lie admissible. From Theorem 1.2, we have ([G2, Theorem 1])

THEOREM 4.1 . Let S is a nonassociative k-algebra, then S is sl_2-Lie admissible if and only if

(4.1) $(p, q, rs) + (p, sr, q) + (rs, p, q) = 0$

for all $p, q, r, s \in S$.

PROOF. For any $(R, -)$, we have $\overline{(a, b, c)} = -(\bar{c}, \bar{b}, \bar{a})$. Letting

$$z = (a, b, \langle c, d \rangle) + (\langle c, d \rangle, a, b) + (a, \langle d, c \rangle, b),$$

we see that Theorem 1.2 (ii) is equivalent to $z = \bar{z}$. For $(R, -) = (S \oplus S^{\mathrm{op}}, ex)$, if we take $a \in S \cup S^{\mathrm{op}}$, then $z \in S \cup S^{\mathrm{op}}$. So $z = \bar{z}$ is equivalent to $z = 0$. Thus, in this case, we can replace (ii) with

(ii') $(a, b, \langle c, d \rangle) + (\langle c, d \rangle, a, b) + (a, \langle d, c \rangle, b) = 0$.

We denote the S-component of (i) by (i)$_S$, etc., and claim (i)$_S$, (ii')$_S$, (iii)$_S$ $\Leftrightarrow$ (4.1) for S. Again we can take a, b, c, etc. in $S \cup S^{\mathrm{op}}$. If the S-component of some term of (i) is nonzero, we can permute the variables to assume it is $(a, \bar{b}, c)$, so $a, \bar{b}, c \in S$, Except for $(c, \bar{b}, a)$, the other terms are in S^{op}, so (i)$_S$ is equivalent to

(4.2) $(p, q, r) + (r, p, q) = 0$

for $p, q, r \in S$. The terms in (ii')$_S$ are zero unless $a, b \in S$. Since $\langle S, S^{\mathrm{op}} \rangle = 0$ and $\langle c, d \rangle = -\langle \bar{d}, \bar{c} \rangle$, we can take $c, d \in S$. Thus, (ii')$_S$ is equivalent to (4.1) for S. Similarly, (iii)$_S$ is equivalent to

(4.3) $(rq, ts, p) + (p, qr, st) + (rq, p, st) = (ts, rq, p) + (p, st, qr) + (ts, p, qr)$

for S. [G2, Section 2] shows (4.1) implies (4.2) and (4.3), so the claim holds. Since $(S^{\mathrm{op}})^{\mathrm{op}} = S$, we can reserve the roles of S and S^{op} to see that (i)$_{S^{\mathrm{op}}}$, (ii')$_{S^{\mathrm{op}}}$, (iii)$_{S^{\mathrm{op}}}$ $\Leftrightarrow$ (4.1) for S^{op}. However, since $(a, b, c)^{\mathrm{op}} = -(c, b, a)$, we see that (4.1)for $S \Leftrightarrow$(4.1) for S^{op}, and the result follows.   □

Example 4.1 By the above Theorem, let S be a sl_2-Lie admissible k-algebra which is not associative, then we have $(S \oplus S^{\mathrm{op}}, ex)$ is a $\mathfrak{eu}_2$-Lie admissible k-algebra with involution which is also not associative. In [G2, Section 4], a class of sl_2-Lie admissible algebras $R(n)$ which are not associative is given, so $(R(n) \oplus R(n)^{\mathrm{op}}, ex)$ are examples of the $\mathfrak{eu}_2$-Lie admissible algebras which are not associative. Since for finite n, the dimension of $R(n) \oplus R(n)^{\mathrm{op}}$ is even, the algebra W given in Example 1.1 can not be isomorphic to one of these algebras.

Suppose that S is an sl_2-Lie admissible k-algebra. If we take $e = (1, -1) \in R$ which satisfies the Assumption 1.1, then we have $\mathfrak{eu}_2(R, -, \gamma) \cong sl_2(S)$ is perfect. Straightforwardly, we can also obtain it since the generators

$$e_{12}(s) = \frac{1}{2}[[e_{12}(1), e_{21}(1)], e_{12}(s)]$$

and $e_{21}(s) = \frac{1}{2}[[e_{12}(1), e_{21}(1)], e_{21}(s)]$ are contained in $[sl_2(S), sl_2(S)]$.

THEOREM 4.2 . Suppose that S is an sl_2-Lie admissible k-algebra. Let Δ' be the subspace of $S^{\otimes 2}$ spanned by all

(4.4)
$$a \otimes b + b \otimes a$$

$$(a(bc) + (cb)a) \otimes d - c \otimes ((ba)d + d(ab))$$

(4.5)
$$+ ((cd)a + a(dc)) \otimes b - a \otimes ((dc)b + (bc)d).$$

Then,
$$H_2(sl_2(S)) \simeq \ker \partial'/\Delta'$$
where $\partial' : S \otimes S \to S$ is defined by $\partial'(a \otimes b) = [a, b]$.

PROOF. First, since S is an sl_2-Lie admissible, we have $(R, -,) = (S \oplus S^{\mathrm{op}}, \mathrm{ex})$ is an $\mathfrak{eu}_2$-Lie admissible k-algebra with Assumption 1.1 for $e = (1, -1)$. So, $\mathfrak{eu}_2(R, -, \gamma) \simeq sl_2(S)$ and $H_2(sl_2(S)) \simeq \ker \phi$ by Theorem 3.2.

We claim that $\ker \phi \simeq \ker \partial'/\Delta'$. Recall the definition of Δ and ∂ in Theorem 3.2. We take $a = (1, 1), b = (1, -1), c = (r, 0), d = (0, s)$ in (3.2), then

$$\langle (1,1), (1, -1), (r, 0) \rangle \otimes (0, s) - (r, 0) \otimes \langle (1, -1), (1, 1), (0, s) \rangle$$
$$+ \langle (r, 0), (0, s), (1, 1) \rangle \otimes (1, -1) - (1, 1) \otimes \langle (0, s), (r, 0), (1, -1) \rangle$$
$$= 4(1, -1)(r, 0) \otimes (0, s) - (r, 0) \otimes 4(1, -1)(0, s)$$
$$= 8(r, 0) \otimes (0, s) \in \Delta.$$

Thus, $S \otimes S^{\mathrm{op}} \subset \Delta$. Similarly, taking $a = (1, 1), b = (1, -1), c = (0, s), d = (r, 0)$ in (3.2), we have $S^{\mathrm{op}} \otimes S \subset \Delta$. Also, if T is the span of all the $a \otimes b + \bar{b} \otimes \bar{a}$ for $a, b \in S$, then $T \subset \Delta$. It is easy to see that

$$R^{\otimes 2} = S^{\otimes 2} \oplus (S \otimes S^{\mathrm{op}}) \oplus (S^{\mathrm{op}} \otimes S) \oplus T.$$

Denote $\Omega = (S \otimes S^{\mathrm{op}}) \oplus (S^{\mathrm{op}} \otimes S) \oplus T$ and let π be the projection from $R^{\otimes 2}$ onto $S^{\otimes 2}$. Since $\Omega \subset \Delta \subset \tilde{\Delta}$, using Corollary 3.1, we have

$$\ker \phi \simeq \ker \partial/\tilde{\Delta} \simeq \pi(\ker \partial)/\pi(\tilde{\Delta}).$$

Clearly, π applied on (3.1) is 0. We can take a, b, c, d in (3.2) to be in $S \cup S^{\mathrm{op}}$. If $a \in S, b \in S^{\mathrm{op}}$ or vice versa, the π applied to (3.2) is 0. Since $\langle a, b, c \rangle = -\langle \bar{b}, \bar{a}, c \rangle$ and

$$\pi(\langle c, d, a \rangle \otimes b) = -\pi(\bar{b} \otimes \overline{\langle c, d, a \rangle}) = -\pi(\bar{b} \otimes \langle c, d, \bar{a} \rangle),$$

we can take $a, b \in S$. Similarly, we can take $c, d \in S$. In this case, π applied to (3.2) gives us (4.5). Thus, $\pi(\Delta)$ is spanned by (4.5). Letting $b = d = 1$ in (4.5) gives

(4.6)
$$(ac + ca) \otimes 1 - c \otimes a - a \otimes c \in \pi(\Delta).$$

Since $\pi((r_1, r_2) \otimes (s_1, s_2)) = r_1 \otimes s_1 - s_2 \otimes r_2$, we see that

(4.7)
$$\pi((1, 1) \otimes (a, -a)) = 1 \otimes a + a \otimes 1.$$

Since $\tilde{\Delta} = \Delta \oplus (1 \otimes R_-)$, $\pi(\tilde{\Delta})$ is spanned by (4.5) and all $1 \otimes a + a \otimes 1$, for $a \in S$. We have $\pi(\tilde{\Delta}) \subset \Delta'$. On the other hand, taking $c = 1$ in (4.6) shows $a \otimes 1 - 1 \otimes a \in \pi(\Delta)$. Thus, $a \otimes 1 \in \pi(\tilde{\Delta})$ for $a \in S$. Then, we have $c \otimes a + a \otimes c \in \pi(\tilde{\Delta})$ for any $a, c \in S$ and $\pi(\tilde{\Delta}) \supset \Delta'$. So, $\pi(\tilde{\Delta}) = \Delta'$.

Since $\Omega \subset \Delta \subset \ker \partial$, we see that $\pi(\ker \partial) = \pi(\ker \partial \cap S^{\otimes 2})$. Since $\partial((a, 0) \otimes (b, 0)) = ([a, b], -[b, a])$, we have $\pi(\ker \partial) = \ker \partial'$.

Finally, $\ker \phi \simeq \pi(\ker \partial)/\pi(\tilde{\Delta}) \simeq \ker \partial'/\Delta'$.   □

For any nonassociative k-algebra S, the Steinberg Lie algebra $\mathfrak{st}_2(S)$ is defined by generators $X_{12}(r), X_{21}(r), T(r,s), r, s \in S$, subject to the relations(See [G2])

$$r \mapsto X_{ij}(r) \text{ is a } k\text{-linear mapping,}$$
$$T(r,s) = [X_{12}(r), X_{21}(s)],$$
$$[T(r,s), X_{12}(t)] = X_{12}((rs)t + t(sr)),$$
$$[T(r,s), X_{21}(t)] = -X_{21}((sr)t + t(rs)),$$
$$[X_{ij}(r), X_{ij}(s)] = 0,$$

where $r, s, t \in S$ and $(i,j) = (1,2)$ or $(2,1)$. For $(R, -, \gamma) = (S \oplus S^{\mathrm{op}}, \mathrm{ex}, \mathbf{1})$, we also have an isomorphism $\psi : \mathfrak{stu}_2(R, -, \gamma) \to \mathfrak{st}_2(S)$ given by $\psi(u_{12}((r,s))) = X_{12}(r) - X_{21}(s)$ and $\psi(h((r_1,r_2),(s_1,s_2))) = T(r_1,s_1) - T(r_2,s_2)$. If S is sl_2-Lie admissible, there also exists a Lie algebra epimorphism $\phi' : \mathfrak{st}_2(S) \to sl_2(S)$ satisfying

$$\phi'(X_{ij}(s)) = e_{ij}(s), \quad \phi'(T(s,t)) = e_{11}(st) - e_{22}(ts),$$

for $s, t \in S, 1 \le i \ne j \le 2$. The following diagram commutes,

$$
\begin{array}{ccccccccc}
0 & \to & \ker\phi & \to & \mathfrak{stu}_2(R,-,\gamma) & \xrightarrow{\phi} & eu_2(R,-,\gamma) & \to & 0 \\
 & & & & \downarrow \wr \psi & & \downarrow \wr \varphi & & \\
0 & \to & \ker\phi' & \to & \mathfrak{st}_2(S) & \xrightarrow{\phi'} & sl_2(S) & \to & 0 .
\end{array}
$$

Thus, the above commutative diagram induces the isomorphism $\psi|_{\ker\phi} : \ker\phi \to \ker\phi'$ by the Five Lemma. By Theorem 4.2, we have the following result,

COROLLARY 4.1 . $\ker\phi' \simeq \ker\partial'/\Delta'$.   □

Remark 4.1 If S is associative, the structure of $\ker\phi'$ is given by as $hC_1(S)$ in [G2, Section 3]. Comparing it with $\ker\partial'/\Delta'$, we define Δ' with less generators in (4.4)-(4.5).

Finally, we give the remark which is similar with [G1, Remark 2.64].
Remark 4.2 Let $\mathfrak{est}_2(R, -, \gamma)$ be the subalgebra of $\mathfrak{st}_2(R)$ generated by the elements $X_{ij}(a) - \gamma_i\gamma_j^{-1}X_{ji}(\bar{a}), a \in R, 1 \le i \ne j \le 2$. It is easy to see that $X_{ij}(a) - \gamma_i\gamma_j^{-1}X_{ji}(\bar{a})$ satisfies the defining relations (2.1)-(2.4), thus by the universal property of $\mathfrak{stu}_2(R, -, \gamma)$, there exists a Lie algebra epimorphism $\varphi : \mathfrak{stu}_2(R, -, \gamma) \to \mathfrak{est}_2(R, \ , \gamma)$ such that $\varphi(u_{ij}(a)) = X_{ij}(a) - \gamma_i\gamma_j^{-1}X_{ji}(\bar{a})$. Moreover, φ is an isomorphism from $\mathfrak{stu}_2(R, -, \gamma)$ to $\mathfrak{est}_2(R, -, \gamma)$ when R is associative.

References

[AF] B. N. Allison and J. R. Faulkner, *Nonassociative coefficient algebras for Steinberg unitary Lie algebras*, J. Algebra 161 (1993) 1–19.

[AG] B. N. Allison and Y. Gao, *Central quotients and coverings of Steinberg unitary Lie algebras*, Canad. J. Math, 48 (1996) 449–482.

[BK] S. Berman and Y. S. Krylyuk, *Universal central extensions of twisted and untwisted Lie algebras extended over commutative rings*, J. Algebra 173(1995), 302–347.

[Bl] S. Bloch, *The dilogarithm and extensions of Lie algebras*, Alg. K-theory, Evanston 1980, Springer Lecture Notes in Math 854 (1981) 1–23.

[F] J. R. Faulkner, *Barbilian planes*, Geom. Dedicata 30 (1989) 125–181.

[G1] Y. Gao, *Steinberg Unitary Lie algebras and Skew-Dihedral Homology*, J.Algebra,17 (1996),261-304.

[G2] Y. Gao, *On the Steinberg Lie algebras st₂(R)*, Comm. in Alg. 21 (1993) 3691–3706.

[Ga] H. Garland, *The arithmetic theory of loop groups*, Publ. Math. IHES 52 (1980) 5–136.

[H] T. Hodge, *Lie triple systems, restricted Lie triple systems, and algebra groups*, J. Algebra, 244 (2001),533–380.

[Ka] C. Kassel, *Kähler differentials and coverings of complex simple Lie algebras extended over a commutative ring*, J. Pure and Appl. Alg. 34 (1984) 265–275.

[KL] C. Kassel and J-L. Loday, *Extensions centrales d'algèbres de Lie*, Ann. Inst. Fourier 32 (4) (1982) 119–142.

[L1] J-L. Loday, *Homologies diédrale et quaternionique*, Adv. in Math 66 (1987) 119–148.

[L2] J-L. Loday, *Cyclic homology*, Grundlehren der mathematischen Wissenschaften 301, Springer 1992.

[S] O. N. Smirnov , *Imbedding of Lie systems in Lie algebras*, to appear.

[St] R. Steinberg, *Lectures on Chevalley groups,*(notes by J. Faulkner & R. Wilson), Yale Univ. Lect. Notes 1967.

DEPARTMENT OF MATHEMATICS, UNIVERSITY OF SCIENCE AND TECHNOLOGY OF CHINA, HEFEI, ANHUI P.R.CHINA 230026

E-mail address: `skshang@mail.ustc.edu.cn`

DEPARTMENT OF MATHEMATICS AND STATISTICS, YORK UNIVERSITY, TORONTO, ONTARIO, CANADA M3J 1P3

E-mail address: `ygao@yorku.ca`

Contemporary Mathematics
Volume **478**, 2009

Lusztig's conjecture for finite classical groups with even characteristic

Toshiaki Shoji

ABSTRACT. The determination of scalars involved in Lusztig's conjecture concerning the characters of finite reductive groups was achieved by Waldspurger in the case of finite classical groups $Sp_{2n}(\mathbf{F}_q)$ or $O_n(\mathbf{F}_q)$ when p, q are large enough. Here p is the characteristic of the finite field $\mathbf{F}_q$. In this paper, we determine the scalars in the case of $Sp_{2n}(\mathbf{F}_q)$ with $p = 2$, by applying the theory of symmetric spaces over a finite field due to Kawanaka and Lusztig. We also obtain a weaker result for $SO_{2n}(\mathbf{F}_q)$ with $p = 2$, of split type.

0. Introduction

Let G be a connected reductive group defined over a finite field $\mathbf{F}_q$ of characteristic p with Frobenius map F. Lusztig's conjecture asserts that, under a suitable parametrization, almost characters of the finite reductive group G^F coincide with the characteristic functions of character sheaves of G up to scalar. Once Lusztig's conjecture is settled, and the scalars involved there are determined, one obtains a uniform algorithm of computing irreducible characters of G^F. Lusztig's conjecture was solved in [S1] in the case where the center of G is connected. In [S2], the scalars in question were determined in the case where G is a classical group with connected center, when p is odd, and the scalars are related to the unipotent characters of G^F. By extending the method there, Waldspurger [W] proved Lusztig's conjecture (or its appropriate generalization) for Sp_{2n} and O_n assuming that p, q are large enough. He also determined the scalars involved in the conjecture. But these methods cannot be applied to the case of classical groups with even characteristic.

In this paper we take up the problem of determining the scalars in the case of classical groups with $p = 2$. We show that the scalars are determined explicitly in the case where $G = Sp_{2n}$ with $p = 2$. We also obtain a somewhat weaker result for the case SO_{2n} of split type, when $p = 2$, containing the case related to the unipotent characters. The main ingredient for the proof is the theory of symmetric spaces over finite fields due to Kawanaka [K] and Lusztig [L4]. They determined the multiplicity of irreducible representations of G^{F^2} occurring in the induced module

1991 *Mathematics Subject Classification.* Primary 20G40; Secondary 20G05.
Key words and phrases. finite classical groups, representation theory.

$\operatorname{Ind}_{G^F}^{G^{F^2}} 1$ in the case where the center of G is connected (for arbitrary characteristic). Using this, one can determine the scalars for G^{F^2} in many cases for a connected classical group with connected center, with arbitrary characteristic.

On the other hand, it was shown in [S3] that there exists a good representatives of C^F for a unipotent class C in Sp_{2n} or SO_N for arbitrary characteristic. This implies that the generalized Green functions of G^{F^m} turn out to be polynomials in q (more precisely, rational functions in q if $p = 2$) for various extension field $\mathbf{F}_{q^m}$, and so certain values of almost characters are also rational functions in q. This makes it possible to apply some sort of specialization argument for the character values of G^{F^m} for any $m \geq 1$, and one can determine the scalars of G^F which are related to the unipotent characters, from the result for G^{F^2}. Thus we rediscover the results in [S2]. But this method works also for $p = 2$, and from this we can deduce the result for G^F.

The main result of this paper was announced in the 4th International Conference on Representation Theory, Lhasa, 2007.

1. Lusztig's conjecture

1.1. Let k be an algebraic closure of a finite field $\mathbf{F}_q$ of characteristic p. Let G be a connected reductive algebraic group defined over k. We fix a Borel subgroup B of G, and a maximal torus T contained in B, and a Weyl group $W = N_G(T)/T$ of G with respect to T. Let $\mathcal{D}G$ be the bounded derived category of constructible $\bar{\mathbf{Q}}_l$-sheaves on G, and let $\mathcal{M}G$ be the full subcategory of $\mathcal{D}G$ consisting of perverse sheaves. Let $\mathcal{S}(T)$ be the set of isomorphism classes of tame local systems on T, i.e., the local systems $\mathcal{L}$ of rank 1 such that $\mathcal{L}^{\otimes n} \simeq \bar{\mathbf{Q}}_l$ for some integer $n \geq 1$, invertible in k. Take a local system $\mathcal{L} \in \mathcal{S}(T)$ such that $w^*\mathcal{L} \simeq \mathcal{L}$ for some $w \in W$. Then one can construct a complex $K_w^{\mathcal{L}} \in \mathcal{D}G$ as in [L2, III, 12.1]. For each $\mathcal{L} \in \mathcal{S}(T)$ we denote by $\widehat{G}_{\mathcal{L}}$ the set of isomorphism classes of irreducible perverse sheaves A on G such that A is a constituent of the i-th perverse cohomology sheaf $^pH^i(K_w^{\mathcal{L}})$ of $K_w^{\mathcal{L}}$ for any i, w. The set $\widehat{G}$ of character sheaves on G is defined as $\widehat{G} = \bigcup_{\mathcal{L} \in \mathcal{S}(T)} \widehat{G}_{\mathcal{L}}$.

1.2. We consider the $\mathbf{F}_q$-structure of G, and assume that G is defined over $\mathbf{F}_q$ with Frobenius map F. We assume that B and T are both F-stable. We assume further that the center of G is connected. Let G^* be the dual group of G and T^* a maximal torus of G^* dual to T. By fixing an isomorphism $\iota : k^* \simeq \mathbf{Q}'/\mathbf{Z}$ ($\mathbf{Q}'$ is the subring of $\mathbf{Q}$ consisting of elements whose numerator is invertible in k), we have an isomorphism $f : T^* \simeq \mathcal{S}(T)$ (see e.g., [S1, II, 1.4, 3.1]). Let $W^* = N_G(T^*)/T^*$ be the Weyl group of G^*. Then W^* may be identified with $W = N_G(T)/T$, compatible with f. F acts naturally on $\mathcal{S}(T)$, via $F^{-1} : \mathcal{L} \mapsto F^*\mathcal{L}$, and the action of F on $\mathcal{S}(T)$ corresponds to the action of F^{-1} on T^* via f.

For each $s \in T^*$ such that the conjugacy class $\{s\}$ in G^* is F-stable, we put

$$W_s = \{w \in W^* \mid w(s) = s\},$$
$$Z_s = \{w \in W^* \mid F(s) = w(s)\}.$$

Then Z_s is non-empty, and one can write $Z_s = z_1 W_s$ for some element $z_1 \in Z_s$. Since the center of G is connected, $Z_{G^*}(s)$ is connected reductive, and W_s is a Weyl group of $Z_{G^*}(s)$. We choose z_1 so that $\gamma = \gamma_s = z_1^{-1}F : W_s \to W_s$ leaves invariant the set of simple roots of $Z_{G^*}(s)$ determined naturally from B and T.

Similarly, for any $\mathcal{L} \in \mathcal{S}(T)$ such that $F^*\mathcal{L} \simeq \mathcal{L}$, we define

$$W_{\mathcal{L}} = \{w \in W \mid w^*\mathcal{L} \simeq \mathcal{L}\},$$
$$Z_{\mathcal{L}} = \{w \in W \mid F^*\mathcal{L} \simeq (w^{-1})^*\mathcal{L}\}.$$

Then $W_{\mathcal{L}}$ (resp. $Z_{\mathcal{L}}$) is naturally identified with W_s (resp. Z_s).

1.3.   Let $\operatorname{Irr} G^F$ be the set of irreducible characters of G^F. Then $\operatorname{Irr} G^F$ is partitioned into a disjoint union of subsets $\mathcal{E}(G^F, \{s\})$, where $s \in T^*$ and $\{s\}$ runs over all the F-stable semisimple classes in G^*. According to [L1], two parameter sets $X(W_s, \gamma)$ and $\overline{X}(W_s, \gamma)$ are attached to $\mathcal{E}(G^F, \{s\})$, and a non-degenerate pairing $\{\ ,\ \} : \overline{X}(W_s, \gamma) \times X(W_s, \gamma) \to \bar{\mathbf{Q}}_l$ is defined. Here $\overline{X}(W_s, \gamma)$ is a finite set, and $X(W_s, \gamma)$ is an infinite set with a free action of the group M of all roots of unity in $\bar{\mathbf{Q}}_l^*$. More precisely, there exists a set $\overline{X}(W_s)$ with γ-action, and a natural map $X(W_s, \gamma) \to \overline{X}(W_s)$ whose image coincides with $\overline{X}(W_s)^\gamma$, the set of γ-fixed points in $\overline{X}(W_s)$. In the case where γ acts trivially on W_s, $X(W_s, \gamma)$ coincides with $\overline{X}(W_s, \gamma) \times M$. In general, the orbits set $X(W_s, \gamma)/M$ is in bijection with $\overline{X}(W_s)^\gamma$.

Now the set $\mathcal{E}(G^F, \{s\})$ is parametrized by $\overline{X}(W_s, \gamma)$. We denote by ρ_y the irreducible character in $\mathcal{E}(G^F, \{s\})$ corresponding to $y \in \overline{X}(W_s, \gamma)$. In turn, for each $x \in X(W_s, \gamma)$, an almost character R_x is defined as

$$(1.3.1) \qquad R_x = (-1)^{l(z_1)} \sum_{y \in \overline{X}(W_s, \gamma)} \{y, x\} \Delta(y) \rho_y,$$

where $\Delta(y) = \pm 1$ is a certain adjustment in the case of exceptional groups E_7, E_8. If c is the order of γ on W_s, the almost characters R_x are determined, up to a c-th root unity multiple, by the M-orbit of x in $X(W_s, \gamma)$.

1.4.   It is known by [L2, V], that the set $\widehat{G}_{\mathcal{L}}$ is parametrized by $\overline{X}(W_{\mathcal{L}}) = \overline{X}(W_s)$ under the identification $W_{\mathcal{L}} \simeq W_s$. For each $y \in \overline{X}(W_s)$, we denote by A_y the corresponding character sheaf in $\widehat{G}_{\mathcal{L}}$. Let $\widehat{G}^F$ be the set of F-stable character sheaves, i.e., the set of $A \in \widehat{G}$ such that $F^*A \simeq A$. Then $\widehat{G}^F = \bigcup_{\mathcal{L}} \widehat{G}_{\mathcal{L}}^F$, where $\mathcal{L}$ runs over the elements in $\mathcal{S}(T)$ such that $(Fw)^*\mathcal{L} \simeq \mathcal{L}$ for some $w \in W$. The set $\widehat{G}_{\mathcal{L}}^F$ is parametrized by $\overline{X}(W_s)^\gamma$. For each $A \in \widehat{G}_{\mathcal{L}}^F$, we fix an isomorphism $\phi_A : F^*A \xrightarrow{\sim} A$ as in [L2, V, 25.1]. Then ϕ_A is unique up to a root of unity multiple. We define a class function $\chi_A = \chi_{A, \phi_A}$ as the characteristic function $G^F \to \bar{\mathbf{Q}}_l$ of A. In the case of classical groups, we have the following theorem, which is a (partial) solution to the Lusztig' conjecture.

THEOREM 1.5 ([S1,II, Theorem 3.2]). *Assume that G is a (connected) classical group with connected center. Then for each $x \in X(W_s, \gamma)$, there exists an algebraic number ζ_x of absolute value 1 such that*

$$R_x = \zeta_x \chi_{A_{\bar{x}}},$$

where $\bar{x}$ is the image of x under the map $X(W_s, \gamma) \to \overline{X}(W_s)^\gamma$.

1.6.   Assume that G is a connected classical group with connected center. Let P be an F-stable parabolic subgroup of G containing B, and L be an F-stable Levi subgroup of P containing T, U_P the unipotent radical of P. Then $W_L = N_L(T)/T$ is a Weyl subgroup of W, and $B_L = B \cap L$ is a Borel subgroup of L containing T. Let $\widehat{L}$ be the set of character sheaves on L. We assume that $\widehat{L}_{\mathcal{L}}$ contains a cuspidal character sheaf A_0 for $\mathcal{L} \in \mathcal{S}(T)$, where $\mathcal{L}$ is Fw-stable for some $w \in W$. Then A_0

may be expressed by the intersection cohomology complex as $A_0 = \mathrm{IC}(\overline{\Sigma}, \mathcal{E})[\dim \Sigma]$, where Σ is the inverse image of a conjugacy class in $\overline{G} = G/Z^0(G)$ under the natural map $\pi : G \to \overline{G}$, and $\mathcal{E}$ is a cuspidal local system on Σ. The pair $(\Sigma, \mathcal{E})$, or its restriction on the conjugacy class, is called a cuspidal pair on G. Then either L is the maximal torus or L has the same type as G, and A_0 is a unique cuspidal character sheaf contained in $\widehat{L}_{\mathcal{L}}$. Consider the induced complex $K = \mathrm{ind}_P^G A_0$ on G. Then K is a semisimple perverse sheaf on G whose components are contained in $\widehat{G}$. By Lemma 5.9 in [S1,I], the endomorphism algebra $\mathrm{End}_{\mathcal{M}G} K$ is isomorphic to the group algebra $\bar{\mathbf{Q}}_l[\mathcal{W}_{\mathcal{E}}]$ of $\mathcal{W}_{\mathcal{E}}$, where

$$\mathcal{W}_{\mathcal{E}} = \{n \in N_G(L) \mid n\Sigma n^{-1} = \Sigma, \mathrm{ad}(n)^*\mathcal{E} \simeq \mathcal{E}\}/L,$$
$$\mathcal{Z}_{\mathcal{E}} = \{n \in N_G(L) \mid F(n\Sigma n^{-1}) = \Sigma, (Fn)^*\mathcal{E} \simeq \mathcal{E}\}/L.$$

On the other hand, if we choose a positive integer r large enough, the set $\mathcal{E}(L^{F^r}, \{s\})$ contains a unique cuspidal character δ of L^{F^r}, where $s \in T^*$ corresponds to $\mathcal{L}$ under f. We define

$$W_\delta = \{w \in N_W(W_L) \mid wB_Lw^{-1} = B_L, {}^w\delta \simeq \delta\},$$
$$Z_\delta = \{w \in N_W(W_L) \mid wB_Lw^{-1} = B_L, {}^{Fw}\delta \simeq \delta\}.$$

Since $\widehat{L}_{\mathcal{L}}$ contains a unique cuspidal character sheaf, we have $W_\delta \simeq \mathcal{W}_{\mathcal{E}}, Z_\delta \simeq \mathcal{Z}_{\mathcal{E}}$ by [S1, I, (5.16.1)]. Moreover there exists $w_1 \in Z_\delta$ such that $Z_\delta = w_1W_\delta$ and that $\gamma_1 = Fw_1 : W_\delta \to W_\delta$ gives rise to an automorphism of the Coxeter group W_δ. We denote by $(W_\delta)^\wedge$ the set of irreducible characters of W_δ, and $(W_\delta)^\wedge_{\mathrm{ex}}$ the subset of $(W_\delta)^\wedge$ consisting of γ_1-stable characters. Then A_0 is Fw_1-stable, and for each $E \in (W_\delta)^\wedge_{\mathrm{ex}}$, there exists $x_E \in X(W_s, \gamma)$ and $\bar{x}_E \in \overline{X}(W_s)^\gamma$ such that $A_E = A_{\bar{x}_E}$ is an F-stable character sheaf in $\widehat{G}_{\mathcal{L}}$ which is a simple component of K corresponding to $E \in \mathrm{End}_{\mathcal{M}} K$. Moreover, $\rho_{x_E} \in \mathcal{E}(G^{F^r}, \{s\})$ is an F-stable irreducible character which is a constituent of the Harish-Chandra induction $\mathrm{Ind}_{PF^r}^{G^{F^r}} \delta$ corresponding to $E \in (W_\delta)^\wedge_{\mathrm{ex}}$, and the image of the Shintani descent $\mathrm{Sh}_{F^r/F}$ of ρ_{x_E} determines the almost character R_{x_E} of G^F. (For the Shintani descent, see [S1]).

1.7.   Let L_{w_1} be an F-stable Levi subgroup twisted by $F(w_1)$, i.e., $L_{w_1} = \alpha L\alpha^{-1}$ for $\alpha \in G$ such that $\alpha^{-1}F(\alpha) = F(\dot{w}_1)$ for a representative $\dot{w}_1 \in N_G(L)$ of $w_1 \in \mathcal{Z}_{\mathcal{E}}$. Then by $\mathrm{ad}(\alpha^{-1}) : L_{w_1} \xrightarrow{\sim} L$, $\mathrm{ad}(\alpha^{-1})^*A_0$ gives rise to an F-stable cuspidal character sheaf on L_{w_1} which we denote by A_0'. If we fix an isomorphism $\varphi_0 : (F\dot{w}_1)^*\mathcal{E} \xrightarrow{\sim} \mathcal{E}$, φ_0 induces an isomorphism $\varphi_0^{w_1} : F^*A_0' \xrightarrow{\sim} A_0'$ on L_{w_1}. We choose $\phi_{A_0'} : F^*A_0' \xrightarrow{\sim} A_0'$ as $\phi_{A_0'} = \varphi_0^{w_1}$. Then by Theorem 1.5, we have

$$(1.7.1) \hspace{3cm} R_0^{L_{w_1}} = \zeta_0\chi_{A_0'}$$

for some $\zeta_0 \in \bar{\mathbf{Q}}_l^*$ of absolute value 1, where $R_0^{L_{w_1}}$ is the almost character of $L_{w_1}^F$ corresponding to $A_0' \in \widehat{L}_{w_1}$. The following result was proved in [S1] in the course of the proof of the main theorem. (Note that in [S1], the constants $\varepsilon_0\xi_{A_0}$ and $\varepsilon_0\xi_{A_E}$ are used. But the proof shows that these constants are indeed given by $\zeta_0 = \varepsilon_0\xi_{A_0}$.)

LEMMA 1.8 ([S1, II, Lemma 3.7]). *Let ζ_0 be as in (1.7.1). Then we have*

$$R_{x_E} = (-1)^{\dim \Sigma}\zeta_0\chi_{A_E}$$

for any $E \in (W_\delta)^\wedge_{\mathrm{ex}}$.

1.9.   Lemma 1.8 shows that the determination of the scalars ζ_x appeared in Theorem 1.5 is reduced to the case of cuspidal character sheaves. We note that it is further reduced to the case of adjoint groups. In fact, let A_0 be an F-stable cuspidal character sheaf contained in $\widehat{G}_{\mathcal{L}}$. Let $\pi : G \to \bar{G}$ be as before. Then A_0 can be written as $A_0 \simeq \mathcal{E}_0 \otimes \pi^* \bar{A}_0[d]$, where $d = \dim Z(G)$, and $\mathcal{E}_0$ is a local system on G which is the inverse image of $\mathcal{E}_0' \in \mathcal{S}(G/G_{\mathrm{der}})$ under the natural map $G \to G_{\mathrm{der}}$ (G_{der} is the derived subgroup of G), and $\bar{A}_0$ is a cuspidal character sheaf on $\bar{G}$. Since $\bar{A}_0$ is a unique cuspidal character sheaf in $\widehat{\bar{G}}_{\mathcal{L}}$, $\bar{A}_0$ is F-stable. Then $\pi^* \bar{A}_0$ is F-stable and so $\mathcal{E}_0$ is also F-stable. Then $\phi_{A_0} : F^* A_0 \xrightarrow{\sim} A_0$ is given by $\phi_{A_0} = \varphi_0 \otimes \pi^* \phi_{\bar{A}_0}$, where $\phi_{\bar{A}_0} : F^* \bar{A}_0 \xrightarrow{\sim} \bar{A}_0$ is the map chosen for $\bar{A}_0$, and φ_0 is the pull-back of the canonical isomorphism $F^* \mathcal{E}_0' \xrightarrow{\sim} \mathcal{E}_0'$. Hence χ_{A_0} is written as $\chi_{A_0} = \theta_0 \otimes \pi^* \chi_{\bar{A}_0}$, where $\pi^* \chi_{\bar{A}_0}$ is the pull-back of $\chi_{\bar{A}_0}$ under the induced map $\pi : G^F \to \bar{G}^F$, and θ_0 is a linear character of G^F corresponding to $\mathcal{E}_0$. A similar description works also for almost characters. Let R_0 (resp. $\bar{R}_0$) be the almost character of G^F (resp. $\bar{G}^F$) corresponding to A_0 (resp. $\bar{A}_0$). Then we have $R_0 = \theta_0 \otimes \pi^* \bar{R}_0$. (This follows from the fact that if δ is a cuspidal irreducible character of G^{F^r} corresponding to A_0 for sufficiently large r, then δ can be written as $\delta = \theta \otimes \bar{\delta}$, where $\bar{\delta}$ is a cuspidal irreducible character of $\bar{G}^{F^r}$ corresponding to $\bar{A}_0$, and θ is an F-stable linear character of G^{F^r}, and by applying the Shintani descent on δ.) Thus ζ_0 for A_0 coincides with ζ_0 for $\bar{A}_0$.

2. Generalized Green functions

2.1.   Under the setting in 1.6, we further assume that G^F is of split type. Let L be as before. Assume that A_0 is a cuspidal character sheaf on L of the form $A_0 = \mathrm{IC}(\bar{\Sigma}, \mathcal{E})[\dim \Sigma]$, where $\Sigma = Z^0(L) \times C$ with a unipotent class C in L and $\mathcal{E} = \bar{\mathbf{Q}}_l \boxtimes \mathcal{E}'$ for a cuspidal local system $\mathcal{E}'$ on C. Then $\mathcal{W}_{\mathcal{E}} = W = N_G(L)/L$. For each $w \in W$, let L_w be an F-stable Levi subgroup of G obtained from L by twisting w as in 1.7, i.e., $L_w = \alpha L \alpha^{-1}$ with $\alpha \in G$ such that $\alpha^{-1} F(\alpha) = F(\dot{w})$ for a representative $\dot{w} \in N_G(L)$ of w. Put $\Sigma_w = \alpha \Sigma \alpha^{-1}, \mathcal{E}_w = \mathrm{ad}(\alpha^{-1})^* \mathcal{E}$, a local system on Σ_w. We assume that the pair $(C, \mathcal{E}')$ is F-stable, and fix an isomorphism $\varphi_0 : F^* \mathcal{E}' \xrightarrow{\sim} \mathcal{E}'$. Then one can construct an isomorphism $(\varphi_0)_w : F^* \mathcal{E}_w \xrightarrow{\sim} \mathcal{E}_w$ as in [L2, II, 10.6], and this induces an isomorphism $\varphi_w : F^* K_w \xrightarrow{\sim} K_w$, where K_w is a complex induced from the pair $(\Sigma_w, \mathcal{E}_w)$. Note that K_w is isomorphic to $\mathrm{ind}_P^G A_0$, with a specific mixed structure twisted by $w \in W$. We denote by χ_{K_w, φ_w} the characteristic function of K_w with respect to φ_w.

Since L is of the same type as G, and F is of split type, $\gamma_1 : W \to W$ is identity. Let $K = \mathrm{ind}_P^G A_0 = \bigoplus_{E \in \mathcal{W}^\wedge} V_E \otimes A_E$ be the decomposition of K into simple components, where A_E is a character sheaf corresponding to $E \in \mathcal{W}^\wedge$, and V_E is the multiplicity space of A_E which has a natural structure of irreducible $\mathcal{W}$-module corresponding to E. Then there exists a unique isomorphism $\phi_{A_E} : F^* A_E \xrightarrow{\sim} A_E$ for each $E \in \mathcal{W}^\wedge$ such that

$$(2.1.1) \qquad \chi_{K_w, \varphi_w} = \sum_{E \in \mathcal{W}^\wedge} \mathrm{Tr}\,(w, V_E) \chi_{A_E}.$$

Let G_{uni} be the unipotent variety of G, and G_{uni}^F be the set of F-fixed points in G_{uni}. The restriction of χ_{K_w, φ_w} on G_{uni}^F is the generalized Green function $Q_{L_w, C_w, \mathcal{E}_w', (\varphi_0)_w'}^G$ ([L2, II, 8.3]), where $C_w = \alpha C \alpha^{-1}, \mathcal{E}_w' = \mathrm{ad}(\alpha^{-1})^* \mathcal{E}'$, local system on C_w, and

$(\varphi_0)'_w$ is the restriction of $(\varphi_0)_w$ on $\mathcal{E}'_w$. On the other hand, by the generalized Springer correspondence, for each $E \in \mathcal{W}^\wedge$, there exists a pair $(C_1, \mathcal{E}_1)$, where C_1 is a unipotent class in G and $\mathcal{E}_1$ is a G-equivariant simple local system on C_1, such that

$$(2.1.2) \qquad A_E|_{G_{\mathrm{uni}}} \simeq \mathrm{IC}(\overline{C}_1, \mathcal{E}_1)[\dim C_1 + \dim Z^0(L)].$$

Now the pair $(C_1, \mathcal{E}_1)$ is F-stable, and $\phi_{A_E} : F^* A_E \xrightarrow{\sim} A_E$ determines an isomorphism $\psi_{\mathcal{E}_1} : F^* \mathcal{E}_1 \xrightarrow{\sim} \mathcal{E}_1$ via (2.1.2) (cf. [L2, V, 24.2]). In other words, the choice of $\varphi_0 : F^* \mathcal{E}' \xrightarrow{\sim} \mathcal{E}'$ determines $\psi_{\mathcal{E}_1}$. Let $\chi_{(C_1, \mathcal{E}_1)} : C_1^F \to \bar{\mathbf{Q}}_l$ be the characteristic function of $\mathcal{E}_1$ with respect to $\psi_{\mathcal{E}_1}$. Lusztig ([L2, V, 24]) gave an algorithm of computing $\chi_{A_E} = \chi_{A_E, \phi_{A_E}}$ on G_{uni}^F. Here $\chi_{A_E}|_{G_{\mathrm{uni}}^F}$ is expressed in terms of a linear combination of various $\chi_{(C'_1, \mathcal{E}'_1)}$, where $(C'_1, \mathcal{E}'_1)$ is a pair as above corresponding to some $E' \in \mathcal{W}^\wedge$. Let $p_{E,E'}$ be the coefficient of $\chi_{(C'_1, \mathcal{E}'_1)}$ in the expansion of χ_{A_E}. Then $p_{E,E'}$ satisfies the following property; if we replace F by F^m for any integer $m > 0$, we obtain a similar coefficient $p_{E,E'}$ (with respect to G^{F^m}) starting from $\varphi_0^{(m)} : (F^m)^* \mathcal{E}' \xrightarrow{\sim} \mathcal{E}'$ induced naturally from φ_0, which we denote by $p_{E,E'}^{(m)}$. Then there exists a rational function $P_{E,E'}(x)$ such that $p_{E,E'}^{(m)} = P_{E,E'}(q^m)$. (Note : It is shown in [L2, V] that $P_{E,E'}$ turns out to be a polynomial if p is good.)

Now $\chi_{(C_1, \mathcal{E}_1)}$ is described as follows; take $u \in C_1^F$ and put $A_G(u) = Z_G(u)/Z_G^0(u)$. Then F acts naturally on $A_G(u)$, and in our setting F acts trivially on it. The set of G^F-conjugacy classes in C_1^F is in 1:1 correspondence with the set $A_G(u)$ (note: $A_G(u)$ is abelian). We denote by a_u a representative of a G^F-class in C_1^F corresponding to $a \in A_G(u)$. On the other hand, the set of G-equivariant simple local systems on C_1 is in 1:1 correspondence with the set $A_G(u)^\wedge$ of irreducible characters of $A_G(u)$. For each $\rho \in A_G(u)^\wedge$, we define a function f_ρ on G_{uni}^F by

$$(2.1.3) \qquad f_\rho(v) = \begin{cases} \rho(a) & \text{if } v = u_a \in C_1^F, \\ 0 & \text{if } v \notin C_1^F \end{cases}$$

for $v \in G_{\mathrm{uni}}^F$. Let $\rho \in A_G(u)^\wedge$ be the character corresponding to $\mathcal{E}_1$. Then there exists $\eta_E \in \bar{\mathbf{Q}}_l^*$ of absolute value 1 such that

$$(2.1.4) \qquad \chi_{(C_1, \mathcal{E}_1)} = \eta_E f_\rho.$$

Note that η_E depends on the choice of $\varphi_0 : F^* \mathcal{E}' \xrightarrow{\sim} \mathcal{E}'$ and on the choice of $u \in C_1^F$. We have the following theorem.

THEOREM 2.2 ([S3]). *Let G be a classical group, simple modulo center. Assume that the derived subgroup of G does not contain the Spin group. Further assume that G is of split type. Then for each unipotent class C_1 in G, there exists $u \in C_1^F$ (called a split unipotent element) satisfying the following; Let $(C, \mathcal{E}')$ be the pair in L as in 2.1 and $u_0 \in C^F$ be a split element. Choose $\varphi_0 : F^* \mathcal{E}' \xrightarrow{\sim} \mathcal{E}'$ so that the isomorphism $(\varphi_0)_{u_0} : \mathcal{E}'_{u_0} \to \mathcal{E}'_{u_0}$ induced on the stalk $\mathcal{E}'_{u_0}$ of $\mathcal{E}'$ at u_0 is identity. Choose a split element $u \in C_1^F$ for defining f_ρ in (2.1.3). Then $\eta_E = 1$ for any $E \in \mathcal{W}^\wedge$.*

2.3. Returning to the setting in 2.1, we choose a split element $u \in C_1^F$ for each unipotent class C_1 of G. Then $u \in C_1^{F^m}$ for any integer $m > 0$, (in fact, it is a split element with respect to G^{F^m}), and we choose $u_a^{(m)} \in C_1^{F^m}$, a representative of G^{F^m}-class in $C_1^{F^m}$ for each $a \in A_G(u)$. For later discussion, we prepare a notation.

Assume given a family of functions $h = \{h^{(m)}\}_{m>0}$, where $h^{(m)}$ is a class function on $G_{\text{uni}}^{F^m}$. Then we say that h is a rational function in q if there exists a rational function $H_{C_1,a}(x)$ for each pair (C_1, a) such that $h^{(m)}(u_a^{(m)}) = H_{C_1,a}(q^m)$.

For each $E \in \mathcal{W}^\wedge$, we have an isomorphism $\phi_{A_E}^{(m)} : (F^m)^* A_E \xrightarrow{\sim} A_E$, and one can define a function $\chi_{A_E}^{(m)} = \chi_{A_E, \phi_{A_E}^{(m)}}$ on G^{F^m}. Then in view of Theorem 2.2, Lusztig's algorithm implies that $\{\chi_{A_E}^{(m)}|_{G_{\text{uni}}^{F^m}}\}_{m>0}$ is a rational function in q.

Thus, by (2.1.1) we have the following corollary.

COROLLARY 2.4. *Assume that G is as in Theorem 2.2, and that G^F is of split type. The generalized Green function $Q_{L_w, C_w, \mathcal{E}_w', (\varphi_0)_w'}^G$ can be expressed as a rational function in q.*

2.5. More generally, if there exists a family of values $h = \{h^{(m)} \in \bar{\mathbf{Q}}_l\}_{m>0}$ such that $h^{(m)} = H(q^m)$ for some rational function $H(x)$, we say that h is a rational function in q.

3. Cuspidal character sheaves

3.1. Let G be an adjoint simple group of classical type. We assume that G is of split type over $\mathbf{F}_q$. Let $\widehat{G}^0$ be the set of cuspidal character sheaves on G. $A \in \widehat{G}^0$ is given in the form $A = \text{IC}(\overline{C}, \mathcal{E})[\dim C]$, where C is a conjugacy class in G, and $\mathcal{E}$ is a simple G-equivariant local system on C. We shall describe the cuspidal character sheaves on G (cf. [L2, V, 22.2, 23.2], see also [S2, 6.6]) and their mixed structures.

(a) $G = PSp_{2n}$ ($n \geq 1$) with p: odd. $\widehat{G}^0$ is empty if n is even. Assume that n is odd. Then for each pair (N_1, N_2) such that $N_i = d_i^2 + d_i$ for some integers $d_i \geq 0$ and that $n = N_1 + N_2$, one can associate cuspidal character sheaves on G as follows. Let C be a conjugacy class of $g = su = us$, where s is a semisimple element of G such that $Z_G^0(s)$ is isomorphic to $H = (Sp_{2N_1} \times Sp_{2N_2})/\{\pm 1\}$, and u is a unipotent element of $Z_G^0(s) \simeq H$ such that the unipotent class C_0 containing u gives a unique cuspidal pair $(C_0, \mathcal{E}_0)$ with unipotent support of H. Here $(C_0, \mathcal{E}_0)$ is described as follows. There exists a cuspidal pair $(C_i, \mathcal{E}_i)$ for Sp_{2N_i} such that $C_0 = C_1 \times C_2$ and $\mathcal{E}_0 = \mathcal{E}_1 \boxtimes \mathcal{E}_2$. Choose $u = (u_1, u_2) \in C_0$ such that $u_i \in C_i$. Let $\rho_i \in A_{H_i}(u_i)^\wedge$ corresponding to $\mathcal{E}_i$, where $H_i = Sp_{2N_i}$. Then $\rho_1 \boxtimes \rho_2 \in (A_{H_1}(u_1) \times A_{H_2}(u_2))^\wedge$ factors through $A_H(u)$ and defines an irreducible character ρ_0 of $A_H(u)$ corresponding to $\mathcal{E}_0$.

Now assume that $N_1 \neq N_2$. Then $Z_G(s)$ is connected, and so $A_G(g) = A_H(u)$, and ρ_0 gives an irreducible character $\rho \in A_G(g)^\wedge$ which determines a local system $\mathcal{E}$ on C, and we denote by A_{N_1,N_2} the character sheaf corresponding to $(C, \mathcal{E})$. Next assume that $N_1 = N_2$. Then $A_G(s) \simeq \mathbf{Z}/2\mathbf{Z}$ and $A_H(u)$ is a subgroup of $A_G(g)$ of index 2. We have $\text{Ind}_{A_H(u)}^{A_G(g)} \rho_0 = \rho + \rho'$, where ρ, ρ' are linear characters of $A_G(u)$. If we write $\mathcal{E}, \mathcal{E}'$ the simple local system on C corresponding to ρ, ρ', then the pairs $(C, \mathcal{E}), (C, \mathcal{E}')$ are both cuspidal pairs of G. We denote by $A_{N_1,N_2}, A'_{N_1,N_2}$ the cuspidal character sheaves on G corresponding to $(C, \mathcal{E}), (C, \mathcal{E}')$, respectively. The set $\widehat{G}^0$ consists of these elements.

We shall fix a mixed structure on $(C, \mathcal{E})$. Since $s \in G^F$, H is F-stable, and so $(C_0, \mathcal{E}_0)$ is also F-stable. Choose $u = (u_1, u_2) \in C_0^F$ such that u_i are split elements in Sp_{2N_i}, and fix $s \in T^F$ appropriately. We choose $\varphi_0 : F^*\mathcal{E} \xrightarrow{\sim} \mathcal{E}$ so that the induced isomorphism $(\varphi_0)_g : \mathcal{E}_g \to \mathcal{E}_g$ on the stalk $\mathcal{E}_g$ at g is identity. Then φ_0 induces an isomorphism $\varphi : F^*A_{N_1,N_2} \xrightarrow{\sim} A_{N_1,N_2}$. We define $\phi_A = \phi_{A_{N_1,N_2}}$ by $\phi_A = q^{(\dim G - \dim C)/2}\varphi$. A similar construction is applied also for $(C, \mathcal{E}')$.

(b) $G = PSO_m$ ($m \geq 3$) with p: odd. $\widehat{G}^0$ is empty unless m is either odd or divisible by 8. Note that $PSO_m = SO_m$ if m is odd. To each pair (N_1, N_2) such that $N_i = d_i^2$ for some $d_i \geq 1$ and that $m = N_1 + N_2$, one can associate cuspidal character sheaves A associated to $(C, \mathcal{E})$ as follows. Let C be the conjugacy class of G containing $g = su = us$, where s is a semisimple element such that $H = Z_G^0(s)$ is isomorphic to $SO_{N_1} \times SO_{N_2}$ if m is odd, and to $(SO_{N_1} \times SO_{N_2})/\{\pm 1\}$ if m is even, and u is a unipotent element in $Z_G^0(s) \simeq H$ such that the unipotent class C_0 containing u gives a unique cuspidal pair $(C_0, \mathcal{E}_0)$ with unipotent support on H. Here $C_0 = C_1 \times C_2$, $\mathcal{E}_0 \simeq \mathcal{E}_1 \boxtimes \mathcal{E}_2$ with the cuspidal pair $(C_i, \mathcal{E}_i)$ on SO_{N_i}. Choose $u = (u_1, u_2) \in C_0$ such that $u_i \in C_i$. Let $\rho_i \in A_{H_i}(u_i)^\wedge$ corresponding to $\mathcal{E}_i$, where $H_i = SO_{N_i}$. Then $\rho_1 \boxtimes \rho_2 \in (A_{H_1}(u_1) \times A_{H_2}(u_2))^\wedge$ factors through $A_H(u)$ and gives an irreducible character $\rho_0 \in A_H(u)^\wedge$ corresponding to $\mathcal{E}_0$. Depending on the structure of $A_G(s)$, the three cases occur.

(i) The case where $N_1 = 0$ or $N_2 = 0$. In this case, $Z_G(s)$ is connected and so $A_G(u) \simeq A_H(u)$. ρ_0 gives $\rho \in A_G(g)^\wedge$, which determines a local system $\mathcal{E}$ on C, and $(C, \mathcal{E})$ corresponds to the cuspidal character sheaf A_{N_1,N_2}.

(ii) The case where $N_1 > 0, N_2 > 0, N_1 \neq N_2$. Then $A_G(s) \simeq \mathbf{Z}/2\mathbf{Z}$, and $A_H(u)$ is regarded as an index 2 subgroup of $A_G(g)$. We have $\mathrm{Ind}_{A_H(u)}^{A_G(g)} \rho_0 = \rho + \rho'$ for $\rho, \rho' \in A_G(g)^\wedge$. If we write $\mathcal{E}, \mathcal{E}'$ the simple local system corresponding to ρ, ρ', $(C, \mathcal{E}), (C, \mathcal{E}')$ are both cuspidal pairs for G. We denote by $A_{N_1,N_2}, A'_{N_1,N_2}$ the cuspidal character sheaves corresponding to them.

(iii) The case where $N_1 = N_2$. In this case $A_G(s) \simeq \mathbf{Z}/2\mathbf{Z} \times \mathbf{Z}/2\mathbf{Z}$ and so $A_G(g)/A_H(u) \simeq \mathbf{Z}/2\mathbf{Z} \times \mathbf{Z}/2\mathbf{Z}$. $\mathrm{Ind}_{A_H(u)}^{A_G(g)} \rho_0$ decomposes into 4 irreducible (linear) characters, $\rho, \rho', \rho'', \rho'''$ of $A_G(g)$. Correspondingly, we have simple local systems $\mathcal{E}, \mathcal{E}', \mathcal{E}'', \mathcal{E}'''$ on C, and all of them give cuspidal pairs on G. We denote by $A_{N_1,N_2}, A'_{N_1,N_2}, A''_{N_1,N_2}, A'''_{N_1,N_2}$ the cuspidal character sheaves corresponding to them.

All of the above three cases give the set $\widehat{G}^0$. We shall fix a mixed structure on cuspidal character sheaves. Since $s \in G^F$, H is F-stable, and so $(C_0, \mathcal{E}_0)$ is F-stable. Take $u = (u_1, u_2) \in C_0^F$ such that u_i are split elements in SO_{N_i}, and fix $s \in T^F$. We choose $\varphi_0 : F^*\mathcal{E} \xrightarrow{\sim} \mathcal{E}$ so that the induced isomorphism $(\varphi_0)_g : \mathcal{E}_g \to \mathcal{E}_g$ on the stalk $\mathcal{E}_g$ at g is identity. φ_0 induces an isomorphism $\varphi : F^*A \xrightarrow{\sim} A$ for $A = A_{N_1,N_2}$. We define ϕ_A by $\phi_A = q^{(\dim G - \dim C)/2}\varphi$. We define similarly for $A'_{N_1,N_2}, A''_{N_1,N_2}, A'''_{N_1,N_2}$.

(c) $G = Sp_{2n}$ ($n \geq 1$) with $p = 2$. $\widehat{G}^0$ is empty unless $n = d^2 + d$ for some $d \geq 1$. Assume that $n = d^2 + d$. Then G contains a unique cuspidal pair $(C, \mathcal{E})$. The set $\widehat{G}^0$ consists of a single character sheaf A associated to $(C, \mathcal{E})$. We fix a mixed structure of A. C is an F-stable unipotent class of G, and we take a split element $u \in C^F$. We fix an isomorphism $\varphi_0 : F^*\mathcal{E} \xrightarrow{\sim} \mathcal{E}$ so that the induced isomorphism $(\varphi_0)_u : \mathcal{E}_u \to \mathcal{E}_u$ on the stalk $\mathcal{E}_u$ of $\mathcal{E}$ at u is identity. φ_0 induces $\varphi : F^*A \xrightarrow{\sim} A$. We define ϕ_A by $\phi_A = q^{(\dim G - \dim C)/2}\varphi$.

(d) $G = SO_{2n}$ with $p = 2$ $(n \geq 1)$. $\widehat{G}^0$ is empty unless $n = 4d^2$ for some $d \geq 1$. Assume that $n = 4d^2$. Then G contains a unique cuspidal pair $(C, \mathcal{E})$. The set $\widehat{G}^0$ consists of a single character sheaf A associated to $(C, \mathcal{E})$. C is an F-stable unipotent class of G, and we take a split element $u \in C^F$. We define ϕ_A in a similar way as in the case (c), by $\phi_A = q^{(\dim G - \dim C)/2} \varphi$.

We show the following lemma.

LEMMA 3.2. *Let ρ be the irreducible character of $A_G(g)$ corresponding to the local system $\mathcal{E}$ on C, as in 3.1. Then ρ is a linear character such that $\rho^2 = 1$. A similar fact holds also for ρ', ρ'', ρ''' if there exists any.*

PROOF. Let ρ be one of the characters $\rho, \rho', \rho'', \rho'''$ if there exists any. It is enough to show that $\rho(a^2) = 1$ for any $a \in A_G(g)$. By investigating the structure of $A_G(g)$, we see that $A_G(g)$ is an elementary abelian 2-group if $N_1 \neq N_2$. Thus in this case, $\rho(a^2) = 1$. We assume that $N_1 = N_2$. Then p is odd, and G is PSp_{2n} or PSO_{2n}. Assume that $G = PSp_{2n}$. We have $A_G(g) \simeq \langle \sigma \rangle \ltimes A_H(u)$, where σ is an element of order 2 permuting two factors of $A_H(u)$. In this case $a \in A_G(g)$ is of order 2 or 4. If a has order 2, there is nothing to prove. Assume that a has order 4. Then we have $a^2 \in A_H(u)$. Put $\theta = \mathrm{Ind}_{A_H(u)}^{A_G(g)} \rho_0$. Since $A_H(u)$ is an elementary abelian 2-group, and ρ_0 is σ-stable, we see that $\theta(a^2) = |A_G(g)|/|A_H(u)|$. This shows that $\rho(a^2) = 1$ for any irreducible factor ρ of θ. Next assume that $G = SO_{2n}$. In this case, $A_G(g) \simeq \langle \sigma \rangle \ltimes \widetilde{A}_H(u)$, where $\widetilde{A}_H(u)$ is an elementary abelian 2-group containing $A_H(u)$ as an index 2 subgroup, and σ is an element of order 2 acting on $\widetilde{A}_H(u)$. σ stabilizes $A_H(u)$ permuting their two factors, and ρ_0 is σ-stable. Thus a similar argument shows that $\rho(a^2) = 1$. The lemma is proved. $\quad\square$

4. Symbols and unipotent characters

4.1. Irreducible characters contained in $\mathcal{E}(G^F, \{1\})$ are called unipotent characters. In the case of classical groups, unipotent characters are parametrized by a combinatorial object called symbols. In this section, we review unipotent characters of classical groups.

Let G be a classical group over $\mathbf{F}_q$ of type B_n, C_n or D_n. We assume that G^F is of split type if G is of type D_n. The set of unipotent characters of G^F is parameterized by symbols. A symbol is an (unordered) pair $\binom{S}{T}$ of finite subsets of $\{0, 1, 2, \ldots\}$ modulo the shift operation $\binom{S}{T} \sim \binom{S'}{T'}$ with $S' = \{0\} \cup (S + 1)$, $T' = \{0\} \cup (T + 1)$. The rank of a symbol $\Lambda = \binom{S}{T}$ is defined by

$$r(\Lambda) = \sum_{\lambda \in S} \lambda + \sum_{\mu \in T} \mu - \left[\left(\frac{|S| + |T| - 1}{2} \right)^2 \right],$$

where $[z]$ denotes the largest integer which does not exceed z. The defect $d(\Lambda)$ of Λ is defined by the absolute value of $|S| - |T|$. The rank and the defect are independent of the shift operation.

For each integer $d \geq 0$, we denote by Φ_n^d the set of symbols of rank n and defect d. In the case where $\Lambda = \binom{S}{T}$ is defect 0, Λ is said to be degenerate if $S = T$, and is said to be non-degenerate otherwise. We denote by $\widetilde{\Phi}_n^0$ the set of symbols of rank

n and defect 0, where the degenerate symbols are counted twice. We put

$$\Phi_n = \coprod_{d:\text{odd}} \Phi_n^d, \qquad \Phi_n^+ = \tilde{\Phi}_n^0 \coprod \Big(\coprod_{d \equiv 0 \pmod 4} \Phi_n^d \Big).$$

Then the unipotent characters of G^F of type B_n or C_n (resp. D_n of split type) are parametrized by Φ_n (resp. Φ_n^+). In the notation of 1.3, $W_s = W$ and $\gamma = 1$ since F is of split type, and Φ_n or Φ_n^+ is nothing but $\overline{X}(W_s, \gamma) = \overline{X}(W, 1)$. We denote by ρ_Λ the unipotent character of G^F corresponding to $\Lambda \in \Phi_n$ or Φ_n^+. The unipotent cuspidal character exists if and only if $n = d^2 + d$ (resp. $n = 4d^2$) for some integer $d \geq 1$ if G is of type B_n or C_n (resp. D_n). In these cases, the symbol Λ_c (the cuspidal symbol) corresponding to the (unique) cuspidal unipotent character is given as follows.

$$\Lambda_c = \begin{pmatrix} 0, 1, 2, \ldots, 2d \\ - \end{pmatrix} \in \Phi_n^{2d+1} \qquad (G : \text{type } B_n \text{ or } C_n, n = d^2 + d),$$

$$\Lambda_c = \begin{pmatrix} 0, 1, 2, \ldots, 4d-1 \\ - \end{pmatrix} \in \Phi_n^{4d} \qquad (G : \text{type } D_n, n = 4d^2).$$

4.2. We introduce a notion of families in Φ_n or Φ_n^+. Two symbols Λ, Λ' belong to the same family if Λ, Λ' are represented by $\binom{S}{T}, \binom{S'}{T'}$ such that $S \cup T = S' \cup T'$ and that $S \cap T = S' \cap T'$. Families give a partition of Φ_n or Φ_n^+. A symbol $\Lambda \in \Phi_n$ of defect 1 is called a special symbol if $\Lambda = \binom{S}{T}$ with $S = \{a_0, a_1, \ldots, a_m\}$ and $T = \{b_1, \ldots, b_m\}$ such that $a_0 \leq b_1 \leq a_1 \leq \cdots \leq b_m \leq a_m$. Similarly, a symbol $\Lambda \in \Phi_n^+$ of defect 0 is called a special symbol if $\Lambda = \binom{S}{T}$ with $S = \{a_1, \ldots, a_m\}$, $T = \{b_1, \ldots, b_m\}$ such that $a_1 \leq b_1 \leq \cdots \leq a_m \leq b_m$. Each family contains a unique special symbol. Let $\mathcal{F}$ be a (non-degenerate) family. Then any symbol $\Lambda \in \mathcal{F}$ can be expressed as

$$\Lambda = \Lambda_M = \begin{pmatrix} Z_2 \coprod (Z_1 - M) \\ Z_2 \coprod M \end{pmatrix},$$

for some M, where Z_1, Z_2 are determined by $\mathcal{F}$; Z_2 is the set of elements which appear in both rows of Λ, Z_1 is the set of singles in Λ, and M is a subset of Z_1. The map $M \mapsto \Lambda_M$ gives a bijective correspondence between the set of subsets M of Z_1 such that $|M| \equiv d_1 \pmod 2$ and $\mathcal{F}$, where $|Z_1| = 2d_1 + 1$ (resp. $|Z_1| = 2d_1$) if $\mathcal{F} \subset \Phi_n$ (resp. $\mathcal{F} \subset \Phi_n^+$) for some integer $d_1 \geq 1$. (In the case of $\mathcal{F} \subset \Phi_n^+$, we further assume that the smallest element in M is bigger that that of $Z_1 - M$.) In particular, the special symbol in $\mathcal{F}$ can be written as Λ_{M_0} for some $M_0 \subset Z_1$ such that $|M_0| = d_1$. For $M \in Z_1$, put $M^\sharp = M_0 \cup M - M_0 \cap M$. We define a pairing $\{ \, , \, \} : \mathcal{F} \times \mathcal{F} \to \mathbf{Q}$ by

(4.2.1) $$\{\Lambda_M, \Lambda_{M'}\} = \frac{1}{2^f} (-1)^{|M^\sharp \cap M'^\sharp|},$$

where $f = d_1$ (resp. $f = d_1 - 1$) if $\mathcal{F} \subset \Phi_n$ (resp. $\mathcal{F} \subset \Phi_n^+$). We extend this pairing to the pairing on Φ_n or Φ_n^+ by requiring that $\mathcal{F}$ and $\mathcal{F}'$ are orthogonal if $\mathcal{F} \neq \mathcal{F}'$, which we denote by the same symbol. Note that the pairing $\{ \, , \, \}$ on Φ_n or Φ_n^+ coincides with the pairing $\{ \, , \, \}$ on $\overline{X}(W, 1)$ given in 1.3. Hence, for each $\Lambda \in \mathcal{F}$, the almost character R_Λ is given as

(4.2.2) $$R_\Lambda = \sum_{\Lambda' \in X_n} \{\Lambda, \Lambda'\} \rho_{\Lambda'},$$

where $X_n = \Phi_n$ or Φ_n^+ according to the cases G is of type B_n or C_n, or G is of type D_n. By the property of the pairing $\{\ ,\ \}$, one can also write, for each $\Lambda \in \mathcal{F}$,

$$(4.2.3) \qquad\qquad \rho_\Lambda = \sum_{\Lambda' \in X_n} \{\Lambda, \Lambda'\} R_{\Lambda'}.$$

4.3.   Assume that G^F contains a cuspidal unipotent character, and denote by $\mathcal{F}_c$ the family containing the cuspidal symbol Λ_c. Then the special symbol Λ_0 contained in $\mathcal{F}_c$ is given as follows.

$$\Lambda_0 = \begin{pmatrix} 0,2,4,\ldots,2d \\ 1,3,\ldots,2d-1 \end{pmatrix} \qquad (G : \text{type } B_n \text{ or } C_n, n = d^2 + d),$$

$$\Lambda_0 = \begin{pmatrix} 0,2,\ldots,4d-2 \\ 1,3,\ldots,4d-1 \end{pmatrix} \qquad (G : \text{ type } D_n, n = 4d^2).$$

In the case where G is of type B_n or C_n, we have $Z_1 = \{0,1,\ldots,2d\}$ and $M_0 = \{1,3,\ldots,2d-1\}$. In the case where G is of type D_n, we have $Z_1 = \{0,1,\ldots,4d-1\}$ and $M_0 = \{1,3,\ldots,4d-1\}$. In both cases, Λ_c is given by $\Lambda_c = \Lambda_M$ with $M = \emptyset$. We denote by $R_0 = R_{\Lambda_c}$ the cuspidal almost character.

4.4.   Let A_0 be the cuspidal character sheaf of G corresponding to $R_0 = R_{\Lambda_c}$. Let $(C, \mathcal{E})$ be the cuspidal pair corresponding to A_0. Then it is contained in the list in 3.1. If $p = 2$, it is uniquely determined since $\widehat{G}^0$ consits of a single element. In the case where $p \neq 2$, the explicit correspondence is known by [S2, Prop. 6.7], see also [L3]. The conjugacy class is given as follows (though we don't need it in later discussions). We use the notation in 3.1. Let $g = su = us \in C$ with $H = Z_G^0(s)$. Assume that $G = PSp_{2n}$ with $n = d^2 + d$. Then H is isogeneous to $Sp_{d^2+d} \times Sp_{d^2+d}$. Assume that $G = PSO_{2n+1}$ with $n = d^2 + d$. Then H is isogenous to $SO_{(d+1)^2} \times SO_{d^2}$. Assume that $G = PSO_{2n}$ with $n = 4d^2$. Then H is isogeneous to $SO_{4d^2} \times SO_{4d^2}$.

5. Symmetric space over finite fields

In this section, we apply the theory of symmetric space over finite fields to the problem of determining the scalars ζ_x occurring in Lusztig's conjecture (Theorem 1.5)] for G^{F^2}.

5.1.   Let G be a connected reductive group over a finite field $\mathbf{F}_q$ with Frobenius map F. We consider the symmetric space G^{F^2}/G^F. For a class function f on G^{F^2}, we define $m_2(f)$ by

$$m_2(f) = \langle \mathrm{Ind}_{G^F}^{G^{F^2}} 1, f \rangle_{G^{F^2}} = \frac{1}{|G^F|} \sum_{x \in G^F} f(x).$$

In the case where G has a connected center, $m_2(\rho)$ is determined by Kawanaka [K], Lusztig [L4] for any irreducible character ρ of G^{F^2}.

5.2.   Let C be an F-stable conjugacy class in G. Take $x \in C^F$ and let $A_G(x)$ be the component group of $Z_G(x)$ as before. F acts naturally on $A_G(x)$. We assume that F acts trivially on $A_G(x)$. Then the set of G^F-conjugacy classes in C^F is in bijection with the set $A_G(x)/\sim$ of conjugacy classes in $A_G(x)$. The correspondence is given as follows; for each $a \in A_G(x)$, take a representative $\dot{a} \in Z_G(x)$. There exists $h_a \in G$ such that $h_a^{-1} F(h_a) = \dot{a}$. Then $x_a = h_a x h_a^{-1}$ in contained in C^F, and the set $\{x_a \mid a \in A_G(x)/\sim\}$ gives a complete set of representatives of the G^F-conjugacy classes in C^F.

The above description works also for the case of C^{F^2}. We denote by $\{y_a \mid a \in A_G(x)/\sim\}$ the set of G^{F^2}-conjugacy classes in C^{F^2}. We define a class function f_τ on G^{F^2} for each $\tau \in A_G(x)^\wedge$ as follows.

$$(5.2.1) \qquad f_\tau(g) = \begin{cases} \tau(a) & \text{if } g \text{ is } G^{F^2}\text{-conjugate to } y_a, \\ 0 & \text{if } g \notin C^{F^2}. \end{cases}$$

We have the following lemma.

LEMMA 5.3. *Let $\tau \in A_G(x)^\wedge$ be a linear character such that $\tau^2 = 1$. Then we have*

$$m_2(f_\tau) = |C^F|/|G^F|.$$

PROOF. Take $g \in C^F$. Then g is G^F-conjugate to an $x_a \in C^F$ for some $a \in A_G(x)$. Then there exists $h \in G$ such that $g = hxh^{-1}$ and that $h^{-1}F(h) = \dot{a} \in Z_G(x)$. Since F acts trivially on $A_G(x)$, we may choose $\dot{a} \in Z_G(x)^F$. We have

$$h^{-1}F^2(h) = h^{-1}F(h) \cdot F(h^{-1}F(h)) = \dot{a}F(\dot{a}) = \dot{a}^2.$$

It follows that any element $g \in C^F$ is G^{F^2}-conjugate to g_{a^2} for some $a \in A_G(x)$. Hence we have $f_\tau(g) = \tau(a^2) = 1$ by our assumption. We have

$$m_2(f_\tau) = \frac{1}{|G^F|} \sum_{g \in C^F} f_\tau(g) = |C^F|/|G^F|$$

as asserted. The lemma is proved. □

5.4.   Assume that G is as in 3.1, and we use the notation there. Let $A = \mathrm{IC}(\overline{C}, \mathcal{E})[\dim C]$ be the cuspidal character sheaf on G. (Here $\mathcal{E}$ represents one of the simple local systems $\mathcal{E}, \mathcal{E}', \ldots$ on C if there exist more than one). Then A is F-stable. Let $\rho \in A_G(g)^\wedge$ be the irreducible character corresponding to $\mathcal{E}$. Note that F acts trivially on $A_G(g)$. We consider the class function f_ρ on $G_{\mathrm{uni}}^{F^2}$ defined as in (5.2.1) for $\tau = \rho$. We show that

$$(5.4.1) \qquad m_2(f_\rho) = q^{-(\dim G - \dim C)}.$$

In fact thanks to Lemma 3.2, one can apply Lemma 5.3 and we have $m_2(f_\rho) = |C^F|/|G^F|$. Since F acts trivially on $A_G(g)$, C^F splits into several G^F-conjugacy classes, which are parametrized by $A_G(g)/\sim$. Let $a_1, \ldots, a_r$ be the representatives of the conjugacy classes in $A_G(g)$, and let $\mathcal{C}_i$ the G^F-conjugacy classes in C^F corresponding to a_i. We choose $g_i \in \mathcal{C}_i$. Since $|Z_G(g_i)^F| = |A_A(a_i)||Z_G^0(g_i)^F|$, where $A = A_G(g)$, we have

$$|C^F|/|G^F| = \sum_{i=1}^r |\mathcal{C}_i|/|G^F| = \sum_{i=1}^r |Z_A(a_i)|^{-1}|Z_G^0(g_i)^F|^{-1}.$$

Here $Z_G^0(g_i)^F \simeq Z_{H_i}^0(u_i)^F$, where $g_i = s_i u_i = u_i s_i$ and $H_i = Z_G^0(s_i)$. Note that $Z_{H_i}^0(u_i) \simeq Z_H^0(u)$, and since $u \in C_0$, where $(C_0, \mathcal{E}_0)$ is a cuspial pair with unipotent support, it is known by [L2,I, Prop.3.12] that $Z_H^0(u)$ is a unipotent group. It follows that

$$|Z_G^0(g_i)^F| = |Z_{H_i}^0(u_i)^F| = q^{\dim H - \dim C_0} = q^{\dim G - \dim C}.$$

Hence we have

$$|C^F|/|G^F| = q^{-(\dim G - \dim C)} \sum_{i=1}^{r} |Z_A(a_i)|^{-1} = q^{-(\dim G - \dim C)}.$$

This proves (5.4.1).

Let $\varphi_0 : (F^2)^* \mathcal{E} \xrightarrow{\sim} \mathcal{E}$, $\varphi : (F^2)^* A \xrightarrow{\sim} A$ and $\phi_A : (F^2)^* A \xrightarrow{\sim} A$ be as in 3.1, but replacing F by F^2. Then the characteristic function $\chi_{\mathcal{E}, \varphi_0}$ on C^{F^2} coincides with $f_\rho|_{C^{F^2}}$. Since A is clean, the function $\chi_{A, \varphi}$ coincides with $f_\rho|_{\overline{C}^{F^2}}$. It follows that χ_A coincides with $q^{\dim G - \dim C} f_\rho$. This implies that

LEMMA 5.5. *Let A be a cuspidal character sheaf of G. Then we have*

$$m_2(\chi_A) = 1.$$

5.6. Let $s \in T^*$ be such that the class $\{s\}$ is F^2-stable. Then there exists $s_0 \in \{s\}$ such that $F^2(s_0) = s_0$. Let $H = Z_G^0(s_0)$. Then H is an F^2-stable reductive subgroup of G. It is known, since the center of G is connected, that there exists a natural bijection $\mathcal{E}(G^{F^2}, \{s\}) \leftrightarrow \mathcal{E}(H^{F^2}, \{1\})$, $\rho \leftrightarrow \rho_{\mathrm{uni}}$. Concerning the values of $m_2(\rho)$ and $m_2(\rho_{\mathrm{uni}})$, the following result is known. (For unipotent characters, we follow the notation in Section 4.)

THEOREM 5.7 ([K], [L4]). *Let G be a connected classical group with connected center. Then*

(i) *If there does not exist $s_0 \in \{s\}^{F^2}$ such that $F(s_0) = s_0^{-1}$, then $m_2(\rho) = 0$ for any $\rho \in \mathcal{E}(G^{F^2}, \{s\})$. If there exists such s_0, then under the notation of 5.6, $m_2(\rho) = m_2(\rho_{\mathrm{uni}})$ for any $\rho \in \mathcal{E}(G^{F^2}, \{s\})$.*

(ii) *Assume that G is of type B_n or C_n. Let $\mathcal{F}$ be a family in Φ_n such that $|Z_1| = 2d_1 + 1$ (cf. 4.2). Then we have*

$$m_2(\rho_\Lambda) = \begin{cases} 2^{d_1} & \text{if } \Lambda \text{ is special,} \\ 0 & \text{otherwise.} \end{cases}$$

(iii) *Assume that G is of type D_n. Let $\mathcal{F}$ be a non-degenerate family in Φ_n^+ such that $|Z_1| = 2d_1$ (cf. 4.2). Then we have*

$$m_2(\rho_\Lambda) = \begin{cases} 2^{d_1 - 1} & \text{if } \Lambda \text{ is special,} \\ 0 & \text{otherwise.} \end{cases}$$

If $\mathcal{F} = \{\Lambda, \Lambda'\}$ is a degenerate family, then we have

$$m_2(\rho_\Lambda) = m_2(\rho_{\Lambda'}) = \begin{cases} 1 & \text{if } F \text{ is of split type,} \\ 0 & \text{otherwise.} \end{cases}$$

In view of (4.2.2), we have the following corollary.

COROLLARY 5.8 ([L4]). *Let G be a classical group of split type. Assume that $\mathcal{F}$ is a non-degenerate family. Then for any $\Lambda \in \mathcal{F}$, we have $m_2(R_\Lambda) = 1$.*

5.9. Let A be as in 5.4 and assume that $A \in \widehat{G}_{\mathcal{L}}$. Let $s \in T^*$ be such that the class $\{s\}$ corresponds to $\mathcal{L}$ via f in 1.2. Then $s^2 = 1$. Let R_0 be the almost character of G^F corresponding to A as given in Theorem 1.5. Then by Lemma 5.5, we have $m_2(R_0) \neq 0$. Note that R_0 is a linear combination of irreducible characters contained in $\mathcal{E}(G^{F^2}, \{s\})$. Thus by Theorem 5.7, (i), there exists $s_0 \in \{s\}$ such that $s_0 \in G^{F^2}$ and that $F(s_0) = s_0^{-1} = s_0$. Then $H = Z_G^0(s_0)$ is an F-stable reductive subgroup of G, and so H is split over $\mathbf{F}_{q^2}$. Recall that the almost character R_0 is given as in (1.3.1). Then it is known that $(-1)^{l(z_1)} = \sigma(H)\sigma(G)$, where $\sigma(G)$ is a split rank of G with respect to $\mathbf{F}_{q^2}$, and vice versa for H. Since G, H are split, we have $l(z_1) = \sigma(H)\sigma(G) = 1$. Let R_0^H be the almost character of H^{F^2} obtained from R_0 under the correspondence $\rho \leftrightarrow \rho_{\mathrm{uni}}$. Then we see that

$$(5.9.1) \qquad\qquad m_2(R_0) = m_2(R_0^H) = 1.$$

In fact, the first equality follows from Theorem 5.7, (i), together with the fact that $l(z_1) = 1$. The second equality follows from Theorem 5.7, (ii), (iii).

By Theorem 1.5, we know that $R_0 = \zeta\chi_A$ with some scalar ζ, if R_0 is the almost character corresponding to the character sheaf A. Since $m_2(R_0) = \zeta m_2(\chi_A)$, we have the following theorem by combining Lemma 5.5 with (5.9.1).

THEOREM 5.10. *Let G be an adjoint simple group of classical type. Let A be a cuspidal character sheaf, and $\chi_A = \chi_{A,\phi_A}$ be the characteristic function of A on G^{F^2} (defined as in 3.1). Let R_0 be the almost character of G^{F^2} corresponding to A. Then we have*

$$R_0 = \chi_A.$$

As a corollary we have the following result, which holds without any restriction on p nor q.

COROLLARY 5.11. *Let G be a connected classical group with connected center. Then the constants ζ_x appearing in Lusztig's, conjecture (Theorem 1.5) can be determined for G^{F^2} in the following cases; under the notation of 1.6, assume that $W_\delta = Z_\delta$. Then we have*

$$R_{x_E} = (-1)^{\dim \Sigma} \chi_{A_E}$$

for any $E \in W_\delta^\wedge$. In other words, we have $\zeta_{x_E} = (-1)^{\dim \Sigma}$.

PROOF. Lemma 1.8 together with the argument in 1.9 shows that the determination of ζ_x is reduced to the case of ζ_0 (the one corresponding to the cuspidal character sheaf) in the case of adjoint simple groups. We know that $\zeta_0 = 1$ by Theorem 5.10. Then the corollary follows from Lemma 1.8, together with 1.9 since $L_{w_1}^{F^2} = L^{F^2}$. $\qquad\square$

REMARK 5.12. In the case where $\overline{G} = PSp_{2n}$ or SO_{2n+1}, Corollary 5.11 gives a complete answer for the determination of constants ζ_x for G^{F^2} since we have always $W_\delta = Z_\delta$ in that case. In the case where $\overline{G} = PSO_{2n}$, the corollary holds if R_{x_E} is a linear combination of unipotent characters, i.e., if $A_{x_E} \in \widehat{G}_{\overline{\mathbf{Q}}_l}$. But it happens that $W_\delta \neq Z_\delta$ for some δ.

6. From G^{F^2} to G^F

6.1.   The results Theorem 5.10 and Corollary 5.11 in the previous section are only valid for the group G^{F^2}. In this section, by using a certain specialization argument, we extend those results to the group G^F as far as χ_A are concerned with unipotent characters. In the case of $p = 2$, this implies the extension of Theorem

5.10 and Corollary 5.11 for the group G^F of split type. We have the following theorem.

THEOREM 6.2. *Let G be a classical group of split type over $\mathbf{F}_q$. Let A be a cuspidal character sheaf contained in $\widehat{G}_{\bar{\mathbf{Q}}_l}$, and $\chi_A = \chi_{A,\phi_A}$ be the characteristic function of A over G^F. Let R_0 be the almost character of G^F corresponding to A. Then we have*

$$R_0 = \chi_A.$$

As a corollary, we have

COROLLARY 6.3. *Let $G = Sp_{2n}$ or SO_{2n} with $p = 2$. Assume that G^F is of split type. Then the constants ζ_x appearing in Lusztig's conjecture (Theorem 1.5) can be determined completely for Sp_{2n}, and partly for SO_{2n}. More precisely, under the notation of 1.6, assume that $W_\delta = Z_\delta$. Then we have*

(6.3.1) $$R_{x_E} = (-1)^{\dim \Sigma} \chi_{A_E}$$

for any $E \in W_\delta^\wedge$. In particular, (6.3.1) holds if $A_E \in \widehat{G}_{\bar{\mathbf{Q}}_l}$ in the case where $G = SO_{2n}$.

PROOF. As in the proof of Corollary 5.11, the determination of ζ_x is reduced to that of ζ_0. Assume that $G = Sp_{2n}$ or SO_{2n} with $p = 2$. In this case, it is known by 3.1, (c), (d), that a unique cuspidal character sheaf (if it exists) is always contained in $\widehat{G}_{\bar{\mathbf{Q}}_l}$. Hence Theorem 6.2 can be applied, and the corollary follows from Lemma 1.8 (see also Remark 5.12). ☐

REMARK 6.4. A similar argument as in Corollary 6.3 works also for the case where $p \neq 2$. In particular the formula (6.3.1) holds for the case where $A_E \in \widehat{G}_{\bar{\mathbf{Q}}_l}$. Thus we rediscover the results in Theorem 6.2 in [S2]. Although the argument given in [S2] can not be applied to the case where $p = 2$, the proof here works simultaneously for arbitrary p.

6.5. The remainder of this paper is devoted to the proof of Theorem 6.2. In what follows, we assume that G is a classical group containing a cuspidal unipotent character. Hence G is of type B_n or C_n with $n = d^2 + d$, or of type D_n with $n = 4d^2$. We assume further that G^F is of split type. We follow the notation in Section 4.

Let I_q be the G^F-module $\mathrm{Ind}_{B^F}^{G^F} 1$ induced from B^F to G^F. Then the irreducible component of I_q is in bijective correspondence with $W^\wedge$. We denote by ρ_E the irreducible G^F-module occurring in V_q corresponding to $E \in W^\wedge$. ρ_E gives a unipotent character, which we denote by ρ_{Λ_E} with $\Lambda_E \in X_n$. Let $\mathcal{H}$ be the Iwahori-Hecke algebra over $\mathbf{Q}[u^{1/2}, u^{-1/2}]$ associated to the Coxeter system (W, S) with generators $\{T_s \mid s \in S\}$. $\mathcal{H}$ has a basis $\{T_w \mid w \in W\}$, where T_w is defined as $T_w = T_{s_{i_1}} \ldots T_{s_{i_k}}$ for a reduced expression $w = s_{i_1} \cdots s_{i_k}$. $\mathcal{H}$ is characterized by the following properties;

$$\begin{cases} (T_s - u)(T_s + 1) = 0, \\ T_s T_w = T_{sw} \quad \text{if } l(sw) = l(w) + 1, \end{cases}$$

where $l : W \to \mathbf{Z}_{>0}$ is the length function of W.

In the case of type B_n or D_n, the generator set S of W is described as follows. Assume that W is the Weyl group of type B_n. Then W is realized as a group of signed permutations of $I = \{1, \bar{1}, 2, \bar{2}, \ldots, n, \bar{n}\}$. The set S of generators is given as $S = \{s_0, s_1, \ldots, s_{n-1}\}$ with $s_0 = (1, \bar{1}), s_1 = (1, 2), \ldots, s_{n-1} = (n-1, n)$, and we denote by T_i the generator of $\mathcal{H}$ corresponding to s_i. Note that the subalgebra of $\mathcal{H}$ generated by $T_1, \ldots, T_{n-1}$ is isomorphic to the Iwahori-Hecke algebra of type A_{n-1}. Next assume that W is of tyep D_n. Then W is a subgroup of the Weyl group of type B_n, generated by $s_0' = (1, \bar{1})(2, \bar{2}), s_1, \ldots, s_{n-1}$. We denote by $T_0', T_1, \ldots, T_{n-1}$ the corresponding generators of $\mathcal{H}$.

The endomorphism algebra $\operatorname{End}_{G^F} I_q$ is isomorphic to the specialized algebra $\bar{\mathbf{Q}}_l \otimes \mathcal{H}$ via the algebra homomorphism $\mathbf{Q}[u^{1/2}, u^{-1/2}] \to \bar{\mathbf{Q}}_l$ by $u \mapsto q$, which we denote by $\mathcal{H}_q$. We denote by E_q the irreducible representation of $\mathcal{H}_q$ corresponding to $E \in W^\wedge$. Now I_q has a structure of $G^F \times \mathcal{H}_q$-module, and the trace for $g \in G^F, T_w \in \mathcal{H}_q$ is written as

$$(6.5.1) \qquad \operatorname{Tr}((g, T_w), I_q) = \sum_{E \in W^\wedge} \operatorname{Tr}(g, \rho_E) \operatorname{Tr}(T_w, E_q).$$

By replacing $\rho_E = \rho_{\Lambda_E}$ by R_Λ by using (4.2.3), we have

$$(6.5.2) \qquad \operatorname{Tr}((g, T_w), I_q) = \sum_{\Lambda \in X_n} f_\Lambda(w) R_\Lambda(g),$$

where

$$(6.5.3) \qquad f_\Lambda(w) = \sum_{E \in W^\wedge} \{\Lambda, \Lambda_E\} \operatorname{Tr}(T_w, E_q).$$

It is known that $\operatorname{Tr}(T_w, E_q)$ is a polynomial in q in the sense of 2.3. Hence $f_\Lambda(w)$ is also a polynomial in q. We are interested in $f_\Lambda(w)$ in the case where Λ is the cuspidal symbol Λ_c, and we want to find some special $w \in W$ such that $f_{\Lambda_c}(w) \neq 0$. Let W be the Weyl group of type B_n or D_n. Then any element of W can be expressed as a product of positive cycles and negative cycles, where the number of negative cycles is even if W is of type D_n. We have the following proposition.

PROPOSITION 6.6. *There exists an element $w \in W$ such that $f_{\Lambda_c}(w) \neq 0$, where either w is a Coxeter element in W, or w contains a positive cycle of length ≥ 2.*

6.7. The proof of the proposition will be given in Section 7. Here assuming the proposition, we continue the proof of the theorem. We prove the theorem by induction on the semisimple rank of G, and so we assume that the theorem holds for the classical groups of the smaller semisimple rank. Let A be the cuspidal character sheaf on G as in the theorem, and let C be the conjugacy class which is the support of A. We choose $g = su = us \in C^F$ as in 3.1. We choose $w \in W$ as in Proposition 6.6. We consider the equation (6.5.2) simultaneously for the groups G^{F^m} for any integer $m \geq 1$. Note that $g \in C^{F^m}$ and it has a uniform description for any $m \geq 1$ since the split unipotent element for G^F is split for any extended group G^{F^m}. Then we can write (6.5.2) as

$$(6.7.1) \qquad \operatorname{Tr}((g, T_w), I_{q^m}) = \sum_{\Lambda \neq \Lambda_c} f_\Lambda(w)(q^m) R_\Lambda^{(m)}(g) + f_{\Lambda_c}(w)(q^m) R_{\Lambda_c}^{(m)}(g),$$

where T_w is an element of $\mathcal{H}_{q^m}$, and $R_\Lambda^{(m)}$ denotes the almost character of G^{F^m}. By induction hypothesis and by Remark 6.4, the formula (6.3.1) holds for $R_\Lambda^{(m)}$ if $\Lambda \neq \Lambda_c$. We show the following lemma.

LEMMA 6.8. *Assume that $\Lambda \neq \Lambda_c$. Then under the induction hypothesis, $R_\Lambda^{(m)}(g)$ is a rational function in q.*

PROOF. Let A_Λ the character sheaf corresponding to R_Λ, and denote by $\chi_{A_\Lambda}^{(m)}$ the characteristic function on G^{F^m} associated to A_Λ. In view of (6.3.1), it is enough to show that $\chi_{A_\Lambda}^{(m)}(g)$ ia a rational function in q. Since $A \in \hat{G}_{\bar{\mathbf{Q}}_l}$, by applying Corollary 4.10 in [S2] ($a_0 = 1$ in the notation there since F is of split type), the computation of $\chi_{A_\Lambda}^{(m)}(g)$ is reduced to that of $\chi_{A_\Lambda}^{(m)}(u)$ for a subgroup H^{F^m}, where H is a connected centralizer of some semisimple element in G. So we consider $\chi_{A_\Lambda}^{(m)}$ on $G_{\mathrm{uni}}^{F^m}$. A_Λ is a direct summand of a certain complex $K = \mathrm{ind}_P^G A_0$, where $A_0 \in \hat{L}_{\bar{\mathbf{Q}}_l}$ is a cuspidal character sheaf of a Levi subgroup L. Then the computation of $\chi_{A_\Lambda}^{(m)}|_{G_{\mathrm{uni}}^{F^m}}$ is reduced to the computation of the generalized Green functions of G^{F^m} associated to L^{F^m}. Hence by Corollary 2.4, we obtain the lemma. $\qquad\square$

6.9. Next we consider the left hand side of (6.7.1). The G^F-module I_q is a permutation representation of G^F on G^F/B^F. Since the action of G^F on G^F/B^F is independent of the isogeny, we may assume that $G = Sp_N$ or SO_N and that $g \in G^F$. Let V be the vector space over $\mathbf{F}_q$ of $\dim V = N$, equipped with a non-degenerated alternating form (resp. symmetric bilinear form) f on V if $G = Sp_N$ (resp. SO_N). In the case where $G = SO_N$ with $p = 2$ and N is even, we also consider the quadratic form Q on V. Now the set G^F/B^F may be identified with the set of flags $\mathcal{F}_q$ as follows; A flag $\mathbf{F} = (V_0 \subset V_1 \subset \cdots \subset V_n)$ is a sequence of subspaces of V such that $\dim V_i = i$ and that V_i are isotropic with respect to f, where $N = 2n, 2n+1$ or $2n$ according to the cases where $G = Sp_{2n}, SO_{2n+1}$ or SO_{2n}. In the case where $p = 2$ and $G = SO_{2n}$, we assume further that the restriction of Q on V_n is zero. Now in the case where $G = Sp_{2n}$ or SO_{2n+1}, $\mathcal{F}_q$ consists of all flags on V. G^F acts naturally on $\mathcal{F}_q$ via $x : (V_0 \subset \cdots \subset V_n) \mapsto (xV_0 \subset \cdots \subset xV_n)$ for $x \in G^F$, and the G^F-set $\mathcal{F}_q$ is identified with the G^F-set G^F/B^F. In the case where $G = SO_{2n}$, we define $\tilde{\mathcal{F}}_q$ as the set of all flags on V as above, then G^F acts on $\tilde{\mathcal{F}}_q$ with two G^F-orbits, $\mathcal{F}_q$ and $\mathcal{F}_q'$. Either of them can be identified with G^F/B^F, and we have a natural bijection between $\mathcal{F}_q$ and $\mathcal{F}_q'$, which is given in the form $(V_0 \subset V_1 \subset \cdots \subset V_n) \mapsto (V_0 \subset V_1 \subset \cdots \subset V_n')$, (only the term V_n is changed to V_n').

We consider the vector space $\mathcal{J}_q$ over $\bar{\mathbf{Q}}_l$ with basis $\mathcal{F}_q$, which is identified with I_q. By the identification $I_q \simeq \mathcal{J}_q$, $\mathcal{H}_q$ acts on $\mathcal{J}_q$, whose action is given as follows; let $\mathbf{F} = (V_0 \subset \cdots \subset V_n)$. For $i = 1, \ldots, n-1$, we have

$$\mathbf{F}T_{n-i} = \sum_{W \neq V_i} (V_0 \subset \cdots \subset V_{i-1} \subset W \subset V_{i+1} \subset \cdots \subset V_n),$$

where the sum is taken over all the isotopic subspaces W such that $V_{i-1} \subset W \subset V_{i+1}$ and that $\dim W = i$, $W \neq V_i$. In the case of B_n, we have

$$\mathbf{F}T_0 = \sum_{W \supset V_{n-1}} (V_0 \subset \cdots \subset V_{n-1} \subset W),$$

where the sum is taken over all the isotropic subspaces W such that $\dim W = n$ and $W \neq V_n$. In the case of type D_n, we have

$$\mathbf{F}T_0' = \sum_{W \supset W' \supset V_{n-2}} (V_0 \subset \cdots \subset V_{n-2} \subset W' \subset W),$$

where the sum is taken over the isotropic subspaces $W \supset W'$ such that $\dim W = n, \dim W' = n - 1$ and W' contains V_{n-2} and some more conditions.

It follows from the description of the action of $\mathcal{H}$ on $\mathcal{I}_q$, we see that

(6.9.1) Assume that $w \in W$ contains a positive cycle of length ≥ 2. Then there exists $k \geq 1$ such that for any $\mathbf{F} = (V_0 \subset \cdots \subset V_n)$, $\mathbf{F}T_w$ is a linear combination of $\mathbf{F}' = (V_0' \subset \cdots \subset V_n')$ such that $V_k = V_k'$.

We show the following lemma.

LEMMA 6.10. *Under the induction hypothesis,* $\mathrm{Tr}\,((g, T_w), I_{q^m})$ *is a rational function in* q.

PROOF. The following argument was inspired by [HR2], where the combinatorial properties of $\mathrm{Tr}\,((u, T_w), I_q)$ is discussed in the case of $GL_n(\mathbf{F}_q)$ with a unipotent element u. First assume that w contains a positive cycle of length ≥ 2. Then by (6.9.1), for any flag $\mathbf{F} = (V_0 \subset \cdots \subset V_n)$, $\mathbf{F}T_w$ is a linear combination of $\mathbf{F}' = (V_0' \subset \cdots \subset V_n')$ such that $V_k = V_k'$. We now prepare a notation. If $v \in \mathcal{J}_q$, and $\mathbf{F} \in \mathcal{F}_q$, we denote by $v|_\mathbf{F}$ the coefficient of $\mathbf{F}$ in the expression of v as a linear combination of base vectors. Let $\mathbf{F} = (V_0 \subset \cdots \subset V_n) \in \mathcal{F}_q$, and assume that $g\mathbf{F}T_w|_\mathbf{F} \neq 0$. Since $g\mathbf{F}T_w|_\mathbf{F} = \mathbf{F}T_w|_{g^{-1}\mathbf{F}}$, $g^{-1}\mathbf{F} = (V_0' \subset \cdots \subset V_n')$ is of the form that $V_k = V_k' = g^{-1}V_k$. It follows that V_k is stabilized by g. Thus we have

$$\mathrm{Tr}\,((g, T_w), I_q) = \sum_{\mathbf{F} \in \mathcal{F}_q} g\mathbf{F}T_w|_\mathbf{F}$$

$$= \sum_W \sum_{\mathbf{F} = (V_0 \subset \cdots \subset W \subset \cdots \subset V_n)} g\mathbf{F}T_w|_\mathbf{F}$$

$$= \sum_W \sum_{\mathbf{F}' = (V_0 \subset \cdots \subset W)} g\mathbf{F}'T_w|_{\mathbf{F}'} \sum_{\mathbf{F}'' = (W \subset \cdots)} g\mathbf{F}''T_w|_{\mathbf{F}''}$$

where W runs over all the isotropic subspaces in V such that $\dim W = k$ and that $gW = W$. Let $H' = GL(W)$, and H'' be the group of isometries $Sp(\overline{W})$ or $SO(\overline{W})$ for $\overline{W} = W^\perp/W$. Let $I_q^W \simeq \mathcal{J}_q^W$ be the corresponding induced modules for H'^F, and similarly define $I_q^{\overline{W}} \simeq \mathcal{J}_q^{\overline{W}}$ for H''^F. g acts naturally on W (resp. on $\overline{W}$), and we denote by $g_W \in H'^F$ (resp. $g_{\overline{W}} \in H''^F$) the corresponding elements. Also the action of T_w on $\mathcal{I}_q$ induces an action on $\mathcal{J}_q^W$ (resp. on $\mathcal{J}_q^{\overline{W}}$) which is given by $T_{w'}$ (resp. $T_{w''}$) with an element w' (resp. w'') in the Weyl group of H' (resp. H''). Then the last sum can be written as

(6.10.1) $$\mathrm{Tr}\,((g, T_w), I_q) = \sum_W \mathrm{Tr}\,((g_W, T_{w'}), I_q^W)\,\mathrm{Tr}\,((g_{\overline{W}}, T_{w''}), I_q^{\overline{W}}).$$

Let $\mathbf{F}_0 = (V_0^0 \subset \cdots \subset V_n^0)$ be the standard flag whose stabilizer in G^F is B^F, and put $W_k = V_k^0$. Then there exists an F-stable maximal parabolic subgroup P of G containing B such that P^F is the stabilizer of W_k in G^F. Let L be an F-stable Levi subgroup of P containing T. Then L is isomorphic to $L' \times L''$, where $L' = GL_k$ and L'' is a similar group as G of rank $n - k$. Let $g_1', \ldots, g_r'$ (resp.

$g_1'', \dots, g_s''$) be representatives of the conjugacy classes in L'^F (resp. L''^F) such that $g_{ij} = (g_i', g_j'') \in L^F$ is conjugate to g under G^F. We denote by X_q^{ij} the set of W such that $W = xW_k$ and that $x^{-1}gx$ is conjugate to g_{ij} in L^F. Then (6.10.1) implies that

$$(6.10.2) \qquad \mathrm{Tr}\,((g, T_w), I_q) = \sum_{i=1}^{r} \sum_{j=1}^{s} |X_q^{ij}|\, \mathrm{Tr}\,((g_i', T_{w'}), I_q^{L'})\, \mathrm{Tr}\,((g_j'', T_{w''}), I_q^{L''}),$$

where $I_q^{L'}, I_q^{L''}$ are corresponding induced modules for L'^F, L''^F. Choose $x_{ij} \in G^F$ such that $g = x_{ij}g_{ij}x_{ij}^{-1}$ for i, j. Then X_q^{ij} is in bijection with the set $Z_G(g)^F x_{ij} L^F / L^F$, hence $|X_q^{ij}| = |Z_G(g)^F|/|Z_L(g_{ij})^F|$.

We now consider (6.10.2) for any G^{F^m}. Then the choice of representatives $g_{ij} \in G^{F^m}$ does not depend on m, and we see that $|X_{q^m}^{ij}|$ is a rational function in q. On the other hand, one can write as in (6.5.1)

$$\mathrm{Tr}\,((g_j'', T_{w''}), I_q^{L''}) = \sum_{E \in W_{L''}{}^{\wedge}} \mathrm{Tr}\,(g_j'', \rho_E)\, \mathrm{Tr}\,(T_{w''}, E_q),$$

where $W_{L''}$ is the Weyl group of L''. By induction hypothesis, $R_\Lambda^{(m)}(g_j'')$ is a rational function in q for any almost character R_Λ of L''^F. Hence $\mathrm{Tr}\,(g_j'', \rho_E^{(m)})$ is a rational function in q. It follows that $\mathrm{Tr}\,((g_j'', T_{w''}), I_{q^m}^{L''})$ is a rational function in q. Similarly, and as it is known since $L' = GL_k$, $\mathrm{Tr}\,((g_i', T_{w'}), I_{q^m}^{L'})$ is a rational function in q. Thus we conclude that $\mathrm{Tr}\,((g, T_w), I_{q^m})$ is a rational function in q as asserted.

Next assume that w is a Coxeter element of W. We note that

$$(6.10.3) \qquad \mathrm{Tr}\,((x, T_w), I_q) = \begin{cases} q^r & \text{if } x_u \text{ is regular unipotent}, \\ 0 & \text{otherwise}, \end{cases}$$

for $x \in G^F$, where x_u is the unipotent part of x and r is the semisimple rank of G. In fact (6.10.3) is discussed in [HR2, Prop. 3.2] in the case where $G = GL_n$. The argument there works in general if we notice that $Z_G(v) = Z_U(v)$ for a regular unipotent element $v \in U^F$ and that $|Z_U^F(v)| = q^r$, where U is the unipotent radical of B. (6.10.3) implies that $\mathrm{Tr}\,((g, T_w), I_{q^m})$ is a polynomial in q. Hence the lemma holds.   □

6.11 We now prove the theorem. In the formula (6.7.1) the left hand side is a rational function in q by Lemma 6.10. For for $\Lambda \neq \Lambda_c$, $R_\Lambda^{(m)}(g)$ is a rational function in q by Lemma 6.8. Since $f_\Lambda(w)$ is a polynomial in q, and $f_{\Lambda_c}(w) \neq 0$ by Proposition 6.6, we see that $R_{\Lambda_c}^{(m)}(g)$ is a rational function in q. By Theorem 1.5, one can write $R_{\Lambda_c}^{(m)}(g) = \zeta^{(m)} \chi_{A_0}^{(m)}(g)$ with some $\zeta^{(m)} \in \bar{\mathbf{Q}}_l^*$ of absolute value 1, where $A_0 = A_{\Lambda_c}$ is the cuspidal character sheaf. We know that $\chi_{A_0}^{(m)}(g)$ is a non-zero polynomial in q. We also know by Theorem 5.10 that $\zeta^{(m)} = 1$ for even m. It follows that $R_{\Lambda_c}^{(m)}(q)/\chi_{A_0}^{(m)}(g)$ is a rational function in q, and takes the value 1 for any power of q^2. Hence $R_{\Lambda_c}^{(m)}(g) = \chi_{A_0}^{(m)}(g)$ for any m, and we have $\zeta^{(m)} = 1$. This shows that $R_{\Lambda_c} = \chi_{A_0}$, and the theorem is proved (modulo Proposition 6.6).

7. Proof of Proposition 6.6

7.1. Recall that Φ_n^1 is the set of symbols of rank n and defect 1 as in 4.1. In the case where W is the Weyl group of type B_n, the set $W^\wedge$ is in bijection with Φ_n^1. The correspondence is given as follows; let $\mathcal{P}_n$ be the set of of double partitions (λ, μ) such that $|\lambda| + |\mu| = n$. Then $W^\wedge$ is parametrized by $\mathcal{P}_n$. For a double partition $(\lambda, \mu) \in \mathcal{P}_n$, we write $\lambda = (\lambda_0 \leq \lambda_1 \leq \cdots \leq \lambda_m)$ and $\mu = (\mu_1 \leq \mu_2 \leq \cdots \leq \mu_m)$ with $\lambda_i, \mu_i \geq 0$ for some integer m. We put $a_i = \lambda_i + i$, $b_i = \mu_i + (i-1)$, and define the sets S, T by $S = \{a_0, a_1, \ldots, a_m\}$ and $T = \{b_1, \ldots, b_m\}$. Then $\Lambda = \binom{S}{T} \in \Phi_n^1$, and this gives a bijective correspondence between Φ_n^1 and $\mathcal{P}_n$, and so gives a bijection between Φ_n^1 and $W^\wedge$. As in 6.5, we denote by Λ_E the symbol in Φ_n^1 corresponding to $E \in W^\wedge$. Assume that $n = d^2 + d$, and let Λ_0 be the special symbol in $\mathcal{F}_c$. Then $\Lambda_0 = \Lambda_E$, where E corresponds to $(\lambda, \mu) \in \mathcal{P}_n$ such that $\lambda = (0 \leq 1 \leq 2 \leq \cdots \leq d)$, $\mu = (1 \leq 2 \leq \cdots \leq d)$.

Recall that $\widetilde{\Phi}_n^0$ is the set of symbols of rank n and defect 0 as in 4.1. In the case where W is the Weyl group of type D_n, $W^\wedge$ is in bijection with $\widetilde{\Phi}_n^0$, which is given as follows; let $\widetilde{\mathcal{P}}_n$ be the set of unordered partitions (λ, μ) such that $|\lambda| + |\mu| = n$, where (λ, λ) is counted twice. Then $W^\wedge$ is parametrized by $\widetilde{\mathcal{P}}_n$. For $(\lambda, \mu) \in \widetilde{\mathcal{P}}_n$, we write $\lambda = (\lambda_1 \leq \cdots \leq \lambda_m)$, $\mu = (\mu_1 \leq \cdots \leq \mu_m)$ with $\lambda_i, \mu_j \geq 0$ for some integer $m \geq 1$. We put $a_i = \lambda_i + (i-1)$, $b_i = \mu_i + (i-1)$, and define the sets S, T by $S = \{a_1, \ldots, a_m\}$, $T = \{b_1, \ldots, b_m\}$. Then $\Lambda = \binom{S}{T} \in \widetilde{\Phi}_n^0$ and this gives a bijective correspondence between $\widetilde{\Phi}_n^0$ and $W^\wedge$. As in 6.5, we denote by Λ_E the symbol in $\widetilde{\Phi}_n^0$ corresponding to $E \in W^\wedge$. Assume that $n = 4d^2$ and let Λ_0 be the special symbol in $\mathcal{F}_c$. Then $\Lambda_0 = \Lambda_E$, where E corresponds to $(\lambda, \mu) \in \widetilde{\mathcal{P}}_n$ such that $\lambda = (0 \leq 1 \leq \cdots \leq 2d-1)$, $\mu = (1 \leq 2 \leq \cdots \leq 2d)$

First we show the following lemma.

LEMMA 7.2. *Assume that $E \in W^\wedge$ is such that $\Lambda_E \in \mathcal{F}_c$. If E corresponds to $(\lambda, \mu) \in \mathcal{P}_n$ (resp. $(\lambda, \mu) \in \widetilde{\mathcal{P}}_n$), in the case where W is of type B_n (resp. D_n), then we have*

$$
\{\Lambda_c, \Lambda_E\} = \begin{cases} \dfrac{1}{2^d}(-1)^{|\lambda| + d(d+1)/2} & \text{if } G \text{ is of type } B_n, \\[3mm] \dfrac{1}{2^{2d-1}}(-1)^{|\lambda| + d(2d-1)} & \text{if } G \text{ is of type } D_n. \end{cases}
$$

PROOF. In the case where $\mathcal{F} = \mathcal{F}_c$ is the cuspidal family, $Z_1 = \{0, 1, \ldots, 2d\}$ (resp. $Z_1 = \{0, 1, \ldots, 4d-1\}$) and $M_0 = \{1, 3, \ldots, 2d-1\}$ (resp. $M_0 = \{1, 3, \ldots, 4d-1\}$) if G is of type B_n (resp. D_n) by 4.3. Moreover $\Lambda_c = \Lambda_M$ with $M = \emptyset$. Then $M^\sharp = M_0$, and for any $M' \subset Z_1$ such that $|M'| = d$ (resp. $|M'| = 2d$), we have $M'^\sharp \cap M^\sharp = M_0 - M'$. Now M' is written as $M' = M'_{\text{odd}} \coprod M'_{\text{ev}}$, where M'_{odd} (resp. M'_{ev}) is the subset of M' consisting of odd numbers (resp. even numbers). We have $M' \cap M_0 = M'_{\text{odd}}$ and so $M_0 - M' = M_0 - M'_{\text{odd}}$. Moreover we have $Z_1 - M' = ((Z_1)_{\text{ev}} - M'_{\text{ev}}) \coprod (M_0 - M'_{\text{odd}})$, where $(Z_1)_{\text{ev}}$ is defined similarly. Assume that M' corresponds to $(\lambda, \mu) \in \mathcal{P}_n$ (resp. $(\lambda, \mu) \in \widetilde{\mathcal{P}}_n$). Let $\gamma = \sum_{i=1}^{d} i = d(d+1)/2$

(resp. $\gamma = \sum_{i=1}^{2d-1} i = d(2d-1)$). Then

$$|\lambda| = (\sum_{a \in Z_1 - M'} a) - \gamma$$

$$\equiv (\sum_{a \in M_0 - M'_{\text{odd}}} a) - \gamma \pmod 2$$

$$\equiv |M_0 - M'_{\text{odd}}| - \gamma \pmod 2.$$

This proves the lemma. $\qquad\qquad\qquad\qquad\qquad\qquad\qquad\qquad\qquad\qquad$ □

7.3. We consider $f_{\Lambda_c}(w) \in \mathbb{Z}[u^{1/2}, u^{-1/2}]$ for some $w \in W$, and compute it by making use of the Murnaghan-Nakayama formula for $\mathcal{H}$ due to Halverson and Ram [HR1]. We use the notation in 6.5.

First assume that W is of type B_n and $\mathcal{H}$ is the corresponding Hecke algebra. For each $1 \le k \le n$, we define $L_k \in \mathcal{H}$ inductively as $L_1 = T_0$ and $L_k = T_{k-1}L_{k-1}T_{k-1}$ for $k = 2, \ldots, n$. For $1 \le k < l \le n$, we define

$$R_{kl} = T_k T_{k+1} \cdots T_{l-1}, \qquad R_{k\bar{l}} = L_k T_k T_{k+1} \cdots T_{l-1}.$$

For an element i or $\bar{i}$ in I, we put $|i| = |\bar{i}| = i$. For a sequence $\mathbf{r} = (r_1, \ldots, r_k)$ of elements of I such that $|r_1| < |r_2| < \cdots < |r_k|$, we define $T_{\mathbf{r}} \in \mathcal{H}$ by

$$T_{\mathbf{r}} = R_{1,r_1} R_{|r_1|+1,r_2} \cdots R_{|r_{k-1}|+1,r_k}.$$

Then $T_{\mathbf{r}}$ coincides with $T_{w_{\mathbf{r}}}$, where $w_{\mathbf{r}} \in W$ is given by a cyclic notation of the signed permutation,

$$(7.3.1) \qquad \begin{aligned} w_{\mathbf{r}} = &(1, 2, \ldots, |r_1| - 1, r_1)(|r_1| + 1, |r_1| + 2, \ldots, |r_2| - 1, r_2) \\ &\cdots (|r_{k-1}| + 1, \ldots, |r_k| - 1, r_k). \end{aligned}$$

We also use the following cycle type expression of $w_{\mathbf{r}}$

$$(7.3.2) \qquad w_{\mathbf{r}} = [l_1, \ldots, l_r],$$

where $l_i \in I$ is such that $|l_i| = |r_i| - |r_{i-1}|$ and l_i is barred if r_i is barred. For example, if $\mathbf{r} = (\bar{1}, \bar{4}, 7, 12)$, then $w_{\mathbf{r}} = [\bar{1}, \bar{3}, 3, 5]$.

We now prepare some notation related to the skew diagram. Let λ be a double partition of size n. Apart from the notation in 7.1, we express it as $\lambda = (\lambda^\alpha, \lambda^\beta)$, where $\lambda^\alpha, \lambda^\beta$ are partitions. For $\mu \in \mathcal{P}_n$, we write $\mu \subseteq \lambda$ if $\mu^\alpha \subseteq \lambda^\alpha$ and $\mu^\beta \subseteq \lambda^\beta$. The Young diagram of λ is defined as a pair of Young diagrams of λ^α and λ^β. We often identify the double partition and the corresponding Young diagram. For double partitions $\mu \subseteq \lambda$, the skew diagram $\lambda/\mu = ((\lambda/\mu)^\alpha, (\lambda/\mu)^\beta)$ is defined naturally. For each node x in the skew diagram λ/μ, the content $ct(x)$ is defined as follows;

$$(7.3.3) \qquad ct(x) = \begin{cases} u^{j-i+1} & \text{if } x \text{ is in position } (i,j) \text{ in } (\lambda/\mu)^\alpha, \\ -u^{j-i} & \text{if } x \text{ is in position } (i,j) \text{ in } (\lambda/\mu)^\beta. \end{cases}$$

The skew diagram X is called a border strip if it is connected and does not contain any 2×2 block of nodes ("connected" means that two nodes are connected horizontally or vertically). The skew diagram X is called a broken border strip if its connected components are border strip. Note that a double partition (α, β) with both α, β non-empty consists of two connected components. For a border strip X, a sharp corner is a node with no node above it and no node to its left. A dull corner in a border strip is a node which has a node to its left and a node above it (and so has no node directly northwest of it).

For a skew diagram X, let $\mathcal{C}$ be the set of connected components of X, and put $m = |\mathcal{C}|$, the number of connected components of X. We define $\Delta(X), \overline{\Delta}(X) \in \mathbb{Z}[u^{1/2}, u^{-1/2}]$ as follows;

$$
\Delta(X) = \begin{cases} (u^{1/2} - u^{-/2})^{m-1} \displaystyle\prod_{Y \in \mathcal{C}} (u^{1/2})^{c(Y)-1}(-u^{-1/2})^{r(Y)-1} \\ \qquad\qquad\qquad\qquad \text{if } X \text{ is a broken border strip,} \\ 0 \quad \text{otherwise.} \end{cases}
$$

$$
\overline{\Delta}(X) = \begin{cases} (u^{1/2})^{c(X)-1}(-u^{-1/2})^{r(X)-1} \displaystyle\prod_{y \in DC} ct(y)^{-1} \prod_{z \in SC} ct(z) \\ \qquad\qquad\qquad\qquad \text{if } X \text{ is a (connected) border strip,} \\ 0 \quad \text{otherwise,} \end{cases}
$$

where SC and DC denote the set of sharp corners and dull corners in a border strip, and $r(X)$ (resp. $c(X)$) is the number of rows (resp. columns) in the border strip X.

The Murnaghan-Nakayam formula for $\mathcal{H}$ by Halverson-Ram is given as follows. Note that in the formula below, $l'(w)$ denotes the number of $s_1, \ldots, s_{n-1}$ (excluding s_0) occurring in the reduced expression of $w \in W$

THEOREM 7.4 ([HR1, Theorem 2.20]). *Assume that W is of type B_n. Let E_u^λ be the irreducible representation of $\mathcal{H}$ associated to $\lambda \in \mathcal{P}_n$. Then*

$$
\mathrm{Tr}\,(T_{w_{\mathbf{r}}}, E_u^\lambda) = u^{l'(w_{\mathbf{r}})/2} \sum_{\emptyset = \mu^{(0)} \subseteq \mu^{(1)} \subseteq \cdots \subseteq \mu^{(k)} = \lambda} \Delta(\mu^{(1)})\Delta(\mu^{(2)}/\mu^{(1)}) \cdots \Delta(\mu^{(k)}/\mu^{(k-1)}),
$$

where the sum is taken over all the sequences $\emptyset = \mu^{(0)} \subseteq \mu^{(1)} \subseteq \cdots \subseteq \mu^{(k)} = \lambda$ such that $|\mu^{(k)}/\mu^{(k-1)}| = |r_k| - |r_{k-1}|$ and the factor $\Delta(\mu^{(k)}/\mu^{(k-1)})$ is barred if r_k in $\mathbf{r}$ is barred.

7.5. Next assume that W is the Weyl group of type D_n and $\mathcal{H}$ is the corresponding Hecke algebra. Then under the notation of 6.5, we define L_k' by $L_1' = 1, L_2' = T_0'T_1$, and $L_k' = T_{k-1}L_{k-1}'T_{k-1}$ for $k = 3, \ldots, n$. For $1 \le k < l \le n$, we define

$$
R_{kl} = T_k T_{k+1} \cdots T_{l-1}, \qquad R_{k\bar{l}} = L_k' T_k T_{k+1} \cdots T_{l-1}.
$$

For a sequence $\mathbf{r} = (r_1, \ldots, r_k)$ of elements I such that $|r_1| < |r_2| < \cdots < |r_k|$ and that the even numbers of r_i are barred, we define $T_{\mathbf{r}} \in \mathcal{H}$ by

$$
T_{\mathbf{r}} = R_{1,r_1} R_{|r_1|+1,r_2} \cdots R_{|r_{k-1}|+1,r_k}.
$$

Note that $T_{\mathbf{r}}$ does not always correspond to T_w for some w, but it corresponds to $T_{w_{\mathbf{r}}}$ for $w_{\mathbf{r}} \in W$ as given in (7.3.1) in the following two cases,
  (i) $r_i > 0$ for $i = 1, \ldots, k$,
  (ii) $r_1 = -1, r_2 < 0$ and $r_i > 0$ for $i = 3, \ldots, k$.
In what follows, we only consider $\mathbf{r}$ as above, and so assume that $T_{\mathbf{r}} = T_{w_{\mathbf{r}}}$. Note that $w_{\mathbf{r}}$ is regarded as an element of the Weyl group of type B_n in the notation of 7.3.

As in 7.3, we consider the skew diagram λ/μ, and define $ct'(x)$ by modifying (7.3.2),

$$ct'(x) = \begin{cases} u^{j-i} & \text{if } x \text{ is in position } (i,j) \text{ in } (\lambda/\mu)^{\alpha}, \\ -u^{j-i} & \text{if } x \text{ is in position } (i,j) \text{ in } (\lambda/\mu)^{\beta}. \end{cases}$$

For a skew diagram X, we define $\Delta(X), \overline{\Delta}'(X) \in \mathbb{Z}[u^{1/2}, u^{-1/2}]$ by the formula in 7.3, but for $\overline{\Delta}'(X)$, we modify the definition of $\overline{\Delta}(X)$ by replacing $ct(x)$ by $ct'(x)$. Concerning the irreducible characters of $\mathcal{H}$, we have the following result.

THEOREM 7.6 ([HR1, Theorem 4.21]). *Assume that W is of type D_n. Let E_u^{λ} be the irreducible representation of $\mathcal{H}$ associated to $\lambda \in \mathcal{P}_n$ such that $\lambda^{\alpha} \neq \lambda^{\beta}$. Then $\mathrm{Tr}\,(T_{w_r}, E_u^{\lambda})$ can be computed by the formula in Theorem 7.4 for type B_n, by replacing $\overline{\Delta}(X)$ by $\overline{\Delta}'(X)$.*

7.7. Let $\mathcal{F}_c$ is the cuspidal family as in 4.3, where $\mathcal{F}_c$ is a subset of Φ_n or Φ_n^+. Let $\mathcal{F}_c^1$ (resp. $\mathcal{F}_c^0$) be the set of symbols of defect 1 (resp. defect 0) contained in $\mathcal{F}_c$. First assume that $\mathcal{F}_c \subset \Phi_n$. Let $\mathcal{P}_n^d$ be the set of double partitions λ such that $\Lambda_{E^{\lambda}} \in \mathcal{F}_c$. Then $\lambda \in \mathcal{P}_n^d$ can be written as $\lambda = (\alpha, \beta)$ with $\alpha : \alpha_1 \geq \alpha_2 \geq \cdots \geq \alpha_d \geq \alpha_{d+1} \geq 0$, and $\beta : \beta_1 \geq \beta_2 \geq \cdots \geq \beta_d \geq 0$. Let $\beta^* : \beta_1^* \geq \beta_2^* \geq \cdots$ be the dual partition of β. The following fact is easily checked.

$$(7.7.1) \qquad \mathcal{F}_c^1 \simeq \mathcal{P}_n^d = \{\lambda = (\alpha, \beta) \in \mathcal{P}_n \mid \alpha_i + \beta_{d-i+2}^* = d \text{ for } 1 \leq i \leq d+1\}.$$

Next assume that $\mathcal{F}_c \subset \Phi_n^+$. Let $\widetilde{\mathcal{P}}_n^d$ be the set of double partitions $\lambda \in \widetilde{\mathcal{P}}_n$ such that $\Lambda_{E^{\lambda}} \in \mathcal{F}_c$, where $\lambda = (\alpha, \beta)$ with $\alpha : \alpha_1 \geq \cdots \geq \alpha_d \geq 0$, $\beta : \beta_1 \geq \cdots \geq \beta_d \geq 0$ (in this case always $\alpha \neq \beta$). Let β^* be the dual partition of β. Then we have

$$(7.7.2) \qquad \mathcal{F}_c^0 \simeq \widetilde{\mathcal{P}}_n^d = \{\lambda = (\alpha, \beta) \in \widetilde{\mathcal{P}}_n \mid \alpha_i + \beta_{2d-i+1}^* = 2d \text{ for } 1 \leq i \leq d\}.$$

(7.7.1) shows that $\lambda = (\alpha, \beta) \in \mathcal{P}_{n,2}^d$ is obtained from a diagram $\gamma = (d^{d+1})$ of rectangular shape as follows; take any partition $\alpha \subseteq \gamma$, and let β be the dual of the partition obtained by rearranging the skew diagram γ/α. Similarly, (7.7.2) shows that $\lambda = (\alpha, \beta) \in \widetilde{\mathcal{P}}_n^d$ is obtained from the diagram $\gamma = ((2d)^{2d})$ of rectangular shape by a similar process as above.

For example, in the case of type B_n with $d = 2$, we have $n = d(d+1) = 6$ and

$$\mathcal{P}_6^2 = \{(21; 21), (2^2; 1^2), (1^2; 31), (2^3; -), (2^2 1; 1),$$
$$(21^2; 2), (1^3; 3), (2; 2^2), (1; 32), (-; 3^2)\}.$$

In the case of type D_n with $d = 1$, we have $n = 4d^2 = 4$ and

$$\widetilde{\mathcal{P}}_4^1 = \{(2^2; -), (21^2; 1), (2, 1^2)\}.$$

7.8. Let w_r be an element in W associated to some $\mathbf{r} = (r_1, \ldots, r_k)$ as in (7.3.1). By Lemma 7.2 and by (6.5.3), we have

$$(7.8.1) \qquad f_{\Lambda_c}(w_{\mathbf{r}}) = \delta \sum_{(\alpha, \beta)} (-1)^{|\alpha|} \mathrm{Tr}\,(T_{w_{\mathbf{r}}}, E_u^{(\alpha, \beta)}),$$

where the sum is taken over all $\lambda = (\alpha, \beta)$ in $\mathcal{P}_n^d$ (resp. in $\widetilde{\mathcal{P}}_n^d$), and the constant δ is given as $\delta = (-1)^{d(d+1)/2} 2^{-d}$ (resp. $(-1)^{d(2d-1)} 2^{-2d+1}$) if G is of type B_n (resp. D_n).

We shall compute this sum for some specific choice of $\mathbf{r}$ by applying the Murnaghan-Nakayama formula (Theorem 7.4 or Theorem 7.6). In order to discuss the case B_n and D_n simultaneously, we consider the following setting. Let $\mathcal{P}_n^{a,b}$ be the set of double partitions $\lambda = (\alpha, \beta) \in \mathcal{P}_n$, where $\alpha : \alpha_1 \geq \cdots \geq \alpha_a \geq 0, \beta : \beta_1 \geq \cdots \geq \beta_b \geq 0$ and $n = ab$, such that

$$(7.8.2) \qquad \mathcal{P}_n^{a,b} = \{\lambda = (\alpha, \beta) \in \mathcal{P}_n \mid \alpha_i + \beta_{a-i+1}^* = b \text{ for } 1 \leq i \leq a\}.$$

Hence $\mathcal{P}_n^{a,b}$ is the set of λ contained in the Young diagram $\gamma = (b^a)$ of rectangular shape in the above sense. In particular, $\mathcal{P}_n^{d,d+1}$ coincides with $\mathcal{P}_n^d$, and $\mathcal{P}_n^{2d,2d}$ (under the identification $(\alpha, \beta) = (\beta, \alpha)$) coincides with $\widetilde{\mathcal{P}}_n^d$.

Put

$$(7.8.3) \qquad f_{a,b}(w_{\mathbf{r}}) = \sum_{(\alpha,\beta) \in \mathcal{P}_n^{a,b}} (-1)^{|\alpha|} \operatorname{Tr}(T_{w_{\mathbf{r}}}, E_u^{(\alpha,\beta)}).$$

We take $r_k = n, r_{k-1} = n - (2a + 2b - 6)$ so that in applying Theorem 7.4 or Theorem 7.6, $\mu^{(k)}/\mu^{(k-1)}$ is a broken border strip of length $2a + 2b - 6$. Let X be a broken border strip of length $2a + 2b - 6$ contained in $\lambda = (\alpha, \beta) \in \mathcal{P}_n^{a,b}$. We can write $X = Y \coprod Z$ with $Y \subset \alpha$, $Z \subset \beta$, broken border strips. Since the maximum length of a border strip is $a + b - 1$, we have only to consider the following 5 cases.

    Case I. $|Y| = a + b - 1, |Z| = a + b - 5$,
    Case II. $|Y| = a + b - 2, |Z| = a + b - 4$,
    Case III. $|Y| = a + b - 3, |Z| = a + b - 3$,
    Case IV. $|Y| = a + b - 4, |Z| = a + b - 2$,
    Case V. $|Y| = a + b - 5, |Z| = a + b - 1$.

We consider the diagram $\gamma = (b^a)$ of rectangular shape so that $\alpha \cup \beta^* = \gamma$. In the following discussion, we regard Y and Z as paths in γ, instead of considering α and β separately. For example, the following figure explains an example of the case I, where $\lambda = (\alpha, \beta) = (42^2 1^2, 4^2 21)$ with $n = 20$, Y is a border strip of length 8 and Z is a border strip of length 4. In the figure, $\bullet$ (resp. $\times$) denotes the starting point and the ending point of Y (resp. Z).

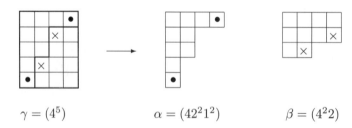

$$\gamma = (4^5) \qquad\qquad \alpha = (42^2 1^2) \qquad\qquad \beta = (4^2 2)$$

In computing $f_{a,b}(w_{\mathbf{r}})$, we use the following cancellation property.

LEMMA 7.9. *Let x be the top rightmost node of γ, and y the west of x, z the south of x. Assume that $x, y \in Y$ and that $z \in Z$. Then $\alpha : \alpha_1 \geq \alpha_2 \geq \cdots, \beta : \beta_1 \geq \beta_2 \geq \cdots$ with $\alpha_1 = a, \beta_1 = b - 1$. Let $\lambda' = (\alpha', \beta') \in \mathcal{P}_n$ be defined by $\alpha_1' = \alpha_1 - 1$, $\beta_1' = \beta_1 + 1$ and $\alpha_j' = \alpha_j, \beta_j' = \beta_j$ for $j \neq 1$. Then $\lambda' \in \mathcal{P}_n^{a,b}$. Put $Y' = Y - \{x\}$ and $Z' = Z + \{x\}$. Then Y' (resp. Z') is a broken border strip of α' (resp. β'). Let $X' = Y' \coprod Z'$. Then we have*

$$\Delta(X) = \Delta(X').$$

Moreover, the double partition $(\alpha - Y, \beta - Z)$ coincides with $(\alpha' - Y', \beta - Z')$. Hence in the computation of $f_{a,b}(w_{\mathbf{r}})$, the broken border strip X of this type may be ignored. Similar situations occur also for the cases, such as x is the bottom leftmost node of γ and y is the north of x, z is the east of x, and $x, y \in Y$, $z \in Z$.

PROOF. Since $y \in Y$ and $z \in Z$, the number of border strips in X is the same as the numbers in X'. Since $c(Y') = c(Y) - 1, c(Z') = c(Z) + 1$, and $r(Y') = r(Y), r(Z') = r(Z)$, we see that $\Delta(X) = \Delta(X')$. Clearly, we have $(\alpha - Y, \beta - Z) = (\alpha' - Y', \beta' - Z')$. It follows that in the sequence $\emptyset \subset \mu^{(1)} \subset \cdots \subset \mu^{(k)} = \lambda$ in Theorem 7.4 or Theorem 7.6, $\emptyset \subset \cdots \subset \mu^{(k-1)}$ is common for λ' if $\mu^{(k)}/\mu^{(k-1)} = X$. Since $|\alpha'| = |\alpha| - 1$, two terms starting from X and from X' are canceled in the computation of $f_{a,b}(w_{\mathbf{r}})$ by using these theorems. □

7.10. We shall classify the broken border strip $X = Y \coprod Z$ which is needed for the computation of $f_{a,b}(w_{\mathbf{r}})$. Let x_t be the top rightmost node and x_b be the bottom leftmost node of the Young diagram α. Let y_t be the top rightmost node of β and y_b be the bottom leftmost node of β. Note that we embed β in γ by using β^*, and regard y_t, y_b as a box in γ.

First we consider the Case I. We have $\alpha_1 = b, \alpha_a \neq 0$, and x_t is the top rightmost node and x_b is the bottom leftmost node of γ. Y is a unique border strip connecting x_t and x_b. Assume that $\alpha_2 < b$ and $\alpha_a = 1$. Then y_t is the south of x_t. If Z contains y_t or y_b, then Lemma 7.9 can be applied, and we can ignore this X. Hence Z does not contain y_t, y_b. In this case, Z is a unique border strip connecting the node west of y_t and the north of y_b. Next assume that $\alpha_2 = b$ and $\alpha_a = 1$. If $y_b \in Z$, then Lemma 7.9 can be applied. So we may assume that $y_b \notin Z$. In this case Z is a unique border strip connecting the node north of y_b and y_t which is the two node south of x_t. Next assume that $\alpha_2 < b$ and $\alpha_a = 2$. Then y_t is the south of x_t, and if $y_t \in Z$, the lemma can be applied. So we may assume that $y_t \notin Z$, and Z is a unique border strip connecting the node west of y_t and y_b. Finally, assume that $\alpha_2 = b$ and $\alpha_a = 2$. In this case, Z is a unique border strip connecting y_t and y_b.

Thus Case I is divided into 4 classes, and in each case, Y and Z are determined uniquely by $\lambda = (\alpha, \beta)$. In a similar way, one can classify all the possible broken border strips X for Case II and Case III. Case IV is symmetric to Case II, and each class is obtained from the class in Case II, by rotating the diagram γ by the angle $180°$, and then replacing Y and Z. Similarly, Case V is obtained from Case I. We shall list up all the possible cases for the cases I, II, III, in the list below, assuming that $a, b \geq 4$. Case I is divided into 4 classes, Case II into 6 classes, and Case III into 12 classes, (Case IV : 6 classes, Case V: 4 classes). Here • (resp. ×) denotes the starting node and the ending node of the border strip Y (resp. Z). In each case, Y and Z are determined uniquely by $\lambda = (\alpha, \beta)$. Or alternately, if we draw in the diagram γ the path connecting nodes marked by •, so that it is compatible with the path connecting boxes marked by ×, then it determines $\lambda = (\alpha, \beta)$ uniquely.

Note that in each case, $\Delta(X)$ does not depend on λ belonging to the class, and has a common value. We have listed those $\Delta(X)$ for each class, where $U = (u^{1/2} - u^{-1/2})$. For Case IV or Case V, $\Delta(X)$ is obtained from the corresponding $\Delta(X)$ for Case II or Case I, by replacing $u^{1/2} \leftrightarrow -u^{-1/2}$.

Case I. $|Y| = a + b - 1, |Z| = a + b - 5.$

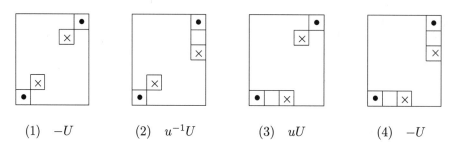

$$(1) \quad -U \qquad\qquad (2) \quad u^{-1}U \qquad\qquad (3) \quad uU \qquad\qquad (4) \quad -U$$

Case II. $|Y| = a + b - 2, |Z| = a + b - 4.$

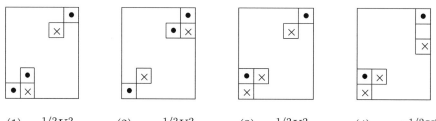

$$(1) \quad u^{1/2}U^2 \qquad (2) \quad -u^{1/2}U^2 \qquad (3) \quad u^{1/2}U^2 \qquad (4) \quad -u^{-1/2}U^2$$

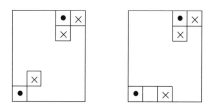

$$(5) \quad -u^{-1/2}U^2 \qquad (6) \quad u^{1/2}U^2$$

Case III. $|Y| = a + b - 3, |Z| = a + b - 3.$

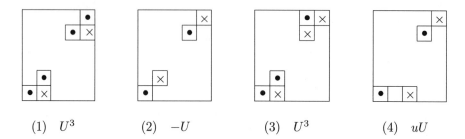

$$(1) \quad U^3 \qquad\qquad (2) \quad -U \qquad\qquad (3) \quad U^3 \qquad\qquad (4) \quad uU$$

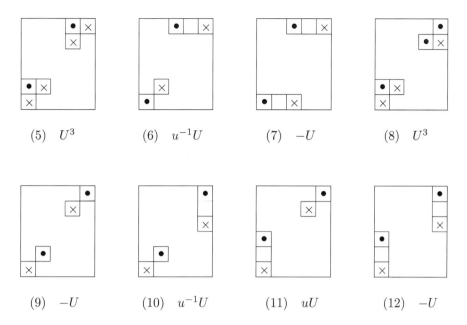

(5) U^3	(6) $u^{-1}U$	(7) $-U$	(8) U^3

(9) $-U$	(10) $u^{-1}U$	(11) uU	(12) $-U$

7.11.   In each case listed in 7.10, one obtains a unique $\lambda' \in \mathcal{P}_{n'}$ ($n' = n - (2a + 2b - 6)$) from λ by removing the broken border strip $X = Y \coprod Z$, where $\lambda' = (\alpha', \beta')$ with $\alpha' = \alpha - Y, \beta' = \beta - Z$. Let J be the set of classes in the list in 7.10, and let $\mathcal{Q}_j$ be the set of $\lambda' \in \mathcal{P}_{n'}$ satisfying the condition in each case j. For example we consider the case $j = \text{I-}(1)$. Then $\mathcal{Q} = \mathcal{Q}_{I-(1)}$ is the set of $\lambda' = (\alpha', \beta')$ satisfying the following conditions.

$$\begin{cases} \alpha' = (\alpha'_1, \alpha'_2, \ldots, \alpha'_{a-3}), \\ \beta'^* = (\beta'^*_1, \beta'^*_2, \ldots, \beta'^*_{a-1}) \quad \text{with} \quad \beta'^*_1 = b - 1, \beta'^*_{a-1} = 1, \\ \alpha_i + \beta'^*_{a-1-i} = b - 2 \text{ for } i = 1, \ldots, a - 3. \end{cases}$$

For each $\lambda' = (\alpha', \beta') \in \mathcal{Q}$, we consider a broken border strip X' of length $2a + 2b - 10$. Then X' is unique, and is given as $X' = Y' \coprod Z'$, where Y' is a unique border strip in α' of length $a + b - 7$, and Z' is a unique border strip in β' of length $a + b - 3$. By removing X' from λ', one obtains $\lambda'' \in \mathcal{P}_{n''}$, where $n'' = n' - (2a + 2b - 10) = (a - 4)(b - 4)$. Then it is easy to check that

(7.11.1) $$\{\lambda'' \in \mathcal{P}_{n''} \mid \lambda' \in \mathcal{Q}\} = \mathcal{P}_{n''}^{a-4, b-4},$$

and one recovers the original set $\mathcal{P}_n^{a-4, b-4}$ by replacing a, b by $a - 4, b - 4$. We can compute that $\Delta(X') = -(u^{1/2} - u^{-1/2})$, which is independent from $\lambda' \in \mathcal{Q}$. For example, let $a = 7, b = 6$ with $n = 42$. Take $\lambda = (6432^2 1^2, 6^2 542) \in \mathcal{P}_{42}^{7,6}$. Then $\lambda' = (\alpha', \beta') = (321^2, 6431^2)$, and under an appropriate rearrangement, $\gamma' = \alpha' \cup \beta'^*$ can be drawn as in the following figure. Here Y' (resp. Z') is a unique border strip of length 6, (resp. length 10), and $\bullet$ (resp. $\times$) denotes the starting point and the ending point of Y' (resp. Z'). From this, we obtain $\gamma'' = (2^3)$, and $\lambda'' = (\alpha'', \beta'') = (1, 32) \in \mathcal{P}_6^{3,2}$.

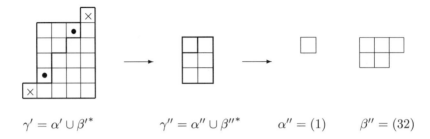

$$\gamma' = \alpha' \cup \beta'^* \qquad\qquad \gamma'' = \alpha'' \cup \beta''^* \qquad \alpha'' = (1) \qquad \beta'' = (32)$$

In fact, similar arguments work for all other cases, and the set Q_j is described in a similar way. In particular, $\gamma' = \alpha' \cup \beta'^*$ is of the shape obtained from a rectangle by attaching two nodes, one on the above or right of the northeast corner, and the other on the below or left of the southwest corner of the rectangle. In all the cases, the broken border strip $X' = Y' \coprod Z'$ of length $2a + 2b - 10$ is determined uniquely, and we always find the set $\mathcal{P}_{n''}^{a-4,b-4}$ after removing the border strips X'. Moreover, $\Delta(X')$ takes the common value $-(u^{1/2} - u^{-1/2})$ for all the cases through Case I $\sim$ Case V.

Recall that $w_{\mathbf{r}} = (r_1, \dots, r_k)$ with $r_k = n$. Assume that

$$r_k - r_{k-1} = 2a + 2b - 6, \quad r_{k-1} - r_{k-2} = 2a + 2b - 10.$$

Thus $r_{k-2} = n''$ with $n'' = (a-4)(b-4)$. We put $\mathbf{r}'' = (r_1, \dots, r_{k-2})$ and consider $w_{\mathbf{r}''} \in W_{n''}$. By investigating the above list, we have the following lemma.

LEMMA 7.12. *Under the notation above, there exists a non-zero (Laurent) polynomial $h(u)$ such that*

$$f_{a,b}(w_{\mathbf{r}}) = h(u) f_{a-4,b-4}(w_{\mathbf{r}''}).$$

PROOF. For each $j \in J$, we denote by $\Delta_j(u)$ the Laurent polynomial $\Delta(X)$ attached to the broken border strip X for Q_j as in the list. We put $\varepsilon_j = (-1)^{|\alpha|}$ for $\lambda = (\alpha, \beta) \in Q_j$. (Note that ε_j is independent of the choice of $\lambda \in \mathcal{Q}_j$.) By (7.11.1), the cardinality of Q_j coincides with the cardinality of $\mathcal{P}_{n''}^{a-4,b-4}$, hence is independent of $j \in J$. Then the investigation in 7.11 shows that

$$(7.12.1) \qquad f_{a,b}(w_{\mathbf{r}}) = \left\{ (u^{1/2} - u^{-1/2}) | \mathcal{P}_{n''}^{a-4,b-4} | \sum_{j \in J} \varepsilon_j \Delta_j(u) \right\} f_{a-4,b-4}(w_{\mathbf{r}''}).$$

Thus in order to show the lemma, it is enough to see that $\sum_j \varepsilon_j \Delta_j(u) \neq 0$. For this we compare the highest degree term $u^{3/2}$ in Δ_j. It follows from the list in 7.10, Δ_j contains the term $u^{2/3}$ in the following cases, where the coefficients are always 1.

Case I.   (3),
Case II.   (1), (3), (6),
Case III. (1), (3), (4), (5), (8), (11),
Case IV.   (2), (4), (5),
Case V.   (2),

where the numbering in Case IV and V is given by the bijective correspondence with Case II and Case I through the rotation of γ. Note that ε_j takes the constant value for each case, I $\sim$ V. They have the common value for Case I, III, or V, and have a different common value for Case II or IV. This shows that the coefficient of $u^{2/3}$ in $\sum_j \varepsilon_j \Delta_j \neq 0$. Hence $h(u) \neq 0$ as asserted. $\square$

Returning to the original setting, we show the following two propositions, which give the proof of Proposition 6.6.

PROPOSITION 7.13. *Assume that W is of type B_n with $n = d^2 + d$. We define an element $w_{\mathbf{r}} \in W$ by the cycle type expression given in (7.3.2) as follows.*

$[\overline{2}, 12, 16, 12 + 16, 16 + 16, \ldots, 12 + 16k, 16 + 16k]$,	*if $d \equiv 1 \pmod 4$,*
$[6, 16, 20, 16 + 16, 20 + 16, \ldots, 16 + 16k, 20 + 16k]$,	*if $d \equiv 2 \pmod 4$,*
$[4, 8, 4 + 16, 8 + 16, \ldots, 4 + 16k, 8 + 16k]$,	*if $d \equiv 3 \pmod 4$,*
$[8, 12, 8 + 16, 12 + 16, \ldots, 8 + 16k, 12 + 16k]$,	*if $d \equiv 0 \pmod 4$,*

for some $k \geq 1$, where the last term is equal to $4d - 4$, and the next term is equal to $4d - 8$, and so on. Then we have $f_{\Lambda_c}(w_{\mathbf{r}}) \neq 0$.

PROOF. We apply Lemma 7.12 with $a = d + 1, b = d$. Then $f_{\Lambda_c}(w_{\mathbf{r}}) = h(u) f_{\Lambda_c}(w_{\mathbf{r}''})$ for some non-zero $h(u)$, where $\mathbf{r} = (r_1, \ldots, r_k)$ and $\mathbf{r}'' = (r_1, \ldots, r_{k-2})$ with $r_k = n = d(d+1)$, $r_k - r_{k-1} = 4d - 4$, $r_{k-1} - r_{k-2} = 4d - 8$, $r_{k-2} = (d-4)(d-3)$. Thus the computation of $f_{\Lambda_c}(w_{\mathbf{r}})$ is reduced to the case where $d = 1, 2, 3$. Assume that $d = 1$, then $n = 2$. One can check by using the formula for $\overline{\Delta}(X)$ that $f_{\Lambda_c}(w) \neq 0$ for $w = (\overline{2})$. (Note that $f_{\Lambda_c}(w) = 0$ for $w = (2)$). Next assume that $d = 2$, then $n = 6$. The direct computation shows that $f_{\Lambda_c(w)} \neq 0$ for $w = (6)$. Finally assume that $d = 3$, then $n = 12$. By using a similar method as in the proof of Lemma 7.12, one can show that $f_{\Lambda_c}(w) \neq 0$ for $w = (4, 8)$ (we obtain a similar list, but some classes in the list in 7.10 don't appear for this case). This proves the proposition. $\square$

PROPOSITION 7.14. *Assume that W is of type D_n with $n = 4d^2$. We define an element $w_{\mathbf{r}} \in W$ as follows.*

$[\overline{1}, \overline{3}, 14, 18, 14 + 16, 18 + 16, \ldots, 14 + 16k, 18 + 16k]$,	*if $d \equiv 1 \pmod 2$,*
$[6, 10, 6 + 16, 10 + 16, \ldots, 6 + 16k, 10 + 16k]$,	*if $d \equiv 0 \pmod 2$,*

for some $k \geq 1$, where the last term is equal to $4d - 6$ and the next term is equal to $4d - 10$, and so on. Then we have $f_{\Lambda_c}(w_{\mathbf{r}}) \neq 0$.

PROOF. We apply Lemma 7.12 with $a = b = 2d$. Note that since $\alpha \neq \beta$ for any $\lambda = (\alpha, \beta) \in \mathcal{P}_n^{2d, 2d}$, the argument for type B_n can be applied without change. Then $f_{\Lambda_c}(w_{\mathbf{r}}) = h(u) f_{\Lambda_c}(w_{\mathbf{r}''})$ for some non-zero $h(u)$, where $\mathbf{r} = (r_1, \ldots, r_k)$ and $\mathbf{r}'' = (r_1, \ldots, r_{k-2})$ with $r_k = n = 4d^2$, $r_k - r_{k-1} = 4d - 6$, $r_{k-1} - r_{k-2} = 4d - 10$, $r_{k-2} = 4(d-2)^2$. Thus the computation of $f_{\Lambda_c}(w_{\mathbf{r}})$ is reduced to the case where $d = 1, 2$. Assume that $d = 2$. Then $n = 16$. Since $a = b = 4$, Lemma 7.12 can be applied, and we see that $f_{\Lambda_c}(w) \neq 0$ for $w = (6, 10)$. Next assume that $d = 1$. Then $n = 4$. One can check by using $\overline{\Delta}'(X)$ that $f_{\Lambda_c}(w) \neq 0$ for $w = (\overline{1}, \overline{3})$. This proves the proposition. $\square$

REMARK 7.15. As the formula (7.12.1) shows, our element $f_{\Lambda_c}(w_{\mathbf{r}})$ turns out to be 0 if $u \mapsto 1$. So our computation cannot be performed in the level of Weyl groups. On the other hand, Lusztig showed in [L5] that there exists an element $w \in W$ such that $f_{\Lambda_c}(w)|_{u=1} \neq 0$. Thus $f_{\Lambda_c}(w) \neq 0$ as a polynomial. This w has a simpler form than ours, but since it is a product of negative cycles of the full length (i.e., the sum of the lengths of negative cycles is equal to n), it is not appropriate for the computation of the bitrace on the flags (cf. Lemma 6.10).

References

[HR1]   T. Halverson and A. Ram; Murnaghan-Nakayama rues for characters of Iwahori-Hecke algebras of classical type, Trans. Amer. Math. Soc. **348** (1996), 3976 - 3995.

[HR2]   T. Halverson and A. Ram; Bitraces for $GL_n(\mathbb{F}_q)$ and the Iwahori-Hecke algebra of type A_{n-1}, Indag. Math. New Series **10** (1999), 247 - 268.

[K]   N. Kawanaka; On subfield symmetric spaces over a finite field, Osaka J. Math. **28** (1991) 759 - 791.

[L1]   G. Lusztig; "Characters of Reductive groups over a finite field", Ann. of Math. Studies, Vol. **107**, Princeton Univ. Press, Princeton, 1984.

[L2]   G. Lusztig; Character sheaves, I Adv. in Math. **56** (1985), 193–237, II Adv. in Math. **57** (1985), 226–265, III, Adv. in Math. **57** (1985), 266–315, IV, Adv. in Math. **59** (1986), 1–63, V, Adv. in Math. **61** (1986), 103–155.

[L3]   G. Lusztig; Remarks on computing irreducible characters, J. of Amer. Math. Soc., **5** (1992), 971–986.

[L4]   G. Lusztig; $G(F_q)$-invariants in irreducible $G(F_{q^2})$-modules, Represent. Theory, **4** (2000), 446 - 465.

[L5]   G. Lusztig; Rationality properties of unipotent representations, J. Algebra **258** (2002), 1 - 22.

[S1]   T. Shoji; Character sheaves and almost characters of reductive groups, Adv. in Math. **111** (1995), 244 - 313, II, Adv. in Math. **111** (1995), 314 - 354.

[S2]   T. Shoji; Unipotent characters of finite classical groups, in "Finite reductive groups; related structures and representations," Progress in Math., Vol.**141** (1997), 373 - 413.

[S3]   T. Shoji; Generalized Green functions and unipotent classes for finite reductive groups, II. To appear in Nagoya Math. Journal.

[W]   J.-L. Waldspurger; "Une conjecture de Lusztig pour les groupes classiques", Mémoires de la Soc. Math. France, No. **96**, Soc. Math. Fracne, 2004.

GRADUATE SCHOOL OF MATHEMATICS, NAGOYA UNIVERSITY, CHIKUSA-KU, NAGOYA 464-8602, JAPAN

Contemporary Mathematics
Volume **478**, 2009

A survey on quasifinite representations
of Weyl type Lie algebras

Yucai Su

Dedicated to my teacher Professor Zhexian Wan for his 80*th birthday*

ABSTRACT. The purpose of this survey article, mainly based on materials from the papers by the author and B. Xin [**26, 30, 31**], and also by other authors [**5, 14, 16, 37**], is to present some results on classifications of quasifinite representations and unitary representations of Lie algebras related to W-infinity algebras, especially, $\mathcal{W}_{1+\infty}$, $\mathcal{W}_\infty$ and $\widehat{\mathcal{D}^N}$. In particular, one main result states that an irreducible quasifinite module over a Lie algebra of this kind is either a highest/lowest weight module or a module of the intermediate series.

Introduction

The W-infinity algebras naturally arise in various physical theories, such as conformal field theory, the theory of the quantum Hall effect and integrable systems, etc. and receive intensive studies in the literature (e.g., [**1, 2, 3, 4, 5, 8, 9, 15, 16, 17, 20, 34**]). The $\mathcal{W}_{1+\infty}$ algebra, also known as the unique nontrivial central extension $\widehat{\mathcal{D}}$ of the Lie algebra $\mathcal{D}$ of differential operators on the circle [**15**], is the most fundamental one among these algebras. A systematic study of representation theory of the Lie algebra $\mathcal{W}_{1+\infty}$ was initiated in [**16**]. In that paper, irreducible quasifinite highest weight representations of $\mathcal{W}_{1+\infty}$ were classified and it was shown that they can be realized in terms of irreducible highest weight representations of the central extension $\widehat{gl}_\infty$ of the Lie algebra $\widetilde{gl}_\infty$ of infinite matrices with only finitely many nonzero diagonals. Furthermore, unitary representations were described. This study was continued in [**9, 17**] in the framework of vertex algebra theory. In particular, character formulas for primitive representations of $\mathcal{W}_{1+\infty}$ with central charge N were established, and the vertex operator algebra structures on the vacuum modules of $\mathcal{W}_{1+\infty}$ and the W-algebra $\mathcal{W}(gl_N)$ were studied. Later on, Ref. [**5**] gave a complete description of the irreducible quasifinite highest weight

---

1991 *Mathematics Subject Classification.* Primary 17B68; Secondary 17B10, 17B65, 17B66.
Key words and phrases. W-infinity algebras, Weyl type Lie algebras, quasifinite representations.
The author was supported Supported by a NSF grant 10471091 of China and "One Hundred Talents Program" from USTC, China.

modules over the Lie algebra $\widehat{\mathcal{D}^N}$ (the unique nontrivial central extension of the Lie algebra $\mathcal{D}^N$ of $N \times N$-matrix differential operators on the circle), and classified the unitary ones, and obtained them in terms of representation theory of the central extension $\widehat{gl}_\infty(R^m)$ of the Lie algebra $\widetilde{gl}_\infty(R^m)$ of infinite matrices with only finitely many nonzero diagonals over the algebras $R^m = \mathbb{C}[t]/(t^{m+1})$ for various $m \in \mathbb{Z}_+$.

In [14], all irreducible quasifinite highest weight modules of the subalgebras $\mathcal{W}_{1+\infty}^{(k)}$ of $\mathcal{W}_{1+\infty}$ were classified, where $\mathcal{W}_{1+\infty}^{(k)}$ is the unique nontrivial central extension of the Lie algebra $\mathcal{D}\partial_t^k$ of differential operators on the circle that annihilate all polynomials of degree less than k (the most important of these subalgebras is another W-infinity algebra $\mathcal{W}_\infty = \mathcal{W}_{1+\infty}^{(1)}$). Furthermore, unitary modules over $\mathcal{W}_\infty$ are also considered. In particular, it is obtained surprisingly that the list of unitary modules over $\mathcal{W}_\infty$ is much richer than that over $\mathcal{W}_{1+\infty}$.

In [26, 30, 31], irreducible quasifinite modules over Weyl type Lie algebras $\mathcal{W}(\Gamma, n)$, $\widehat{\mathcal{W}}(\Gamma, 1)$, $\mathcal{W}(\Gamma, n)^{(1)}$, $\widehat{\mathcal{W}}(\Gamma, 1)^{(1)}$, $\widehat{\mathcal{D}^N}$ were classified. Here $\mathcal{W}(\Gamma, n)$ is the Lie algebra of generalized differential operators, which can be defined as the tensor product space of the group algebra $\mathbb{C}[\Gamma]$ with basis $\{t^\alpha \,|\, \alpha \in \Gamma\}$ (for any nondegenerate additive subgroup Γ of $\mathbb{C}^n$) and the polynomial algebra $\mathbb{C}[D_1, ..., D_n]$ (where D_i's are degree operators of $\mathbb{C}[\Gamma]$), while $\mathcal{W}(\Gamma, n)^{(1)}$ is the Lie subalgebra of $\mathcal{W}(\Gamma, n)$ of generalized differential operators of degree at least 1, and $\widehat{\mathcal{W}}(\Gamma, n)$, $\widehat{\mathcal{W}}(\Gamma, 1)$ are respectively the universal central extensions of $\mathcal{W}(\Gamma, 1)$, $\mathcal{W}(\Gamma, 1)^{(1)}$. Thus in particular, $\mathcal{W}_{1+\infty} = \widehat{\mathcal{W}}(\mathbb{Z}, 1)$, $\mathcal{W}_\infty = \widehat{\mathcal{W}}(\mathbb{Z}, 1)^{(1)}$, and so, Lie algebras $\mathcal{W}(\Gamma, n)$, $\widehat{\mathcal{W}}(\Gamma, 1)$, $\mathcal{W}(\Gamma, n)^{(1)}$, $\widehat{\mathcal{W}}(\Gamma, 1)^{(1)}$ can be regarded as multi-variable generalizations of W-infinity algebras. A particular result of [26, 30, 31] states that an irreducible quasifinite module over a Lie algebra of this kind is either a highest/lowest weight module or a module of the intermediate series.

As is pointed in [14, 16], when studying the representation theory of a Lie algebra of this kind, we encounter the difficulty that though it admits a $\mathbb{Z}$-gradation, each of the graded subspaces is still infinite dimensional, and thus the study of modules which satisfy the quasifiniteness condition that its graded subspaces have finite dimensions, becomes a nontrivial problem. The aim of this survey article is to give an overview of the above mentioned papers, and to present the main techniques and results of these papers.

1. Weyl type Lie algebras and some subalgebras

1.1. Weyl type Lie algebras. A (classical) *Weyl algebra of rank n* is the associative algebra

$$A_n^+ = \mathbb{C}[t_1, ..., t_n, \frac{\partial}{\partial t_1}, ..., \frac{\partial}{\partial t_n}] \quad \text{or} \quad A_n = \mathbb{C}[t_1^{\pm 1}, ..., t_n^{\pm 1}, \frac{\partial}{\partial t_1}, ..., \frac{\partial}{\partial t_n}]$$

of differential operators over $\mathbb{C}$. Under the commutator, A_n is a *Lie algebra of Weyl type*, denoted by $\mathcal{W}(n)$. Then $\mathcal{W}(n) = \oplus_{\alpha \in \mathbb{Z}^n} \mathcal{W}(n)_\alpha$ is a $\mathbb{Z}^n$-graded Lie algebra such that the graded subspace $\mathcal{W}(n)_\alpha$ for $\alpha = (\alpha_1, ..., \alpha_n) \in \mathbb{Z}^n$ has basis elements

$$(1.1) \qquad t^\alpha D^\mu = t_1^{\alpha_1} \cdots t_n^{\alpha_n} D_1^{\mu_1} \cdots D_n^{\mu_n}, \quad \mu = (\mu_1, ..., \mu_n) \in \mathbb{Z}_+^n,$$

where $D_i = t_i \frac{\partial}{\partial t_i}$. It is known [8, 22] that $\mathcal{W}(n)$ has a nontrivial universal central extension if and only if $n = 1$. The *W-infinity algebra* $\mathcal{W}_{1+\infty}$ is the universal central extension of $\mathcal{W}(1)$ (cf. (1.4)).

Based on a classification (in [**32**]) of pairs $(\mathcal{A}, \mathcal{D})$ of a commutative associative algebra $\mathcal{A}$ and a finite-dimensional locally finite Abelian derivation subalgebra $\mathcal{D}$ such that $\mathcal{A}$ is $\mathcal{D}$-simple, a large class of Lie algebras of Weyl type was constructed in [**33**], whose structure theory was studied in [**34, 36**]. This type of Lie algebras of Weyl type is essentially a generalization of the Lie algebra $\mathcal{W}(n)$.

Now we introduce a special subclass of such Lie algebras, which can be regarded as a multi-variable generalization of W-infinity algebras. Consider the n-dimensional vector space $\mathbb{C}^n$. Denote an element in $\mathbb{C}^n$ by

$$\alpha = (\alpha_1, ..., \alpha_n).$$

Let Γ be an additive subgroup of $\mathbb{C}^n$ such that Γ is *nondegenerate*, i.e., it contains a $\mathbb{C}$-basis of $\mathbb{C}^n$. Let $\mathbb{C}[\Gamma]$ be the *group algebra* of the group Γ with basis element written as t^α, $\alpha \in \Gamma$. We define the *degree operators* D_i to be the derivations

$$D_i : t^\alpha \mapsto \alpha_i t^\alpha \quad \text{for} \ \alpha \in \Gamma,$$

where $i \in [1, n]$. Then the *generalized Weyl algebra* $\mathcal{A}(\Gamma, n)$ *of rank* n is the associative subalgebra of $\text{End}_{\mathbb{C}}(\mathbb{C}[\Gamma])$ generated by D_i's and left multiplications by t^α, $\alpha \in \Gamma$. Thus $\mathcal{A}(\Gamma, n)$ is a tensor product space of $\mathbb{C}[\Gamma]$ and the polynomial algebra $\mathbb{C}[D_1, ..., D_n]$, with basis $\{t^\alpha D^\mu \,|\, \alpha \in \Gamma, \mu \in \mathbb{Z}_+^n\}$, where D^μ stands for $\prod_{i=1}^n D_i^{\mu_i}$, and product:

$$(1.2) \qquad t^\alpha D^\mu \cdot t^\beta D^\nu = \sum_{\lambda \in \mathbb{Z}_+^n} \binom{\mu}{\lambda} \beta^\lambda t^{\alpha+\beta} D^{\mu+\nu-\lambda},$$

where $\beta^\lambda = \prod_{i=1}^n \beta_i^{\lambda_i}$, $\binom{\mu}{\lambda} = \prod_{i=1}^n \binom{\mu_i}{\lambda_i}$, and for $i, j \in \mathbb{C}$, the binomial coefficient $\binom{i}{j} = \prod_{k=0}^{j-1}(i-k)/j!$ if $j \in \mathbb{Z}_+$, or 0 otherwise.

Under the commutator, $\mathcal{A}(\Gamma, n)$ is a *Lie algebra of generalized Weyl type of rank* n (also called a *Lie algebra of generalized differential operators*) [**26**], denoted by $\mathcal{W}(\Gamma, n)$, whose Lie bracket is thus given by

$$(1.3) \qquad [t^\alpha D^\mu, t^\beta D^\nu] = t^\alpha D^\mu \cdot t^\beta D^\nu - t^\beta D^\nu \cdot t^\alpha D^\mu.$$

The classical Lie algebra $\mathcal{W}(n)$ of Weyl type is simply the Lie algebra $\mathcal{W}(\mathbb{Z}^n, n)$.

The Lie algebra $\mathcal{W}(\Gamma, n)$ has a nontrivial universal central extension if and only if $n = 1$ [**22**]. The universal central extension $\widehat{\mathcal{W}}(\Gamma, 1)$ of $\mathcal{W}(\Gamma, 1)$ is defined by: the bracket (1.3) is replaced as

$$[t^\alpha [D]_\mu, t^\beta [D]_\nu] = t^\alpha [D]_\mu \cdot t^\beta [D]_\nu - t^\beta [D]_\nu \cdot t^\alpha [D]_\mu$$
$$(1.4) \hspace{4cm} + \delta_{\alpha, -\beta} (-1)^\mu \mu! \nu! \binom{\alpha+\mu}{\mu+\nu+1} \kappa,$$

where κ is a central element of $\widehat{\mathcal{W}}(\Gamma, 1)$, $D = t\frac{d}{dt}$ and $[D]_\mu = \prod_{i=0}^{\mu-1}(D-i)$ (here we have to use this new notation in order to have a simple expression of the coefficient of κ).

In case $\Gamma = \mathbb{Z}$, (1.4) can be rewritten into the better form

$$(1.5) \quad [t^i f(D), t^j g(D)] = t^{i+j}\big(f(D-j)g(D) - f(D)g(D+i)\big) + \varphi\big(t^i f(D), t^j g(D)\big)\kappa,$$

where, the 2-cocycle φ, which seems to appear first in [**15**], is defined by

$$\varphi\big(t^i f(D), t^j g(D)\big) = \begin{cases} \displaystyle\sum_{-i \le k \le -1} f(k)g(k+j) & \text{if } i = -j > 0, \\ 0 & \text{if } i+j \ne 0 \text{ or } i = j = 0, \end{cases}$$

for $i, j \in \mathbb{Z}$ and all polynomials $f(D), g(D)$.

Let $k > 0$. Denote the Lie subalgebra of $\mathcal{W}(\Gamma, n)$ of (generalized) differential operators of degree at least k by $\mathcal{W}(\Gamma, n)^{(k)}$, namely, $\mathcal{W}(\Gamma, n)^{(k)}$ is the Lie subalgebra with basis elements

$$(1.6) \qquad t^\alpha D^\mu \text{ for } \alpha \in \Gamma, \ |\mu| \geq k,$$

where $|\mu| = \sum_{i=1}^n \mu_i$ is the *level* of μ. Similarly, one can define $\widehat{\mathcal{W}}(\Gamma, 1)^{(k)}$ to be the universal central extension of $\mathcal{W}(\Gamma, 1)^{(k)}$. In particular, we denote $\mathcal{W}_{1+\infty}^{(k)} = \widehat{\mathcal{W}}(\mathbb{Z}, 1)^{(k)}$. Then another W-*infinity algebra* $\mathcal{W}_\infty = \mathcal{W}_{1+\infty}^{(1)}$ is the universal central extension of $\mathcal{W}(\mathbb{Z}, 1)^{(1)}$.

A Lie algebra $\mathcal{L}$ is called

  * Γ-*graded* if $\mathcal{L} = \oplus_{\alpha \in \Gamma} \mathcal{L}_\alpha$ is decomposed as a direct sum of graded subspaces such that the Lie bracket respects the gradation, namely, $[\mathcal{L}_\alpha, \mathcal{L}_\beta] \subset \mathcal{L}_{\alpha+\beta}$ for $\alpha, \beta \in \Gamma$;
  * *finitely* Γ-*graded* if all graded subspaces $\mathcal{L}_\alpha$ are finite dimensional;
  * *not-finitely* Γ-*graded* if $\dim L_\alpha = \infty$ for some $\alpha \in \Gamma$.

Thus, $\mathcal{W}(\Gamma, n)$, $\mathcal{W}(\Gamma, n)^{(k)}$, and their central extensions are not-finitely Γ-graded Lie algebras such that the basis element $t^\alpha D^\mu$ has degree $\alpha \in \Gamma$ (cf. (1.1)), and κ has degree 0. For the case $\mathcal{W}(\Gamma, n)$, we can write $\mathcal{W}(\Gamma, n) = \oplus_{\alpha \in \Gamma} \mathcal{W}(\Gamma, n)_\alpha$ such that each graded subspace $\mathcal{W}(\Gamma, n)_\alpha$ has basis elements

$$(1.7) \qquad t^\alpha D^\mu \text{ for } \mu \in \mathbb{Z}_+^n.$$

Thus $\mathcal{W}(\Gamma, n)_0$ is commutative which is also a *Cartan subalgebra* (i.e., a maximal self-normalized nilpotent subalgebra), and it contains the (unique) *maximal torus* (i.e., a maximal commutative subalgebra with each element being ad-semi-simple) $\mathcal{H}$ with basis elements $1 := t^0 D^0$, D_i for $i \in [1, n]$ (the uniqueness can be see from [**34, 36**]).

Next, we introduce the W-infinity algebra $\widehat{\mathcal{D}^N}$ (e.g., [**5, 30**]). Denote by $\mathcal{D}_{as}^N$ the associative algebra which is the tensor product of two associative algebras $\mathcal{W}(\mathbb{Z}, 1)$ and gl_N. Thus $\mathcal{D}_{as}^N$ has basis elements

$$(1.8) \qquad t^i D^j A \text{ for } i \in \mathbb{Z}, \ j \in \mathbb{Z}_+, \ A \in gl_N,$$

with product

$$(1.9) \qquad t^i D^j A \cdot t^k D^l B = t^{i+k} (D+k)^j D^l AB = \sum_{s=0}^j \binom{j}{s} k^s t^{j+k} D^{j+l-s} AB.$$

Under the commutator, $\mathcal{D}_{as}^N$ is the *Lie algebra of* $N \times N$-*matrix differential operators on the circle*, denoted by $\mathcal{D}^N$, whose Lie bracket is thus given by

$$(1.10) \qquad [t^i D^j A, t^k D^l B] = t^i D^j A \cdot t^k D^l B - t^k D^l B \cdot t^i D^j A.$$

The Lie algebra $\mathcal{D}^N$ has a unique nontrivial central extension [**8, 22**]. The W-*infinity algebra* $\widehat{\mathcal{D}^N}$ is the universal central extension of $\mathcal{D}^N$, defined by [**5, 9**]: the bracket (1.10) is replaced as

$$(1.11) \qquad \begin{aligned} [t^i [D]_j A, t^k [D]_l B] &= t^i [D]_j A \cdot t^k [D]_l B - t^k [D]_l B \cdot t^i [D]_j A \\ &\quad + \delta_{i,-k} (-1)^j j! l! \binom{i+j}{j+l+1} \mathrm{tr}(AB) \kappa. \end{aligned}$$

Thus $\widehat{\mathcal{D}^1}$ is simply the Lie algebra $\mathcal{W}_{1+\infty}$.

Let $\{E_{p,q} \,|\, p, q \in [1, N]\}$ be the standard basis of gl_N, where $E_{p,q}$ is the matrix with entry 1 at (p, q) and 0 otherwise. Then $\widehat{\mathcal{D}^N}$ has the *principal $\mathbb{Z}$-gradation* $\widehat{\mathcal{D}^N} = \oplus_{i \in \mathbb{Z}} \widehat{\mathcal{D}^N}_i$ with the graded subspace [5]

$$(1.12) \qquad \widehat{\mathcal{D}^N}_i = \text{span}\{t^k D^j E_{p,q} \,|\, k \in \mathbb{Z},\ j \in \mathbb{Z}_+,\ p, q \in [1, N],\ kN + p - q = i\} \oplus \delta_{i,0} \mathbb{C} \kappa.$$

Note that t has degree N in this case. In particular, $\widehat{\mathcal{D}^N}_0$ with basis $\{D^j E_{p,p}, \kappa \,|\, j \in \mathbb{Z}_+, p \in [1, N]\}$ is commutative (a Cartan subalgebra), and $\mathcal{H}$ with basis $\{E_{pp}, DE_{pp}, \kappa \,|\, p \in [1, N]\}$ is a maximal torus.

1.2. Some subalgebras. It might be worth mentioning that Weyl type Lie algebras contain some interesting Lie subalgebras. It is well known [19] that the *Virasoro algebra* Vir is a finitely $\mathbb{Z}$-graded Lie algebra which can be defined as the universal central extension of the Lie algebra of linear differential operators on the circle. Thus Vir has a basis $\{L_i, \kappa \,|\, i \in \mathbb{Z}\}$ such that κ is central and

$$(1.13) \qquad\qquad [L_i, L_j] = (j - i)L_{i+j} + \frac{i^3 - i}{12} \kappa.$$

Ref. [21, 35] generalized the notion of the Virasoro algebra to *generalized Virasoro algebras* Vir(Γ), for any nonzero additive subgroup Γ of $\mathbb{C}$, which is the finitely Γ-graded Lie algebra with basis $\{L_\alpha, \kappa \,|\, \alpha \in \Gamma\}$ and relation

$$(1.14) \qquad\qquad [L_\alpha, L_\beta] = (\alpha - \beta)L_{\alpha+\beta} + \frac{\alpha^3 - \alpha}{12} \kappa.$$

When Γ is finitely generated additive subgroup of $\mathbb{C}$, Vir(Γ) is also called a *higher rank Virasoro algebra* [21]. One immediately sees that $\widehat{\mathcal{W}}(\Gamma, 1)^{(k)}$ contains a generalized Virasoro subalgebra if and only if $k \leq 1$, and when $k \leq 1$, the subspace of $\widehat{\mathcal{W}}(\Gamma, 1)$ with basis $\{L_\alpha := t^\alpha D, \kappa \,|\, \alpha \in \Gamma\}$ is the generalized Virasoro algebra Vir(Γ). One also sees that $\mathcal{W}(\Gamma, n)^{(k)}$, $k \leq 1$ contains the finitely Γ-graded Lie subalgebra, denoted by Witt(Γ, n) and called a *generalized Witt algebra* (e.g., [32, 38]), which has basis $\{t^\alpha D_i \,|\, \alpha \in \Gamma, i \in [1, n]\}$ and relation

$$(1.15) \qquad\qquad [t^\alpha D_i, t^\beta D_j] = t^{\alpha+\beta}(\beta_i D_j - \alpha_j D_i).$$

Since any finitely Γ-graded Cartan type Lie algebra can be realized as a subalgebra of Witt(Γ, n) for some n (e.g., [38]), we see that $\mathcal{W}(\Gamma, n)^{(k)}$, $k \leq 1$ for various n also contain all finitely Γ-graded Cartan type Lie algebras. Moreover, when $k \geq 2$, $\mathcal{W}(\Gamma, n)^{(k)}$ does not contain a generalized Witt algebra.

Next, we see that $\widehat{\mathcal{D}^N}$ not only contains a Virasoro subalgebra with basis $\{L_i := t^i DI, \kappa \,|\, i \in \mathbb{Z}\}$, where I is the $N \times N$ identity matrix, but also contains the finitely $\mathbb{Z}$-graded Lie subalgebra linearly spanned by $\{t^i A, \kappa, D \,|\, i \in \mathbb{Z}, A \in gl_N\}$, which is in fact the *affine gl_N Lie algebra*, whose nontrivial Lie brackets are given by

$$(1.16) \qquad [t^i A, t^j B] = t^{i+j}[A, B] + \delta_{i,-j} i \operatorname{tr}(AB)\kappa, \quad [D, t^i A] = it^i A.$$

Since every affine Kac-Moody algebra can be embedded into the affine gl_N Lie algebra for some N, we obtained that $\widehat{\mathcal{D}^N}$ for various N contain all affine Kac-Moody algebras.

1.3. Quasifinite representations. Let $\mathcal{L} = \oplus_{\alpha \in \Gamma} \mathcal{L}_\alpha$ be a Γ-graded Lie algebra such that $\mathcal{L}_0$ is commutative, and in general we assume dim $\mathcal{L}_\alpha = \infty$ for $\alpha \in \Gamma$. We always assume that $\mathcal{L}_0$ is a Cartan subalgebra containing a finite-dimensional maximal torus $\mathcal{H}$ such that $\mathcal{L} = \oplus_{\alpha \in \Gamma} \mathcal{L}_\alpha$ is also the root space decomposition with respect to $\mathcal{H}$, namely, there exists $\varphi_\alpha \in \mathcal{H}^*$ (the dual space of $\mathcal{H}$) such that

$$(1.17) \qquad \mathcal{L}_\alpha = \{x \in \mathcal{L} \,|\, [h, x] = \varphi_\alpha(h)x \ \text{ for all } \ h \in \mathcal{H}\} \ \text{ for } \ \alpha \in \Gamma.$$

In this case, φ_α is called a *root* of $\mathcal{H}$.

We combine several concepts related to the representation theory into the following definition.

DEFINITION 1.1. An $\mathcal{L}$-module V is called a

(1) *generalized weight module* if $V = \oplus_{\lambda \in \mathcal{H}^*} V^\lambda$ is decomposed as a direct sum of generalized weight spaces, where

$$V^\lambda = \{v \in V \,|\, \forall h \in \mathcal{H}, \exists n \in \mathbb{Z}_+ \text{ with } \big(h - \lambda(h)\big)^n v = 0\}$$

is the *generalized weight space* of weight λ;

(2) *weight module* if $V = \oplus_{\lambda \in \mathcal{H}^*} V^{0,\lambda}$ is decomposed as a direct sum of weight spaces, where $V^{0,\lambda} = \{v \in V \,|\, hv = \lambda(h)v \text{ for } h \in \mathcal{H}\}$ is the *weight space* of weight λ;

(3) Γ-*graded module* if $V = \oplus_{\alpha \in \Gamma} V_\alpha$ is decomposed as a direct sum of graded subspaces such that the action of $\mathcal{L}$ respects the gradation, namely, $\mathcal{L}_\alpha V_\beta \subset V_{\alpha+\beta}$ for $\alpha, \beta \in \Gamma$;

(4) *quasifinite module* [16] if V is Γ-graded such that all graded subspaces are finite dimensional, i.e., dim $V_\alpha < \infty$ for $\alpha \in \Gamma$;

(5) *uniformly bounded module* if V is quasifinite and the dimensions of graded subspaces are uniformly bounded, i.e., there exists an integer $K > 0$ such that dim $V_\alpha \leq K$ for $\alpha \in \Gamma$;

(6) *module of the intermediate series* if it is quasifinite and dim $V_\alpha \leq 1$ for $\alpha \in \Gamma$;

(7) *highest/lowest weight module* if there exists a total ordering $\succ$ on Γ compatible with its group structure (i.e., if $\alpha \succ \beta$, $\lambda \succeq \mu$ for $\alpha, \beta, \lambda, \mu \in \Gamma$ then $\alpha + \lambda \succ \beta + \mu$), and there exist $\lambda \in \mathcal{L}_0^*$ (the dual space of $\mathcal{L}_0$) and nonzero vector $v_\lambda \in V$ (in this case λ is called the *highest/lowest weight* of V and v_λ a *highest/lowest weight vector* of V) such that V is generated by v_λ satisfying

$$\mathcal{L}_\pm v_\lambda = 0, \ \text{ and } \ xv_\lambda = \lambda(x)v_\lambda \ \text{ for } \ x \in \mathcal{L}_0,$$

where $\mathcal{L}_\pm = \oplus_{\pm \alpha \succ 0} \mathcal{L}_\alpha$ (so that $\mathcal{L} = \mathcal{L}_- \oplus \mathcal{L}_0 \oplus \mathcal{L}_+$ has a triangular decomposition);

(8) *Verma module* if $V = V(\lambda)$, where $V(\lambda)$ is defined as follows: there exist a total ordering $\succ$ on Γ compatible with its group structure, and $\lambda \in \mathcal{L}_0^*$ and $0 \neq v_\lambda \in V(\lambda)$ such that $\mathcal{L}_+ v_\lambda = 0$, $xv_\lambda = \lambda(x)v_\lambda$ for $x \in \mathcal{L}_0$, and

$$(1.18) \qquad V(\lambda) = \mathrm{Ind}^{\mathcal{L}}_{\mathcal{L}_0 \oplus \mathcal{L}_+} \mathbb{C}v_\lambda = U(\mathcal{L}) \otimes_{U(\mathcal{L}_0 \oplus \mathcal{L}_+)} \mathbb{C}v_\lambda \cong U(\mathcal{L}_-) \otimes_\mathbb{C} \mathbb{C}v_\lambda,$$

where, $U(\mathcal{L})$ denotes the universal enveloping algebra of $\mathcal{L}$.

REMARK 1.2.   (1) A quasifinite module V may not be a weight module. But V is necessarily a generalized weight module since each finite-dimensional homogenous subspace V_α is an $\mathcal{H}$-module, thus has a generalized weight space decomposition. However V_α may contain more than one weights.

(2) By (1), the category of quasifinite modules is equivalent to the category of generalized weight modules with finite-dimensional generalized weight spaces.

(3) A highest/lowest weight module V is Γ-graded: $V = \oplus_{\alpha \in \Gamma} V_\alpha$ if we define v_λ to have degree zero, but it may not be quasifinite. Also, V is necessarily a weight module, whose highest weight vector v_λ is a common eigenvector for both $\mathcal{H}$ and $\mathcal{L}_0$ (a weight vector may not be a common eigenvector for $\mathcal{L}_0$; for instance, in the adjoint module $\mathcal{W}(\Gamma, n)$, the vector $v = t^\alpha$ with $\alpha \in \Gamma \backslash \{0\}$ is a weight vector but not a common eigenvector for $\mathcal{W}(\Gamma, n)_0$).

(4) By (1.17), one can prove that any submodule of a generalized weight module is again a generalized weight module.

(5) Let $V(\lambda)$ be the Verma module defined in (1.18), and M a proper submodule of $V(\lambda)$. Then $v_\lambda \notin M$. By (4), the sum of all proper submodules is still proper, thus it is the maximal proper submodule, denoted by $M(\lambda)$. Then

$$(1.19) \qquad\qquad L(\lambda) = V(\lambda)/M(\lambda),$$

is the (unique) irreducible highest weight module with highest weight λ.

Note that Verma modules for finitely $\mathbb{Z}$-graded Lie algebras (in particular, for all finite dimensional Lie algebras [10], affine Kac-Moody Lie algebras [12], the Virasoro algebra [19], etc) are always quasifinite in the sense of Definition 1.1(4). However, when $\Gamma \not\cong \mathbb{Z}$, Verma modules for finitely Γ-graded Lie algebras might not be quasifinite. In fact, it is proved in [18, 27] that every nontrivial highest weight module for $Vir(\Gamma)$ or $Witt(\Gamma, n)$ is not quasifinite when $\Gamma \not\cong \mathbb{Z}$. Furthermore, one sees immediately from (1.18) that every Verma module $V(\lambda)$ for a not-finitely Γ-graded Lie algebra is not quasifinite (in fact all graded subspaces of $V(\lambda)$ are infinite dimensional except the one with degree 0 which is one dimensional spanned by the highest weight vector v_λ). Thus, as stated in the introduction, when we study the representation theory of a not-finitely Γ-graded Lie algebra $\mathcal{L}$, we encounter the difficulty that although it admits a Γ-gradation, each of the graded subspaces is still infinite dimensional, and therefore the study of quasifinite modules becomes a nontrivial and interesting problem.

Some natural problems arise (where a *singular vector* in a highest weight module is a nonzero weight vector v other than the highest weight vector such that $\mathcal{L}_+ v = 0$).

PROBLEM 1.3. Let Γ be a torsion-free Abelian group such that $\Gamma \not\cong \mathbb{Z}$, and let $\mathcal{L} = \oplus_{\alpha \in \Gamma} \mathcal{L}_\alpha$ be a finitely Γ-graded Lie algebra such that $\mathcal{L}_\alpha \neq 0$ for $\alpha \in \Gamma$. Is every Verma module $V(\lambda)$ over $\mathcal{L}$ not quasifinite? (We conjecture that the answer is "yes".)

PROBLEM 1.4. Let $\mathcal{W}$ be one of Weyl type Lie algebras $\mathcal{W}(\Gamma, n)^{(k)}$, $\widehat{\mathcal{W}}(\Gamma, 1)^{(k)}$, $\widehat{\mathcal{D}^N}$.

(1) Determine when the Verma $\mathcal{W}$-module $V(\lambda)$ is irreducible.

(2) If the Verma $\mathcal{W}$-module $V(\lambda)$ is not irreducible, determine all singular vectors (or equivalently, all composition factors).

(3) Classify irreducible quasifinite highest weight $\mathcal{W}$-modules.

(4) Classify irreducible quasifinite $\mathcal{W}$-modules.

(5) Classify indecomposable uniformly bounded (not necessarily weight) $\mathcal{W}$-modules.

(6) Classify indecomposable quasifinite (not necessarily weight) $\mathcal{W}$-modules.

PROBLEM 1.5. Consider similar problems for $\mathcal{W}(\Gamma, n)^{(k)} \otimes gl_N$ or its central extension.

In the following sections, we shall present partial results for the problems listed in Problem 1.4. The proofs of some obtained results heavily depend on the representation theories of the Virasoro algebra and Witt algebra (cf. (1.13)–(1.15)) which are contained as subalgebras in the Weyl type Lie algebras. Since when $k \geq 2$, the Lie algebras $\mathcal{W}(\Gamma, n)^{(k)}$ and $\widehat{\mathcal{W}}(\Gamma, 1)^{(k)}$ do not contain a Virasoro subalgebra or Witt subalgebra, the techniques used for the Virasoro algebra and Witt algebra cannot be applied, most problems concerning them remain unsolved. For convenience, we list below what has been done for the problems listed in Problem 1.4:

(1) is done for $\mathcal{W} = \widehat{\mathcal{W}}(\Gamma, 1)$ [**37**], thus also for $\mathcal{W} = \mathcal{W}(\Gamma, 1)$.

(2) is not done in any case.

(3) is done for $\mathcal{W} = \mathcal{W}(\Gamma, n)^{(k)}, \widehat{\mathcal{W}}(\Gamma, 1)^{(k)}$ with $k \leq 1$ or $\Gamma \cong \mathbb{Z}$, and $\widehat{\mathcal{D}^N}$ [**5, 9, 14, 16, 17, 26, 30, 31**].

(4) is done for $\mathcal{W} = \mathcal{W}(\Gamma, n)^{(k)}, \widehat{\mathcal{W}}(\Gamma, 1)^{(k)}$ with $k \leq 1$, and $\widehat{\mathcal{D}^N}$ [**26, 30, 31**].

(5) is done for $\mathcal{W} = \mathcal{W}(\Gamma, n), \widehat{\mathcal{W}}(\Gamma, 1)$ [**26, 30**].

(6) is done for $\mathcal{W} = \mathcal{W}(\Gamma, n), \widehat{\mathcal{W}}(\Gamma, 1)$ with $\Gamma \not\cong \mathbb{Z}$ [**26, 30**].

2. Irreducible and unitary quasifinite highest weight representations

Highest weight representations are the most important objects in the study of representation theory of Lie algebras; for instance, finite dimensional simple Lie algebras [**10**], the Virasoro algebra [**15, 19, 23**], affine Kac-Moody Lie algebras [**12**]. We shall present some results of [**5, 14, 16, 37**] which give characterizations of irreducible (including unitary) quasifinite highest weight $\mathcal{W}$-modules $L(\lambda)$ for $\mathcal{W} = \mathcal{W}_{1+\infty}, \mathcal{W}_\infty$ or $\widehat{\mathcal{D}^N}$.

Note that $L(\lambda)$ is uniquely determined by its highest weight λ. Thus the following definition of generating series is natural.

DEFINITION 2.1.   (1) If $\mathcal{W} = \mathcal{W}_{1+\infty}$, an element $\lambda \in (\mathcal{W}_{1+\infty})_0^*$ is characterized by the *central charge* $c_\lambda = \lambda(\kappa)$ and its *labels* $\Delta_n = -\lambda(D^n)$, $n \geq 0$. We define the *generating series* to be

(2.1)
$$\Delta_\lambda(x) = \sum_{n=0}^{\infty} \frac{x^n}{n!} \Delta_n.$$

(2) If $\mathcal{W} = \mathcal{W}_\infty$, we define the *labels* of $\lambda \in (\mathcal{W}_\infty)_0^*$ to be $\Delta_n = -\lambda(D^{n+1})$, $n \geq 0$, and define the *generating series* as in (2.1).

(3) If $\mathcal{W} = \widehat{\mathcal{D}^N}$, we define the *labels* of $\lambda \in (\widehat{\mathcal{D}^N})_0^*$ to be $\Delta_{p,n} = -\lambda(D^n E_{pp})$, $p \in [1, N]$, $n \geq 0$ and define the *generating series*

$$\Delta_{p,\lambda}(x) = \sum_{n=0}^{\infty} \frac{x^n}{n!} \Delta_{p,n}.$$

A function is called a *quasipolynomial* if it is a linear combination of functions of the form $p(x)e^{ax}$, where $p(x)$ is a polynomial, $a \in \mathbb{C}$. Then a well-known fact [**9, 14, 16**] states that a formal power series is a quasipolynomial if and only if it satisfies a nontrivial linear differential equation with constant coefficients.

THEOREM 2.2.   (1) *Suppose* $\mathcal{W} = \mathcal{W}_{1+\infty}$. *Then*

$$L(\lambda) \text{ is quasifinite} \iff \Delta_\lambda(x) = \frac{\phi_\lambda(x)}{e^x - 1} \text{ (see [\textbf{16}])}$$
$$\iff L(\lambda) \text{ is a proper quotient of } V(\lambda) \text{ (see [\textbf{37}])},$$

for some quasipolynomial $\phi_\lambda(x)$ *with* $\phi_\lambda(0) = 0$.
(2) *Suppose* $\mathcal{W} = \mathcal{W}_\infty$. *Then*

$$L(\lambda) \text{ is quasifinite} \iff \Delta_\lambda(x) + \frac{d}{dx}\left((e^x - 1)\Delta_\lambda(x)\right) \text{ is a quasipolynomial}$$
$$\iff \Delta_\lambda(x) = \frac{d}{dx}\left(\frac{\phi_\lambda(x)}{e^x - 1}\right)$$

for some quasipolynomial $\phi_\lambda(x)$ *with* $\phi_\lambda(0) = 0$ *(see [\textbf{14}])*.
(3) *Suppose* $\mathcal{W} = \widehat{\mathcal{D}^N}$. *Then*

$$L(\lambda) \text{ is quasifinite} \iff \Delta_{N,\lambda}(x) = \frac{\phi_{N,\lambda}(x)}{e^x - 1}, \text{ and}$$
$$\Delta_{p,\lambda}(x) = \phi_{p,\lambda}(x) + \Delta_{N,\lambda}(x), \quad p = 1, 2, ..., N-1,$$

for some quasipolynomials $\phi_{p,\lambda}(x)$ *with* $\phi_{N,\lambda}(0) = 0$ *(see [\textbf{5}])*.

Thus the last equivalence in Theorem 2.2(1) in particular shows that the Verma module $V(\lambda)$ over $\mathcal{W}_{1+\infty}$ is irreducible if and only if $(e^x - 1)\Delta_\lambda(x)$ is a quasipoly-nomial. For the case $\widehat{\mathcal{W}}(\Gamma, 1)$ with $\Gamma \not\cong \mathbb{Z}$, Ref. [**37**] obtained the following.

THEOREM 2.3. *Let* $\mathcal{L} = \widehat{\mathcal{W}}(\Gamma, 1)$, *and let* $\lambda \in \mathcal{L}_0^*$.
(1) *Suppose the order* $\succ$ *on* Γ *is* **dense** *in the sense that for each* $\alpha \in \Gamma_+ = \{\beta \in \Gamma \mid \beta \succ 0\}$, *the subset* $B(\alpha) = \{\beta \in \Gamma \mid \alpha \succ \beta \succ 0\}$ *of* Γ *is infinite. Then the Verma* $\mathcal{L}$-*module* $V(\lambda)$ *is irreducible if and only if* $\lambda \neq 0$.
(2) *Suppose the order* $\succ$ *on* Γ *is* **discrete** *in the sense that there exists* $a \in \Gamma_+$, *called the* **minimal element** *of* Γ_+ *such that* $B(a) = \emptyset$. *Then the Verma* $\mathcal{L}$-*module* $V(\lambda)$ *is irreducible if and only if the Verma module with highest weight* λ *over the Lie algebra* $\widehat{\mathcal{W}}(a\mathbb{Z}) \cong \mathcal{W}_{1+\infty}$ *is irreducible.*

The main techniques in obtaining Theorem 2.2 are the descriptions of parabolic subalgebras and generalized Verma modules. We briefly overview below.

For convenience, we assume $\mathcal{W} = \mathcal{W}_{1+\infty}$ or $\mathcal{W}_\infty$ (we refer to [5] if $\mathcal{W} = \widehat{\mathcal{D}^N}$). A *parabolic subalgebra* of $\mathcal{W}$ is a subalgebra $\mathcal{P}$ such that

$$\mathcal{P} \underset{\neq}{\supseteq} \mathcal{W}_0 \oplus \mathcal{W}_+, \text{ where } \mathcal{W}_+ = \underset{i > 0}{\oplus} \mathcal{W}_i.$$

Then $\mathcal{P} = \oplus_{i \in \mathbb{Z}} \mathcal{P}_i$ such that $\mathcal{P}_i = \mathcal{W}_i$ for $i \geq 0$ and $\mathcal{P}_i \neq \{0\}$ for some $i < 0$. Let $\lambda \in \mathcal{W}_0^*$ be such that $\lambda|_{\mathcal{W}_0 \cap [\mathcal{P}, \mathcal{P}]} = 0$. The 1-dimensional $(\mathcal{W}_0 \oplus \mathcal{W}_+)$-module $\mathbb{C}v_\lambda$

(cf. Definition 1.1(8)) extends to a $\mathcal{P}$-module by letting $\mathcal{P}_i$ act as zero for $i < 0$. The *generalized Verma module* is the highest weight module

$$V(\mathcal{P}, \lambda) = U(\mathcal{W}) \otimes_{U(\mathcal{P})} \mathbb{C}_\lambda.$$

For a nonzero element $a \in \mathcal{W}_{-1}$, denote $\mathcal{P}^a = \oplus_{i \in \mathbb{Z}} \mathcal{P}_i^a$ to be the *minimal parabolic subalgebra containing* a. Then $\mathcal{P}_i^a = \mathcal{W}_i$ for $i \geq 0$, and

$$\mathcal{P}_{-1}^a = \mathrm{span}\{[...[[a, \mathcal{W}_0], \mathcal{W}_0], ...,]\}, \quad \mathcal{P}_{-i-1}^a = [\mathcal{P}_{-1}^a, \mathcal{P}_{-i}^a],$$

furthermore, $\mathcal{W}_0^a := [\mathcal{P}^a, \mathcal{P}^a] \cap \mathcal{W}_0 = [a, \mathcal{W}_1]$. A parabolic subalgebra $\mathcal{P}$ is *nondegenerate* if $\mathcal{P}_{-i}$ has finite codimension in $\mathcal{W}_{-i}$ for all $i > 0$. A nonzero element $a \in \mathcal{W}_{-1}$ is *nondegenerate* if $\mathcal{P}^a$ is nondegenerate. Then the following results can be proved without too much difficulty [**5, 14, 16**].

LEMMA 2.4. *The Lie algebra $\mathcal{W}$ satisfies the following properties.*

(1) *If $a \in \mathcal{W}_{-\beta}$ for some $\beta > 0$ and $[a, \mathcal{W}_1] = 0$, then $a = 0$.*
(2) *If $\mathcal{P}$ is a nondegenerate parabolic subalgebra of $\mathcal{W}$, then there exists a nondegenerate element a such that $\mathcal{P}^a \subset \mathcal{P}$.*
(3) *Any nonzero element $a \in \mathcal{W}_{-1}$ is nondegenerate.*
(4) *Any parabolic subalgebra of $\mathcal{W}$ is nondegenerate.*

LEMMA 2.5. *The following conditions on $\lambda \in \mathcal{W}_0^*$ are equivalent:*

(1) *$V(\lambda)$ contains a singular vector $a \cdot v_\lambda$ in $V(\lambda)_{-1}$, where a is nondegenerate;*
(2) *there exists a nondegenerate element $a \in \mathcal{W}_{-1}$ such that $\lambda(\mathcal{W}_0^a) = 0$;*
(3) *$L(\lambda)$ is quasifinite;*
(4) *there exists a nondegenerate element $a \in \mathcal{W}_{-1}$ such that $L(\lambda)$ is the irreducible quotient of the generalized Verma module $V(\mathcal{P}^a, \lambda)$.*

Now suppose $L(\lambda)$ is quasifinite. By Lemma 2.5(2) and (4), there exists some monic polynomial $f_\lambda(x)$ of minimal degree, uniquely determined by the highest weight λ, called the *characteristic polynomial* of $L(\lambda)$, such that

(2.2) $$\left(t^{-1} f_\lambda(D)\right) v_\lambda = 0.$$

Denote

(2.3) $$\phi_\lambda(x) = \begin{cases} (1 - e^x)\Delta_\lambda(x) & \text{if } \mathcal{W} = \mathcal{W}_{1+\infty}, \\ \Delta_\lambda(x) + \frac{d}{dx}\left((e^x - 1)\Delta_\lambda(x)\right) & \text{if } \mathcal{W} = \mathcal{W}_\infty. \end{cases}$$

Then Theorem 2.2 is obtained by proving that (2.2) is equivalent to the following differential equation [**14, 16**]

(2.4) $$f_\lambda\left(\frac{d}{dx}\right)\phi_\lambda(x) = 0,$$

which in particular shows that $\phi_\lambda(x)$ is a quasipolynomial. For example, let us prove (2.4) for the case $\mathcal{W} = \mathcal{W}_{1+\infty}$. Write $f_\lambda(x) = \sum_{j=0}^m f_{\lambda,j} x^j$ for some $f_{\lambda,j} \in \mathbb{C}$ with $f_{\lambda,m} = 1$. For $k \geq 0$, applying the element $S_k := t(D+1)^k$ of $\mathcal{W}_1$ (the graded subspace of $\mathcal{W}$ of degree 1) to (2.2), using (1.5), we obtain

$$\lambda\left(D^k f_\lambda(D) - (D+1)^k f_\lambda(D+1) + f_\lambda(0)\delta_{k,0}\kappa\right) = 0,$$

which is equivalent to

(2.5) $$\sum_{j=0}^m f_{\lambda,j} F_{j+k} = 0 \text{ for all } k = 0, 1, ...,$$

where

$$(2.6) \qquad F_j = \delta_{j,0}c + \sum_{s=0}^{j-1} \binom{j}{s}\Delta_s.$$

Introducing the generating series $F(x) = \sum_{s=0}^{\infty} \frac{x^s}{s!}F_s$, we can rewrite (2.5) into the form

$$(2.7) \qquad \Big(\sum_{j=0}^{m} f_{\lambda,j}\big(\frac{d}{dx}\big)^j\Big)F(x) = 0.$$

Thus $L(\lambda)$ is quasifinite if and only if $F(x)$ is a quasipolynomial. But (2.6) in fact means that $c_\lambda - F(x) = (e^x - 1)\Delta_\lambda(x) = \phi_\lambda(x)$, and (2.4) follows from (2.7) (note that in case $\mathcal{W} = \mathcal{W}_\infty$, since the element S_k is not in $\mathcal{W}_\infty$, we need to replace S_k by $t(D+1)^k D \in \mathcal{W}_\infty$ and in this case we need to define $\phi_\lambda(x)$ as the second case of (2.3) in order to have the differential equation (2.4), see [14] for detail).

REMARK 2.6. For all not-finitely $\mathbb{Z}$-graded Lie algebras $\mathcal{L}$ which have properties of Lemmas 2.4 and 2.5, one can consider quasifinite highest weight $\mathcal{L}$-modules as above (e.g., [28, 29]).

Now let us describe unitary irreducible quasifinite highest weight modules. First we introduce the *semi-linear anti-involution* ω on $\widehat{\mathcal{D}^N}$ (for $\mathcal{W}_{1+\infty}$, it is simply the special case $\widehat{\mathcal{D}^1}$), i.e., ω is $\mathbb{R}$-linear satisfying

$$\omega^2 = 1, \quad \omega(ax) = \bar{a}\omega(x), \quad \omega([x,y]) = [\omega(y),\omega(x)],$$

for $a \in \mathbb{C}$, $x,y \in \widehat{\mathcal{D}^N}$, where the overbar denotes the conjugate number. Thus ω is uniquely determined by $\omega(t) = t^{-1}$, $\omega(D) = D$ and $\omega(E_{ij}) = E_{ji}$. Precisely, ω is defined by

$$(2.8) \qquad \omega(t^i f(D)E_{kl}) = \bar{f}(D)t^{-i}E_{kl} = t^{-i}\bar{f}(D-i)E_{kl}, \quad \omega(\kappa) = \kappa,$$

where $\bar{f}(D) = \sum_{j\geq 0} \bar{f}_j D^j$ for $f(D) = \sum_{j\geq 0} f_j D^j$, $f_j \in \mathbb{C}$. For the case $\mathcal{W}_\infty$, since the element $t^{-i}\bar{f}(D-i)$ is in general not in $\mathcal{W}_\infty$, we need to define ω to satisfy

$$\omega(t^i f(D)D) = \bar{f}(D)t^{-i}D = t^{-i}\bar{f}(D-i)D, \quad \omega(\kappa) = \kappa.$$

A module V over $\mathcal{W} = \mathcal{W}_{1+\infty}, \mathcal{W}_\infty, \widehat{\mathcal{D}^N}$ is *unitary* (with respect to the anti-involution ω) if there exists a positive definite Hermitian form $H(\cdot,\cdot)$ on V such that it is *contravariant*, i.e., $\omega(a)$ and a are adjoint operators on V with respect to the Hermitian form H:

$$(2.9) \qquad H(av_1, v_2) = H(v_1, \omega(a)v_2) \quad \text{for} \quad v_1, v_2 \in V \text{ and } a \in \mathcal{W}.$$

For the irreducible highest weight module $L(\lambda)$ such that the central charge $c_\lambda = \lambda(\kappa)$ and all labels are real numbers, there exists a unique contravariant Hermitian form $H(\cdot,\cdot)$ on $L(\lambda)$ such that $H(v_\lambda, v_\lambda) = 1$, which is defined by

$$H(av_\lambda, bv_\lambda) = \langle \omega(a)bv_\lambda \rangle \quad \text{for} \quad a,b \in U(\mathcal{W}),$$

where $\langle v \rangle$ is the coefficient of v_λ in the decomposition of v with respect to the gradation of $L(\lambda)$ for $v \in L(\lambda)$.

We have the following (note that Theorem 2.7(1) is the special case of (3) for $N = 1$).

THEOREM 2.7.   (1) (see [16]) *The irreducible quasifinite highest weight* $\mathcal{W}_{1+\infty}$-*module* $L(\lambda)$ *is unitary if and only if* $\phi_\lambda(x) = \sum_i n_i(e^{r_i x} - 1)$ *for some positive integers* n_i *and distinct real numbers* r_i, *such that* $c_\lambda = \sum_i n_i$, *and* $\phi_\lambda(x)$ *is as in Theorem 2.2.*

(2) (see [14]) *The nontrivial* $\mathcal{W}_\infty$-*module* $L(\lambda)$ *is unitary if and only if the following conditions are satisfied:*

    (a) $\phi_\lambda(x) + c = \sum_i n_i e^{r_i x}$ *for some nonzero integers* n_i *and distinct real numbers* r_i.

    (b) *For each* i, $n_i > 0$ *whenever* $r_i \in \mathbb{Z}$.

    (c) *For* $0 < \alpha < 1$, *let* $E_\alpha = \{i \,|\, r_i \in \alpha + \mathbb{Z}\}$. *Then for each* α *with* $E_\alpha \neq \emptyset$, *the set* $\{n_i \,|\, i \in E_\alpha\}$ *contains exactly one negative number, and* $m_\alpha = -\sum_{i \in E_\alpha} n_i > 0$ *and* $r_i - r_j \leq m_\alpha$ *for all* $i, j \in E_\alpha$.

(3) (see [5]) *A nontrivial* $\widehat{\mathcal{D}^N}$-*module* $L(\lambda)$ *is unitary if and only if* c *is a positive integer and there exist* $r_1, ..., r_c \in \mathbb{R}$ *and a partition* $\{1, ..., c\} = I_0 \cup \cdots \cup I_{N-1}$, *such that*

$$(e^x - 1)\Delta_{p,\lambda}(x) = \sum_{i \in I_0 \cup \cdots \cup I_{p-1}} e^{r_i x} + \sum_{i \in I_p \cup \cdots \cup I_{N-1}} e^{(r_i+1)x} - c.$$

The main technique to obtain the results is the description of relations of $\mathcal{W}_{1+\infty}$, $\mathcal{W}_\infty$ and $\widehat{\mathcal{D}^N}$ to the central extension $\widehat{gl}_\infty(R^m)$ of the Lie algebra $\widetilde{gl}_\infty(R^m)$ of infinite matrices with only finitely many nonzero diagonals over the algebras $R^m = \mathbb{C}[t]/(t^{m+1})$ for $m \in \mathbb{Z}_+$. A key in obtaining Theorem 2.7(1) is the following (similar results also hold for $\mathcal{W}_\infty$, $\widehat{\mathcal{D}^N}$, and even for other not-finitely $\mathbb{Z}$-graded Lie algebras, e.g., Lie algebras considered in [28, 29]).

LEMMA 2.8. *If* $\mathcal{W}_{1+\infty}$-*module* $L(\lambda)$ *is unitary, then the characteristic polynomial* $f_\lambda(x)$ *has only simple real roots.*

This can be proved as follows. Let $T = -\frac{1}{2}(D^2 - \Delta_2 - 1) \in \mathcal{W} = \mathcal{W}_{1+\infty}$, one can check $T^j(t^{-1}v_\lambda) = (t^{-1}D^j)v_\lambda$ for $j \geq 0$. Since $\big(t^{-1}f_\lambda(D)\big)v_\lambda = 0$, we see $\{(t^{-1}D^j)v_\lambda \,|\, 0 \leq j < m\}$ is a basis of $L(\lambda)_{-1}$ (the graded subspace of $L(\lambda)$ of degree -1, cf. Remark 1.2(3)), where $m = \deg f_\lambda(x)$. Thus $f_\lambda(T)(t^{-1}v_\lambda) = 0$ and $\{T^j(t^{-1}v_\lambda) \,|\, 0 \leq j < m\}$ is a basis of $L(\lambda)_{-1}$. This shows that $f_\lambda(x)$ is the characteristic polynomial of the linear transformation T on $L(\lambda)_{-1}$. Since T is self-adjoint, hence roots of $f_\lambda(x)$ are real. Let α be a root of $f_\lambda(x)$ with multiplicity k, so that $f_\lambda(x) = g(x)(x-\alpha)^k$, $g(x) \in \mathbb{C}[x]$. Then $v = (T-\alpha)^{k-1}g(T)(t^{-1}v)$ is a nonzero vector in $L(\lambda)_{-1}$, but by (2.8) and (2.9),

$$H(v,v) = H\big(g(T)(t^{-1}v), (T-\alpha)^{2k-2}g(T)(t^{-1}v)\big) = 0 \quad \text{if } k \geq 2.$$

Thus $k = 1$, which proves the lemma.

By Lemma 2.8, one sees that a solution to the differential equation (2.4) must have the form $\phi_\lambda(x) = \sum_i n_i e^{r_i x}$ for some $n_i \in \mathbb{C}$ and some real numbers r_i. From this and the unitary condition, one can then deduce Theorem 2.7(1).

3. Classification of quasifinite modules

The classification of irreducible modules in some category is definitely one of the most important problems in the representation theory of Lie algebras. In this section, we shall consider the problem of classification of quasifinite modules over Weyl type Lie algebras. Before we state the main results, let us present some related results for some well-known Lie algebras (some results may be used to prove the results for Weyl type Lie algebras).

3.1. Some results for the Virasoro algebra and affine Kac-Moody algebras.

It is well known [**11, 19, 23, 25, 27**] that a module of the intermediate series over Vir (cf. (1.13)) is a subquotient of modules $A_{a,b}$, $A(\alpha)$, $B(\alpha)$ for $a, b, \alpha \in \mathbb{C}$, where $A_{a,b}$, $A(\alpha)$, $B(\alpha)$ all have basis $\{v_i \mid i \in \mathbb{Z}\}$ such that the central element κ acts trivially and

$$(3.1) \qquad A_{a,b}: \; L_i v_j \;=\; (a + j + bi) v_{i+j},$$

$$(3.2) \qquad A(\alpha): \; L_i v_j \;=\; (i + j) v_{i+j} \; (j \neq 0), \quad L_i v_0 = i(i + \alpha) v_i,$$

$$(3.3) \qquad B(\alpha): \; L_i v_j \;=\; j v_{i+j} \; (j \neq -i), \qquad L_i v_{-i} = -i(i + \alpha) v_0.$$

The following famous theorem was originally conjectured by Kac [**11**] in 1982, and proved by Mathieu [**19**] in 1992 (with partial results proved in [**23**]).

THEOREM 3.1. *A **Harish-Chandra** Vir-module (i.e., an irreducible quasifinite Vir-module) is either a module of the intermediate series or a highest/lowest weight module.*

The same theorem also holds for the super-Virasoro algebras [**24**]. It is also obtained in [**25**] that a uniformly bounded Harish-Chandra module over a higher rank Virasoro algebra is a module of the intermediate series. Furthermore, Theorem 3.1 was generalized to the higher hank Virasoro algebras [**18, 27**]:

THEOREM 3.2. *Any Harish-Chandra module over the higher rank Virasoro algebra $Vir(\Gamma)$ is*

  (1) *a module of the intermediate series or a highest/lowest weight module if $\Gamma \cong \mathbb{Z}$;*
  (2) *a module of the intermediate series if $\operatorname{rank} \Gamma = 1$ and $\Gamma \not\cong \mathbb{Z}$;*
  (3) *a module of the intermediate series or isomorphic to $L(a, b, \Gamma', \alpha)$ for some $a, b \in \mathbb{C}$, $\alpha \in \Gamma$ and a subgroup Γ' of Γ with $\Gamma = \Gamma' \oplus \mathbb{Z}\alpha$ if $\operatorname{rank} \Gamma > 1$.*

Here the $Vir(\Gamma)$-module $V(a, b, \Gamma', \alpha)$ is defined as follows: Suppose $\Gamma = \Gamma' \oplus \mathbb{Z}\alpha$ for some subgroup Γ' of Γ and some $\alpha \in \Gamma$. Let $a, b \in \mathbb{C}$, and let $M_{a,b}$ be the (unique) nontrivial irreducible quotient of the $Vir(\Gamma')$-module of the intermediate series of the form $A_{a,b}$ (cf. (3.1)). Write $Vir(\Gamma) = \oplus_{i \in \mathbb{Z}} Vir(\Gamma)_i$ such that $Vir(\Gamma)_i$ has basis $\{L_{\beta + i\alpha}, \delta_{i,0}\kappa \mid \beta \in \Gamma'\}$ (thus $Vir(\Gamma') = Vir(\Gamma)_0$). Then

$$Vir(\Gamma) \;=\; Vir(\Gamma)_- \oplus Vir(\Gamma)_0 \oplus Vir(\Gamma)_+$$

is a triangular decomposition of $Vir(\Gamma)$, where $Vir(\Gamma)_\pm = \oplus_{\pm i > 0} Vir(\Gamma)_i$. Put $Vir(\Gamma)_+ M_{a,b} = 0$ and define the induced module (sometimes also called a *generalized Verma module*)

$$
\begin{aligned}
V(M_{a,b}) &= \operatorname{Ind}_{Vir(\Gamma)_+ \oplus Vir(\Gamma)_0}^{Vir(\Gamma)} M_{a,b} \\
&= U(Vir(\Gamma)) \otimes_{U(Vir(\Gamma)_+ \oplus Vir(\Gamma)_0)} M_{a,b} \cong U(Vir(\Gamma)_-) \otimes_{\mathbb{C}} M_{a,b}.
\end{aligned}
$$

Then $L(a, b, \Gamma', \alpha) = \widetilde{V}(M_{a,b})/J$ is the (unique) irreducible quotient of $V(M_{a,b})$, where J is the maximal proper submodule of $V(M_{a,b})$.

The following is also obtained in [**27**].

THEOREM 3.3. *Suppose there is a group injection $\mathbb{Z} \times \mathbb{Z} \to \Gamma$. Then a Harish-Chandra module over $\mathrm{Witt}(\Gamma, n)$ is either a uniformly bounded module with all nonzero weights having the same multiplicity, or a finitely-dense module (see [27] for definition).*

Earlier than the time when Theorem 3.1 was proved, an analogous result of Theorem 3.1 was obtained in [6, 7] in 1986 and 1988 for integrable irreducible modules over (nontwisted and twisted) affine Kac-Moody algebras. Here a module over a Kac-Moody algebra $\mathfrak{g}$ is called *integrable* [12] if it is diagonalizable with respect to a Cartan subalgebra $\mathfrak{h}$ of $\mathfrak{g}$ and if, as a representation of each of the sl_2-subalgebras of $\mathfrak{g}$ corresponding to the Chevalley generators of $\mathfrak{g}$, it breaks up as a direct sum of finite-dimensional representations (such representations are interesting because they automatically lift to representations of the corresponding Kac-Moody group). The main results of [6, 7] can be stated as the following (for convenience we only state the result for the nontwisted case).

THEOREM 3.4. *Let V be an integrable irreducible module over the nontwisted affine Kac-Moody algebra $\mathfrak{g}$. Then the center κ' (here κ' is a proper scalar multiple of κ such that in (1.16), κ is replaced by κ' and $\mathrm{tr}(AB)$ is replaced by minus the Killing form of the corresponding finite dimensional simple Lie algebra $\mathfrak{g}_0$) acts on V by an integer $l(V)$ (called the **level** of V), and there are three cases:*

(1) *if $l(V) > 0$ then V is a highest weight module (usually called a **standard module**);*

(2) *if $l(V) < 0$ then V is a lowest weight module;*

(3) *if $l(V) = 0$ then V is a loop module (see [6] for definition).*

In the following subsections, we shall see that Theorem 3.1 can be generalized to Weyl type Lie algebras. We would like to mention that similar results might also hold for some other Γ-graded Lie algebras which contain the Virasoro algebra (or Witt algebra) as a subalgebra; for example, Block type Lie algebras [28, 29].

3.2. Classification of quasifinite modules over $\mathcal{W}(\Gamma, n)$ and $\widehat{\mathcal{W}}(\Gamma, 1)$.
Denote the set of n-tuples of commuting $p \times p$ matrices by

$$gl_p^n = \{G = (G_1, ..., G_n) \mid G_i \in gl_p, \ G_i G_j = G_j G_i \ \text{for} \ i, j = 1, ..., n\}.$$

Let $G = (G_1, ..., G_n) \in gl_p^n$. The tensor space $\mathbb{C}[\Gamma] \otimes \mathbb{C}^p$ can be defined as a module over the associative algebra $\mathcal{A}(\Gamma, n)$ (cf.(1.2)), thus also a $\mathcal{W}(\Gamma, n)$-module (denoted by $A_{p,G}$), such that t acts as the multiplication and D_i acts as the derivation on $\mathbb{C}[\Gamma]$ and as G_i on $\mathbb{C}^p$. Namely,

$$(3.4) \quad A_{p,G}: \quad t^\alpha D^\mu \cdot t^\beta v = t^{\alpha+\beta} [\beta \cdot \mathbb{1} + G]^\mu v \quad \text{for} \quad \alpha, \beta \in \Gamma, \ \mu \in \mathbb{Z}_+^n, \ v \in \mathbb{C}^p,$$

where $\mathbb{1} = (\mathbf{1}_p, ..., \mathbf{1}_p) \in gl_p^n$ with $\mathbf{1}_p$ being the $p \times p$ identity matrix, $[\beta \cdot \mathbb{1} + G]^\mu = \prod_{i=1}^n (\beta_i \cdot \mathbf{1}_p + G_i)^{\mu_i}$ and $\beta_i \cdot \mathbf{1}_p$ denotes the scalar multiplication of the identity matrix, and where the action of a matrix on v is defined by the matrix-vector multiplication (by regarding v as a column vector). Obviously, $A_{p,G}$ is a uniformly bounded $\mathcal{W}(\Gamma, n)$-module.

By [34], there exists a Lie algebra isomorphism (involution) $\sigma : \mathcal{W}(\Gamma, n) \cong \mathcal{W}(\Gamma, n)$ such that

$$(3.5) \qquad \sigma(t^\alpha D^\mu) = (-1)^{|\mu|+1} D^\mu \cdot t^\alpha \quad \text{for} \quad \alpha \in \Gamma, \ \mu \in \mathbb{Z}_+,$$

where $|\mu| = \sum_{i=1}^{n} \mu_p$ is as in (1.6), and the "$\cdot$" in the right-hand side means the product defined by (1.2). Using this isomorphism, we have another $\mathcal{W}(\Gamma, n)$-module $\overline{A}_{p,G}$, called the *twisted module of* $A_{p,G}$, for the pair (p, G), defined by

$$(3.6) \qquad \overline{A}_{p,G} : \qquad t^{\alpha} D^{\mu} \cdot t^{\beta} v = (-1)^{|\mu|+1} t^{\alpha+\beta} [(\alpha + \beta) \cdot \mathbb{1} + G]^{\mu} v,$$

for $\alpha, \beta \in \Gamma$, $\mu \in \mathbb{Z}_+^n$, $v \in \mathbb{C}^p$. Clearly, $A_{p,G}$ or $\overline{A}_{p,G}$ is decomposable if and only if there exist some invertible matrix P and some integer p_1 with $1 \leq p_1 < p$ such that every $P^{-1} G_i P$ has the form $\mathrm{diag}(G_{i,1}, G_{i,2})$ for some $p_1 \times p_1$ matrix $G_{i,1}$ and $(p - p_1) \times (p - p_1)$ matrix $G_{i,2}$, and

$$(3.7) \qquad A_{p,G} \text{ or } \overline{A}_{p,G} \text{ is irreducible } \iff p = 1,$$

(in this case $G \in \mathbb{C}^n$, and $A_{1,G}$, $\overline{A}_{1,G}$ are modules of the intermediate series).

Now we state the result obtained in [**26**].

THEOREM 3.5.   (1) *Let* $\mathcal{W}$ *be the Lie algebra* $\mathcal{W}(\mathbb{Z}, 1)$ *or* $\mathcal{W}_{1+\infty}$.
    (i) *A uniformly bounded* $\mathcal{W}$-*module is a direct sum of a trivial module, a module* $A_{p,G}$ *and a module* $\overline{A}_{p',G'}$ *for some* $p, p' \in \mathbb{Z}_+$, $G \in gl_p$, $G' \in gl_{p'}$ *(in the central extension case, the central element* κ *acts trivially).*
    (ii) *An irreducible quasifinite* $\mathcal{W}$-*module is a highest/lowest weight module or a module of the intermediate series (i.e.,* $A_{1,G}$ *or* $\overline{A}_{1,G}$ *for some* $G \in \mathbb{C}$).
(2) *Suppose* $\mathcal{W} = \mathcal{W}(\Gamma, n)$, $\widehat{\mathcal{W}}(\Gamma, 1)$ *with* $\Gamma \not\cong \mathbb{Z}$.
    (i) *A quasifinite* $\mathcal{W}$-*module is a direct sum of a trivial module and a uniformly bounded module.*
    (ii) *A uniformly bounded* $\mathcal{W}$-*module is a direct sum of a trivial module, a module* $A_{p,G}$ *and a module* $\overline{A}_{p',G'}$ *for some* $p, p' \in \mathbb{Z}_+$, $G \in gl_p^n$, $G' \in gl_{p'}^n$.
    (iii) *A nontrivial irreducible quasifinite* $\mathcal{W}$-*module is a module of the intermediate series (i.e.,* $A_{1,G}$ *or* $\overline{A}_{1,G}$ *for some* $G \in \mathbb{C}^n$).

REMARK 3.6.   (1) (cf. Remark 1.2(1)) If V is a weight module, each G_i in (3.4) is diagonalizable, and all uniformly bounded modules are completely reducible.
(2) In Theorem 3.5(2), if a module have infinite number of the trivial composition factors, then it may not be uniformly bounded since any Γ-graded vector space can be defined as a trivial module.

The main techniques in obtaining Theorem 3.5 are developed from the techniques of Mathieu [**19**] used in his proof of Theorem 3.1. These techniques seem to be useful in obtaining analogous results for some $\mathbb{Z}$-graded Lie algebras (e.g., [**28, 29**]). The main idea is to make use of results of representations of the Virasoro subalgebra (cf. (1.13)).

Let us first consider $\mathcal{W} = \mathcal{W}_{1+\infty}$, and denote (cf. (1.4)):

$$L_{i,j} = t^i [D]_j = t^{i+j} (\frac{d}{dt})^j \text{ for } i \in \mathbb{Z}, j \in \mathbb{Z}_+.$$

Using Theorems 3.1 and 3.2, one can obtain the following 2 lemmas, from which, Theorem 3.5(1) follows.

LEMMA 3.7.   (1) *Suppose V is a uniformly bounded* Vir-*module generated by a vector v such that there exists some $k_0 \in \mathbb{Z}$ satisfying $L_i v = 0$ for all $i \in \mathbb{Z}k_0$. Then $V = \mathbb{C}v$ is trivial.*

(2) *Suppose V is an irreducible quasifinite $\mathcal{W}$-module without a highest/lowest weight. Then*

$$L_{1,0} : V_i \to V_{i+1} \quad and \quad L_{-1,0} : V_i \to V_{i-1}$$

are injective and thus bijective for all $i \in \mathbb{Z}$. In particular, V is uniformly bounded.

LEMMA 3.8. *Suppose $V = \oplus_{i\in\mathbb{Z}} V_i$ is an indecomposable uniformly bounded $\mathcal{W}$-module without the trivial composition factor. Then*

(1) *$L_{0,0}$ acts as a constant, and κ acts as zero.*
(2) *$L_{i,0}$ acts nondegenerately on V for all $i \neq 0$.*
(3) *V is a module of the form $A_{p,G}$ or $\overline{A}_{p,G}$.*

The proof of Lemma 3.7(1) is straightforward, while Lemma 3.7(2) can be obtained by following the arguments in the proof of Lemma 1.6 in [19]. To prove Lemma 3.8, by Lemma 3.7(2), there exists $p \in \mathbb{Z}_+$ such that $\dim V_i = p$ for $i \in \mathbb{Z}$, and hence we can choose a basis $Y_0 = (y_0^{(1)}, ..., y_0^{(p)})$ of V_0, and define a basis $Y_i = (y_i^{(1)}, ..., y_i^{(p)})$ of V_i satisfying $Y_i = L_{1,0}Y_{i-1}$. Then by supposing $L_{i,j}Y_k = Y_{i+k}P_{i+j,j,k}$ for some $P_{i+j,j,k} \in gl_p$, and using relation (1.4), one can easily deduce

$$P_{i,0,n} = P_i, \qquad\qquad P_{i,1,n} = \bar{n}P_i + Q_i,$$
$$P_{i,2,n} = [\bar{n}]_2 P_i + 2\bar{n}Q_i + R_i, \qquad P_{i,3,n} = [\bar{n}]_3 P_i + 3[\bar{n}]_2 Q_i + 3\bar{n}R_i + S_i,$$

for some $P_i, Q_i, R_i, S_i \in gl_p$, where $\bar{n}+G$ for some fixed $G \in M_{p\times p}$ (here we identify a scalar with the corresponding $p \times p$ scalar matrix), $[\bar{n}]_j$ is a similar notation to $[D]_j$, and $Q_1 = 0$ (we use notation $\bar{n} + G$ in order to be able to take $Q_1 = 0$; note from $[L_{0,1}, L_{i,j}] = iL_{i,j}$ that G commutes with all other matrices involved in the discussion). Then by further use of relation (1.4), we can deduce $P_i = \pm 1$. Thus by (3.5), we can suppose $P_i = 1$, and then deduce $Q_i = R_i = 0$. ¿From this, the lemma follows.

For the case $\mathcal{W} = \mathcal{W}(\Gamma, n)$ with $\Gamma \not\cong \mathbb{Z}$, let $Witt(\Gamma, n)$ be the generalized Witt algebra defined in (1.15). Using Theorem 3.3, one can obtain the following result, from which, Theorem 3.5(2) can be then deduced.

LEMMA 3.9.   (1) *Any quasifinite $\mathcal{W}$-module V with a finite number of the trivial composition factor is a uniformly bounded module.*

(2) *Let V be a uniformly bounded $\mathcal{W}$-module without the trivial composition factor. Then $t^\alpha \cdot v \neq 0$ for all $\alpha \in \Gamma\backslash\{0\}$, $v \in V\backslash\{0\}$.*

3.3. Classification of quasifinite representations of $\mathcal{W}(\Gamma, n)^{(1)}$ and $\widehat{\mathcal{W}}(\Gamma, 1)^{(1)}$. Although $\mathcal{W}(\Gamma, n)^{(1)}$ is a subalgebra of $\mathcal{W}(\Gamma, n)$, the representations of $\mathcal{W}(\Gamma, n)^{(1)}$ seem to be more complicated than those of $\mathcal{W}(\Gamma, n)$, due to the crucial fact that the elements $t^\beta = t^\beta D^0$ for $\beta \in \Gamma$ do not appear in $\mathcal{W}(\Gamma, n)^{(1)}$ (thus we do not have some nice results such as Lemma 3.8(1) and (2)). Because of this, we are unable to work out indecomposable uniformly bounded $\mathcal{W}(\Gamma, n)^{(1)}$-module. But we can still determine its irreducible modules, and the proof of the results seems to be more technical than that of $\mathcal{W}(\Gamma, n)$ [31].

For $\alpha \in \mathbb{C}^n$, we have two families of modules $A_\alpha, \overline{A}_\alpha$ of the intermediate series over $\mathcal{W}(\Gamma, n)^{(1)}$ or $\widehat{\mathcal{W}}(\Gamma, 1)^{(1)}$ as follows (which correspond to (3.4), (3.6) with $p = 1$): They have basis $\{y_\beta \mid \beta \in \Gamma\}$ such that the central element κ acts trivially and

$$A_\alpha : (t^\beta D^\mu) y_\gamma = (\alpha + \gamma)^\mu y_{\beta+\gamma},$$

$$\overline{A}_\alpha : (t^\beta D^\mu) y_\gamma = (-1)^{|\mu|+1} (\alpha + \beta + \gamma)^\mu y_{\beta+\gamma},$$

for $\beta, \gamma \in \Gamma$, $\mu \in \mathbb{Z}_+^n \setminus \{0\}$ (where $(\alpha + \gamma)^\mu$ is a notation as β^λ in (1.2)). Obviously, A_α or $\overline{A}_\alpha$ is irreducible if and only if $\alpha \notin \Gamma$.

REMARK 3.10. It is straightforward to prove that $A_\alpha, \overline{A}_\alpha$ are the only $\mathcal{W}(\Gamma, n)^{(1)}$-modules of the intermediate series. In particular, unlike the Virasoro algebra case (cf. (3.2) and (3.3)), $\mathcal{W}(\Gamma, n)^{(1)}$-modules $A_\alpha, \overline{A}_\alpha$ do not have any deformation.

The main result of this subsection is the following result obtained in [**31**].

THEOREM 3.11.   (1) *Suppose* $\mathcal{W} = \mathcal{W}(\mathbb{Z}, 1)^{(1)}$ *or* $\mathcal{W}_\infty$.
  (i) *An irreducible quasifinite* $\mathcal{W}$-*module is a highest/lowest weight module, or a module of the intermediate series.*
  (ii) *A nontrivial uniformly bounded indecomposable weight* $\mathcal{W}$-*module is a module of the intermediate series.*
 (2) *Suppose* $\mathcal{W} = \mathcal{W}(\Gamma, n)^{(1)}$ *or* $\widehat{\mathcal{W}}(\Gamma, n)^{(1)}$ *such that* $\Gamma \not\cong \mathbb{Z}$.
  (i) *A nontrivial irreducible quasifinite* $\mathcal{W}$-*module is a module of the intermediate series.*
  (ii) *A nontrivial quasifinite indecomposable weight* $\mathcal{W}$-*module is a module of the intermediate series.*

This theorem mainly follows from the following 2 lemmas, where $\mathcal{W} = \mathcal{W}_\infty$.

LEMMA 3.12. *Let* S *be a subspace of* $\mathcal{W}_0$ *with finite codimension. Given* $i_0 > 0$, *let* $M_{i_0, S}$ *denote the subalgebra of* $\mathcal{W}$ *generated by* $t^{i_0} D, t^{i_0+1} D, t^{i_0} D^2$ *and* S. *Then there exists some integer* $K > 0$ *such that* $\mathcal{W}_{[K, \infty)} := \oplus_{i \geq K} \mathcal{W}_i \subset M_{i_0, S}$.

LEMMA 3.13. *Assume that* V *is an irreducible quasifinite* $\mathcal{W}$-*module without a highest/lowest weight. For any* $i, j \in \mathbb{Z}$, $i \neq 0, -1$, *the linear map*

$$t^i D|_{V_j} \oplus t^{i+1} D|_{V_j} \oplus t^i D^2|_{V_j} : \quad V_j \to V_{i+j} \oplus V_{i+j+1} \oplus V_{i+j}$$

is injective. In particular, $\dim V_j \leq 2 \dim V_0 + \dim V_1$ *for* $j \in \mathbb{Z}$, *thus* V *is uniformly bounded.*

3.4. Classification of quasifinite representations of $\widehat{\mathcal{D}^N}$. We define 2 families of $\widehat{\mathcal{D}^N}$-modules $V(\alpha), \overline{V}(\alpha)$, $\alpha \in \mathbb{C}$, of the intermediate series below. For a fixed $\alpha \in \mathbb{C}$, the obvious representation of $\widehat{\mathcal{D}^N}$ (with trivial action of the central element κ) on the vector space $V(\alpha) = t^\alpha \mathbb{C}^N[t, t^{-1}]$ defines an irreducible module $V(\alpha)$. Let $\{\varepsilon_p = (\delta_{p1}, ..., \delta_{pN})^\mathrm{T} \mid p \in [1, N]\}$ be the standard basis of $\mathbb{C}^N$, where the superscript "T" means the transpose of vectors or matrices (thus the elements of $\mathbb{C}^N$ are column vectors). Then an element $t^i D^j E_{p,q} \in \widehat{\mathcal{D}^N}$ acts on a vector $t^{k+\alpha} \varepsilon_r \in V(\alpha)$ by

(3.8) $$(t^i D^j E_{p,q})(t^{k+\alpha} \varepsilon_r) = \delta_{q,r}(k + \alpha)^j t^{i+k+\alpha} \varepsilon_p,$$

for $i, k \in \mathbb{Z}$, $j \in \mathbb{Z}_+$, $p, q, r \in [1, N]$. For $j \in \mathbb{Z}$, let $V(\alpha)_j = \mathbb{C}t^{k+\alpha} \varepsilon_r$, where $k \in \mathbb{Z}$, $r \in [1, N]$ are unique such that $j + 1 = kN + r$, then $V(\alpha) = \oplus_{j \in \mathbb{Z}} V(\alpha)_j$ is a $\mathbb{Z}$-graded space such that $\dim V(\alpha)_j = 1$ for $j \in \mathbb{Z}$. Thus $V(\alpha)$ is a module of the intermediate series.

For $v \in \mathbb{C}^N$, $k \in \mathbb{Z}$, denote $v_k = t^{k+\alpha}v \in V(\alpha)$ (note that v_k is in general not a homogeneous vector). For $A \in gl_N$, define $Av_k = (Av)_k$, where Av is the natural action of A on v defined linearly by $E_{p,q}\varepsilon_r = \delta_{q,r}\varepsilon_p$ (i.e., the action is defined by the matrix-vector multiplication). Then (3.8) can be rewritten as

$$(3.9) \qquad (t^i D^j A)v_k = (k+\alpha)^j Av_{i+k},$$

for $i, k \in \mathbb{Z}$, $j \in \mathbb{Z}_+$, $A \in gl_N$, $v \in \mathbb{C}^N$. Clearly $V(\alpha)$ is also a $\mathcal{D}_{as}^N$-module (cf. (1.8) and (1.9)). Note that (cf. (3.5)) there exists a Lie algebra isomorphism $\sigma : \mathcal{D}^N \cong \mathcal{D}^N$ such that

$$(3.10) \qquad \sigma(t^i D^j A) = (-1)^{j+1} t^i (D+i)^j A^{\mathrm{T}},$$

for $i \in \mathbb{Z}$, $j \in \mathbb{Z}_+$, $A \in gl_N$. Using this isomorphism, we have another $\widehat{\mathcal{D}^N}$-module $\overline{V}(\alpha)$ (again with trivial action of κ), called the *twisted module of* $V(\alpha)$, defined by

$$(3.11) \qquad (t^i D^j A)v_k = (-1)^{j+1}(i+k+\alpha)^j A^{\mathrm{T}} v_{i+k},$$

for $i, k \in \mathbb{Z}$, $j \in \mathbb{Z}_+$, $A \in gl_N$, $v \in \mathbb{C}^N$. In fact, $\overline{V}(\alpha)$ is the *dual module* of $V(-\alpha)$: If we define a nondegenerate bilinear form on $\overline{V}(\alpha) \times V(-\alpha)$ by $\langle t^{i+\alpha}\varepsilon_p, t^{j-\alpha}\varepsilon_q \rangle = \delta_{i+j,0}\delta_{p,q}$, then

$$\langle x\overline{v}, v \rangle = -\langle \overline{v}, xv \rangle \quad \text{for} \quad x \in \widehat{\mathcal{D}^N}, \; \overline{v} \in \overline{V}(\alpha), \; v \in V(-\alpha).$$

Obviously, $\overline{V}(\alpha)$ is not a $\mathcal{D}_{as}^N$-module.

Now we can generalize the above modules $V(\alpha)$ and $\overline{V}(\alpha)$ as follows: Let α be an indecomposable linear transformation on $\mathbb{C}^m$ (thus up to equivalences, α is uniquely determined by its unique eigenvalue λ). Let gl_N and α act on $\mathbb{C}^N \otimes \mathbb{C}^m$ defined by

$$A(u \otimes v) = Au \otimes v, \; \alpha(u \otimes v) = u \otimes \alpha v \quad \text{for} \quad A \in gl_N, \; u \in \mathbb{C}^N, \; u \in \mathbb{C}^m.$$

Then in (3.9) and (3.11), by allowing v to be in $\mathbb{C}^N \otimes \mathbb{C}^m$, we obtain 2 families of indecomposable uniformly bounded modules $V(m, \alpha)$, $\overline{V}(m, \alpha)$.

The main result of this subsection is the following theorem obtained in [**30**].

THEOREM 3.14.   (1) *An irreducible quasifinite module over* $\widehat{\mathcal{D}^N}$ *is a highest/lowest weight module or else a module of the intermediate series.*

(2) *A nontrivial module of the intermediate series over* $\widehat{\mathcal{D}^N}$ *is a module* $V(\alpha)$ *or* $\overline{V}(\alpha)$ *for some* $\alpha \in \mathbb{C}$.

(3) *A nontrivial indecomposable uniformly bounded module over* $\widehat{\mathcal{D}^N}$ *is a module* $V(m, \alpha)$ *or* $\overline{V}(m, \alpha)$ *for some* $m \in \mathbb{Z}_+ \backslash \{0\}$ *and some indecomposable linear transformation* α *of* $\mathbb{C}^m$.

Thus in particular, a nontrivial indecomposable uniformly bounded module over $\mathcal{D}^N$ is simply a $\mathcal{D}_{as}^N$-module or its twist, and there is an equivalence between the category of uniformly bounded $\mathcal{D}_{as}^N$-modules without the trivial composition factor and the category of linear transformations on finite-dimensional vector spaces. Since irreducible quasifinite highest weight modules have been classified in [**5**] (Theorem 2.2(3)) and irreducible lowest weight modules are simply dual modules of irreducible highest weight modules, Theorem 3.14 in fact classifies all irreducible quasifinite modules over $\widehat{\mathcal{D}^N}$ and over $\mathcal{D}^N$.

Note that there is a one to one correspondence between Lie conformal algebras and maximal formal distribution Lie algebras, and the Lie algebra $\mathcal{D}^N$ is simply the formal distribution Lie algebra associated to the general Lie conformal algebra gc_N. Thus in the language of conformal algebras, this theorem in particular also gives proofs of Theorems 6.1 and 6.2 of [13] on the classification of finite indecomposable modules over the conformal algebras Cend_N and gc_N.

The proof of Theorem 3.14 mainly follows from Theorem 3.5 and the following result.

LEMMA 3.15. *Suppose V is an irreducible quasifinite $\widehat{\mathcal{D}^N}$-module without highest and lowest weight vectors. Then*

$$t|_{V_i} : V_i \to V_{i+N} \quad and \quad t^{-1}|_{V_i} : V_i \to V_{i-N}$$

are injective and thus bijective for all $i \in \mathbb{Z}$ (recall from (1.12) that t has degree N). In particular, by letting $K = \max\{\dim V_p \,|\, p \in [1, N]\}$, we have $\dim V_i \leq K$ for $i \in \mathbb{Z}$; thus V is uniformly bounded.

Acknowledgement. The author would like to thank Professor Zongzhu Lin and the referee for useful comments and suggestions.

References

[1] H. Awata, M. Fukuma, Y. Matsuo, S. Odake, *Character and determinant formulae of quasifinite representations of the $\mathcal{W}_{1+\infty}$ algebra*, Commun. Math. Phys. **172** (1995), 377–400.

[2] S. Bloch, *Zeta values and differential operators on the circle*, J. Algebra **182** (1996), 476–500.

[3] R. Blumenhagen, W. Eholzer, A. Honecker, K. Hornfeck and R. Hübel, *Unifying W-algebras*, Phys. Lett. B. **332** (1994), 51–60.

[4] P. Bouwknegt, K. Schoutens, *W-symmetry in conformal field theory*, Phys. Rep. **223** (1993), 183–276.

[5] C. Boyallian, V. Kac, J. Liberati, C. Yan, *Quasifinite highest weight modules of the Lie algebra of matrix differential operators on the circle*, J. Math. Phys. **39** (1998), 2910–2928.

[6] V. Chari, *Integrable representations of affine Lie algebras*, Invent. Math. **85** (1986), 317–335.

[7] V. Chari, A. Pressley, *Integrable representations of twisted affine Lie algebras*, J. Algebra **113** (1988), 438–464.

[8] B. Feigin, *The Lie algebra $gl(\lambda)$ and the cohomology of the Lie algebra of differential operators*, Uspechi Math. Nauk **35** (1988), 157–158.

[9] E. Frenkel, V. Kac, R. Radul, W. Wang, $\mathcal{W}_{1+\infty}$ and $\mathcal{W}(gl_N)$ with central charge N, Commun. Math. Phys. **170** (1995), 337–357.

[10] J. Humphreys, *Introduction to Lie algebras and representation theory*, Graduate Texts in Mathematics 9, Springer-Verlag, New York-Berlin, 1978.

[11] V. Kac, *Some problems of infinite-dimensional Lie algebras and their representations*, Lecture Notes in Mathematics, vol. 933, Springer, 1982, pp. 117–126.

[12] V. Kac, *Infinite-dimensional Lie algebras*, 3rd edition, Cambridge University Press, Cambridge, 1990.

[13] V. Kac, *Formal distribution algebras and conformal algebras*, a talk at the Brisbane, in Proc. XIIth International Congress of Mathematical Physics (ICMP'97) (Brisbane), 80–97.

[14] V. Kac, J. Liberati, *Unitary quasi-finite representations of $\mathcal{W}_\infty$*, Lett. Math. Phys. **53** (2000), 11–27.

[15] V. Kac, D. Peterson, *Spin and wedge representations of infinite dimensional Lie algebras and groups*, Proc. Nat. Acad. Sci. U. S. A. **78** (1981), 3308–3312.

[16] V. Kac, A. Radul, *Quasi-finite highest weight modules over the Lie algebra of differential operators on the circle*, Commun. Math. Phys. **157** (1993), 429–457.

[17] V. Kac, A. Radul, *Representation theory of the vertex algebra $\mathcal{W}_{1+\infty}$*, Trans. Groups **1** (1996), 41–70.

[18] R. Lü, K. Zhao, *Classification of irreducible weight modules over higher rank Virasoro algebras*, Adv. Math. **206** (2006), 630–656.

[19] O. Mathieu, *Classification of Harish-Chandra modules over the Virasoro Lie algebra*, Invent. Math. **107** (1992), 225–234.

[20] Y. Matsuo, *Free fields and quasi-finite representations of $\mathcal{W}_{1+\infty}$*, Phys. Lett. B. **326** (1994), 95–100.

[21] J. Patera, H. Zassenhaus, *The higher rank Virasoro algebras*, Commun. Math. Phys. **136** (1991), 1–14.

[22] G. Song, Y. Su, *2-cocycles on the Lie superalgebras of Weyl type*, Commun. Algebra **33** (2005), 2991–3007.

[23] Y. Su, *A classification of indecomposable $sl_2(\mathbb{C})$-modules and a conjecture of Kac on irreducible modules over the Virasoro algebra*, J. Algebra **161** (1993), 33–46.

[24] Y. Su, *Classification of Harish-Chandra modules over the super-Virasoro algebras*, Commun. Algebra **23** (1995), 3653–3675.

[25] Y. Su, *Simple modules over the high rank Virasoro algebras*, Commun. Algebra **29** (2001), 2067–2080.

[26] Y. Su, *Classification of quasifinite modules over the Lie algebras of Weyl type*, Adv. Math. **174** (2003), 57–68.

[27] Y. Su, *Classification of Harish-Chandra modules over the higher rank Virasoro algebras*, Commun. Math. Phys. **240** (2003), 539–551.

[28] Y. Su, *Quasifinite representations of a Lie algebra of Block type*, J. Algebra **276** (2004), 117–128.

[29] Y. Su, *Quasifinite representations of a family of Lie algebras of Block type*, J. Pure Appl. Algebra **192** (2004), 293–305.

[30] Y. Su, *Classification of quasifinite modules over Lie algebras of matrix differential operators on the circle*, Proc. Amer. Math. Soc. **133** (2005), 1949–1957.

[31] Y. Su, B. Xin, *Classification of quasifinite $\mathcal{W}_\infty$-modules*, Israel J. Math. **151** (2006), 223–236.

[32] Y. Su, X. Xu, H. Zhang, *Derivation-simple algebras and the structures of Lie algebras of Witt type*, J. Algebra **233** (2000), 642–662.

[33] Y. Su, K. Zhao, *Simple algebras of Weyl type*, Science in China A **44** (2001), 419–426.

[34] Y. Su, K. Zhao, *Isomorphism classes and automorphism groups of algebras of Weyl type*, Science in China A **45** (2002), 953–963.

[35] Y. Su, K. Zhao, *Generalized Virasoro and super-Virasoro algebras and modules of the intermediate series*, J. Algebra **252** (2002), 1–19.

[36] Y. Su, K. Zhao, *Structure of algebras of Weyl type*, Commun. Algebra **32** (2004), 1051–1059.

[37] B. Xin, Y. Wu, *Generalized Verma modules over Lie algebras of Weyl type*, Algeba Colloquium, in press.

[38] X. Xu, *New generalized simple Lie algebras of Cartan type over a field with characteristic 0*, J. Algebra **224** (2000), 23–58.

DEPARTMENT OF MATHEMATICS, UNIVERSITY OF SCIENCE AND TECHNOLOGY OF CHINA, HEFEI 230026, CHINA

Contemporary Mathematics
Volume **478**, 2009

Maximal and Primitive Elements
in Baby Verma Modules for Type B_2

Nanhua Xi

The purpose of this paper is to find maximal and primitive elements of baby Verma modules for a quantum group of type B_2. As a consequence the composition factors of the baby Verma modules are determined. A similar approach can be used to find maximal and primitive elements of Weyl modules for type B_2. In principle the results can be used to determine the module structure of a baby Verma module, but the calculations are rather involved, much more complicated than the case of type A_2.

For type A_2, submodule structure of a Weyl module has been determined in [DS1, I, K] and by Cline (unpublished). For type B_2, the socle series of Weyl modules was determined in [DS2]. In [X2] we determine the maximal and primitive elements in Weyl modules for type A_2, so that the Weyl modules are understood more explicitly. This paper is a sequel to [X2], but less complete, since submodule structure of a baby Verma module is not determined. In this paper we only work with quantized enveloping algebras at roots of 1 (Lusztig version). For hyperalgebras the approach is completely similar, actually simpler.

The contents of the paper are as follows. In section 1 we recall some definitions and results about maximal and primitive elements. In section 2 we recall some facts about a quantized enveloping algebra of type B_2. In section 3 we determine the maximal and primitive elements in a Verma module of the (slightly enlarged) Frobenius kernel of type B_2. In section 4 we indicate that the maximal and primitive elements in a Weyl module for type B_2 can be worked out similarly, but we omit the results. To avoid complicated expressions and for simplicity we assume that the order of the involved root of 1 is odd and greater than 3 and we only work with some special weights. The approach for general cases is completely similar.

1. Maximal and Primitive Elements

In this section we fix notation and recall the definition and some results for maximal and primitive elements. We refer to [L1-4, X1-2] for additional information.

2000 Mathematics Subject Classification: Primary 17B37; Secondary 20G05.

Key words: maximal element, primitive element, baby Verma module .

The author was supported in part by the National Natural Science Foundation of China (No. 10671193).

1.1. Let U_ξ be a quantized enveloping algebra (over $\mathbf{Q}(\xi)$) at a root ξ of 1 (Lusztig version). We assume that the rank of the associated Cartan matrix is n and the order of $\xi \geq 3$. As usual, the generators of U_ξ are denoted by $e_i^{(a)}, f_i^{(a)}, k_i, k_i^{-1}$, etc. Let $\mathbf{u}$ be the Frobenius kernel and $\tilde{\mathbf{u}}$ the subalgebra of U_ξ generated by all elements in $\mathbf{u}$ and in the zero part of U_ξ. For $\lambda \in \mathbf{Z}^n$ and a U_ξ-module (or $\tilde{\mathbf{u}}$-module M) we denote by M_λ the λ-weight space of M. A nonzero element in M_λ will be called a vector of weight λ or a weight vector. Let m be a weight vector of a U_ξ-module (resp. $\tilde{\mathbf{u}}$-module) M. We call m **maximal** if $e_i^{(a)}m = 0$ for all i and $a \geq 1$ (resp. $e_\alpha m = 0$ for all root vectors e_α in the positive part of $\tilde{\mathbf{u}}$). We call m a **primitive element** if there exist two submodules $M_2 \subset M_1$ of M such that $m \in M_1$ and the image in M_1/M_2 of m is maximal. Obviously, maximal elements are primitive. We have (see [X2]):

(a) Let $m \in M$ be a weight vector and let P_1 be the submodule of M generated by m. Then m is primitive if and only if the image in P_1/P_2 of m is maximal for some proper submodule P_2 of P_1.

We shall write $L(\lambda)$ (resp. $\tilde{L}(\lambda)$) for an irreducible U_ξ-module (resp. $\tilde{\mathbf{u}}$-module) of highest weight λ.

(b) If m is a primitive element of weight λ, then $L(\lambda)$ (or $\tilde{L}(\lambda)$) is a composition factor of M (depending whether M is a U_ξ-module or a $\tilde{\mathbf{u}}$-module).

(c) Let M and N be modules and $\phi : M \to N$ a homomorphism. Let m be a weight vector in M. If $\phi(m)$ is a primitive element of N, then m is a primitive element of M.

(d) Let M, N, ϕ, m be as in (c) and assume $\phi(m) \neq 0$. If m is a primitive element of M, then either $\phi(m)$ is a primitive element of N or $\phi(P_1) = \phi(P_2)$, where P_1 is the submodule of M generated by m and P_2 is any submodule of P_1 such that the image in P_1/P_2 of m is maximal.

(e) Let M, N, ϕ, m be as in (c) and assume $\phi(m) \neq 0$. If m is a maximal element of M, then $\phi(m)$ is a maximal element of N.

We shall denote by $\tilde{Z}(\lambda)$ the (baby) Verma module of $\tilde{\mathbf{u}}$ with highest weight λ and denote by $\tilde{1}_\lambda$ a nonzero element in $\tilde{Z}(\lambda)_\lambda$. Recall that to define U_ξ we need to choose $d_i \in \{1, 2, 3\}$ such that $(d_i a_{ij})$ is symmetric, where (a_{ij}) is the concerned $n \times n$ Cartan matrix. Let l_i be the order of ξ^{2d_i}. For $\lambda = (\lambda_1, ..., \lambda_n) \in \mathbf{Z}^n$ we set $l\lambda = (l_1\lambda_1, ..., l_n\lambda_n)$. We call λ l-restricted if $0 \leq \lambda_i \leq l_i - 1$ for all i. Denote by $\mathbf{N}_l^n$ the set of all l-restricted elements in $\mathbf{Z}^n$. The following fact is well known and follows easily from the commutation formula for $e_j^{(a)} f_i^{(b)}$ (see [L1, 4.1(a)]).

(f) Let $\lambda \in \mathbf{N}_l^n$, $\lambda' \in \mathbf{Z}^n$. Set $\mu = \lambda + l\lambda' \in \mathbf{Z}^n$. Then $f_i^{(\lambda_i+1)}\tilde{1}_\mu$ is maximal in $\tilde{Z}(\mu)$ if $\lambda_i \neq l_i - 1$.

2. Some basic facts

2.1. From now on we assume that U_ξ is of type B_2. In this section we recall some basic facts about U_ξ and the Verma modules $\tilde{Z}(\lambda)$. For completeness and in order to fix notations, we give the definition of U_ξ and $\tilde{Z}(\lambda)$.

Let $a_{ii} = 2$, $a_{12} = -2$, $a_{21} = -1$. Let U be the associative algebra over $\mathbf{Q}(v)$ (v an indeterminate) generated by e_i, f_i, k_i, k_i^{-1} ($i = 1, 2$) with relations

$$k_1 k_2 = k_2 k_1, \qquad k_i k_i^{-1} = k_i^{-1} k_i = 1,$$

$$k_i e_j = v^{i a_{ij}} e_j k_i, \qquad k_i f_j = v^{-i a_{ij}} f_j k_i,$$

$$e_i f_j - f_j e_i = \delta_{ij} \frac{k_i - k_i^{-1}}{v_i - v_i^{-1}},$$

$$e_1 e_2^2 - (v^2 + v^{-2}) e_2 e_1 e_2 + e_2^2 e_1 = 0,$$

$$e_1^3 e_2 - (v^2 + 1 + v^{-2}) e_1^2 e_2 e_1 + (v^2 + 1 + v^{-2}) e_1 e_2 e_1^2 - e_2 e_1^3 = 0,$$

$$f_1 f_2^2 - (v^2 + v^{-2}) f_2 f_1 f_2 + f_2^2 f_1 = 0,$$

$$f_1^3 f_2 - (v^2 + 1 + v^{-2}) f_1^2 f_2 f_1 + (v^2 + 1 + v^{-2}) f_1 f_2 f_1^2 - f_2 f_1^3 = 0,$$

where $v_1 = v$ and $v_2 = v^2$. Let U' be the $A = \mathbf{Z}[v, v^{-1}]$-subalgebra of U generated by all $e_i^{(a)} = e_i^a/[a]_i!, f_i^{(a)} = f_i^a/[a]_i!, k_i, k_i^{-1}, \ a \in \mathbf{N}, \ i = 1, 2$, where $[a]_i! = \prod_{h=1}^a \frac{v^{ih} - v^{-ih}}{v^i - v^{-i}}$ if $a \geq 1$ and $[0]_i! = 1$. Note that the element

$$\begin{bmatrix} k_i, c \\ a \end{bmatrix} = \prod_{h=1}^a \frac{v_i^{c-h+1} k_i - v_i^{-c+h-1} k_i^{-1}}{v_i^h - v_i^{-h}}$$

is in U' for all $c \in \mathbf{Z}, \ a \in \mathbf{N}$. We understand that $\begin{bmatrix} k_i, c \\ a \end{bmatrix} = 1$ if $a = 0$. Note that $f_{12}'^{(a)} = (f_1 f_2 - v^2 f_2 f_1)^a/[a]!$ and $f_{12}^{(a)} = (f_2 f_1 - v^2 f_1 f_2)^a/[a]!$ are in U' for all $a \in \mathbf{N}$. Also

$$f_{112}'^{(a)} = \frac{(f_1 f_{12}' - f_{12}' f_1)^a}{(v + v^{-1})^a [a]_2!} \quad \text{and} \quad f_{112}^{(a)} = \frac{(f_{12} f_1 - f_1 f_{12})^a}{(v + v^{-1})^a [a]_2!}$$

are in U' for all $a \in \mathbf{N}$. Regard $\mathbf{Q}(\xi)$ as an A-algebra by specializing v to ξ. Then $U_\xi = U' \otimes_A \mathbf{Q}(\xi)$. See [L3].

For convenience, the images in U_ξ of $e_i^{(a)}, f_i^{(a)}, f_{12}'^{(a)}, f_{12}^{(a)}, f_{112}'^{(a)}, f_{112}^{(a)}, k_i, k_i^{-1}, \begin{bmatrix} k_i, c \\ a \end{bmatrix}$ etc. will be denoted by the same notation respectively. Let l be the order of ξ and l_i be the order of ξ^{2i}. In U_ξ we have $e_i^{l_i} = f_i^{l_i} = 0$. **For simplicity in this paper we assume that l is odd.** Then $l_1 = l_2 = l$. The Frobenius kernel $\mathbf{u}$ of U_ξ is the subalgebra of U_ξ generated by all $e_i, f_i, k_i, k_i^{-1}, \ i = 1, 2$. Its negative part $\mathbf{u}^-$ is generated by all f_i. Note that $f_{12}'^{(a)}, f_{12}^{(a)}, f_{112}'^{(a)}, f_{112}^{(a)}$, are in $\mathbf{u}^-$ if $0 \leq a \leq l - 1$. The subalgebra $\tilde{\mathbf{u}}$ of U_ξ is generated by all $e_i, f_i, k_i, k_i^{-1}, \begin{bmatrix} k_i, c \\ a \end{bmatrix}, \ i = 1, 2; c \in \mathbf{Z}, a \in \mathbf{N}$.

For $\lambda = (\lambda_1, \lambda_2) \in \mathbf{Z}^2$, we denote by $\tilde{I}_\lambda$ the left ideal of U_ξ generated by all $e_i^{(a)} \ (a > 0), k_i - \xi^{i\lambda_i}, \begin{bmatrix} k_i, c \\ a \end{bmatrix} - \begin{bmatrix} \lambda_i + c \\ a \end{bmatrix}_{\xi^i}$. (We denote by $\begin{bmatrix} b \\ a \end{bmatrix}_{\xi^i}$ the value at ξ^i of $\prod_{h=1}^a \frac{v^{b-h+1} - v^{b+h-1}}{v^h - v^{-h}}$ for any $b \in \mathbf{Z}, a \in \mathbf{N}$ and $i = 1, 2$.) The Verma module $Z(\lambda)$ of U_ξ is defined to be $U_\xi/\tilde{I}_\lambda$. Let $\tilde{1}_\lambda$ be the image in $Z(\lambda)$ of 1. The Verma module $\tilde{Z}(\lambda)$ of $\tilde{\mathbf{u}}$ is defined to be the $\tilde{\mathbf{u}}$-submodule of $Z(\lambda)$ generated by $\tilde{1}_\lambda$. Given non-negative integers a and b, we set

$$x_{a,b} = f_1^{(a)} f_2^{(a+b)} f_1^{(a+2b)} f_2^{(b)} = f_2^{(b)} f_1^{(a+2b)} f_2^{(a+b)} f_1^{(a)}.$$

Recall that l is the order of ξ. The following result is a special case of [X1, 4.2 (ii)].
(a) Assume $0 \leq a, b \leq l - 1, c, d \in \mathbf{Z}$, and let $\mu = (lc - 1 + a, ld - 1 + b)$. Then the element $x_{a,b}$ is in $\mathbf{u}^-$ and $x_{a,b} \tilde{1}_\mu$ is maximal in $\tilde{Z}(\mu)$ and generates the unique irreducible submodule of $\tilde{Z}(\mu)$. The irreducible submodule is isomorphic to $\tilde{L}(lc - 1 - a, ld - 1 - b)$.

The argument for [X1, 4.4(iv)] also gives the following result.

(b) Keep the assumption and notations in (a). Let $p, q, s, t \in \mathbf{N}$ such that $x = f_1^{(a+2b-pl)} f_2^{(a+b-ql)} f_1^{(a)}$ and $y = f_2^{(a+b-sl)} f_1^{(a+2b-tl)} f_2^{(b)}$ are nonzero elements, then $e_i x \tilde{1}_\mu = e_i y \tilde{1}_\mu = 0$ for $i = 1, 2$. If x and y are further in $\mathbf{u}^-$, then $x \tilde{1}_\mu$ and $y \tilde{1}_\mu$ are maximal in $\tilde{Z}(\mu)$.

We shall need a few formulas, which are due to Lusztig (see [L3, L4]). In U_ξ we have

(c) $f_i^{(a)} f_i^{(b)} = \begin{bmatrix} a+b \\ a \end{bmatrix}_{\xi^i} f_i^{(a+b)}$,

(d) $f_{12}^{(i)} f_2^{(j)} = \xi^{2ij} f_2^{(j)} f_{12}^{(i)}$,

(e) $f_{112}^{(i)} f_{12}^{(j)} = \xi^{2ij} f_{12}^{(j)} f_{112}^{(i)}$,

(f) $f_1^{(i)} f_{112}^{(j)} = \xi^{2ij} f_{112}^{(j)} f_1^{(i)}$,

(g) $f_2^{(i)} f_{112}^{(j)} = \sum_{\substack{r,s,t\in\mathbf{N} \\ r+s=j \\ s+t=i}} \xi^{-2rs-2st} \prod_{h=1}^{s} (\xi^{-4h+2} - 1) f_{112}^{(r)} f_{12}^{(2s)} f_2^{(t)}$,

(h) $f_{12}^{(i)} f_1^{(j)} = \sum_{\substack{r,s,t\in\mathbf{N} \\ r+s=j \\ s+t=i}} \xi^{-rs-st+s} \prod_{h=1}^{s} (\xi^{-2h} + 1) f_1^{(r)} f_{112}^{(s)} f_{12}^{(t)}$,

(i) $f_2^{(i)} f_1^{(j)} = \sum_{\substack{r,s,t,u\in\mathbf{N} \\ s+t+u=i \\ r+2s+t=j}} \xi^{2ru+2su+rt} f_1^{(r)} f_{112}^{(s)} f_{12}^{(t)} f_2^{(u)}$.

(j) $f'^{(i)}_{112} f_2^{(j)} = \sum_{\substack{r,s,t\in\mathbf{N} \\ r+s=j \\ s+t=i}} \xi^{-2rs-2st} \prod_{h=1}^{s} (\xi^{-4h+2} - 1) f_2^{(r)} f'^{(2s)}_{12} f'^{(t)}_{112}$,

(k) $f_1^{(i)} f'^{(j)}_{12} = \sum_{\substack{r,s,t\in\mathbf{N} \\ r+s=j \\ s+t=i}} \xi^{-rs-st+s} \prod_{h=1}^{s} \xi^{-2h} + 1) f'^{(r)}_{12} f'^{(s)}_{112} f_1^{(t)}$,

(l) $f_1^{(i)} f_2^{(j)} = \sum_{\substack{r,s,t,u\in\mathbf{N} \\ r+s+t=j \\ s+2t+u=i}} \xi^{2ru+2rt+us} f_2^{(r)} f'^{(s)}_{12} f'^{(t)}_{112} f_1^{(u)}$.

(m) Assume $0 \le a_0, b_0 \le l - 1$ and $a_1, b_1 \in \mathbf{Z}$. We have

$$\begin{bmatrix} a_0 + a_1 l \\ b_0 + b_1 l \end{bmatrix}_{\xi^i} = \begin{bmatrix} a_0 \\ b_0 \end{bmatrix}_{\xi^i} \begin{pmatrix} a_1 \\ b_1 \end{pmatrix},$$

where $\begin{pmatrix} a_1 \\ b_1 \end{pmatrix}$ is the ordinary binomial coefficient.

Using (i), (l) and (m) we get

(n) If $0 \le a \le l - 1$, then $f_1^{(l)} f_2^{(a)} - f_2^{(a)} f_1^{(l)}$ and $f_2^{(l)} f_1^{(a)} - f_1^{(a)} f_2^{(l)}$ are in $\mathbf{u}^-$.

(o) Let $0 \le a, b, c \le l - 1$. Then $f_1^{(a)} f_2^{(b)} f_1^{(c)} = 0$ and $f_1^{(a)} f_2^{(l+b)} f_1^{(c)}$ is in $\mathbf{u}^-$, if $a + c - 2b \ge l$. Similarly $f_2^{(a)} f_1^{(b)} f_2^{(c)} = 0$ and $f_2^{(a)} f_1^{(l+b)} f_2^{(c)}$ is in $\mathbf{u}^-$, if $a + c - b \ge l$.

The assertions (n) and (o) will be frequently used in computations.

Let $\alpha_1 = (2, -1), \alpha_2 = (-2, 2) \in \mathbf{Z}^2$. The set of positive roots is $R^+ = \{\alpha_1, \alpha_2, \alpha_1 + \alpha_2, 2\alpha_1 + \alpha_2\}$. Let W be the Weyl group generated by the simple reflections s_i corresponding to α_i. Assume that $l \geq 5$. Then $\langle \rho, \beta^\vee \rangle < l$ for all $\beta \in R^+$, where $\rho = (1,1)$. For $\lambda, \mu \in \mathbf{Z}^2$, we write that $\lambda \leq \mu$ if $\mu - \lambda = a\alpha_1 + b\alpha_2$ for some non-negative integers a, b.

We say that $x \in \mathbf{u}^-$ is homogeneous (of degree β) if there exists $\beta \in \mathbf{Z}^2$ such that $\begin{bmatrix} k_i, c \\ a \end{bmatrix} x = x \begin{bmatrix} k_i, c + \langle \beta, \alpha_i^\vee \rangle \\ a \end{bmatrix}$ and $k_i x = \xi^{i\langle \beta, \alpha_i^\vee \rangle} x k_i$ for all $c \in \mathbf{Z}$ and $a \in \mathbf{N}$.

2.2. The W-orbit of $\lambda = (0,0)$ (dot action) consists of the following 8 elements: λ, $s_1.\lambda = \lambda - \alpha_1$, $s_2.\lambda = \lambda - \alpha_2$, $s_2s_1.\lambda = \lambda - \alpha_1 - 2\alpha_2$, $s_1s_2.\lambda = \lambda - 3\alpha_1 - \alpha_2$, $s_1s_2s_1.\lambda = \lambda - 4\alpha_1 - 2\alpha_2$, $s_2s_1s_2.\lambda = \lambda - 3\alpha_1 - 3\alpha_2$, $s_1s_2s_1s_2.\lambda = \lambda - 4\alpha_1 - 3\alpha_2$.

Let a, b be integers and $\lambda = (la, lb)$. Using 1.1 (e-f) and 2.1 (a-b) we get

(1) The following elements are maximal in $\tilde{Z}(\lambda)$:

$$\tilde{1}_\lambda, \qquad f_1\tilde{1}_\lambda, \qquad f_2\tilde{1}_\lambda, \qquad f_2^{(2)}f_1\tilde{1}_\lambda, \qquad f_1^{(3)}f_2\tilde{1}_\lambda,$$
$$f_2^{(2)}f_1^{(3)}f_2\tilde{1}_\lambda, \qquad f_1^{(3)}f_2^{(2)}f_1\tilde{1}_\lambda, \qquad f_1f_2^{(2)}f_1^{(3)}f_2\tilde{1}_\lambda.$$

(2) Let $\mu = \lambda + (l-1)\alpha_1$. The following elements are maximal in $\tilde{Z}(\mu)$:

$$\tilde{1}_\mu, \qquad f_1^{(l-1)}\tilde{1}_\mu, \qquad f_2^{(2)}\tilde{1}_\mu, \qquad f_2f_1^{(l-1)}\tilde{1}_\mu, \qquad f_1^{(3)}f_2^{(2)}\tilde{1}_\mu,$$
$$f_2f_1^{(3)}f_2^{(2)}\tilde{1}_\mu, \qquad f_1^{(3)}f_2^{(l+1)}f_1^{(l-1)}\tilde{1}_\mu,$$
$$f_2^{(2)}f_1^{(l+3)}f_2^{(l+1)}f_1^{(l-1)}\tilde{1}_\mu.$$

(3) Let $\mu = \lambda + (l-1)\alpha_2$. The following elements are maximal in $\tilde{Z}(\mu)$:

$$\tilde{1}_\mu, \qquad f_1^{(3)}\tilde{1}_\mu, \qquad f_2^{(l-1)}\tilde{1}_\mu, \qquad f_2^{(2)}f_1^{(3)}\tilde{1}_\mu, \qquad f_1f_2^{(l-1)}\tilde{1}_\mu,$$
$$f_1f_2^{(2)}f_1^{(3)}\tilde{1}_\mu, \qquad f_2^{(2)}f_1^{(l+1)}f_2^{(l-1)}\tilde{1}_\mu, \qquad f_1^{(3)}f_2^{(l+2)}f_1^{(2l+1)}f_2^{(l-1)}\tilde{1}_\mu.$$

(4) Let $\mu = \lambda + (l-1)\alpha_1 + (l-2)\alpha_2$. The following elements are maximal in $\tilde{Z}(\mu)$:

$$\tilde{1}_\mu, \qquad f_1^{(3)}\tilde{1}_\mu, \qquad f_2^{(l-2)}\tilde{1}_\mu, \qquad f_2f_1^{(3)}\tilde{1}_\mu,$$
$$f_1^{(l-1)}f_2^{(l-2)}\tilde{1}_\mu, \qquad f_2f_1^{(l-1)}f_2^{(l-2)}\tilde{1}_\mu.$$
$$f_1^{(l-1)}f_2^{(l+1)}f_1^{(3)}\tilde{1}_\mu, \qquad f_1^{(3)}f_2^{(l+1)}f_1^{(2l-1)}f_2^{(l-2)}\tilde{1}_\mu.$$

(5) Let $\mu = \lambda + (l-3)\alpha_1 + (l-1)\alpha_2$. The following elements are maximal in $\tilde{Z}(\mu)$:

$$\tilde{1}_\mu, \qquad f_1^{(l-3)}\tilde{1}_\mu, \qquad f_2^{(2)}\tilde{1}_\mu, \qquad f_2^{(l-1)}f_1^{(l-3)}\tilde{1}_\mu,$$
$$f_1f_2^{(2)}\tilde{1}_\mu, \qquad f_1f_2^{(l-1)}f_1^{(l-3)}\tilde{1}_\mu,$$
$$f_2^{(l-1)}f_1^{(l+1)}f_2^{(2)}\tilde{1}_\mu, \qquad f_2^{(2)}f_1^{(l+1)}f_2^{(l-1)}f_1^{(l-3)}\tilde{1}_\mu.$$

(6) Let $\mu = \lambda + (l-4)\alpha_1 + (l-2)\alpha_2$. The following elements are maximal in $\tilde{Z}(\mu)$:

$$\tilde{1}_\mu, \qquad f_1^{(l-3)}\tilde{1}_\mu, \qquad f_2\tilde{1}_\mu, \qquad f_2^{(l-2)}f_1^{(l-3)}\tilde{1}_\mu,$$
$$f_1^{(l-1)}f_2\tilde{1}_\mu, \qquad f_1^{(l-1)}f_2^{(l-2)}f_1^{(l-3)}\tilde{1}_\mu.$$
$$f_2^{(l-2)}f_1^{(l-1)}f_2\tilde{1}_\mu, \qquad f_2f_1^{(l-1)}f_2^{(l-2)}f_1^{(l-3)}\tilde{1}_\mu.$$

(7) Let $\mu = \lambda + (l-3)\alpha_1 + (l-3)\alpha_2$. The following elements are maximal in $\tilde{Z}(\mu)$:

$$\tilde{1}_\mu, \qquad f_1\tilde{1}_\mu, \qquad f_2^{(l-2)}\tilde{1}_\mu, \qquad f_2^{(l-1)}f_1\tilde{1}_\mu,$$
$$f_1^{(l-3)}f_2^{(l-2)}\tilde{1}_\mu, \qquad f_1^{(l-3)}f_2^{(l-1)}f_1\tilde{1}_\mu.$$
$$f_2^{(l-1)}f_1^{(2l-3)}f_2^{(l-2)}\tilde{1}_\mu, \qquad f_2^{(l-2)}f_1^{(2l-3)}f_2^{(l-1)}f_1\tilde{1}_\mu.$$

(8) Let $\mu = \lambda + (l-4)\alpha_1 + (l-3)\alpha_2$. The following elements are maximal in $\tilde{Z}(\mu)$:

$$\tilde{1}_\mu, \qquad f_1^{(l-1)}\tilde{1}_\mu, \qquad f_2^{(l-1)}\tilde{1}_\mu, \qquad f_2^{(l-2)}f_1^{(l-1)}\tilde{1}_\mu,$$

$$f_1^{(l-3)}f_2^{(l-1)}\tilde{1}_\mu, \qquad f_1^{(l-3)}f_2^{(l-2)}f_1^{(l-1)}\tilde{1}_\mu.$$

$$f_2^{(l-2)}f_1^{(2l-3)}f_2^{(l-1)}\tilde{1}_\mu, \qquad f_2^{(l-1)}f_1^{(3l-3)}f_2^{(2l-2)}f_1^{(l-1)}\tilde{1}_\mu.$$

3. maximal and primitive elements of $\tilde{Z}(\lambda)$ for type B_2

In this section we determine the maximal and primitive elements in $\tilde{Z}(\lambda)$ (or equivalently in any highest weight module of $\tilde{\mathbf{u}}$). To avoid complicated expressions we only work with some weights in the W-orbit of $(0,0)$. For general cases the approach is completely similar. From now on we assume that the odd integer l is at least 5.

THEOREM 3.1. *Let a, b be integers and $\lambda = (la, lb)$. Then*
(i) The following 8 elements are maximal in $\tilde{Z}(\lambda)$:

$$\tilde{1}_\lambda, \qquad f_1\tilde{1}_\lambda, \qquad f_2\tilde{1}_\lambda, \qquad f_2^{(2)}f_1\tilde{1}_\lambda, \qquad f_1^{(3)}f_2\tilde{1}_\lambda,$$

$$f_2^{(2)}f_1^{(3)}f_2\tilde{1}_\lambda, \qquad f_1^{(3)}f_2^{(2)}f_1\tilde{1}_\lambda, \qquad f_1f_2^{(2)}f_1^{(3)}f_2\tilde{1}_\lambda.$$

(ii) The following 12 elements are primitive elements in $\tilde{Z}(\lambda)$ but not maximal:

$$[f_1^{(3)}, f_2^{(l)}]f_2\tilde{1}_\lambda, \qquad \frac{x_{l-1,2}}{f_1^{(l-1)}}\tilde{1}_\lambda, \qquad [f_2^{(2)}, f_1^{(l)}]f_1^{(3)}f_2\tilde{1}_\lambda$$

$$[f_2^{(2)}, f_1^{(l)}]f_1\tilde{1}_\lambda, \qquad \frac{x_{3,l-1}}{f_2^{(l-1)}}\tilde{1}_\lambda, \qquad \frac{x_{3,l-1}}{f_1^{(l)}f_2^{(l-1)}}\tilde{1}_\lambda, \qquad [f_1^{(3)}, f_2^{(l)}]f_2^{(2)}f_1\tilde{1}_\lambda,$$

$$\frac{x_{3,l-2}}{f_1^{(l-1)}f_2^{(l-2)}}\tilde{1}_\lambda, \qquad f_2^{(l-1)}f_1^{(l)}f_2\tilde{1}_\lambda, \qquad f_1f_2^{(l-1)}f_1^{(l)}f_2\tilde{1}_\lambda,$$

$$\frac{x_{l-1,l-1}}{f_1^{(l-3)}f_2^{(l-2)}f_1^{(l-1)}}f_2\tilde{1}_\lambda, \qquad \frac{x_{l-1,l-1}}{f_2^{(l-2)}f_1^{(2l-3)}f_2^{(l-1)}}f_1\tilde{1}_\lambda.$$

(See 2.1 for the definition of $x_{a,b}$. Convention: $[x,y] = xy - yx$ and $\frac{x}{y}$ stands for an arbitrary homogeneous element z in $\mathbf{u}^-$ such that $zy = x$.)

Moreover no maximal element in $\tilde{Z}(\lambda)$ has the same weight as any of the above 12 elements.

(iii) The maximal and primitive elements in (i-ii) provide 20 composition factors of $\tilde{Z}(\lambda)$, which are

$$\tilde{L}(\lambda), \qquad \tilde{L}(\lambda - \alpha_1), \qquad \tilde{L}(\lambda - \alpha_2),$$

$$\tilde{L}(\lambda - \alpha_1 - 2\alpha_2), \qquad \tilde{L}(\lambda - 3\alpha_1 - \alpha_2)$$

$$\tilde{L}(\lambda - 3\alpha_1 - 3\alpha_2), \qquad \tilde{L}(\lambda - 4\alpha_1 - 2\alpha_2), \qquad \tilde{L}(\lambda - 4\alpha_1 - 3\alpha_2),$$

$$\tilde{L}(\lambda - 3\alpha_1 - (l+1)\alpha_2), \qquad \tilde{L}(\lambda - (l+3)\alpha_1 - (l+3)a_2), \qquad \tilde{L}(\lambda - (l+3)\alpha_1 - 3\alpha_2),$$

$$\tilde{L}(\lambda - (l+1)\alpha_1 - 2\alpha_2), \qquad \tilde{L}(\lambda - (2l+4)\alpha_1 - (l+2)\alpha_2), \qquad \tilde{L}(\lambda - (l+4)\alpha_1 - (l+2)\alpha_2),$$

$$\tilde{L}(\lambda - 4\alpha_1 - (l+2)\alpha_2), \qquad \tilde{L}(\lambda - (l+3)\alpha_1 - (l+1)\alpha_2), \qquad \tilde{L}(\lambda - l\alpha_1 - l\alpha_2),$$

$$\tilde{L}(\lambda - (l+1)\alpha_1 - l\alpha_2), \qquad \tilde{L}(\lambda - 2l\alpha_1 - 2l\alpha_2), \qquad \tilde{L}(\lambda - 2l\alpha_1 - l\alpha_2).$$

Moreover, $\tilde{Z}(\lambda)$ has only the 20 composition factors.

PROOF. (i) According to 1.1 (e-f), we see that (i) is true.

(ii) Now we argue for (ii).

(1) Consider the homomorphism:

$$\varphi_1:\ \tilde{Z}(\lambda) \to \tilde{Z}(\lambda + (l-1)\alpha_1)), \quad \tilde{1}_\lambda \to t_1 = f_1^{(l-1)}\tilde{1}_{\lambda+(l-1)\alpha_1}.$$

Let

$$x_1 = (f_1^{(3)}f_2^{(l)} - f_2^{(l)}f_1^{(3)})f_2 \in \mathbf{u}^-.$$

Note that

$$f_1^{(3)}f_2^{(l+1)}t_1 = f_1^{(3)}f_2^{(l+1)}f_1^{(l-1)}\tilde{1}_{\lambda+(l-1)\alpha_1} = x_1 t_1.$$

Using 1.1 (c) we see that $x_1\tilde{1}_\lambda$ is a primitive element of weight $\gamma_{38} = \lambda - 3\alpha_1 - (l+1)\alpha_2$.

Let

$$y_1 = (f_2^{(2)}f_1^{(l)} - f_1^{(l)}f_2^{(2)})x_1 \in \mathbf{u}^-.$$

Note that $f_2^{(2)}f_1^{(3)}f_2^{(l+1)}f_1^{(l-1)} = 0$. We then can check that

$$f_2^{(2)}f_1^{(l+3)}f_2^{(l+1)}t_1 = f_2^{(2)}f_1^{(l+3)}f_2^{(l+1)}f_1^{(l-1)}\tilde{1}_{\lambda+(l-1)\alpha_1} = y_1 t_1.$$

Using 1.1 (c) we see that $y_1\tilde{1}_\lambda$ is a primitive element of weight $\gamma_{48} = \lambda - (l+3)\alpha_1 - (l+3)\alpha_2$. Note that we have $y_1 = \frac{x_{l-1,2}}{f_1^{(l-1)}}$.

Note that $f_2^{(l+1)}f_1^{(l-1)} = f_2 f_2^{(l)}f_1^{(l-1)}$, so we have $xf_2^{(l+1)}f_1^{(l-1)} = 0$ if $xf_2 = 0$ and $x \in \mathbf{u}^-$. Thus we have a homomorphism (recall that $\tilde{1}_{\lambda+(l-1)\alpha_1+l\alpha_2}$ is also an element of the Verma module $Z(\lambda + (l-1)\alpha_1 + l\alpha_2)$ of U_ξ):

$$\psi_1:\ \tilde{\mathbf{u}}f_2\tilde{1}_\lambda \to \tilde{\mathbf{u}}f_2^{(l+1)}f_1^{(l-1)}\tilde{1}_{\lambda+(l-1)\alpha_1+l\alpha_2},$$

$$f_2\tilde{1}_\lambda \to f_2^{(l+1)}f_1^{(l-1)}\tilde{1}_{\lambda+(l-1)\alpha_1+l\alpha_2}.$$

Note that $\psi_1(f_1^{(3)}f_2\tilde{1}_\lambda) = f_1^{(3)}f_2^{(l+1)}f_1^{(l-1)}\tilde{1}_{\lambda+(l-1)\alpha_1+l\alpha_2}$ and $f_1^{(3)}f_2^{(l+1)}f_1^{(l-1)}$ is in $\mathbf{u}^-$.

Let $z_1 = f_2^{(2)}f_1^{(l)} - f_1^{(l)}f_2^{(2)} \in \mathbf{u}^-$. Using 1.1 (c) we see that $z_1 f_1^{(3)}f_2\tilde{1}_\lambda$ is a primitive element of weight $\gamma_{47} = \lambda - (l+3)\alpha_1 - 3\alpha_2$. Note that $z_1 f_1^{(3)}f_2 = \frac{x_{l-1,2}}{f_2^{(l)}f_1^{(l-1)}}$.

(2) Now we consider the homomorphism:

$$\varphi_2:\ \tilde{Z}(\lambda) \to \tilde{Z}(\lambda + (l-1)\alpha_2)), \quad \tilde{1}_\lambda \to t_2 = f_2^{(l-1)}\tilde{1}_{\lambda+(l-1)\alpha_2}.$$

Let

$$x_2 = (f_2^{(2)}f_1^{(l)} - f_1^{(l)}f_2^{(2)})f_1 \in \mathbf{u}^-.$$

Note that

$$f_2^{(2)}f_1^{(l+1)}t_2 = f_2^{(2)}f_1^{(l+1)}f_2^{(l-1)}\tilde{1}_{\lambda+(l-1)\alpha_2} = x_2 t_2.$$

Using 1.1 (c) we see that $x_2\tilde{1}_\lambda$ is a primitive element of weight $\gamma_{31} = \lambda - (l+1)\alpha_1 - 2\alpha_2$.

Let y_2 be homogeneous in $\mathbf{u}^-$ such that $y_2 t_2 = f_1^{(3)}f_2^{(l+2)}f_1^{(2l+1)}f_2^{(l-1)}\tilde{1}_\mu$, here $\mu = \lambda + (l-1)\alpha_2$. According to 1.1 (c) we know that $y_2\tilde{1}_\lambda$ is a primitive element of weight $\gamma_{41} = \lambda - (2l+4)\alpha_1 - (l+2)\alpha_2$. Note that $y_2 = \frac{x_{3,l-1}}{f_2^{(l-1)}}$.

As the reason for ψ_1, we have a homomorphism:

$$\psi_2:\ \tilde{\mathbf{u}}f_1\tilde{1}_\lambda \to \tilde{\mathbf{u}}f_1^{(l+1)}f_2^{(l-1)}\tilde{1}_{\lambda+l\alpha_1+(l-1)\alpha_2},$$

$$f_1\tilde{1}_\lambda \to f_1^{(l+1)}f_2^{(l-1)}\tilde{1}_{\lambda+l\alpha_1+(l-1)\alpha_2}.$$

Note that $\psi_2(f_2^{(2)} f_1 \tilde{1}_\lambda) = t_2' = f_2^{(2)} f_1^{(l+1)} f_2^{(l-1)} \tilde{1}_{\lambda+l\alpha_1+(l-1)\alpha_2}$ and the element $f_2^{(2)} f_1^{(l+1)} f_2^{(l-1)}$ is in $\mathbf{u}^-$. Let z_2 be homogeneous in $\mathbf{u}^-$ such that

$$z_2 t_2' = f_1^{(3)} f_2^{(l+2)} f_1^{(2l+1)} f_2^{(l-1)} \tilde{1}_{\lambda+l\alpha_1+(l-1)\alpha_2}.$$

Using 1.1 (c) we see that $z_2 f_2^{(2)} f_1 \tilde{1}_\lambda$ is a primitive element of weight $\gamma_{42} = \lambda - (l + 4)\alpha_1 - (l+2)\alpha_2$. Note that $z_2 f_2^{(2)} f_1 = \frac{x_{3,l-1}}{f_1^{(l)} f_2^{(l-1)}}$.

We also have a homomorphism (recall that $\tilde{1}_{\lambda+2l\alpha_1+(l-1)\alpha_2}$ is also an element of the Verma module $Z(\lambda + 2l\alpha_1 + (l-1)\alpha_2)$ of U_ξ) :

$$\theta_2 : \tilde{\mathbf{u}} f_1 \tilde{1}_\lambda \to \tilde{\mathbf{u}} f_1^{(2l+1)} f_2^{(l-1)} \tilde{1}_{\lambda+2l\alpha_1+(l-1)\alpha_2},$$

$$f_1 \tilde{1}_\lambda \to t_2'' = f_1^{(2l+1)} f_2^{(l-1)} \tilde{1}_{\lambda+2l\alpha_1+(l-1)\alpha_2}.$$

Let $w_2 = [f_1^{(3)}, f_2^{(l)}] \in \mathbf{u}^-$. Then

$$w_2 f_2^{(2)} t_2'' = f_1^{(3)} f_2^{(l+2)} f_1^{(2l+1)} f_2^{(l-1)} \tilde{1}_{\lambda+2l\alpha_1+(l-1)\alpha_2}.$$

Using 1.1 (c) we see that $w_2 f_2^{(2)} f_1 \tilde{1}_\lambda$ is a primitive element of weight $\gamma_{44} = \lambda - 4\alpha_1 - (l+2)\alpha_2$. Note that $w_2 f_2^{(2)} f_1 = \frac{x_{3,l-1}}{f_1^{(2l)} f_2^{(l-1)}}$.

(3) Now we consider the homomorphism:

$$\tilde{Z}(\lambda) \to \tilde{Z}(\lambda + (l-1)\alpha_1 + (l-2)\alpha_2),$$

$$\tilde{1}_\lambda \to t_3 = f_1^{(l-1)} f_2^{(l-2)} \tilde{1}_{\lambda+(l-1)\alpha_1+(l-2)\alpha_2}.$$

Let x_3 be homogeneous in $\mathbf{u}^-$ such that

$$x_3 t_3 = f_1^{(3)} f_2^{(l+1)} f_1^{(2l-1)} f_2^{(l-2)} \tilde{1}_{\lambda+(l-1)\alpha_1+(l-2)\alpha_2}.$$

Using 1.1 (c) we know that $x_3 \tilde{1}_\lambda$ is a primitive element of weight $\gamma_{37} = \lambda - (l + 3)\alpha_1 - (l+1)\alpha_2$. Note that $x_3 = \frac{x_{3,l-2}}{f_1^{(l-1)} f_2^{(l-2)}}$.

(4) Since $f_1^{(l-3)} f_2^{(l-2)} \in f_2 \mathbf{u}^-$ (see 2.1), we have a surjective homomorphism

$$\varphi_4 : \tilde{\mathbf{u}} f_2 \tilde{1}_\lambda \to \tilde{\mathbf{u}} f_1^{(l-3)} f_2^{(l-2)} \tilde{1}_{\lambda+(l-3)\alpha_1+(l-3)\alpha_2}$$

$$f_2 \tilde{1}_\lambda \to f_1^{(l-3)} f_2^{(l-2)} \tilde{1}_{\lambda+(l-3)\alpha_1+(l-3)\alpha_2}.$$

Let $x_4 = f_2^{(l-1)} f_1^{(l)} - f_1^{(l)} f_2^{(l-1)} \in \mathbf{u}^-$. Then

$$x_4 f_2 \tilde{1}_\lambda = f_2^{(l-1)} f_1^{(l)} f_2 \tilde{1}_\lambda$$

is a primitive element of weight $\gamma_{34} = \lambda - l\alpha_1 - l\alpha_2$.

Let $y_4 = f_1 x_4 f_2$. Then $y_4 \tilde{1}_\lambda$ is a primitive element of weight $\gamma_{45} = \lambda - (l + 1)\alpha_1 - l\alpha_2$.

(5) Consider the homomorphism

$$\tilde{\mathbf{u}} f_2 \tilde{1}_\lambda \to \tilde{\mathbf{u}} f_1^{(l-3)} f_2^{(l-2)} f_1^{(l-1)} \tilde{1}_{\lambda+(2l-4)\alpha_1+(l-3)\alpha_2}$$

$$f_2 \tilde{1}_\lambda \to f_1^{(l-3)} f_2^{(l-2)} f_1^{(l-1)} \tilde{1}_{\lambda+(2l-4)\alpha_1+(l-3)\alpha_2}.$$

Let x_5 be homogeneous in $\mathbf{u}^-$ such that

$$x_5 f_1^{(l-3)} f_2^{(l-2)} f_1^{(l-1)} = f_2^{(l-1)} f_1^{(3l-3)} f_2^{(2l-2)} f_1^{(l-1)}.$$

By 1.1 (c), $x_5 f_2 \tilde{1}_\lambda$ is primitive and is of weight $\gamma_{35} = \lambda - 2l\alpha_1 - 2l\alpha_2$. Note that $x_5 = \frac{x_{l-1,l-1}}{f_1^{(l-3)} f_2^{(l-2)} f_1^{(l-1)}}$.

Consider the surjective homomorphism

$$\tilde{\mathbf{u}} f_1 \tilde{1}_\lambda \to \tilde{\mathbf{u}} f_2^{(l-2)} f_1^{(2l-3)} f_2^{(l-1)} \tilde{1}_{\lambda + (2l-4)\alpha_1 + (2l-3)\alpha_2}$$

$$f_1 \tilde{1}_\lambda \to f_2^{(l-2)} f_1^{(2l-3)} f_2^{(l-1)} \tilde{1}_{\lambda + (2l-4)\alpha_1 + (2l-3)\alpha_2}.$$

Let y_5 be homogeneous in $\mathbf{u}^-$ such that

$$y_5 f_2^{(l-2)} f_1^{(2l-3)} f_2^{(l-1)} = f_2^{(l-1)} f_1^{(3l-3)} f_2^{(2l-2)} f_1^{(l-1)}.$$

By 1.1 (c), $y_5 f_1 \tilde{1}_\lambda$ is primitive and is of weight $\gamma_{32} = \lambda - 2l\alpha_1 - l\alpha_2$. Note that $y_5 = \frac{x_{l-1,l-1}}{f_2^{(l-2)} f_1^{(2l-3)} f_2^{(l-1)}}$.

We may also consider the homomorphism:

$$\tilde{\mathbf{u}} f_2 \tilde{1}_\lambda \to \tilde{\mathbf{u}} f_1^{(l-3)} f_2^{(2l-2)} f_1^{(l-1)} \tilde{1}_{\lambda + (2l-4)\alpha_1 + (2l-2)\alpha_2},$$

$$\tilde{\mathbf{u}} f_1 \tilde{1}_\lambda \to f_1^{(l-3)} f_2^{(2l-2)} f_1^{(l-1)} \tilde{1}_{\lambda + (2l-4)\alpha_1 + (2l-2)\alpha_2}.$$

Using 1.1 (c) we know that $[[f_2^{(l-1)}, f_1^{(l)}], f_1^{(l)}] f_2 \tilde{1}_\lambda = \frac{x_{l-1,l-1}}{f_1^{(l-3)} f_2^{(2l-2)} f_1^{(l-1)}} f_2 \tilde{1}_\lambda$ is also a primitive element of weight $\lambda - 2l\alpha_1 - l\alpha_2$.

The element $f_1 f_2^{(2)} f_1^{(3)} f_2 \tilde{1}_\lambda$ generates the unique irreducible submodule of $\tilde{Z}(\lambda)$. Clearly, the weight of any element in (ii) is not greater than $\lambda - 4\alpha_1 - 3\alpha_2$, therefore no maximal element in $\tilde{Z}(\lambda)$ has the same weight as any of the elements in (ii).

(iii) Using (i), (ii) and 1.1 (b), we see that $\tilde{Z}(\lambda)$ has the 20 composition factors. The dimensions of irreducible $\tilde{\mathbf{u}}$-modules are known (see [APW]). By a comparison of dimensions we know that $\tilde{Z}(\lambda)$ has only the 20 composition factors.

The theorem is proved.   □

THEOREM 3.2. *Let a, b be integers and $\lambda = (la, lb - 3)$. Then*
(i) The following elements are maximal in $\tilde{Z}(\lambda)$:

$$\tilde{1}_\lambda, \qquad f_1 \tilde{1}_\lambda, \qquad f_2^{(l-2)} \tilde{1}_\lambda, \qquad f_2^{(l-1)} f_1 \tilde{1}_\lambda,$$

$$f_1^{(l-3)} f_2^{(l-2)} \tilde{1}_\lambda, \qquad f_1^{(l-3)} f_2^{(l-1)} f_1 \tilde{1}_\lambda.$$

$$f_2^{(l-1)} f_1^{(2l-3)} f_2^{(l-2)} \tilde{1}_\lambda, \qquad f_2^{(l-2)} f_1^{(2l-3)} f_2^{(l-1)} f_1 \tilde{1}_\lambda,$$

$$\frac{f_2^{(l-1)} f_1^{(l-3)}}{f_2^{(2)}} \tilde{1}_\lambda, \qquad \frac{f_1 f_2^{(l-1)} f_1^{(l-3)}}{f_2^{(2)}} \tilde{1}_\lambda, \qquad f_1^{(l-3)} [f_2^{(l-1)}, f_1^{(l)}] f_1 \tilde{1}_\lambda.$$

(ii) The following elements are primitive elements in $\tilde{Z}(\lambda)$ but not maximal:

$$\frac{x_{l-1,l-1}}{f_1^{(l-1)}} \tilde{1}_\lambda, \qquad \frac{x_{l-1,l-1}}{f_2^{(l)} f_1^{(l-1)}} \tilde{1}_\lambda, \qquad [f_2^{(l-1)}, f_1^{(l)}] f_1 \tilde{1}_\lambda,$$

$$\frac{x_{3,l-1}}{f_2^{(2)} f_1^{(3)}} \tilde{1}_\lambda, \qquad \frac{x_{3,l-1}}{f_2^{(l+2)} f_1^{(3)}} \tilde{1}_\lambda, \qquad f_1^{(l-1)} f_2^{(l)} f_1 \tilde{1}_\lambda,$$

$$\frac{x_{3,l-2}}{f_2 f_1^{(3)}} f_1 \tilde{1}_\lambda, \qquad [f_2^{(l-2)}, f_1^{(l)}] \tilde{1}_\lambda, \qquad \frac{x_{l-1,2}}{f_2 f_1^{(3)} f_2^{(2)}} f_1 \tilde{1}_\lambda.$$

Moreover no maximal element in $\tilde{Z}(\lambda)$ has the same weight as any of the above 9 elements.
(iii) The maximal and primitive elements in (i-ii) provide 20 composition factors of $\tilde{Z}(\lambda)$, which are

$$\tilde{L}(\lambda), \qquad \tilde{L}(\lambda - \alpha_1), \qquad \tilde{L}(\lambda - (l-2)\alpha_2),$$

$$\tilde{L}(\lambda - \alpha_1 - (l-1)\alpha_2), \qquad \tilde{L}(\lambda - (l-3)\alpha_1 - (l-2)\alpha_2)$$

$\tilde{L}(\lambda-(l-2)\alpha_1-(l-1)\alpha_2)$, $\tilde{L}(\lambda-(2l-3)\alpha_1-(2l-3)\alpha_2)$, $\tilde{L}(\lambda-(2l-2)\alpha_1-(2l-3)\alpha_2)$,
$\tilde{L}(\lambda-(l-3)\alpha_1-(l-3)\alpha_2)$, $\tilde{L}(\lambda-(l-2)\alpha_1-(l-3)a_2)$, $\tilde{L}(\lambda-(3l-3)\alpha_1-(3l-3)\alpha_2)$,
$\tilde{L}(\lambda-(3l-3)\alpha_1-(2l-3)\alpha_2)$, $\tilde{L}(\lambda-(l+1)\alpha_1-(l-1)\alpha_2)$, $\tilde{L}(\lambda-(2l-2)\alpha_1-(l-1)\alpha_2)$,
$\tilde{L}(\lambda-(2l+1)\alpha_1-(2l-1)\alpha_2)$, $\tilde{L}(\lambda-(2l+1)\alpha_1-(l-1)\alpha_2)$, $\tilde{L}(\lambda-l\alpha_1-l\alpha_2)$,
$\tilde{L}(\lambda-l\alpha_1-(2l-2)\alpha_2)$, $\tilde{L}(\lambda-l\alpha_1-(l-2)\alpha_2)$, $\tilde{L}(\lambda-2l\alpha_1-l\alpha_2)$.

Moreover, $\tilde{Z}(\lambda)$ has only the 20 composition factors.

PROOF. (i) According to 1.1 (e-f), we see that the first 8 elements in (i) are maximal.

Consider the homomorphism:

$$\tilde{Z}(\lambda) \to \tilde{Z}(\lambda + 2\alpha_2), \quad \tilde{1}_\lambda \to t_1 = f_2^{(2)}\tilde{1}_{\lambda+2\alpha_2}.$$

Since $f_2^{(l-1)}f_1^{(l-3)}$ is in $\mathbf{u}^-f_2^{(2)}$, using 1.1 (c) we see that $\frac{f_2^{(l-1)}f_1^{(l-3)}}{f_2^{(2)}}\tilde{1}_\lambda$ and $\frac{f_1f_2^{(l-1)}f_1^{(l-3)}}{f_2^{(2)}}\tilde{1}_\lambda$ are primitive elements of weights $\lambda - (l-3)\alpha_1 - (l-3)\alpha_2$ and $\lambda-(l-2)\alpha_1-(l-3)\alpha_2$ respectively. One can check directly that the two elements are maximal. We will show that the last element in (i) is maximal in part (2) of the argument for (ii).

(ii) Now we argue for (ii).

(1) Consider the homomorphism:

$$\tilde{Z}(\lambda) \to \tilde{Z}(\lambda + (l-1)\alpha_1)), \quad \tilde{1}_\lambda \to t_1 = f_1^{(l-1)}\tilde{1}_{\lambda+(l-1)\alpha_1}.$$

Let $x_1 = \frac{x_{l-1,l-1}}{f_1^{(l-1)}} \in \mathbf{u}^-$. Using 1.1 (c) we see that $x_1\tilde{1}_\lambda$ is a primitive element of weight $\lambda - (3l-3)\alpha_1 - (3l-3)\alpha_2$.

Consider the homomorphism:

$$\tilde{\mathbf{u}}f_2^{(l-2)}\tilde{1}_\lambda \to \tilde{\mathbf{u}}f_2^{(2l-2)}f_1^{(l-1)}\tilde{1}_{\lambda+(l-1)\alpha_1+l\alpha_2},$$
$$f_2^{(l-2)}\tilde{1}_\lambda \to f_2^{(2l-2)}f_1^{(l-1)}\tilde{1}_{\lambda+(l-1)\alpha_1+l\alpha_2}.$$

Let

$$y_1 = \frac{1}{2}[[f_2^{(l-1)}, f_1^{(l)}], f_1^{(l)}]f_1^{(l-3)}.$$

Then $y_1 f_2^{(2l-2)}f_1^{(l-1)} = x_{l-1,l-1}$. Using 1.1 (c) we see that $y_1 f_2^{(l-2)}\tilde{1}_\lambda$ is a primitive element of weight $\lambda-(3l-3)\alpha_1-(2l-3)\alpha_2$. Note that $y_1 f_2^{(l-2)} = x_{l-1,l-1}/f_2^{(l)}f_1^{(l-1)}$ is in $\mathbf{u}^-$.

(2) Now we consider the homomorphism:

$$\tilde{Z}(\lambda) \to \tilde{Z}(\lambda + 2\alpha_2), \quad \tilde{1}_\lambda \to t_2 = f_2^{(2)}\tilde{1}_{\lambda+2\alpha_2}.$$

Let

$$x_2 = (f_2^{(l-1)}f_1^{(l)} - f_1^{(l)}f_2^{(l-1)})f_1 \in \mathbf{u}^-.$$

Using 1.1 (c) we see that $x_2\tilde{1}_\lambda$ is a primitive element of weight $\lambda-(l+1)\alpha_1-(l-1)\alpha_2$.

Let $y_2 = f_1^{(l-3)}x_2 \in \mathbf{u}^-$. According to 1.1 (c) we know that $y_2\tilde{1}_\lambda$ is a primitive element of weight $\lambda-(2l-2)\alpha_1-(l-1)\alpha_2$. It is easy to see that $y_2\tilde{1}_\lambda$ is maximal.

(3) Now we consider the homomorphism:

$$\tilde{Z}(\lambda) \to \tilde{Z}(\lambda + 3\alpha_1 + 2\alpha_2),$$
$$\tilde{1}_\lambda \to t_3 = f_2^{(2)}f_1^{(3)}\tilde{1}_{\lambda+3\alpha_1+2\alpha_2}.$$

Let $x_3 = \frac{x_{3,l-1}}{f_2^{(2)} f_1^{(3)}} \in \mathbf{u}^-$. Using 1.1 (c) we know that $x_3 \tilde{1}_\lambda$ is a primitive element of weight $\lambda - (2l+1)\alpha_1 - (2l-1)\alpha_2$.

Consider the homomorphism:

$$\tilde{Z}(\lambda) \to \tilde{\mathbf{u}} f_2^{(l+2)} f_1^{(3)} \tilde{1}_{\lambda+3\alpha_1+(l+2)\alpha_2},$$

$$\tilde{1}_\lambda \to t_3 = f_2^{(l+2)} f_1^{(3)} \tilde{1}_{\lambda+3\alpha_1+(l+2)\alpha_2}.$$

Let $y_3 = [[f_2^{(l-1)}, f_1^{(l)}], f_1^{(l)}] f_1 = \frac{x_{3,l-1}}{f_2^{(l+2)} f_1^{(3)}} \in \mathbf{u}^-$. Using 1.1 (c) we know that $y_3 \tilde{1}_\lambda$ is a primitive element of weight $\lambda - (2l+1)\alpha_1 - (l-1)\alpha_2$.

(4) We consider the homomorphism:

$$\tilde{\mathbf{u}} f_1 \tilde{1}_\lambda \to \tilde{\mathbf{u}} f_2 f_1^{(3)} \tilde{1}_{\lambda+2\alpha_1+\alpha_2},$$

$$f_1 \tilde{1}_\lambda \to t_4 = f_2 f_1^{(3)} \tilde{1}_{\lambda+2\alpha_1+\alpha_2}.$$

Let $x_4 = f_1^{(l-1)} f_2^{(l)} f_1 \in \mathbf{u}^-$. Using 1.1 (c) we know that $x_4 \tilde{1}_\lambda$ is a primitive element of weight $\lambda - l\alpha_1 - l\alpha_2$.

Let $y_4 = \frac{x_{3,l-2}}{f_2 f_1^{(3)}} \in \mathbf{u}^-$. Using 1.1 (c) we know that $y_4 f_1 \tilde{1}_\lambda$ is a primitive element of weight $\lambda - 2l\alpha_1 - (2l-2)\alpha_2$.

(5) Now we consider the homomorphism:

$$\tilde{Z}(\lambda) \to \tilde{Z}(\lambda+(l+2)\alpha_1+(l+1)\alpha_2), \quad \tilde{1}_\lambda \to t_3 = f_1^{(l-1)} f_2^{(l+1)} f_1^{(3)} \tilde{1}_{\lambda+(l+2)\alpha_1+(l+1)\alpha_2}.$$

Let

$$x_5 = (f_2^{(l-2)} f_1^{(l)} - f_1^{(l)} f_2^{(l-2)}) \in \mathbf{u}^-.$$

Using 1.1 (c) we see that $x_5 \tilde{1}_\lambda$ is a primitive element of weight $\lambda - l\alpha_1 - (l-2)\alpha_2$.

(6) Consider the homomorphism

$$\tilde{\mathbf{u}} f_1 \tilde{1}_\lambda \to \tilde{\mathbf{u}} f_2 f_1^{(3)} f_2^{(2)} \tilde{1}_{\lambda+2\alpha_1+3\alpha_2}$$

$$f_1 \tilde{1}_\lambda \to f_2 f_1^{(3)} f_2^{(2)} \tilde{1}_{\lambda+2\alpha_1+3\alpha_2}.$$

Let $x_6 = \frac{x_{l-1,2}}{f_2 f_1^{(3)} f_2^{(2)}} \in \mathbf{u}^-$ By 1.1 (c), $x_6 f_1 \tilde{1}_\lambda$ is primitive and is of weight $\lambda - 2l\alpha_1 - l\alpha_2$.

Note that the element $m = f_2^{(l-2)} f_1^{(2l-3)} f_2^{(l-1)} f_1 \tilde{1}_\lambda$ generates the unique irreducible submodule of $\tilde{Z}(\lambda)$. By comparing the weights of the following 6 elements with the weight of m,

$$\frac{x_{l-1,l-1}}{f_1^{(l-1)}} \tilde{1}_\lambda, \qquad \frac{x_{l-1,l-1}}{f_2^{(l)} f_1^{(l-1)}} \tilde{1}_\lambda, \qquad \frac{x_{3,l-1}}{f_2^{(2)} f_1^{(3)}} \tilde{1}_\lambda,$$

$$\frac{x_{3,l-1}}{f_2^{(l+2)} f_1^{(3)}} \tilde{1}_\lambda, \qquad f_2^{(l-2)} f_1^{(l-1)} f_2^{(l)} f_1 \tilde{1}_\lambda, \qquad \frac{x_{l-1,2}}{f_2 f_1^{(3)} f_2^{(2)}} f_1 \tilde{1}_\lambda,$$

we see that there are no maximal elements in $\tilde{Z}(\lambda)$ that have the same weight with any of above 6 elements.

Now we show that there are no maximal elements in $\tilde{Z}(\lambda)$ that have the same weight with any of other 3 elements in (ii) by assuming (iii).

By (iii), $\tilde{Z}(\lambda)$ has only one composition factor isomorphic to $\tilde{L}(\lambda - (l+1)\alpha_1 - (l-1)\alpha_2)$. Suppose that there is a maximal element m in $\tilde{Z}(\lambda)$ of weight $\lambda - (l+1)\alpha_1 - (l-1)\alpha_2$. Then m is in $\tilde{\mathbf{u}} y$, here $y = [f_2^{(l-1)}, f_1^{(l)}] f_1 \tilde{1}_\lambda$. It is clear that $\tilde{\mathbf{u}} y \subset \tilde{\mathbf{u}}^- y + \tilde{\mathbf{u}}^- f_1^{(l-3)} f_2^{(l-1)} f_1 \tilde{1}_\lambda$. Thus $m = ay + b f_1^{(3)} f_1^{(l-3)} f_2^{(l-1)} f_1 \tilde{1}_\lambda$ for some a, b

in $\mathbf{Q}(\xi)$. Thus $m = ay$. But y is not maximal. So there are no maximal elements in $\tilde{Z}(\lambda)$ that have the same weight with $[f_2^{(l-1)}, f_1^{(l)}]f_1\tilde{1}_\lambda$.

Similarly, we see that there are no maximal elements in $\tilde{Z}(\lambda)$ that have the same weight with any of $[f_2^{(l-2)}, f_1^{(l)}]\tilde{1}_\lambda$, $f_1^{(l-1)}f_2^{(l)}f_1\tilde{1}_\lambda$.

(iii) Using (i), (ii) and 1.1 (b), we see that $\tilde{Z}(\lambda)$ has the 20 composition factors. By a comparison of dimensions we know that $\tilde{Z}(\lambda)$ has only the 20 composition factors.

The theorem is proved. $\square$

THEOREM 3.3. *Let a, b be integers and $\lambda = (la + l - 2, lb + 1)$. Then*
(i) The following elements are maximal in $\tilde{Z}(\lambda)$:

$$\tilde{1}_\lambda, \qquad f_1^{(l-1)}\tilde{1}_\lambda, \qquad f_2^{(2)}\tilde{1}_\lambda, \qquad f_2 f_1^{(l-1)}\tilde{1}_\lambda,$$

$$f_1^{(3)}f_2^{(2)}\tilde{1}_\lambda, \qquad f_1^{(3)}f_2^{(l+1)}f_1^{(l-1)}\tilde{1}_\lambda.$$

$$f_2 f_1^{(3)}f_2^{(2)}\tilde{1}_\lambda, \qquad f_2^{(2)}f_1^{(l+3)}f_2^{(l+1)}f_1^{(l-1)}\tilde{1}_\lambda,$$

$$\frac{f_1^{(3)}f_2}{f_1}\tilde{1}_\lambda, \qquad \frac{f_2^{(2)}f_1^{(3)}f_2}{f_1}\tilde{1}_\lambda, \qquad \frac{x_{3,l-2}}{f_2^{(l-2)}}\tilde{1}_\lambda.$$

(ii) The following elements are primitive elements in $\tilde{Z}(\lambda)$ but not maximal:

$$[f_2^{(2)}, f_1^{(l)}]\tilde{1}_\lambda, \qquad \frac{x_{l-1,l-1}}{f_2^{(l-2)}f_1^{(2l-3)}f_2^{(l-1)}}\tilde{1}_\lambda, \qquad \frac{f_2^{(l-1)}f_1^{(l)}f_2}{f_1}\tilde{1}_\lambda, \qquad \frac{f_1 f_2^{(l-1)}f_1^{(l)}f_2}{f_1}\tilde{1}_\lambda,$$

$$\frac{[f_2^{(2)}, f_1^{(l)}]f_1^{(3)}f_2}{f_1}\tilde{1}_\lambda, \qquad \frac{x_{3,l-1}}{f_1 f_2^{(l-1)}}\tilde{1}_\lambda, \qquad \frac{x_{3,l-1}}{f_1^{(l+1)}f_2^{(l-1)}}\tilde{1}_\lambda, \qquad \frac{x_{3,l-1}}{f_1^{(2l+1)}f_2^{(l-1)}}\tilde{1}_\lambda,$$

$$\frac{x_{l-1,l-1}}{f_1^{(l-3)}f_2^{(l-1)}}f_2^{(2)}\tilde{1}_\lambda.$$

Moreover there are no maximal elements in $\tilde{Z}(\lambda)$ which have the same weight with any of above 9 elements.
(iii) The maximal and primitive elements in (i-ii) provide 20 composition factors of $\tilde{Z}(\lambda)$, which are

$$\tilde{L}(\lambda), \qquad \tilde{L}(\lambda - (l-1)\alpha_1), \qquad \tilde{L}(\lambda - 2\alpha_2),$$

$$\tilde{L}(\lambda - (l-1)\alpha_1 - \alpha_2), \qquad \tilde{L}(\lambda - 3\alpha_1 - 2\alpha_2)$$

$$\tilde{L}(\lambda - (l+2)\alpha_1 - (l+1)\alpha_2), \qquad \tilde{L}(\lambda - 3\alpha_1 - 3\alpha_2),$$

$$\tilde{L}(\lambda - (2l+2)\alpha_1 - (l+3)\alpha_2) \qquad \tilde{L}(\lambda - 2\alpha_1 - \alpha_2),$$

$$\tilde{L}(\lambda - 2\alpha_1 - 3\alpha_2), \qquad \tilde{L}(\lambda - l\alpha_1 - 2\alpha_2), \qquad \tilde{L}(\lambda - (2l-1)\alpha_1 - l\alpha_2),$$

$$\tilde{L}(\lambda - (l-1)\alpha_1 - l\alpha_2), \qquad \tilde{L}(\lambda - l\alpha_1 - l\alpha_2), \qquad \tilde{L}(\lambda - (l+2)\alpha_1 - 3\alpha_2),$$

$$\tilde{L}(\lambda - (2l+3)\alpha_1 - (l+2)\alpha_2), \qquad \tilde{L}(\lambda - (l+3)\alpha_1 - (l+2)\alpha_2), \qquad \tilde{L}(\lambda - 3\alpha_1 - (l+2)\alpha_2),$$

$$\tilde{L}(\lambda - (2l+2)\alpha_1 - (l+1)\alpha_2), \qquad \tilde{L}(\lambda - (3l-1)\alpha_1 - 2l\alpha_2).$$

Moreover, $\tilde{Z}(\lambda)$ has only the 20 composition factors.

PROOF. (i) According to 1.1 (e-f), we see that the first 8 elements in (i) are maximal.

Consider the homomorphism:

$$\tilde{Z}(\lambda) \to \tilde{Z}(\lambda + \alpha_1), \quad \tilde{1}_\lambda \to t_1 = f_1 \tilde{1}_{\lambda + \alpha_1}.$$

Since $f_1^{(3)} f_2$ is in $\mathbf{u}^- f_1$, using 1.1 (c) we see that $\frac{f_1^{(3)} f_2}{f_1} \tilde{1}_\lambda$ and $\frac{f_2^{(2)} f_1^{(3)} f_2}{f_1} \tilde{1}_\lambda$ are primitive elements of weights $\lambda - 2\alpha_1 - \alpha_2$ and $\lambda - 2\alpha_1 - 3\alpha_2$ respectively.

Now we consider the homomorphism:

$$\tilde{Z}(\lambda) \to \tilde{Z}(\lambda + (l-2)\alpha_2)), \quad \tilde{1}_\lambda \to t_2 = f_2^{(l-2)} \tilde{1}_{\lambda + (l-2)\alpha_2}.$$

Using 1.1 (c) we see that $f_1^{(3)} f_2 [f_2^{(l)}, f_1^{(l)}] f_1^{(l-1)} \tilde{1}_\lambda = \frac{x_{3,l-2}}{f_2^{(l-2)}} \tilde{1}_\lambda$ is a primitive element of weight $\lambda - (2l+2)\alpha_1 - (l+1)\alpha_2$.

One can check directly that the three elements are maximal.

(ii) Now we argue for (ii).

(1) Consider the homomorphism:

$$\tilde{Z}(\lambda) \to \tilde{Z}(\lambda + \alpha_1), \quad \tilde{1}_\lambda \to t_1 = f_1 \tilde{1}_{\lambda + \alpha_1}.$$

Using Theorem 3.1 and 1.1 (c) we see that the first 8 elements are primitive.

(2) Now we consider the homomorphism:

$$\tilde{\mathbf{u}} f_2^{(2)} \tilde{1}_\lambda \to \tilde{\mathbf{u}} f_1^{(l-3)} f_2^{(l-1)} \tilde{1}_{\lambda + (l-3)\alpha_1 + (l-3)\alpha_2},$$

$$f_2^{(2)} \tilde{1}_\lambda \to f_1^{(l-3)} f_2^{(l-1)} \tilde{1}_{\lambda + (l-3)\alpha_1 + (l-3)\alpha_2}.$$

Using 1.1 (c) we know that $\frac{x_{l-1,l-1}}{f_1^{(l-3)} f_2^{(l-1)}} f_2^{(2)} \tilde{1}_\lambda$ is a primitive element of weight $\lambda - (3l-1)\alpha_1 - 2l\alpha_2$.

Consider the homomorphism:

$$\tilde{Z}(\lambda) \to \tilde{Z}(\lambda + \alpha_1)), \quad \tilde{1}_\lambda \to t_1 = f_1 \tilde{1}_{\lambda + \alpha_1}.$$

It is easy to see that all the 9 primitive elements have non-zero image. By Theorem 3.1 (ii) and 1.1 (e) we know that there are no maximal elements in $\tilde{Z}(\lambda)$ which have the same weight with any of the 9 primitive elements.

(iii) Using (i), (ii) and 1.1 (b), we see that $\tilde{Z}(\lambda)$ has the 20 composition factors. By a comparison of dimensions we know that $\tilde{Z}(\lambda)$ has only the 20 composition factors.

The theorem is proved. $\qquad\qquad\qquad\qquad\qquad\qquad\qquad\qquad\qquad\qquad$ $\square$

THEOREM 3.4. *Let* a, b *be integers and* $\lambda = (la + 2, lb + l - 2)$. *Then*
(i) The following elements are maximal in $\tilde{Z}(\lambda)$:

$$\tilde{1}_\lambda, \qquad f_1^{(3)} \tilde{1}_\lambda, \qquad f_2^{(l-1)} \tilde{1}_\lambda, \qquad f_2^{(2)} f_1^{(3)} \tilde{1}_\lambda,$$

$$f_1 f_2^{(l-1)} \tilde{1}_\lambda, \qquad f_1 f_2^{(2)} f_1^{(3)} \tilde{1}_\lambda.$$

$$f_2^{(2)} f_1^{(l+1)} f_2^{(l-1)} \tilde{1}_\lambda, \qquad f_2^{(l-1)} f_1^{(2l+1)} f_2^{(l+2)} f_1^{(3)} \tilde{1}_\lambda,$$

$$\frac{f_2^{(2)} f_1}{f_2} \tilde{1}_\lambda, \qquad \frac{f_1^{(3)} f_2^{(2)} f_1}{f_2} \tilde{1}_\lambda.$$

(ii) The following elements are primitive elements in $\tilde{Z}(\lambda)$ *but not maximal:*

$$[f_2^{(l-1)}, f_1^{(l)}] \tilde{1}_\lambda, \qquad \frac{x_{l-1,l-1}}{f_1^{(l-3)} f_2^{(l-2)} f_1^{(l-1)}} \tilde{1}_\lambda, \qquad \frac{x_{3,l-2}}{f_2 f_1^{(l-1)} f_2^{(l-2)}} \tilde{1}_\lambda, \qquad [f_1^{(3)}, f_2^{(l)}] \tilde{1}_\lambda,$$

$$\frac{x_{3,l-1}}{f_2 f_1^{(2l)} f_2^{(l-1)}}\tilde{1}_\lambda, \qquad \frac{x_{3,l-1}}{f_2 f_1^{(l)} f_2^{(l-1)}}\tilde{1}_\lambda, \qquad f_1[f_2^{(l-1)}, f_1^{(l)}]\tilde{1}_\lambda, \qquad \frac{x_{l-1,2}}{f_2 f_1^{(l-1)}}\tilde{1}_\lambda,$$

$$[f_2^{(2)}, f_1^{(l)}]f_1^{(3)}\tilde{1}_\lambda, \qquad \frac{x_{l-1,l-1}}{f_1^{(l-3)} f_2^{(2l-2)} f_1^{(l-1)}}\tilde{1}_\lambda.$$

Moreover no maximal element in $\tilde{Z}(\lambda)$ has the same weight as any of above 10 elements.

(iii) The maximal and primitive elements in (i-ii) provide 20 composition factors of $\tilde{Z}(\lambda)$, which are

$$\tilde{L}(\lambda), \qquad \tilde{L}(\lambda - 3\alpha_1), \qquad \tilde{L}(\lambda - (l-1)\alpha_2),$$

$$\tilde{L}(\lambda - 3\alpha_1 - 2\alpha_2), \qquad \tilde{L}(\lambda - \alpha_1 - (l-1)\alpha_2)$$

$$\tilde{L}(\lambda - 4\alpha_1 - 2\alpha_2), \qquad \tilde{L}(\lambda - (l+1)\alpha_1 - (l+1)\alpha_2), \qquad \tilde{L}(\lambda - (2l+4)\alpha_1 - (2l+1)\alpha_2),$$

$$\tilde{L}(\lambda - \alpha_1 - \alpha_2), \qquad \tilde{L}(\lambda - 4\alpha_1 - \alpha_2), \qquad \tilde{L}(\lambda - l\alpha_1 - (l-1)\alpha_2),$$

$$\tilde{L}(\lambda - 2l\alpha_1 - (2l-1)\alpha_2), \qquad \tilde{L}(\lambda - (l+3)\alpha_1 - l\alpha_2), \qquad \tilde{L}(\lambda - 3\alpha_1 - l\alpha_2),$$

$$\tilde{L}(\lambda - 4\alpha_1 - (l+1)\alpha_2), \qquad \tilde{L}(\lambda - (l+4)\alpha_1 - (l+1)\alpha_2), \qquad \tilde{L}(\lambda - (l+1)\alpha_1 - (l-1)\alpha_2),$$

$$\tilde{L}(\lambda - (l+3)\alpha_1 - (l+2)\alpha_2), \qquad \tilde{L}(\lambda - (l+3)\alpha_1 - 2\alpha_2), \qquad \tilde{L}(\lambda - 2l\alpha_1 - (l-1)\alpha_2).$$

Moreover, $\tilde{Z}(\lambda)$ has only the 20 composition factors.

PROOF. (i) According to 1.1 (e-f), we see the first 8 elements in (i) are maximal. Consider the homomorphism:

$$\tilde{Z}(\lambda) \to \tilde{Z}(\lambda + \alpha_2), \quad \tilde{1}_\lambda \to t_1 = f_2\tilde{1}_{\lambda+\alpha_2}.$$

Since $f_2^{(2)} f_1$ is in $\mathbf{u}^- f_2$, using 1.1 (c) we see that $\frac{f_2^{(2)} f_1}{f_2}\tilde{1}_\lambda$ and $\frac{f_1^{(3)} f_2^{(2)} f_1}{f_2}\tilde{1}_\lambda$ are primitive elements of weights $\lambda - \alpha_1 - \alpha_2$ and $\lambda - 4\alpha_1 - \alpha_2$ respectively. One can check directly that the two elements are maximal.

(ii) Now we argue for (ii).

(1) Consider the homomorphism:

$$\tilde{Z}(\lambda) \to \tilde{Z}(\lambda + \alpha_2)), \quad \tilde{1}_\lambda \to t_1 = f_2\tilde{1}_{\lambda+\alpha_2}.$$

Using Theorem 3.1 and 1.1 (c) we see that the first 9 elements are primitive.

(2) Now we consider the homomorphism:

$$\tilde{Z}(\lambda) \to \tilde{\mathbf{u}} f_1^{(l-3)} f_2^{(2l-2)} f_1^{(l-1)}\tilde{1}_{\lambda+(2l-4)\alpha_1+(2l-2)\alpha_2},$$

$$\tilde{1}_\lambda \to f_1^{(l-3)} f_2^{(2l-2)} f_1^{(l-1)}\tilde{1}_{\lambda+(2l-4)\alpha_1+(2l-2)\alpha_2}.$$

Using 1.1 (c) we know that $[[f_2^{(l-1)}, f_1^{(l)}], f_1^{(l)}]\tilde{1}_\lambda = \frac{x_{l-1,l-1}}{f_1^{(l-3)} f_2^{(2l-2)} f_1^{(l-1)}}\tilde{1}_\lambda$ is a primitive element of weight $\lambda - 2l\alpha_1 - (l-1)\alpha_2$.

Consider the homomorphism:

$$\tilde{Z}(\lambda) \to \tilde{Z}(\lambda + \alpha_2)), \quad \tilde{1}_\lambda \to t_1 = f_2\tilde{1}_{\lambda+\alpha_2}.$$

It is easy to see that all the 10 primitive elements have non-zero image. By Theorem 3.1 (ii) and 1.1 (e) we know that there are no maximal elements in $\tilde{Z}(\lambda)$ which have the same weight with any of the 10 primitive elements.

(iii) Using (i), (ii) and 1.1 (b), we see that $\tilde{Z}(\lambda)$ has the 20 composition factors. By a comparison of dimensions we know that $\tilde{Z}(\lambda)$ has only the 20 composition factors.

The theorem is proved.   □

4. Weyl Modules for Type B_2

For $\lambda = (\lambda_1, \lambda_2) \in \mathbf{Z}_+^2$ we denote by I_λ the left ideal of U_ξ generated by all $e_i^{(a)}$ $(a \geq 1)$, $f_i^{(a_i)}$ $(a_i \geq \lambda_i + 1)$, $k_i - \xi^{\lambda_i}$, $\begin{bmatrix} k_i, c \\ a \end{bmatrix} - \begin{bmatrix} \lambda_i + c \\ a \end{bmatrix}_{\xi^i}$. The Weyl module $V(\lambda)$ of U_ξ is defined to be U_ξ/I_λ, its dimension is $(\lambda_1 + 1)(\lambda_2 + 1)(\lambda_1 + \lambda_2 + 2)(\lambda_1 + 2\lambda_2 + 3)/6$. Let v_λ be a nonzero element in $V(\lambda)_\lambda$. We can work out the maximal and primitive elements in $V(\lambda)$ as in section 3 (cf. [X2]). We omit the results here.

Acknowledgement: The work was completed during my visit to the University of Sydney. It is a great pleasure to thank Professor G. Lehrer for the invitation. Part of the work was done during my visit to Bielefeld University. I am grateful to the SFB 343 in Bielefeld University for financial support. Finally I would like to thank the referee for helpful comments.

References

[APW] H.H. Andersen, P. Polo, K. Wen, *Representations of quantum algebras.* Invent. Math. 104 (1991), no. 1, 1–59.

[DS1] S.R. Doty and J.B. Sullivan, *The submodule structure of Weyl module for SL_3,* J. Alg. 96 (1985), 78-93.

[DS2] S.R. Doty and J.B. Sullivan, *On the structure of the higher cohomology modules of line bundles on G/B,* J. Alg. 114(1988), 286-332.

[I] R.S. Irving, *The structure of certain highest weight modules for SL_3,* J. Alg. 99 (1986), 438-457.

[K] Kühne-Hausmann, K, *Zur Untermodulstruktur der Weylmoduln für Sl_3,* Bonner Mathematische Schriften 162, Universität Bonn, Mathematisches Institut, Bonn, 1985. vi+190 pp.

[L1] G. Lusztig, *Modular representations and quantum groups,* Contemp. Math. 82 (1989), 59-77.

[L2] G. Lusztig, *Finite dimensional Hopf algebras arising from quantized universal enveloping algebras,* Jour. Amer. Math. Soc. 3 (1990), 257-296.

[L3] G. Lusztig, *Quantum groups at roots of 1,* Geom. Ded. 35 (1990), 89-114.

[L4] G. Lusztig, *Introduction to quantum groups,* Progress in Mathematics 110, Birkhäuser, Boston · Basel · Berlin, 1993.

[X1] N. Xi, *Irreducible modules of quantized enveloping algebras at roots of 1,* Publ. RIMS. Kyoto Univ. 32 (1996), 235-276.

[X2] N. Xi, *Maximal and primitive elements in Weyl modules for type A_2,* J. Alg., 215(1999), 735-756.

INSTITUTE OF MATHEMATICS, CHINESE ACADEMY OF SCIENCES, BEIJING 100080, CHINA
E-mail address: nanhua@math.ac.cn

Contemporary Mathematics
Volume **478**, 2009

Irreducible representations of the special algebras in prime characteristic

Yu-Feng Yao and Bin Shu

ABSTRACT. Let $L = S(m; \mathbf{n})$ be a graded Lie algebra of special type in the Cartan type series over an algebraically closed field of characteristic $p > 0$. In this paper we study irreducible modules of L. For this we adopt the notion of the category $\mathfrak{C}$ which was introduced in [**11**] by Skryabin for the study of representations of generalized Witt algebras. In the generalized restricted Lie algebra setting, a class of induced modules of L of (generalized) p-character χ are proved to belong to $\mathfrak{C}$. All irreducible modules of L of p-character χ are determined when the height of χ is not bigger than $\min\{p^{n_i} - p^{n_i-1} \mid 1 \leq i \leq m\} - 2$. These irreducible modules turn out to be just induced modules in the so-called non-exceptional cases. As to the exceptional cases, irreducible modules, as well as the dimensions, are precisely obtained through a complex of induced modules.

1. Introduction

This paper is a continuation of the work in [**10**]. In the previous paper, we studied irreducible modules of the generalized Jacobson-Witt algebras $W(m; \mathbf{n})$ which is a so-called generalized restricted Lie algebra as introduced in [**8**]. There we determined simple modules of $W(m; \mathbf{n})$ of generalized p-character χ when the height of $\chi < \min\{p^{n_i} - p^{n_i-1} \mid 1 \leq i \leq m\}$ for $\mathbf{n} = (n_1, \cdots, n_m)$, by introducing a modified induced module structure and thereby endowing it with Skryabin's so-called (R, L)-module structure in the generalized χ-reduced module category. We generally took use of techniques of dealing with simple modules, developed in [**7**], [**5**], [**1**] and [**2**].

In this paper, we first establish the corresponding (R, L)-module category which will be replaced by the notion $\mathfrak{C}$-category in the text, for $R = \mathfrak{A}(m; \mathbf{n})$ (the divided power algebra), and $L = S(m; \mathbf{n})$. By definition, L is a graded simple Lie algebra of series S of Cartan type in $W(m; \mathbf{n})$, which is the derived algebra of $\widetilde{S(m; \mathbf{n})} := \{X \in W(m; \mathbf{n}) \mid \mathrm{div}(X) = 0\}$. Here the mapping $\mathrm{div} : W(m; \mathbf{n}) \to R$ is defined by $\mathrm{div}(\sum_{i=1}^{m} f_i D_i) = \sum_{i=1}^{m} D_i(f_i)$, called a divergence (to see §2.1).

The theory on Skryabin's $\mathfrak{C}$-category in [**11**] can be set up for L as he asserted there. This enables us to continue the arguments in [**10**] for study of irreducible modules of L.

2000 *Mathematics Subject Classification.* 17B50, 17B70.

2. Preliminaries

In this paper, we always assume that the ground field F is algebraically closed with prime characteristic, $m \in \mathbb{N}$, $m \geq 3$.

2.1. Graded simple Lie algebras of series S of Cartan type.

Let $a = (a_1, a_2, \cdots, a_m)$, $b = (b_1, b_2, \cdots, b_m) \in \mathbb{Z}^m$, we write $a \leq b$ (resp. $a \geq b$) if $a_i \leq b_i$ (resp. $a_i \geq b_i$) for all $1 \leq i \leq m$ and we write $a < b$ (resp. $a > b$) if $a \leq b$ (resp. $a \geq b$), but $a \neq b$. If $a, b \geq 0$, define $\binom{a}{b} = \prod_i \binom{a_i}{b_i}$, where $\binom{a_i}{b_i}$ is the usual binomial coefficient with the convention $\binom{a_i}{b_i} = 0$ unless $a_i \geq b_i$. Set $A(m; \mathbf{n}) = \{ a = (a_1, a_2, \cdots, a_m) \mid 0 \leq a_i \leq p^{n_i} - 1, \forall i = 1, 2, \cdots m \}$. Denote $\tau = (p^{n_1} - 1, p^{n_2} - 1, \cdots, p^{n_m} - 1)$, then $A(m; \mathbf{n}) = \{ a = (a_1, a_2, \cdots, a_m) \in \mathbb{Z}^m \mid 0 \leq a \leq \tau \}$. The divided power algebra $\mathfrak{A}(m; \mathbf{n})$ is an F-algebra having F-basis $\{ x^{(a)} \mid a \in A(m; \mathbf{n}) \}$ with the multiplication rule $x^{(a)} x^{(b)} = \binom{a+b}{a} x^{(a+b)}$, where $x^{(c)} = 0$ if $c \notin A(m; \mathbf{n})$. For each $i, 1 \leq i \leq m$, denote $\varepsilon_i = (\delta_{i1}, \delta_{i2}, \cdots, \delta_{im})$. The following formula assures that $\mathfrak{A}(m; \mathbf{n})$ is an associative and commutative algebra which can be easily verified by straightforward computation: for any $\alpha, \beta, \gamma \in A(m; \mathbf{n})$,

$$(2.1) \qquad \binom{\gamma}{\alpha} \binom{\gamma - \alpha}{\beta} = \binom{\gamma}{\beta} \binom{\gamma - \beta}{\alpha}.$$

Let D_i denote the derivation of $\mathfrak{A}(m; \mathbf{n})$ uniquely determined by the rule: $D_i(x^{(a)}) = x^{(a - \varepsilon_i)}$. The generalized Jacobson-Witt algebra $W(m; \mathbf{n})$ is a Lie algebra consisting of all special derivations of $\mathfrak{A}(m; \mathbf{n})$. According to [**12**, 4.2], $W(m; \mathbf{n}) = F$-span$\{ x^{(a)} D_i \mid a \in A(m; \mathbf{n}), 1 \leq i \leq m \} \subseteq Der(\mathfrak{A}(m; \mathbf{n}))$. For $a \in A(m; \mathbf{n})$, set $|a| := \sum_{i=1}^{m} a_i$ and $W(m; \mathbf{n})_{[i]} := F$-span$\{ x^{(a)} D_j \mid |a| = i + 1, j = 1, 2, \cdots, m \}$, then $W(m; \mathbf{n}) = \bigoplus_{i=-1}^{s} W(m; \mathbf{n})_{[i]}$ is a natural $\mathbb{Z}$-gradation of $W(m; \mathbf{n})$, where $s = \sum_{i=1}^{m} (p^{n_i} - 1) - 1$. Then $W(m; \mathbf{n})$ has a natural filtration associated with the gradation above:

$$W(m; \mathbf{n}) = W(m; \mathbf{n})_{-1} \supset W(m; \mathbf{n})_0 \supset W(m; \mathbf{n})_1 \supset \cdots$$

where $W(m; \mathbf{n})_i = \bigoplus_{j \geq i} W(m; \mathbf{n})_{[j]}$.

Define a linear map $\mathrm{div} : W(m; \mathbf{n}) \to \mathfrak{A}(m; \mathbf{n})$, $x^{(a)} D_i \mapsto D_i(x^{(a)}) = x^{(a - \varepsilon_i)} \in \mathfrak{A}(m; \mathbf{n})$. Let $\widetilde{S(m; \mathbf{n})} := \mathsf{Ker}\,(\mathrm{div}) = \{ \sum_{i=1}^{m} f_i D_i \in W(m; \mathbf{n}) \mid \sum_{i=1}^{m} D_i(f_i) = 0 \}$. The special algebra $S(m; \mathbf{n})$ is defined by the derived algebra of $\widetilde{S(m; \mathbf{n})}$, i.e. $S(m; \mathbf{n}) = \widetilde{S(m; \mathbf{n})}^{(1)} = [\widetilde{S(m; \mathbf{n})}, \widetilde{S(m; \mathbf{n})}]$. According to [**12**, 4.3], $S(m; \mathbf{n}) = F$-span$\{ D_{ij}(x^{(a)}) \mid a \in A(m; \mathbf{n}), 1 \leq i < j \leq m \}$, where D_{ij} is a linear map from $\mathfrak{A}(m; \mathbf{n})$ to $W(m; \mathbf{n})$ defined via: $D_{ij}(x^{(a)}) = D_j(x^{(a)}) D_i - D_i(x^{(a)}) D_j = x^{(a - \varepsilon_j)} D_i - x^{(a - \varepsilon_i)} D_j$. According to [**12**, 4.2, 4.3], both $W(m; \mathbf{n})$ and $S(m; \mathbf{n})$ are simple Lie algebras. And each of them is restricted if and only if $\mathbf{n} = \mathbf{1} := (1, 1, \cdots, 1)$.

2.2. Generalized restricted Lie algebras and generalized χ-reduced representations.

Although $S(m; \mathbf{n})$ is not restricted whenever $\mathbf{n} \neq \mathbf{1}$, it is always a generalized restricted Lie algebra in the sense as below (cf. [**8**]).

DEFINITION 2.1. A generalized restricted Lie algebra L over F is a Lie algebra associated with an ordered basis $E = (e_i)|_{i \in I}$ and a mapping $\varphi_{\mathbf{s}} : E \to L$ sending $e_i \mapsto e_i^{\varphi_{\mathbf{s}}}$ with $\mathbf{s} = (s_i)|_{i \in I}$, where $s_i \in \mathbb{Z}_+$ such that $\mathrm{ad}\, e_i^{\varphi_{\mathbf{s}}} = (\mathrm{ad}\, e_i)^{p^{s_i}}$ for all $i \in I$.

By a straightforward verification, $L_0 = S(m;\mathbf{n})_0$ is restricted under the mapping $D \mapsto D^{[p]}$, where $D^{[p]}$ is the usual pth power of the derivation D. So for a basis E_1 of $S(m;\mathbf{n})_0$, we have $\mathrm{ad}\, x^{[p]} = (\mathrm{ad}\, x)^p$ for any $x \in E_1$. Take $E = E_1 \cup \{D_1, D_2, \cdots, D_m\}$, then E is a basis of $S(m;\mathbf{n})$. We can assume that $E = (e_i)|_{i \in I}$ such that $e_i = D_i$, $i = 1, 2, \cdots, m$ and $e_j \in E_1$ for $j > m$. Then let $\mathbf{s} = (n_1, n_2, \cdots, n_m, 1, 1, \cdots, 1)$, define $\varphi_{\mathbf{s}} : E \to L$ sending $e_i \mapsto 0$ for $1 \le i \le m$ and $e_j \mapsto e_j^{[p]}$ for $j > m$, then $\mathrm{ad}\, e_i^{\varphi_{\mathbf{s}}} = (\mathrm{ad}\, e_i)^{p^{s_i}}$, $\forall i \in I$. So $S(m;\mathbf{n})$ is a generalized restricted Lie algebra in the sense of Definition 2.1 .

For a generalized restricted Lie algebra over F, by Schur lemma, we have the following fact.

LEMMA 2.2. *If* $(L, \varphi_{\mathbf{s}})$ *is a generalized restricted Lie algebra over F associated with a basis* $E = (e_i)|_{i \in I}$ *and* $\varphi_{\mathbf{s}}$, $\mathbf{s} = (s_i)|_{i \in I}$. *Suppose* (V, ρ) *is an irreducible representation of L, then there exists a unique* $\chi \in L^*$ *such that :*

$$(2.2) \qquad \rho(e_i)^{p^{s_i}} - \rho(e_i^{\varphi_{\mathbf{s}}}) = \chi(e_i)^{p^{s_i}} \,\mathrm{id}_V, \quad e_i \in E.$$

Here the function χ is called a (generalized) p-character of V. A representation (module) of L satisfying (2.2) is called a generalized χ-reduced representation (module). Especially when $\chi = 0$, it is called a generalized restricted representation (module) of L.

Assume that $(L, \varphi_{\mathbf{s}})$ is a generalized restricted Lie algebra associated with a basis $E = (e_i)|_{i \in I}$ and $\varphi_{\mathbf{s}}$, $\mathbf{s} = (s_i)|_{i \in I}$. For a $\chi \in L^*$, we define $U_{p^{\mathbf{s}}}(L, \chi) := U(L)/(e_i^{p^{s_i}} - e_i^{\varphi_{\mathbf{s}}} - \chi(e_i)^{p^{s_i}} \mid e_i \in E)$, where $(e_i^{p^{s_i}} - e_i^{\varphi_{\mathbf{s}}} - \chi(e_i)^{p^{s_i}} \mid e_i \in E)$ denote the ideal in $U(L)$ generated by these central elements $e_i^{p^{s_i}} - e_i^{\varphi_{\mathbf{s}}} - \chi(e_i)^{p^{s_i}}$ for all $e_i \in E$. $U_{p^{\mathbf{s}}}(L, \chi)$ is called the generalized χ-reduced enveloping algebra of L (cf. [8]). When $\chi = 0$, $U_{p^{\mathbf{s}}}(L, 0)$ is often called the generalized restricted enveloping algebra of L, simply denoted by $U_{p^{\mathbf{s}}}(L)$. We have category equivalence between the generalized χ-reduced (resp. generalized restricted) module category of L and $U_{p^{\mathbf{s}}}(L, \chi)$ (resp. $U_{p^{\mathbf{s}}}(L)$)-module category.

REMARK 2.3. A restricted Lie algebra $(g, [p])$ is naturally a generalized restricted Lie algebra associated with an arbitrary given basis E and $\mathbf{s} = \mathbf{1} := (1, 1, \cdots, 1)$. Furthermore, in this case , a generalized χ-reduced module (algebra) coincides with the χ-reduced module (algebra).

2.3. The primitive p-envelope of $S(m;\mathbf{n})$. Recall that associated with a generalized restricted Lie algebra L, there is a special p-envelope consisting of primitive elements of the generalized restricted enveloping algebra of L, called a primitive p-envelope (cf. [9]). We turn to $L = S(m;\mathbf{n})$. Then, there is a natural realization in $\mathrm{Der}(R)$ of the primitive p-envelope of L which is as follows:

$$\mathfrak{L} = S(m;\mathbf{n}) \bigoplus \sum_{i=1}^{m} \sum_{d_i=1}^{n_i-1} F D_i^{p^{d_i}}.$$

For $\chi \in L^*$, we denote by $\bar{\chi} \in \mathfrak{L}^*$ the trivial extension of χ to $\mathfrak{L}^*$, i.e. $\bar{\chi}|_{S(m;\mathbf{n})} = \chi$, while $\bar{\chi}(x) = 0$, for any $x \in \sum_{i=1}^{m} \sum_{d_i=1}^{n_i-1} F D_i^{p^{d_i}}$. The following lemma will be useful.

LEMMA 2.4. ([9] and [10]) *Keep notations as above. There is an algebra iso-morphism:*

$$U_{p^s}(L, \chi) \cong U(\mathfrak{L}, \bar{\chi}) \tag{2.3}$$

and thus

$$U_{p^s}(L, \chi) \cong U_{p^s}(L, \chi^\Phi), \tag{2.4}$$

where $\Phi \in \mathrm{Aut}(L)$ *and* $\chi^\Phi(D) := \chi(\Phi^{-1}(D)), \forall D \in L$.

3. Skryabin's $\mathfrak{C}$-category and independence of differential operators

3.1. In the sequel, set $L = S(m; \mathbf{n})$. As stated above, L admits a natural gradation: $L = \bigoplus_{i=-1}^{s} L_{[i]}$, where $s = \sum_{i=1}^{m}(p^{n_i} - 1) - 2$ and $L_{[t]} = F\text{-span}\{D_{ij}(x^\alpha) \mid |\alpha| = \sum_{k=1}^{m} \alpha_k = t + 2, 1 \le i < j \le m\}$. Then L has a natural filtration associated with the gradation above : $L = L_{-1} \supset L_0 \supset L_1 \supset \cdots$, where $L_i = \sum_{j \ge i} L_{[j]}$. Note that $L_0 = F\text{-span}\{D_{ij}(x^\alpha) \mid |\alpha| = \sum_{k=1}^{m} \alpha_k \ge 2, 1 \le i < j \le m\}$ is a restricted subalgebra of L with the p–mapping being the pth power as usual derivations.

3.2. In [11], Skryabin introduced the $\mathfrak{C}$-category in the case of the generalized Jacobson-Witt algebra. Using the $\mathfrak{C}$-category, Skryabin studied the simple modules of the generalized Jacobson-Witt algebra generalizing the arguments on simple graded modules by Guang-Yu Shen in [7] (also see [5]). One can refer to [11] for details. In the following we set up the $\mathfrak{C}$-category in the case of the special algebra, following Skryabin.

DEFINITION 3.1. (M, σ) is called a discrete L_0-module if for any $x \in M$, there exists some nonnegative integer l such that $\sigma(L_l)x = 0$.

DEFINITION 3.2. Set $R = \mathfrak{A}(m; \mathbf{n})$ and $L = S(m; \mathbf{n})$. Denote by $\mathfrak{C}$ the category whose objects are additive groups M endowed with an R-module structure (M, ρ_R), an L-module structure (M, ρ_L) and an L_0-module structure (M, σ) so that M is discrete as $\sigma(L_0)$ module and the following properties are satisfied :
  (R1) $[\rho_L(D), \rho_R(f)] = \rho_R(Df)$;
  (R2) $[\sigma(D'), \rho_R(f)] = 0$;
  (R3) $[\sigma(D'), \rho_L(D_i)] = 0$;
  (R4) $\rho_L(D_{ij}(f)) = \rho_R(D_j(f)) \circ \rho_L(D_i) - \rho_R(D_i(f)) \circ \rho_L(D_j) +$
    $\sum_{|\alpha| \ge 2} \rho_R(D^\alpha f) \circ \sigma(D_{ij}(x^\alpha))$,

where $f \in R, D \in L, D' \in L_0, i, j = 1, 2, \cdots, m$. The morphisms in $\mathfrak{C}$-category are the mappings admissible with the three module structures. A module M in $\mathfrak{C}$-category is called a $\mathfrak{C}$-module.

DEFINITION 3.3. Let R, L be as above. Denote by (R, L)-**mod** the category whose objects are additive groups M endowed with an R-module structure (M, ρ_R) and an L-module structure (M, ρ_L) which satisfies $(R1)$ above. The morphisms in (R, L)-**mod** are the mappings admissible with the two module structures. A module M in (R, L)-**mod** is called an (R, L)-module.

3.3. In this subsection, we will recall some facts about independent systems of differential operators introduced in [**11**] which we will use to study submodules and homomorphisms later. Assume R is a commutative algebra with unit over F. Endow the endomorphism algebra $\operatorname{End}_F R$ with an R-module structure by putting $(f \cdot \varphi)(g) = f\varphi(g)$, for $f, g \in R$, $\varphi \in \operatorname{End}_F R$.

DEFINITION 3.4. ([**11**]) A system of endomorphisms $\Phi \subseteq \operatorname{End}_F R$ is called independent if for any finite subset $\Phi' = \{\varphi_1, \varphi_2, \cdots, \varphi_n\} \subseteq \Phi$, the submodule $Val\Phi'$ of the free F-module R^n generated by all n-tuples $(\varphi_1(g), \varphi_2(g), \cdots, \varphi_n(g))$ with $g \in R$ coincides with R^n.

PROPOSITION 3.5. ([**11**]) Suppose $\{\partial_i^{p^{r_i}} \mid 1 \leq i \leq m, 0 \leq r_i < n_i\}$ is an independent system of derivations of R. Given a finite subset $A \subseteq A(m;\mathbf{n})$ and an n-tuple $\gamma \in A$, then there exists a finite number of elements $f_1, f_2, \cdots, f_u, g_1, g_2, \cdots, g_u \in R$ such that :

$$(3.1) \qquad \sum_{\nu=1}^{u} f_\nu \partial^\alpha g_\nu = \begin{cases} 1, & if \quad \alpha = \gamma, \\ 0, & if \quad \alpha \in A, \alpha \neq \gamma. \end{cases}$$

REMARK 3.6. (1) Any subsystem of an independent system of endomorphisms is also independent.

(2) Any independent system $\Phi \subseteq \operatorname{End}_F R$ is linear independent. Assume $\sum_{i=1}^{n} f_i \varphi_i = 0$, where $f_i \in R, \varphi_i \in \Phi$. Set $\Phi' = \{\varphi_1, \varphi_2, \cdots, \varphi_n\}$, according to statement (1), Φ' is also independent, then the R-module homomorphism $\psi :$ $R^n \to R$ defined by the rule $\psi(r_1, r_2, \cdots, r_n) = \sum_{i=1}^{n} f_i r_i$ for $r_1, r_2, \cdots, r_n \in R$ vanishes on $Val\Phi'$. Note that $Val\Phi' = R^n$, so $\psi = 0$. We can take value of $(r_1, r_2, \cdots, r_n)$ with $(1, 0, 0, \cdots, 0), (0, 1, 0, \cdots, 0), \cdots, (0, 0, 0, \cdots, 1)$ in the equation $\psi(r_1, r_2, \cdots, r_n) = \sum_{i=1}^{n} f_i r_i$, then we will obtain that $f_i = 0$, $i = 1, 2, \cdots, n$. Hence Φ is linear independent. But generally speaking, a linear independent system is not necessarily independent in the sense of Definition 3.4.

(3) Set $R = \mathfrak{A}(m;\mathbf{n})$, then one can easily see that $\{D_i^{p^{r_i}} \mid 1 \leq i \leq m, 0 \leq r_i < n_i\}$ is independent.

4. Submodules and homomorphisms in the category $\mathfrak{C}$

In the following sequel, we always assume $R = \mathfrak{A}(m;\mathbf{n})$. According to Remark 3.6(3), $\{D_i^{p^{r_i}} \mid 1 \leq i \leq m, 0 \leq r_i < n_i\}$ is independent. Denote $A'(m;\mathbf{n}) = \{(\alpha_1, \alpha_2, \cdots, \alpha_m) \in A(m;\mathbf{n}) \mid \alpha_i < p^{n_i} - p^{n_i-1}, i = 1, 2, \cdots, m\}$.

For objects $M, N \in \mathfrak{C}$ and a mapping $\varphi : M \to N$, denote by $\Gamma(\varphi)$ the graph $\{(m, \varphi(m)) \mid m \in M\} \subseteq M \bigoplus N$ of φ. Then φ respects any of the three module structures if and only if $\Gamma(\varphi)$ is a submodule of $M \bigoplus N$ with respect to the corresponding module structure. So φ is a morphism in $\mathfrak{C}$ if and only if $\Gamma(\varphi)$ is a submodule of $M \bigoplus N$.

DEFINITION 4.1. For an irreducible L_0-module (M, σ), define the height of M be the smallest nonnegative integer l such that M is annihilated by $\sigma(L_l)$.

A restricted simple $L_{[0]} \cong \mathfrak{sl}(m)$-module can be regarded a restricted simple L_0-module with L_1-trivial action. Recall that the set of isomorphism classes of restricted simple $\mathfrak{sl}(m)$-modules can be parameterized by restricted weights of $\mathfrak{sl}(m)$ (cf. [**3**] and [**4**, Part II, Ch2-3]), which play roles as "highest weights". The simple restricted L_0-module corresponding to fundamental weights of $\mathfrak{sl}(m)$ will be called *exceptional (weighted-) module*.

LEMMA 4.2. *(i) Let $M \in \mathfrak{C}$, assume that*

$$(4.1) \quad \sigma(D_{ij}(x^\alpha)) = 0 \ for \ \alpha \in A(m; \mathbf{n}) \backslash A'(m; \mathbf{n}) \ and \ 1 \le i < j \le m.$$

Then any (R, L)-submodule M' of M is a $\mathfrak{C}$-submodule.

(ii) Let $M, N \in \mathfrak{C}$, assume that both of M and N satisfy (4.1). Then any (R, L)-module homomorphism $\varphi : M \to N$ is a morphism in $\mathfrak{C}$.

PROOF. (i) We only need to prove that M' is a $\sigma(L_0)$-submodule. Given $q \in M'$, then there exists $l \ge 0$ such that $\sigma(L_l)q = 0$. The subset $Q = \{m \in M \mid \sigma(L_l)m = 0\}$ is a $\sigma(L_0)$-submodule of M, because L_l is an ideal of L_0. By (R2) and (R3), it is also an R-submodule and a $\mathfrak{D}$-submodule, where $\mathfrak{D} = F$-span$\{D_1, D_2, \cdots, D_m\}$. By the formula (R4), it is then a L-submodule as well. So it is a $\mathfrak{C}$-submodule. Now $Q \cap M'$ is an (R, L)-submodule of Q which is annihilated by $\sigma(L_l)$ for some integer $l \ge 0$. If we can prove that $Q \cap M'$ is a $\mathfrak{C}$-submodule, then it would follow that $\sigma(D')q \in Q \cap M' \subseteq M'$ for all $D' \in L_0$ since $q \in Q \cap M'$.

Hence we have reduced the proof of this lemma to the case when M is annihilated by L_l for some $l \ge 0$. Under this assumption, the subset $A_{ij} = \{\alpha \in A(m; \mathbf{n}) \mid |\alpha| \ge 2$ and $\sigma(D_{ij}(x^\alpha)) \ne 0\}$ is finite for any i, j such that $1 \le i < j \le m$. Note that $A_{ij} \subseteq A'(m; \mathbf{n})$. Applying Proposition 3.5 to the subset $A = A_{ij} \cup \{0\}$ and an n-tuple $\gamma \in A_{ij}$, we can find a finite number of elements $f_\nu, g_\nu \in R$ satisfying (3.1). Then we see that M' is stable under the endomorphism : $\sum_\nu \rho_R(f_\nu)\rho_L(D_{ij}(g_\nu))$.

At the same time,

$$\sum_\nu \rho_R(f_\nu)\rho_L(D_{ij}(g_\nu))$$

$$= \sum_\nu \rho_R(f_\nu)\rho_L(D_j(g_\nu)D_i - D_i(g_\nu)D_j)$$

$$= \sum_\nu \rho_R(f_\nu)(\rho_R(D_j(g_\nu))\rho_L(D_i) - \rho_R(D_i(g_\nu))\rho_L(D_j)) +$$

$$\sum_\nu \sum_\alpha \rho_R(f_\nu)\rho_R(D^\alpha(g_\nu))\sigma(D_{ij}(x^\alpha))$$

$$= \sum_\nu \rho_R(f_\nu)(\rho_R(D_j(g_\nu))\rho_L(D_i) - \rho_R(D_i(g_\nu))\rho_L(D_j)) + \sigma(D_{ij}(x^\gamma))$$

As M' is an (R, L)-submodule, then it is stable under $\sum_\nu \rho_R(f_\nu)(\rho_R(D_j(g_\nu))\rho_L(D_i) - \rho_R(D_i(g_\nu))\rho_L(D_j))$. Hence M' is stable under $\sigma(D_{ij}(x^\gamma))$ for all $\gamma, |\gamma| \ge 2$, $1 \le i < j \le m$. So M' is stable under $\sigma(L_0)$, i.e. M' is a $\mathfrak{C}$-submodule.

(ii) The direct sum $M \bigoplus N$ is an object of $\mathfrak{C}$ satisfying (4.1). The graph $\Gamma(\varphi)$ is its (R, L)-submodule. According to (i), $\Gamma(\varphi)$ is a $\sigma(L_0)$-submodule. So φ respects the $\sigma(L_0)$-module structure. Therefore φ is a morphism in $\mathfrak{C}$. $\square$

THEOREM 4.3. *(i) Let $M \in \mathfrak{C}$. Assume that :*

(4.2) M is completely reducible as $\sigma(L_0)$-module and none of its irreducible

summand is exceptional.

(4.3) $\sigma(D_{ij}(x^\alpha)) = 0$ *for* $\alpha \in A(m; \mathbf{n}) \backslash A'(m; \mathbf{n}), i, j = 1, 2, \cdots, m, i \neq j$.

Then any L-submodule M' of M is a $\mathfrak{C}$-submodule.

(ii) *Let M, N be two objects of $\mathfrak{C}$ satisfying (4.2), (4.3). Then any L-module homomorphism $\varphi : M \to N$ is a morphism in $\mathfrak{C}$.*

PROOF. As in Lemma 4.2, we know that (ii) is a sequence of (i). The same arguments as in the proof of Lemma 4.2 reduce the proof of (i) to the case when M is annihilated by $\sigma(L_l)$ for some $l \geq 0$. From Lemma 4.2, we only need to prove that M' is an R-submodule of M. We use a similar method developed by Skryabin in [11]. Let $P = \{m \in M \mid Rm \subseteq M'\}$ be the largest R-submodule contained in M' and $Q = RM'$ be the smallest R-submodule containing M'. By (R1), P, Q are L-submodules. Hence by Lemma 4.2, P, Q are $\mathfrak{C}$-submodules. Then we can now pass to an object $Q/P \in \mathfrak{C}$ and its L-submodule M'/P. Hence, at the beginning we can impose the additional assumption that M' contains no nonzero R-submodule of M and $RM' = M$. Then it's sufficient for us to prove that $M = 0$.

We use a similar strategy developed by Skryabin in [11]. The main ideal is as follows. We want to seek endomorphisms φ of M lying in the associative algebra generated by the endomorphisms $\sigma(D')$, $D' \in L_0$ with the property that for any $f \in R$, the endomorphism $f\varphi$ belongs to the associative subalgebra generated by the endomorphisms $\rho_L(D)$, $D \in L$. The L-submodule M' is stable under $f\varphi$ for any $f \in R$. Hence, it contains the R-submodule $R\varphi(M')$. By the hypothesis, then $\varphi(M') = 0$. From (R2), φ is an R-module endomorphism , so $\varphi(M) = \varphi(RM') = R\varphi(M') = 0$, i.e. $\varphi = 0$. This gives a certain relations between the endomorphisms $\sigma(D')$, $D' \in L_0$. It appears that there are sufficiently many relations of a similar kind so that they can't satisfy simultaneously unless $M = 0$.

Assume $M \neq 0$ and let l be the least positive integer such that M is annihilated by $\sigma(L_l)$. First of all, consider $l \leq 1$. Then M can be viewed as the factor algebra $L_0/L_1 \cong \mathfrak{sl}(\mathfrak{D})$-module, where $\mathfrak{D} = F\text{-span}\{D_1, D_2, \cdots, D_m\}$. Given indices $s_1, s_2, s_1', s_2' \in \{1, 2, \cdots, m\}$. Let $f_\nu, g_\nu \in R$ be a finite number of elements satisfying (3.1) for $A = \{\alpha \in A(m; \mathbf{n}) \mid |\alpha| \leq 4\}$ and $\gamma = \varepsilon_{s_1} + \varepsilon_{s_2} + \varepsilon_{s_1'} + \varepsilon_{s_2'}$, i.e.

(4.4)
$$\sum_{\nu=1}^{u} f_\nu D^\alpha g_\nu = \begin{cases} 1, & \text{if } \alpha = \varepsilon_{s_1} + \varepsilon_{s_2} + \varepsilon_{s_1'} + \varepsilon_{s_2'}, \\ 0, & \text{otherwise.} \end{cases}$$

Hence for any $f \in R$, $i, j, s, t \in \{1, 2, \cdots, m\}$, we have :

$$\sum_\nu \rho_L(D_{ij}(ff_\nu))\rho_L(D_{st}(g_\nu))$$

$$= \sum_\nu \rho_L(D_j(ff_\nu)D_i - D_i(ff_\nu)D_j)\rho_L(D_t(g_\nu)D_s - D_s(g_\nu)D_t)$$

$$= \sum_\nu \big(\rho_R(D_j(ff_\nu))\rho_L(D_i) - \rho_R(D_i(ff_\nu))\rho_L(D_j) +$$

$$\sum_\alpha \rho_R(D^\alpha(ff_\nu))\sigma(D_{ij}(x^\alpha))\big)\big(\rho_R(D_t(g_\nu))\rho_L(D_s) -$$

$$\rho_R(D_s(g_\nu))\rho_L(D_t) + \sum_\beta \rho_R(D^\beta(g_\nu))\sigma(D_{st}(x^\beta))\big)$$

$$= \sum_{\nu} \big(\rho_R(D_j(ff_\nu))\rho_L(D_i) - \rho_R(D_i(ff_\nu))\rho_L(D_j) +$$

$$\rho_R(D_iD_j(ff_\nu))\sigma(E_{ii} - E_{jj}) + \sum_{l\neq i}\rho_R(D_jD_l(ff_\nu))\sigma(E_{li}) -$$

$$\sum_{q\neq j}\rho_R(D_iD_q(ff_\nu))\sigma(E_{qj})\big)\big(\rho_R(D_t(g_\nu))\rho_L(D_s) -$$

$$\rho_R(D_s(g_\nu))\rho_L(D_t) + \rho_R(D_sD_t(g_\nu))\sigma(E_{ss} - E_{tt}) +$$

$$\sum_{r\neq s}\rho_R(D_tD_r(g_\nu))\sigma(E_{rs}) - \sum_{u\neq t}\rho_R(D_sD_u(g_\nu))\sigma(E_{ut})\big)$$

$$= \rho_R(f)\big(\sum_{\nu}\rho_R(f_\nu D_iD_jD_sD_t(g_\nu))\sigma(E_{ii} - E_{jj})\sigma(E_{ss} - E_{tt}) +$$

$$\sum_{\nu}\sum_{r\neq s}\rho_R(f_\nu D_iD_jD_tD_r(g_\nu))\sigma(E_{ii} - E_{jj})\sigma(E_{rs}) -$$

$$\sum_{\nu}\sum_{u\neq t}\rho_R(f_\nu D_iD_jD_sD_u(g_\nu))\sigma(E_{ii} - E_{jj})\sigma(E_{ut}) +$$

$$\sum_{\nu}\sum_{l\neq i}\rho_R(f_\nu D_jD_lD_sD_t(g_\nu))\sigma(E_{li})\sigma(E_{ss} - E_{tt}) +$$

$$\sum_{\nu}\sum_{l\neq i}\sum_{r\neq s}\rho_R(f_\nu D_jD_lD_tD_r(g_\nu))\sigma(E_{li})\sigma(E_{rs}) -$$

$$\sum_{\nu}\sum_{l\neq i}\sum_{u\neq t}\rho_R(f_\nu D_jD_lD_sD_u(g_\nu))\sigma(E_{li})\sigma(E_{ut}) -$$

$$\sum_{\nu}\sum_{q\neq j}\rho_R(f_\nu D_iD_qD_sD_t(g_\nu))\sigma(E_{qj})\sigma(E_{ss} - E_{tt}) -$$

$$\sum_{\nu}\sum_{q\neq j}\sum_{r\neq s}\rho_R(f_\nu D_iD_qD_tD_r(g_\nu))\sigma(E_{qj})\sigma(E_{rs}) +$$

$$\sum_{\nu}\sum_{q\neq j}\sum_{u\neq t}\rho_R(f_\nu D_iD_qD_sD_u(g_\nu))\sigma(E_{qj})\sigma(E_{ut})\big)$$

By the remark in the first paragraph and (4.4), we obtain the following relation

(4.5)

$$\delta_{\{i,j,s,t\},\{s_1,s_2,s_1',s_2'\}}\sigma(E_{ii} - E_{jj})\sigma(E_{ss} - E_{tt}) +$$

$$\sum_{r\neq s}\delta_{\{i,j,t,r\},\{s_1,s_2,s_1',s_2'\}}\sigma(E_{ii} - E_{jj})\sigma(E_{rs}) -$$

$$\sum_{u\neq t}\delta_{\{i,j,s,u\},\{s_1,s_2,s_1',s_2'\}}\sigma(E_{ii} - E_{jj})\sigma(E_{ut}) +$$

$$\sum_{l\neq i}\delta_{\{j,l,s,t\},\{s_1,s_2,s_1',s_2'\}}\sigma(E_{li})\sigma(E_{ss} - E_{tt}) +$$

$$\sum_{l\neq i}\sum_{r\neq s}\delta_{\{j,l,t,r\},\{s_1,s_2,s_1',s_2'\}}\sigma(E_{li})\sigma(E_{rs}) -$$

$$\sum_{l\neq i}\sum_{u\neq t}\delta_{\{j,l,s,u\},\{s_1,s_2,s_1',s_2'\}}\sigma(E_{li})\sigma(E_{ut}) -$$

$$\sum_{q \neq j} \delta_{\{i,q,s,t\},\{s_1,s_2,s_1',s_2'\}} \sigma(E_{qj})\sigma(E_{ss} - E_{tt}) -$$

$$\sum_{q \neq j} \sum_{r \neq s} \delta_{\{i,q,t,r\},\{s_1,s_2,s_1',s_2'\}} \sigma(E_{qj})\sigma(E_{rs}) +$$

$$\sum_{q \neq j} \sum_{u \neq t} \delta_{\{i,q,s,u\},\{s_1,s_2,s_1',s_2'\}} \sigma(E_{qj})\sigma(E_{ut})$$

$$=0.$$

LEMMA 4.4. *If an irreducible representation σ of the Lie algebra $\mathfrak{sl}(m)$ in a vector space V satisfies (4.5), then V is exceptional.*

PROOF. Set $\mathsf{n}^+ := F\text{-span}\{E_{ij} \mid i < j\}$, $\mathsf{h} := F\text{-span}\{E_{i,i} - E_{i+1,i+1} \mid i = 1, 2, \cdots, m-1\}$, $\mathsf{n}^- := F\text{-span}\{E_{ij} \mid i > j\}$. For any $a, b \in \{1, 2, \cdots, m\}, a \neq b$, take $s_1 = s_2 = s_1' = s_2' = b$, $i = s = a$, $j = t = b$ in (4.5). From (4.5), we get $\sigma(E_{ba})^2 = 0$. So n^+ and n^- act on V nilpotently. Then there exists a maximal vector v in V with respect to the Borel subalgebra $\mathsf{h} + \mathsf{n}^+$. Assume λ is the corresponding maximal weight. Set $\lambda_k = \lambda(E_{k,k} - E_{k+1,k+1}), k = 1, 2, \cdots, m-1$. Take $s_1 = k+1$, $s_1' = k'+1$, $s_2 = k$, $s_2' = k'$. $i = k$, $j = k+1$, $s = k'$, $t = k'+1$ in (4.5). Then from (4.5), we have:

$$(4.6) \quad \sigma(E_{kk} - E_{k+1,k+1})\sigma(E_{k'k'} - E_{k'+1,k'+1}) + (1 - \delta_{k,k'})\sigma(E_{k'k})\sigma(E_{kk'})$$
$$- (1 - \delta_{k'+1,k})\sigma(E_{k'+1,k})\sigma(E_{k,k'+1}) - (1 - \delta_{k+1,k'})\sigma(E_{k',k+1})\sigma(E_{k+1,k'})$$
$$+ (1 - \delta_{k,k'})\sigma(E_{k'+1,k+1})\sigma(E_{k+1,k'+1})$$
$$=0.$$

Taking $k = k'$ in (4.6), and then letting both sides of (4.6) act on v, we get $\lambda_k^2 - \lambda_k = 0$. So $\lambda_k = 0$ or 1 for $k = 1, 2, \cdots, m-1$. If all $\lambda_k = 0$, $k = 1, 2, \cdots, m-1$, then v is an exceptional-weight vector. Otherwise there exists some $k \in \{1, 2, \cdots, m-1\}$ such that $\lambda_k = 1$, and all $\lambda_i = 0$ for $i < k$. Taking $k' \geq k+1$ in (4.6), and also letting both sides of (4.6) act on v, we then get $\lambda_{k'}\lambda_k = 0$, thereby $\lambda_{k'} = 0$. Hence v is an exceptional-weight vector. In conclusion, V is exceptional. $\square$

If V is an irreducible L_0-submodule of M, then V is exceptional, due to Lemma 4.4. By the assumption of (4.2), there occurs a contradiction $M = 0$. Therefore, $l > 1$, so $\sigma(L_l) = 0$. And $\sigma(L_{l-1})$ is a nonzero abelian ideal of $\sigma(L_0)$. Define a total order "$\prec$" on the set of n-tuples $A(m; \mathbf{n})$ by putting $\alpha \prec \beta$ (or $\beta \succ \alpha$) for $\alpha, \beta \in A(m; \mathbf{n})$ if either $|\alpha| < |\beta|$ or $|\alpha| = |\beta|$ and β is greater than α in lexicographical order. For $f, g \in R$, we have:

$$(4.7) \qquad \rho_L(D_{ij}(f))\rho_L(D_{ij}(g))$$
$$= \rho_L(D_j(f)D_i - D_i(f)D_j)\rho_L(D_j(g)D_i - D_i(g)D_j)$$
$$= (\rho_R(D_j(f))\rho_L(D_i) - \rho_R(D_i(f))\rho_L(D_j) +$$
$$\sum_\alpha \rho_R(D^\alpha(f))\sigma(D_{ij}(x^\alpha)))(\rho_R(D_j(g))\rho_L(D_i) -$$
$$\rho_R(D_i(g))\rho_L(D_j) + \sum_\beta \rho_R(D^\beta(g))\sigma(D_{ij}(x^\beta)))$$

$$
\begin{aligned}
= \;& \rho_R(D_j(f)D_j(g))\rho_L(D_i)^2 + \rho_R(D_j(f)D_iD_j(g))\rho_L(D_i) - \\
& \rho_R(D_j(f)D_i(g))\rho_L(D_i)\rho_L(D_j) - \rho_R(D_j(f)D_iD_i(g))\rho_L(D_j) - \\
& \rho_R(D_i(f)D_j(g))\rho_L(D_j)\rho_L(D_i) - \rho_R(D_i(f)D_jD_j(g))\rho_L(D_i) + \\
& \rho_R(D_i(f)D_i(g))\rho_L(D_j)^2 + \rho_R(D_i(f)D_jD_i(g))\rho_L(D_j) + \\
& \sum_\beta \rho_R(D_j(f)D^{\beta+\varepsilon_i}(g))\sigma(D_{ij}(x^\beta)) + \\
& \sum_\beta \rho_R(D_j(f)D^\beta(g))\rho_L(D_i)\sigma(D_{ij}(x^\beta)) - \\
& \sum_\beta \rho_R(D_i(f)D^\beta(g))\rho_L(D_j)\sigma(D_{ij}(x^\beta)) - \\
& \sum_\beta \rho_R(D_i(f)D^{\beta+\varepsilon_j}(g))\sigma(D_{ij}(x^\beta)) + \\
& \sum_\alpha \rho_R(D^\alpha(f)D_j(g))\sigma(D_{ij}(x^\alpha))\rho_L(D_i) - \\
& \sum_\alpha \rho_R(D^\alpha(f)D_i(g))\sigma(D_{ij}(x^\alpha))\rho_L(D_j) + \\
& \sum_\alpha \sum_\beta \rho_R(D^\alpha(f)D^\beta(g))\sigma(D_{ij}(x^\alpha))\sigma(D_{ij}(x^\beta)) \\
= \;& \rho_R(D_j(fD_j(g)))\rho_L(D_i)^2 - \rho_R(fD_jD_j(g))\rho_L(D_i)^2 + \\
& \rho_R(D_j(fD_iD_j(g)))\rho_L(D_i) - \rho_R(fD_iD_jD_j(g))\rho_L(D_i) - \\
& \rho_R(D_j(fD_i(g)))\rho_L(D_i)\rho_L(D_j) + \rho_R(fD_jD_i(g))\rho_L(D_i)\rho_L(D_j) - \\
& \rho_R(D_j(fD_iD_i(g)))\rho_L(D_j) + \rho_R(fD_jD_iD_i(g))\rho_L(D_j) - \\
& \rho_R(D_i(fD_j(g)))\rho_L(D_j)\rho_L(D_i) + \rho_R(fD_iD_j(g))\rho_L(D_j)\rho_L(D_i) - \\
& \rho_R(D_i(fD_jD_j(g)))\rho_L(D_i) + \rho_R(fD_iD_jD_j(g))\rho_L(D_i) + \\
& \rho_R(D_i(fD_i(g)))\rho_L(D_j)^2 - \rho_R(fD_iD_i(g))\rho_L(D_j)^2 + \\
& \rho_R(D_i(fD_jD_i(g)))\rho_L(D_j) - \rho_R(fD_iD_jD_i(g))\rho_L(D_j) + \\
& \sum_\beta \rho_R(D_j(fD^\beta(g)))\rho_L(D_i)\sigma(D_{ij}(x^\beta)) - \\
& \sum_\beta \rho_R(fD^{\beta+\varepsilon_j}(g))\rho_L(D_i)\sigma(D_{ij}(x^\beta)) + \\
& \sum_\beta \rho_R(D_j(fD^{\beta+\varepsilon_i}(g)))\sigma(D_{ij}(x^\beta)) - \\
& \sum_\beta \rho_R(D_i(fD^\beta(g)))\rho_L(D_j)\sigma(D_{ij}(x^\beta)) + \\
& \sum_\beta \rho_R(fD^{\beta+\varepsilon_i}(g))\sigma(D_{ij}(x^\beta)) - \\
& \sum_\beta \rho_R(D_i(fD^{\beta+\varepsilon_j}(g)))\sigma(D_{ij}(x^\beta)) + \\
& \sum_\alpha \sum_{\alpha'+\alpha''=\alpha} (-1)^{|\alpha''|} \binom{\alpha}{\alpha'} \rho_R(D^{\alpha'}(fD^{\alpha''+\varepsilon_j}(g)))\sigma(D_{ij}(x^\alpha))\rho_L(D_i) -
\end{aligned}
$$

$$\sum_{\alpha} \sum_{\alpha'+\alpha''=\alpha} (-1)^{|\alpha''|} \binom{\alpha}{\alpha'} \rho_R(D^{\alpha'}(fD^{\alpha''+\varepsilon_i}(g)))\sigma(D_{ij}(x^\alpha))\rho_L(D_j) +$$

$$\sum_{\alpha} \sum_{\alpha'+\alpha''=\alpha} \sum_{\beta} (-1)^{|\alpha''|} \binom{\alpha}{\alpha'} \rho_R(D^{\alpha'}(fD^{\alpha''+\beta}(g)))\sigma(D_{ij}(x^\alpha))\sigma(D_{ij}(x^\beta)).$$

The third and the fourth equations in (4.7) hold because of the equations

$$D_i(f)g = D_i(fg) - fD_i(g) \text{ and } D^\alpha(f)g = \sum_{\alpha'+\alpha''=\alpha} (-1)^{|\alpha''|}\binom{\alpha}{\alpha'}D^{\alpha'}(fD^{\alpha''}(g)).$$

For a family of R-linear endomorphisms of M: $\Phi = (\varphi_\alpha)_{\alpha \in A(m;\mathbf{n})}$, we define its support $\text{Supp}\Phi = \{\alpha \in A(m;\mathbf{n}) \mid \varphi_\alpha \neq 0\}$. Define the degree $\deg\Phi$ of Φ to be the largest integer t such that there exists $\alpha \in \text{Supp}\Phi$ with $|\alpha| = t$. The next lemma is due to Skryabin ([11]), which is needed to complete the proof of Theorem 4.3.

LEMMA 4.5. ([11]) *Let $t \geq 0$ be an integer and $\Phi = (\varphi_\alpha)_{\alpha \in A(m;\mathbf{n})}$ be a family of R-linear endomorphisms of M satisfying :*
(1) $\deg\Phi \leq t$.
(2) The endomorphisms φ_α with $|\alpha| = t$ are pairwise commuting.
(3) M' is stable under all endomorphisms $\Phi(f) = \sum_{\alpha \in A(m;\mathbf{n})} \rho_R(D^\alpha(f))\varphi_\alpha, \forall f \in R$.
Then the endomorphisms φ_α are nilpotent for all α with $|\alpha| = t$.

For a fixed pair $i, j \in \{1, 2, \cdots, m\}$, $i < j$, put $\varphi_\alpha = 0$ for $|\alpha| \leq 1$, $\varphi_\alpha = \sigma(D_{ij}(x^\alpha))$ for $|\alpha| \geq 2$ and $t = l+1$. The properties (1) and (2) in Lemma 4.5 hold, but (3) is lacking, so we can't directly apply Lemma 4.5. However, fortunately M' is stable under the endomorphism $\rho_L(D_{ij}(f)) = \rho_L(D_j(f)D_i - D_i(f)D_j) = \rho_R(D_j(f))\rho_L(D_i) - \rho_R(D_i(f))\rho_L(D_j) + \Phi(f)$. If $\gamma, \gamma' \in A(m;\mathbf{n})$ and $f_\nu, g_\nu \in R$ are chosen as in the proof of Lemma 4.5 in [11], i.e.

(4.8) $\qquad \sum_{\nu} f_\nu D^\alpha g_\nu = \begin{cases} 1, & \text{when } \alpha = \gamma + \gamma', \\ 0, & \text{when } \alpha \in A(m;\mathbf{n}), |\alpha| \leq 2t, \alpha \neq \gamma + \gamma'. \end{cases}$

From (4.7) and Lemma 4.5, we see that

$$\sum_{\nu} \rho_L(D_{ij}(ff_\nu))\rho_L(D_{ij}(g_\nu)) = \sum_{\nu} \Phi(ff_\nu)\Phi(g_\nu) = \Psi(f)$$

where $\Psi = (\psi_{\alpha'})_{\alpha' \in A(m;\mathbf{n})}$ and $\Psi(f) = \sum_{\alpha'} \rho_R(D^{\alpha'}(f))\psi_{\alpha'}$, $\deg\Psi \leq t - t'$, $t' = |\gamma'|$. Furthermore,

$$\psi_{\alpha'} = \sum_{\alpha} \sum_{\beta} (-1)^{t'} \binom{\alpha}{\alpha'} \varphi_\alpha \varphi_\beta$$

whenever $|\alpha'| = t - t'$, and the sum is taken over all $\alpha, \beta \in A(m;\mathbf{n})$ such that $|\alpha| = |\beta| = t, \alpha \succeq \alpha', \alpha - \alpha' + \beta = \gamma + \gamma'$. So M' is stable under the endomorphisms $\Psi(f)$. The arguments used in the proof of Lemma 4.5 in [11] now work without any change. We deduce as Lemma 4.5 in [11] that $\sigma(D_{ij}(x^\alpha))$ are nilpotent for all $\alpha \in A(m;\mathbf{n})$ with $|\alpha| = l + 1$ and $1 \leq i < j \leq m$, i.e. $\sigma(L_{l-1})$ consists of nilpotent endomorphisms. For any irreducible $\sigma(L_0)$-submodule V of M, the subspace $V' := \{v \in V \mid \sigma(L_{l-1})v = 0\}$ is a nonzero $\sigma(L_0)$-submodule of V, because L_{l-1} is an ideal of L_0. By the simplicity of V, we know $V' = V$. Hence $\sigma(L_{l-1})$ vanish on V. The complete reducibility of M as $\sigma(L_0)$-module now implies that $\sigma(L_{l-1}) = 0$. It contradicts the definition of l. The proof of Theorem 4.3 is completed.   □

5. Simple modules for $S(m; \mathbf{n})$ in non-exceptional cases

5.1. For $\chi \in L^*$, define the height of χ $\mathrm{ht}\chi := \min\{i \geq -1 \mid \chi(L_i) = 0\}$. Let (V, ρ_0) be a $\chi|_{L_0}$-reduced representation of the restricted Lie algebra L_0. Then we have an induced module $\mathcal{V} := \mathbf{Ind}_{U(L_0, \chi)}^{U_{p^{\mathbf{s}}}(L, \chi)} V = U_{p^{\mathbf{s}}}(L, \chi) \bigotimes_{U(L_0, \chi)} V$, where $\mathbf{s}$ is the same one as in §2.2, and $U(L_0, \chi)$ is the so-called χ-reduced enveloping algebra of L_0 which is a quotient of $U(L_0)$ by the ideal generated by $x^p - x^{[p]} - \chi(x)^p$, $\forall x \in L_0$. As a vector space, $\mathcal{V} = \sum_{\beta} FE^{\beta} \otimes V$, where $E^{\alpha} = e_{-1}^{\alpha_1} e_{-2}^{\alpha_2} \cdots e_{-m}^{\alpha_m}, 0 \leq \alpha_i \leq p^{m_i} - 1, e_{-i} = D_i, 1 \leq i \leq m$.

5.2. Below we will introduce some module operators on $\mathcal{V}$, finally present a desired module category after three steps as follows.

Step 1: The R-module structure ρ_R defined via

$$(5.1) \qquad \rho_R(x^{\alpha})E^{\beta} \otimes v = (-1)^{|\alpha|} \binom{\beta}{\alpha} E^{\beta-\alpha} \otimes v.$$

By (2.1), one can easily see that $\mathcal{V}$ becomes an R-module concerning the R-module structure defined by (5.1).

Step 2: The L-module structure ρ_L defined via :

$$(5.2)$$

$$\rho_L(D_{ij}(x^{\alpha}))E^{\beta} \otimes v$$
$$= (-1)^{|\alpha|-1}\left(\binom{\beta + \varepsilon_i}{\alpha - \varepsilon_j} - \binom{\beta + \varepsilon_j}{\alpha - \varepsilon_i}\right) E^{\beta+\varepsilon_i+\varepsilon_j-\alpha} \otimes v +$$
$$\sum_{0 < \gamma \leq \alpha, |\gamma| \geq 2} (-1)^{|\alpha|-|\gamma|} \binom{\beta}{\alpha - \gamma} E^{\beta+\gamma-\alpha} \otimes \rho_0(D_{ij}(x^{\gamma}))v.$$

Denote by $\mathbf{ind}$ the induced representation of L associated with the induced module structure on $\mathcal{V} = \mathbf{Ind}_{U(L_0, \chi)}^{U_{p^{\mathbf{s}}}(L, \chi)} V$ from the L_0-module (V, ρ_0). We have the following proposition.

PROPOSITION 5.1. *The action of L on $\mathcal{V}$ defined by (5.2) coincides with $\mathbf{ind}$. So $\mathcal{V}$ becomes an L-module with the L-module structure defined by (5.2). Furthermore it is a generalized χ-reduced L-module.*

PROOF. We need to prove that the following equation holds for any $\alpha, \beta \in A(m; \mathbf{n}), 1 \leq i < j \leq m$:

$$\rho_L(D_{ij}(x^{\alpha}))E^{\beta} \otimes v = \mathbf{ind}(D_{ij}(x^{\alpha}))E^{\beta} \otimes v.$$

This can be verified by a direct computation:

$$(\text{LHS}) = (-1)^{|\alpha|-1}\left(\binom{\beta+\varepsilon_i}{\alpha-\varepsilon_j} - \binom{\beta+\varepsilon_j}{\alpha-\varepsilon_i}\right)E^{\beta+\varepsilon_i+\varepsilon_j-\alpha}\otimes v +$$

$$\sum_{0<\gamma\le\alpha,\,|\gamma|\ge 2}(-1)^{|\alpha|-|\gamma|}\binom{\beta}{\alpha-\gamma}E^{\beta+\gamma-\alpha}\otimes\rho_0(D_{ij}(x^\gamma))v$$

$$= (-1)^{|\alpha|-1}\left(\left(\binom{\beta+\varepsilon_i}{\alpha-\varepsilon_j} - \binom{\beta}{\alpha-\varepsilon_j}\right) - \right.$$

$$\left.\left(\binom{\beta+\varepsilon_j}{\alpha-\varepsilon_i} - \binom{\beta}{\alpha-\varepsilon_i}\right)\right)E^{\beta+\varepsilon_i+\varepsilon_j-\alpha}\otimes v +$$

$$(-1)^{|\alpha|-1}\binom{\beta}{\alpha-\varepsilon_j}E^{\beta+\varepsilon_i+\varepsilon_j-\alpha}\otimes v -$$

$$(-1)^{|\alpha|-1}\binom{\beta}{\alpha-\varepsilon_i}E^{\beta+\varepsilon_i+\varepsilon_j-\alpha}\otimes v +$$

$$\sum_{\substack{0\le\gamma<\alpha\\ |\alpha-\gamma|\ge 2}}(-1)^{|\gamma|}\binom{\beta}{\gamma}E^{\beta-\gamma}\otimes\rho_0(D_{ij}(x^{\alpha-\gamma}))v$$

$$= (-1)^{|\alpha|-1}\left(\binom{\beta}{\alpha-\varepsilon_i-\varepsilon_j} - \binom{\beta}{\alpha-\varepsilon_i-\varepsilon_j}\right)E^{\beta+\varepsilon_i+\varepsilon_j-\alpha}\otimes v$$

$$+ \operatorname{ind}(D_{ij}(x^\alpha))E^\beta\otimes v$$

$$= \operatorname{ind}(D_{ij}(x^\alpha))E^\beta\otimes v$$

$$= \text{RHS}.$$

The third equation above is because in the generalized χ-reduced algebra $U_{p^s}(L,\chi)$, we have the following formula which will be used frequently:

$$D_{ij}(x^\alpha)E^\beta = \sum_{0\le\gamma\le\beta}(-1)^{|\gamma|}\binom{\beta}{\gamma}E^{\beta-\gamma}D_{ij}(x^{\alpha-\gamma})$$

$\square$

Step 3: The L_0-module structure defined via :

$$(5.3)\qquad \sigma(D')E^\beta\otimes v = E^\beta\otimes\rho_0(D')v \quad D'\in L_0, \beta\in A(m;\mathbf{n}), v\in V.$$

It is clear that $\mathcal{V}$ becomes a $\chi|_{L_0}$-reduced L_0-module with the module structure defined by (5.3) from the assumption that (V,ρ_0) is a $\chi|_{L_0}$-reduced representation of L_0.

Next we will prove that $\mathcal{V}$ belongs to the category $\mathfrak{C}$ for the three module structures defined by (5.1), (5.2) and (5.3). We need the following combinatorial formulae as (2.1) which can be directly verified.

LEMMA 5.2. *For any $\alpha,\beta,\alpha',\gamma\in A(m;\mathbf{n}), 1\le i < j\le m$, then the following combinatorial formulae hold.*

$$(5.4)\qquad \binom{\beta}{\alpha'}\binom{\beta-\alpha'}{\alpha-\gamma} = \binom{\beta}{\alpha-\gamma}\binom{\beta+\gamma-\alpha}{\alpha'}$$

$$(5.5) \quad \binom{\beta}{\alpha'}\left(\binom{\beta - \alpha' + \varepsilon_j}{\alpha - \varepsilon_i} - \binom{\beta - \alpha' + \varepsilon_i}{\alpha - \varepsilon_j}\right) +$$

$$\binom{\beta + \varepsilon_i + \varepsilon_j - \alpha}{\alpha'}\left(\binom{\beta + \varepsilon_i}{\alpha - \varepsilon_j} - \binom{\beta + \varepsilon_j}{\alpha - \varepsilon_i}\right)$$

$$= \binom{\beta}{\alpha + \alpha' - \varepsilon_i - \varepsilon_j}\left(\binom{\alpha + \alpha' - \varepsilon_i - \varepsilon_j}{\alpha - \varepsilon_j} - \binom{\alpha + \alpha' - \varepsilon_i - \varepsilon_j}{\alpha - \varepsilon_i}\right)$$

THEOREM 5.3. *Notations are as above, and the R-module structure, the L-module structure and the L_0-module structure on $\mathcal{V}$ are defined by (5.1), (5.2) and (5.3) respectively. Then the module $\mathcal{V}$ must belong to the category $\mathfrak{C}$.*

PROOF. We need to prove that (R1)–(R4) in Definition 3.2 hold.

(1) $\forall \in D_{ij}(x^\alpha) \in L, x^{\alpha'} \in R, E^\beta \otimes v \in \mathcal{V}$, then :

$$[\rho_L(D_{ij}(x^\alpha)), \rho_R(x^{\alpha'})](E^\beta \otimes v)$$

$$= \rho_L(D_{ij}(x^\alpha)) \circ \rho_R(x^{\alpha'})(E^\beta \otimes v) - \rho_R(x^{\alpha'}) \circ \rho_L(D_{ij}(x^\alpha))(E^\beta \otimes v)$$

$$= \rho_L(D_{ij}(x^\alpha))\left((-1)^{|\alpha'|}\binom{\beta}{\alpha'}E^{\beta-\alpha'} \otimes v\right) -$$

$$\rho_R(x^{\alpha'})\left((-1)^{|\alpha|-1}\left(\binom{\beta + \varepsilon_i}{\alpha - \varepsilon_j} - \binom{\beta + \varepsilon_j}{\alpha - \varepsilon_i}\right)E^{\beta + \varepsilon_i + \varepsilon_j - \alpha} \otimes v\right)$$

$$- \rho_R(x^{\alpha'})\left(\sum_{\substack{0 < \gamma \le \alpha \\ |\gamma| \ge 2}} (-1)^{|\alpha| - |\gamma|}\binom{\beta}{\alpha - \gamma}E^{\beta + \gamma - \alpha} \otimes \rho_0(D_{ij}(x^\gamma))v\right)$$

$$= (-1)^{|\alpha'|}\binom{\beta}{\alpha'}\left((-1)^{|\alpha|-1}\left(\binom{\beta - \alpha' + \varepsilon_i}{\alpha - \varepsilon_j} - \binom{\beta - \alpha' + \varepsilon_j}{\alpha - \varepsilon_i}\right)E^{\beta - \alpha' + \varepsilon_i + \varepsilon_j - \alpha} \otimes v\right) -$$

$$(-1)^{|\alpha|-1}\left(\binom{\beta + \varepsilon_i}{\alpha - \varepsilon_j} - \binom{\beta + \varepsilon_j}{\alpha - \varepsilon_i}\right)(-1)^{|\alpha'|}\binom{\beta + \varepsilon_i + \varepsilon_j - \alpha}{\alpha'}E^{\beta + \varepsilon_i + \varepsilon_j - \alpha - \alpha'} \otimes v +$$

$$\sum_{\substack{0 < \gamma \le \alpha \\ |\gamma| \ge 2}} (-1)^{|\alpha'|}\binom{\beta}{\alpha'}(-1)^{|\alpha| - |\gamma|}\binom{\beta - \alpha'}{\alpha - \gamma}E^{\beta - \alpha' + \gamma - \alpha} \otimes \rho_0(D_{ij}(x^\gamma))v -$$

$$\sum_{\substack{0 < \gamma \le \alpha \\ |\gamma| \ge 2}} (-1)^{|\alpha'|}(-1)^{|\alpha| - |\gamma|}\binom{\beta}{\alpha - \gamma}\binom{\beta + \gamma - \alpha}{\alpha'}E^{\beta + \gamma - \alpha - \alpha'} \otimes \rho_0(D_{ij}(x^\gamma))v$$

$$= (-1)^{|\alpha| + |\alpha'| - 1}\left(\binom{\beta}{\alpha'}\left(\binom{\beta - \alpha' + \varepsilon_i}{\alpha - \varepsilon_j} - \binom{\beta - \alpha' + \varepsilon_j}{\alpha - \varepsilon_i}\right) -$$

$$\binom{\beta + \varepsilon_i + \varepsilon_j - \alpha}{\alpha'}\left(\binom{\beta + \varepsilon_i}{\alpha - \varepsilon_j} - \binom{\beta + \varepsilon_j}{\alpha - \varepsilon_i}\right)\right)E^{\beta + \varepsilon_i + \varepsilon_j - \alpha - \alpha'} \otimes v +$$

$$\sum_{\substack{0 < \gamma \le \alpha \\ |\gamma| \ge 2}} (-1)^{|\alpha| + |\alpha'| - |\gamma|}\left(\binom{\beta}{\alpha'}\binom{\beta - \alpha'}{\alpha - \gamma} -$$

$$\binom{\beta}{\alpha - \gamma}\binom{\beta + \gamma - \alpha}{\alpha'}\right)E^{\beta + \gamma - \alpha - \alpha'} \otimes \rho_0(D_{ij}(x^\gamma))v.$$

By Lemma 5.2, we have:

$$[\rho_L(D_{ij}(x^\alpha)), \rho_R(x^{\alpha'})](E^\beta \otimes v)$$

$$=(-1)^{|\alpha|+|\alpha'|-1}\begin{pmatrix}\beta \\ \alpha+\alpha'-\varepsilon_i-\varepsilon_j\end{pmatrix}(\begin{pmatrix}\alpha+\alpha'-\varepsilon_i-\varepsilon_j \\ \alpha-\varepsilon_i\end{pmatrix}-$$

$$\begin{pmatrix}\alpha+\alpha'-\varepsilon_i-\varepsilon_j \\ \alpha-\varepsilon_j\end{pmatrix})E^{\beta+\varepsilon_i+\varepsilon_j-\alpha-\alpha'} \otimes v.$$

On the other hand:

$$\rho_R(D_{ij}(x^\alpha)(x^{\alpha'}))(E^\beta \otimes v)$$

$$=\rho_R((x^{\alpha-\varepsilon_j}D_i - x^{\alpha-\varepsilon_i}D_j)(x^{\alpha'}))(E^\beta \otimes v)$$

$$=\rho_R(x^{\alpha-\varepsilon_j}x^{\alpha'-\varepsilon_i} - x^{\alpha-\varepsilon_i}x^{\alpha'-\varepsilon_j})(E^\beta \otimes v)$$

$$=(\begin{pmatrix}\alpha+\alpha'-\varepsilon_i-\varepsilon_j \\ \alpha-\varepsilon_j\end{pmatrix} - \begin{pmatrix}\alpha+\alpha'-\varepsilon_i-\varepsilon_j \\ \alpha-\varepsilon_i\end{pmatrix})\rho_R(x^{\alpha+\alpha'-\varepsilon_i-\varepsilon_j})(E^\beta \otimes v)$$

$$=(-1)^{|\alpha|+|\alpha'|-2}(\begin{pmatrix}\alpha+\alpha'-\varepsilon_i-\varepsilon_j \\ \alpha-\varepsilon_j\end{pmatrix}-$$

$$\begin{pmatrix}\alpha+\alpha'-\varepsilon_i-\varepsilon_j \\ \alpha-\varepsilon_i\end{pmatrix})\begin{pmatrix}\beta \\ \alpha+\alpha'-\varepsilon_i-\varepsilon_j\end{pmatrix}E^{\beta+\varepsilon_i+\varepsilon_j-\alpha-\alpha'} \otimes v$$

$$=(-1)^{|\alpha|+|\alpha'|-1}\begin{pmatrix}\beta \\ \alpha+\alpha'-\varepsilon_i-\varepsilon_j\end{pmatrix}(\begin{pmatrix}\alpha+\alpha'-\varepsilon_i-\varepsilon_j \\ \alpha-\varepsilon_i\end{pmatrix}-$$

$$\begin{pmatrix}\alpha+\alpha'-\varepsilon_i-\varepsilon_j \\ \alpha-\varepsilon_j\end{pmatrix})E^{\beta+\varepsilon_i+\varepsilon_j-\alpha-\alpha'} \otimes v.$$

Therefore $[\rho_L(D_{ij}(x^\alpha)), \rho_R(x^{\alpha'})] = \rho_R(D_{ij}(x^\alpha)(x^{\alpha'}))$. Hence (R1) holds.

(2) $\forall D_{ij}(x^\alpha) \in L_0, x^{\alpha'} \in R, E^\beta \otimes v \in \mathcal{V}$, then:

$$[\sigma(D_{ij}(x^\alpha)), \rho_R(x^{\alpha'})](E^\beta \otimes v)$$

$$=\sigma(D_{ij}(x^\alpha)) \circ \rho_R(x^{\alpha'})(E^\beta \otimes v) - \rho_R(x^{\alpha'}) \circ \sigma(D_{ij}(x^\alpha))(E^\beta \otimes v)$$

$$=\sigma(D_{ij}(x^\alpha))((-1)^{|\alpha'|}\begin{pmatrix}\beta \\ \alpha'\end{pmatrix}E^{\beta-\alpha'} \otimes v) - \rho_R(x^{\alpha'})(E^\beta \otimes \rho_0(D_{ij}(x^\alpha))v)$$

$$=(-1)^{|\alpha'|}\begin{pmatrix}\beta \\ \alpha'\end{pmatrix}E^{\beta-\alpha'} \otimes \rho_0(D_{ij}(x^\alpha))v - (-1)^{|\alpha'|}\begin{pmatrix}\beta \\ \alpha'\end{pmatrix}E^{\beta-\alpha'} \otimes \rho_0(D_{ij}(x^\alpha))v$$

$$=0.$$

Therefore $[\sigma(D_{ij}(x^\alpha)), \rho_R(x^{\alpha'})] = 0$. Hence (R2) holds.

(3) $\forall D_k \in L_{[-1]}, k = 1, 2, \cdots, m, D_{ij}(x^\alpha) \in L_0, E^\beta \otimes v \in \mathcal{V}$, then:

$$[\rho_L(D_k), \sigma(D_{ij}(x^\alpha))](E^\beta \otimes v)$$

$$=\rho_L(D_k) \circ \sigma(D_{ij}(x^\alpha))(E^\beta \otimes v) - \sigma(D_{ij}(x^\alpha)) \circ \rho_L(D_k)(E^\beta \otimes v)$$

$$=\rho_L(D_k)(E^\beta \otimes \rho_0(D_{ij}(x^\alpha))v) - \sigma(D_{ij}(x^\alpha))(E^{\beta+\varepsilon_k} \otimes v)$$

$$=E^{\beta+\varepsilon_k} \otimes \rho_0(D_{ij}(x^\alpha))v - E^{\beta+\varepsilon_k} \otimes \rho_0(D_{ij}(x^\alpha))v$$

$$=0.$$

Therefore $[\rho_L(D_k), \sigma(D_{ij}(x^\alpha))] = 0$. Hence (R3) holds.

(4) $\forall\, D_{ij}(x^{\alpha}) \in L,\ E^{\beta} \otimes v \in \mathcal{V}$, then:

$$\rho_L(D_{ij}(x^{\alpha}))(E^{\beta} \otimes v)$$

$$=(-1)^{|\alpha|-1}\left(\binom{\beta + \varepsilon_i}{\alpha - \varepsilon_j} - \binom{\beta + \varepsilon_j}{\alpha - \varepsilon_i}\right)E^{\beta+\varepsilon_i+\varepsilon_j-\alpha} \otimes v +$$

$$\sum_{\substack{0<\gamma\le\alpha \\ |\gamma|\ge 2}} (-1)^{|\alpha|-|\gamma|}\binom{\beta}{\alpha - \gamma}E^{\beta+\gamma-\alpha} \otimes \rho_0(D_{ij}(x^{\gamma}))v.$$

On the other hand:

$$\Big(\rho_R(D_j(x^{\alpha})) \circ \rho_L(D_i) - \rho_R(D_i(x^{\alpha})) \circ \rho_L(D_j) +$$

$$\sum_{\substack{0<\gamma\le\alpha \\ |\gamma|\ge 2}} \rho_R(D^{\gamma}(x^{\alpha})) \circ \sigma(D_{ij}(x^{\gamma}))\Big)(E^{\beta} \otimes v)$$

$$=\rho_R(x^{\alpha-\varepsilon_j})(E^{\beta+\varepsilon_i} \otimes v) - \rho_R(x^{\alpha-\varepsilon_i})(E^{\beta+\varepsilon_j} \otimes v) +$$

$$\sum_{\substack{0<\gamma\le\alpha \\ |\gamma|\ge 2}} \rho_R(x^{\alpha-\gamma})\big(E^{\beta} \otimes \rho_0(D_{ij}(x^{\gamma}))v\big)$$

$$=(-1)^{|\alpha|-1}\binom{\beta + \varepsilon_i}{\alpha - \varepsilon_j}E^{\beta+\varepsilon_i+\varepsilon_j-\alpha} \otimes v - (-1)^{|\alpha|-1}\binom{\beta + \varepsilon_j}{\alpha - \varepsilon_i}E^{\beta+\varepsilon_i+\varepsilon_j-\alpha} \otimes v +$$

$$\sum_{\substack{0<\gamma\le\alpha \\ |\gamma|\ge 2}} (-1)^{|\alpha|-|\gamma|}\binom{\beta}{\alpha - \gamma}E^{\beta+\gamma-\alpha} \otimes \rho_0(D_{ij}(x^{\gamma}))v$$

$$=(-1)^{|\alpha|-1}\left(\binom{\beta + \varepsilon_i}{\alpha - \varepsilon_j} - \binom{\beta + \varepsilon_j}{\alpha - \varepsilon_i}\right)E^{\beta+\varepsilon_i+\varepsilon_j-\alpha} \otimes v +$$

$$\sum_{\substack{0<\gamma\le\alpha \\ |\gamma|\ge 2}} (-1)^{|\alpha|-|\gamma|}\binom{\beta}{\alpha - \gamma}E^{\beta+\gamma-\alpha} \otimes \rho_0(D_{ij}(x^{\gamma}))v.$$

So $\rho_L(D_{ij}(x^{\alpha})) = \rho_R(D_j(x^{\alpha})) \circ \rho_L(D_i) - \rho_R(D_i(x^{\alpha})) \circ \rho_L(D_j) +$
$\sum_{\substack{0<\gamma\le\alpha \\ |\gamma|\ge 2}} \rho_R(D^{\gamma}(x^{\alpha})) \circ \sigma(D_{ij}(x^{\gamma}))$. Therefore (R4) holds.

Summing up, we know that $\mathcal{V}$ belongs to the category $\mathfrak{C}$. The proof is completed. $\qquad\square$

5.3. According to [**12**, 4.3], $L_{[0]} \cong \mathfrak{sl}(m,F)$. Denote $h_i := D_{i,i+1}(x^{\varepsilon_i+\varepsilon_{i+1}}) = x^{\varepsilon_i}D_i - x^{\varepsilon_{i+1}}D_{i+1}, i = 1,2,\cdots m-1$, then $\mathcal{H} := F\text{-span}\{h_i \mid i = 1,2,\cdots,m-1\}$ is a canonical torus of $L_{[0]}$. Let V be an irreducible L_0-module, $\lambda = (\lambda_1, \lambda_2, \cdots, \lambda_{m-1}) \in F^{m-1}$, $0 \ne v \in V$ is called a weight vector of weight λ if $\rho(h_i)v = \lambda_i v, i = 1,2,\cdots m-1$, where ρ is the representation of L_0 corresponding to the L_0-module V. If in addition $\rho(\mathcal{N})v = 0$, where $\mathcal{N} := F\text{-span}\{D_{ij}(x^{\varepsilon_l+\varepsilon_j}) = x^{\varepsilon_l}D_i \mid 1 \le l < i \le m, j \ne i\} \bigoplus L_1$, then v is called a maximal vector of weight λ. Set $\omega_k = \varepsilon_k, 1 \le k \le m-1$, where $\varepsilon_i \in \mathcal{H}^*$ such that $\varepsilon_i(h_j) = \delta_{ij}, i,j = 1,2,\cdots,m-1$. Set $\omega_0 = \omega_m = (0,0,\cdots,0)$. The above ω_k are called exceptional weights, $k = 0,1,\cdots,m-1$. Naturally, a weight λ not exceptional is called a nonexceptional one.

For $\chi \in L^*$, then each simple generalized χ-reduced module of L is some homomorphic image of an induced module $\mathcal{V}$ from a $\chi|_{L_0}$-reduced simple L_0-module V. When $\mathrm{ht}\chi \leq 0$, then L_1 act on V trivially and any simple $L_{[0]}$-module can be regarded as L_0-module via trivial L_1 action. In this case, all simple $L_{[0]}$-module are parameterized by the restricted weight set $\Lambda := \{\lambda \in \mathcal{H}^* \mid \lambda(h_i) \in \mathbb{F}_p, i = 1, 2, \cdots, m-1\} = \mathbb{F}_p^{m-1}$, as "highest weight" (cf. [3]). Conversely each $\lambda \in \Lambda$, as "highest weight", corresponds to a unique simple $L_{[0]}$-module denoted by $L_0(\lambda)$. In this case, we may say "exceptional" or "nonexceptional" for a simple L_0-module, depending on what the "highest weight" is.

Thanks to Theorems 4.3 and 5.3, we have the following result concerning the representations of special algebra $L = S(m; \mathbf{n})$.

THEOREM 5.4. *Let $\chi \in L^*$ with $\mathrm{ht}\chi \leq \min\{p^{n_i} - p^{n_i-1} \mid 1 \leq i \leq m\} - 2$. If V is a nonexceptional irreducible L_0-module with character χ, then $(\mathcal{V}, \rho_L)$ is an irreducible L-module.*

PROOF. Set $R = \mathfrak{A}(m; \mathbf{n}), L = S(m; \mathbf{n})$. By Theorem 5.3, V belongs to the category $\mathfrak{C}$. Set $\mathcal{V}_\theta = F$-span$\{E^\theta \otimes v \mid v \in V\}$ for $\theta \in A(m; \mathbf{n})$. Then $\mathcal{V} = \bigoplus_{\theta \in A(m; \mathbf{n})} \mathcal{V}_\theta$ and $\mathcal{V}_\theta \cong V$ as $\sigma(L_0)$-modules. Therefore $\mathcal{V}_\theta$ is an irreducible submodule of $\mathcal{V}$ as $\sigma(L_0)$-module. Furthermore $\mathcal{V}_\theta$ is not exceptional, so the first condition in Theorem 4.3 holds. The assumption $\mathrm{ht}\chi \leq \min\{p^{n_i} - p^{n_i-1} \mid 1 \leq i \leq m\} - 2$ assures the second condition in Theorem 4.3. So any L-submodule $\mathcal{V}'$ of $\mathcal{V}$ is also an R-submodule of $\mathcal{V}$ by Theorem 4.3.

Assume $\mathcal{V}'$ is a nonzero L-submodule of $\mathcal{V}$. Next we will prove that $\mathcal{V}' = \mathcal{V}$. Assume $0 \neq v = \sum_{i=1}^{n} E^{\theta_i} \otimes v_i \in \mathcal{V}'$, where $\theta_i \in A(m; \mathbf{n}), 0 \neq v_i \in V$. We can define a total order in $A(m; \mathbf{n})$ by lexicographical order, i.e. $\alpha = (\alpha_1, \alpha_2, \cdots, \alpha_m) \prec \beta = (\beta_1, \beta_2, \cdots, \beta_m)$ if and only if $|\alpha| < |\beta|$ or $|\alpha| = |\beta|$ and there exists some $i \in \{1, 2, \cdots, m\}$ such that $\alpha_j = \beta_j$ for $j < i$, while $\alpha_i < \beta_i$. Without loss of generality, we can assume $\theta_1 = \max\{\theta_i \mid i = 1, 2, \cdots, n\}$. Then:

$$\rho_R(x^{\theta_1})v = (-1)^{|\theta_1|} 1 \otimes v_1 \in \mathcal{V}'.$$

Therefore $\mathcal{V}' = \mathcal{V}$ by the simplicity of V as L_0-module. We are done. □

REMARK 5.5. For the case of the restricted Lie algebras, the results in this Theorem have been obtained by Zhang Chao-Wen in [14].

6. Simple modules for $S(m; \mathbf{n})$ with exceptional weights

6.1. Let $\chi \in L^*$ and let M be a simple generalized χ-reduced L-module. As noted in §5.3, M is a homomorphic image of a module $\mathcal{V}$ induced from some $\chi|_{L_0}$-reduced simple L_0-module V. By the arguments in §5, we know that $\mathcal{V}$ is simple (and hence isomorphic to M) if V is not exceptional and the height of χ is no more than $\min\{p^{n_i} - p^{n_i-1} \mid i = 1, 2, \cdots, m\} - 2$. In this section we will take up the case where V is exceptional and construct M as a quotient of $\mathcal{V}$. The dimension of M will be determined as well.

When $\mathrm{ht}\chi = -1$ i.e. $\chi = 0$, the determination of the simple exceptional-weight modules were completed respectively by Shen and Nakano (cf. [7] and [5]). So in this section we always assume $\chi \in L^*, \mathrm{ht}\chi = 0$. Maintain the notations before. In particular $R = \mathfrak{A}(m; \mathbf{n}), L = S(m; \mathbf{n})$.

6.2. We need the following lemma to simplify the sequent argument.

LEMMA 6.1. *For $\chi \in L^*$, $\mathrm{ht}\chi = 0$, there exists $\Phi \in \mathrm{Aut}(L)$ and $l \in \{1, 2, \cdots, m\}$ such that $\chi^\Phi(D_i) := \chi(\Phi^{-1}(D_i)) = 0$ for $i \neq l$. Furthermore one can choose a suitable $\Phi \in \mathrm{Aut}(L)$ such that $\chi^\Phi(D_i) = \delta_{il}$.*

PROOF. Since $S(m; \mathbf{n}) \cong S(m; \mathbf{n}')$ for $\mathbf{n} = (n_1, n_2, \cdots, n_m)$, $\mathbf{n}' = (n_{\sigma(1)}, n_{\sigma(2)}, \cdots, n_{\sigma(m)})$, $\sigma \in S_m$. We might suppose $n_1 \leq n_2 \leq \cdots \leq n_m$. As $\mathrm{ht}\chi = 0$, then there exists $i \in \{1, 2, \cdots, m\}$ such that $\chi(D_i) \neq 0$. Set $l = \max\{i \mid \chi(D_i) \neq 0\}$. Define $\varphi \in \mathrm{Aut}(\mathfrak{A}(m; \mathbf{n}))$ via $\varphi(x_i) = x_i$ if $i \neq l$ and $\varphi(x_l) = x_l - \sum_{i=1}^{l-1} \chi(D_l)^{-1}\chi(D_i)x_i$. Set $\Phi := \tilde{\varphi} : W(m; \mathbf{n}) \to W(m; \mathbf{n})$ sending $X \in W(m; \mathbf{n})$ to $\varphi^{-1}X\varphi \in W(m; \mathbf{n})$. According to [13], Φ is an automorphism of $W(m; \mathbf{n})$. By direct computation, one can easily see that $\Phi(x^\alpha D_i) = \varphi^{-1}(x^\alpha)D_i - \delta_{i<l}\chi(D_l)^{-1}\chi(D_i)\varphi^{-1}(x^\alpha)D_l$, where

$$\delta_{i<l} = \begin{cases} 1, & \text{if } i < l, \\ 0, & \text{if } i \geq l. \end{cases}$$

For any $D_{ij}(x^\beta) \in S(m; \mathbf{n})$, then

$$\Phi(D_{ij}(x^\beta))$$
$$= \varphi^{-1}(x^{\beta-\varepsilon_j})D_i - \delta_{i<l}\chi(D_l)^{-1}\chi(D_i)\varphi^{-1}(x^{\beta-\varepsilon_j})D_l -$$
$$(\varphi^{-1}(x^{\beta-\varepsilon_i})D_j - \delta_{j<l}\chi(D_l)^{-1}\chi(D_j)\varphi^{-1}(x^{\beta-\varepsilon_i})D_l).$$

By direct computation, we have:

$$D_i(\varphi^{-1}(x^{\beta-\varepsilon_j})) - D_j(\varphi^{-1}(x^{\beta-\varepsilon_i})) + \delta_{j<l}\chi(D_l)^{-1}\chi(D_j)D_l(\varphi^{-1}(x^{\beta-\varepsilon_i}))$$
$$- \delta_{i<l}\chi(D_l)^{-1}\chi(D_i)D_l(\varphi^{-1}(x^{\beta-\varepsilon_j}))$$
$$= 0.$$

Therefore $\Phi|_{S(m;\mathbf{n})}$ is an automorphism of $S(m; \mathbf{n})$, we also denote it by Φ.

By straightforward computation, one can obtain that $\chi^{\Phi^{-1}}(D_i) = \chi(\Phi(D_i)) = 0$ for $i \neq l$ and $\chi^{\Phi^{-1}}(D_l) = \chi(\Phi(D_l)) = \chi(D_l) \neq 0$. Set $\psi := \tilde{\varphi_l} \circ \Phi$ where $\varphi_l(x_i) = x_i$ for $i \neq l$ and $\varphi_l(x_l) = \chi(D_l)^{-1}\varphi(x_l)$. Then $\chi^{\psi^{-1}}(D_i) = \delta_{il}$, $1 \leq i \leq m$, as required. $\square$

6.3. Combining Lemma 2.4 and Lemma 6.1, we might as well assume $\chi \in L^*$ with $\mathrm{ht}\chi = 0$ satisfying $\chi(D_i) = \delta_{il}$ for a certain positive integer l, without loss of generality, up to module equivalence.

One can realize exceptional module $L_0(\omega_k)$ by k-fold exterior product of the natural module V of $L_{[0]} \cong \mathfrak{sl}(m)$. i.e. V is an m-dimensional vector space with an ordered basis $\{v_i \mid i = 1, 2, \cdots, m\}$ and the action $x_i D_j := E_{ij}(i \neq j)$: $v_j \mapsto v_i, v_s \mapsto 0$ for $s \neq j$; $D_{i,i+1}(x^{\varepsilon_i+\varepsilon_{i+1}}) := E_{ii} - E_{i+1,i+1}$: $v_i \mapsto v_i, v_{i+1} \mapsto -v_{i+1}, v_s \mapsto 0, s \neq i, i+1$; the action of L_1 is trivial. Then $L_0(\omega_k) \cong V_k := \bigwedge^k V$ with exceptional weight vector $v_1 \wedge v_2 \wedge \cdots \wedge v_k$ of exceptional weight ω_k.

Set $I_k := \mathbf{Ind}V_k = U_{p^s}(L, \chi) \bigotimes_{u(L_0)} V_k$, where $u(L_0)$ is the restricted enveloping algebra of L_0. By Theorem 5.3, I_k is a module in the $\mathfrak{C}$-category. By PBW

Theorem, I_k has an F-basis $\{E^\beta \otimes \wedge^k v_\Gamma \mid \beta \in A(m; \mathbf{n}), E^\beta = \prod\limits_{i=1}^{m} D_i^{\beta_i}, \wedge^k v_\Gamma = v_{\gamma_1} \wedge v_{\gamma_2} \wedge \cdots \wedge v_{\gamma_k}, \Gamma = \{\gamma_1 < \gamma_2 < \cdots < \gamma_k\} \subseteq \{1, 2, \cdots, m\}\}$.

Define a linear mapping : $d_k : I_k \to I_{k+1}$ by $d_k(E^\beta \otimes \wedge^k v_\Gamma) = \sum\limits_{i \notin \Gamma} \rho_L(D_i) E^\beta \otimes \wedge^k v_\Gamma \wedge v_i = \sum\limits_{i \notin \Gamma} E^{\beta+\varepsilon_i} \otimes \wedge^k v_\Gamma \wedge v_i$. In the following sequel, for simplicity, we often omit the subscript k of $\wedge^k v_\Gamma$ if the context is clear.

We will take use of some notations in the sequel, following Holmes in [1]: assume Γ is a subset of $\{1, 2, \cdots, m\}$, then denote $|i\Gamma j| = \#\{k \in \Gamma \mid i < k < j \text{ or } j < k < i\}$, $|\Gamma > k| = \#\{i \in \Gamma \mid i > k\}$ and denote $\Gamma \backslash i \cup j$ a tuple obtained by replacing i by j and reordering which contribute to the sign change of the corresponding exterior product by $(-1)^{|i\Gamma j|}$. For a statement P, Set

$$\delta_P = \begin{cases} 1, & \text{if } P \text{ is true,} \\ 0, & \text{if } P \text{ is false.} \end{cases}$$

We have the following proposition, the proof of which can be achieved with the aid of Theorem 3.9 in [1] if one notes that Ω_k^χ in [1] can be similarly defined for the generalized restricted Cartan-Type series W and S, and there is an L-module isomorphism between I_k and Ω_k^χ.

PROPOSITION 6.2.' *Maintain the notations as above. The following statements are true.*

(1) d_k is an L-module homomorphism.

(2) $d_{k+1} \circ d_k = 0$.

PROOF. As argument in the previous paragraph, one can similarly construct Ω_k^χ as in [1] with the L-module structure similar to the one in Proposition 3.7 in [1]. Then define

$$\Psi : \quad \begin{matrix} I_k & \longrightarrow & \Omega_k^\chi \\ E^\beta \otimes \wedge^k v_\Gamma & \longmapsto & y^{(\tau-\beta)} \otimes dx_\Gamma \end{matrix}$$

where $\tau = (p^{n_1}-1, p^{n_2}-1, \cdots, p^{n_m}-1)$. One can see that Ψ is actually an L-module isomorphism between I_k and Ω_k^χ. Furthermore, by the arguments parallel to the proof of Theorem 3.9 in [1], the same statements also hold for the nonrestricted case in the generalized restricted Lie algebra setting. Having this in mind, one immediately get the results of this proposition, parallel to Theorem 3.9 in [1]. $\square$

By Proposition 6.2, we obtain a complex of L-modules:

$$(6.1) \qquad 0 \to I_0 \xrightarrow{d_0} I_1 \xrightarrow{d_1} I_2 \xrightarrow{d_2} \cdots \xrightarrow{d_{m-1}} I_m \to 0.$$

Set $\bar{I}_k = I_k / \operatorname{Ker} d_k$. By Proposition 6.2, d_k is an L-module homomorphism. So $\operatorname{Ker} d_k$ is a submodule of I_k, therefore $\bar{I}_k$ is an L-module. We need the following lemma which is similar to Proposition 3.10 in [1] to obtain our main theorems.

LEMMA 6.3. *The following statements hold*

(1) For any $k \in \{0, 1, \cdots, m-1\}$, $\bar{I}_k$ has a basis $\{\overline{E^\beta \otimes \wedge v_\Gamma} \mid \beta \in A(m; \mathbf{n}), \Gamma \subseteq \{1, 2, \cdots, m\}, |\Gamma| = k, l \notin \Gamma\}$, where l is the same as in Lemma 6.1, i.e. $\chi(D_i) = \delta_{il}$.

(2) The sequence (6.1) is exact.

We finally obtain the main theorem of this section.

THEOREM 6.4. *Assume* $\mathrm{ht}\chi = 0$. *The following statements hold.*

(1) *If* $0 < k < m$, *then* I_k *has a unique proper L-submodule* $\mathrm{Ker}\, d_k$, *therefore* $\bar{I}_k$ *is simple for* $0 \leq k < m$. *Furthermore* $\bar{I}_k \cong L^\chi(\omega_k)$, *where* $L^\chi(\omega_k)$ *denote irreducible exceptional-weight module with exceptional weight* ω_k.

(2) $I_0 \cong I_m$ *are simple.*

(3) $\dim_F \bar{I}_k = p^{\sum n_i} \binom{m-1}{k}, 0 \leq k < m.$

PROOF. For any simple L-submodule M in $\bar{I}_k$, $0 < k < m - 1$, assume $w = \sum_{\beta \in A} \overline{E^\beta \otimes u_\beta}$ is a maximal vector of M, where $A \subseteq A(m; \mathbf{n})$, $u_\beta = \sum a_{\beta, \Gamma} \wedge v_\Gamma$, i.e. $\rho_L(\mathcal{N})w = 0$, $\rho_L(h_i)w = \lambda(h_i)w$, where $\mathcal{N} = F\text{-span}\{x^{\varepsilon_i}D_j \mid 1 \leq i < j \leq m\} \bigoplus L_1$, $h_i := x^{\varepsilon_i}D_i - x^{\varepsilon_{i+1}}D_{i+1}$, $i = 1, 2, \cdots, m-1$, $\lambda \in \mathbb{F}_p^{m-1}$. In $A(m; \mathbf{n})$, there is an total order $\succ : \alpha \succ \beta$ if $\alpha(l) > \beta(l)$ or $\alpha(l) = \beta(l)$ and α is greater than β in the lexicographical order. In the sense of this order, we choose $\beta_0 = \max\{\beta \mid \beta \in A\}$. We first claim that $\beta_0(l) < p^{n_l} - 1$. Otherwise, $\beta_0(l) = p^{n_l} - 1$, and then there would occur two cases as follows:

Case 1: $\exists i \neq l$ such that $\beta_0(i) = p^{n_i} - 1$. Then

$$0 = \rho_L(x^{(p^{n_i}-1)\varepsilon_i}D_l)w$$
$$= \chi(D_l)^{p^{n_l}}\overline{E^{\beta_0 - (p^{n_i}-1)\varepsilon_i - (p^{n_l}-1)\varepsilon_l} \otimes u_{\beta_0}} + \sum *\overline{E^\beta \otimes u_\beta}$$

where all β appearing in $\sum *\overline{E^\beta \otimes u_\beta}$ are different from $\beta_0 - (p^{n_i}-1)\varepsilon_i - (p^{n_l}-1)\varepsilon_l$. So the above equation contradicts with $\chi(D_l) \neq 0$.

Case 2: $\beta_0(i) < p^{n_i} - 1, \forall i \neq l$.

In this case, take $i \notin \Gamma$ such that $a_{\beta_0, \Gamma} \neq 0$(this is possible, since we have assumed $k < m - 1$), then

$$0 = \rho_L(x^{(p^{n_l}-1)\varepsilon_l}D_i)w$$
$$= \sum_{i \notin \Gamma} a_{\beta_0, \Gamma}\overline{E^{\beta_0 + \varepsilon_i - (p^{n_l}-1)\varepsilon_l} \otimes \wedge v_\Gamma} + \sum *\overline{E^\beta \otimes \wedge v_{\Gamma'}}$$

where all β appearing in $\sum *\overline{E^\beta \otimes \wedge v_{\Gamma'}}$ are different from $\beta_0 + \varepsilon_i - (p^{n_l}-1)\varepsilon_l$. The choice of i makes the above equation a contradiction.

So, we finally obtain that $\beta_0(l) < p^{n_l} - 1$. If there exists $i < l$ such that $\beta_0(i) \geq 1$, then

$$0 = \rho_L(x^{\varepsilon_i}D_l)w$$
$$= \overline{E^{\beta_0 + \varepsilon_l - \varepsilon_i} \otimes u_{\beta_0}} + \sum *\overline{E^\beta \otimes u_\beta}$$

where all β appearing in $\sum *\overline{E^\beta \otimes u_\beta}$ are different from $\beta_0 + \varepsilon_l - \varepsilon_i$. Therefore $\overline{E^{\beta_0 + \varepsilon_l - \varepsilon_i} \otimes u_{\beta_0}} = 0$. It is a contradiction. So $\beta_0(i) = 0, \forall i < l$. If there exists $j > l$ such that $\beta_0(j) \geq 2$, then

$$0 = \rho_L(x^{\beta_0(j)\varepsilon_j}D_l)w$$
$$= (-1)^{\beta_0(j)}\overline{E^{\beta_0 + \varepsilon_l - \beta_0(j)\varepsilon_j} \otimes u_{\beta_0}} + \sum *\overline{E^\beta \otimes u_\beta}$$

where all β appearing in $\sum *\overline{E^\beta \otimes u_\beta}$ are different from $\beta_0 + \varepsilon_l - \beta_0(j)\varepsilon_j$. So $(-1)^{\beta_0(j)}\overline{E^{\beta_0 + \varepsilon_l - \beta_0(j)\varepsilon_j} \otimes u_{\beta_0}} = 0$. It is still a contradiction. If there exists $j, k > l$ such that $\beta_0(j) = \beta_0(k) = 1$, then

$$0 = \rho_L(x^{\varepsilon_j + \varepsilon_k}D_l)w$$
$$= \overline{E^{\beta_0 + \varepsilon_l - \varepsilon_j - \varepsilon_k} \otimes u_{\beta_0}} + \sum *\overline{E^\beta \otimes u_\beta}$$

where all β appearing in $\sum *\overline{E^\beta \otimes u_\beta}$ are different from $\beta_0 + \varepsilon_l - \varepsilon_j - \varepsilon_k$. So $\overline{E^{\beta_0 + \varepsilon_l - \varepsilon_j - \varepsilon_k} \otimes u_{\beta_0}} = 0$, a contradiction. If there exists $k > l$ such that $\beta_0(k) = 1$ and $\beta_0(l) \geq 1$. Assume $i \in \Gamma$ such that $a_{\beta_0, \Gamma} \neq 0$(this is possible, since we have assumed $k > 0$), then

$$0 = \rho_L(x^{(\beta_0(l)+1)\varepsilon_l} D_i)w$$

$$= (-1)^{\beta_0(l)} \sum \delta_{i \in \Gamma} a_{\beta_0, \Gamma}(-1)^{|i\Gamma l|}\overline{D_k \otimes \wedge v_{\Gamma \setminus i \cup l}} + \sum *\overline{E^\beta \otimes u_\beta}$$

where all β appearing in $\sum *\overline{E^\beta \otimes u_\beta}$ are different from ε_k. So

$$(-1)^{\beta_0(l)} \sum \delta_{i \in \Gamma} a_{\beta_0, \Gamma}(-1)^{|i\Gamma l|}\overline{D_k \otimes \wedge v_{\Gamma \setminus i \cup l}} = 0,$$

a contradiction. Therefore $|\beta_0| < 2$. If $|\beta_0| = 1$, then $w = \sum \overline{D_i \otimes u_{\varepsilon_i}} + \overline{1 \otimes u_0}$. If $w = \overline{D_i \otimes v} + \overline{1 \otimes u_0}, i < m$. Then

$$0 = \rho_L(x^{\varepsilon_i} D_m)w$$

$$= -\overline{D_m \otimes v} + \overline{D_i \otimes v'} + \overline{1 \otimes u_0'}$$

which is a contradiction.

So $w = \overline{D_m \otimes v} + \overline{1 \otimes u_0}$. Choose $1 \leq i < j \leq m$, then

$$0 = \rho_L(x^{\varepsilon_i} D_j)w$$

$$= \overline{D_m \otimes \varepsilon_i D_j \cdot v} + \overline{1 \otimes u_0'}$$

which implies that $x^{\varepsilon_i} D_j \cdot v = 0 \forall 1 \leq i < j \leq m$. So $v = av_1$, $a \in F$(for the case that $k = 1$), or $v = av_1 \wedge v_2 \wedge \cdots \wedge \cdots v_{m-1}$, $a \in F$(for the case that $k = m - 1$ and $l = m$). The latter case is excluded by our assumption that $k < m - 1$. So $v = av_1$ for some $a \in F$.

Take $D = D_{m1}(x^{2\varepsilon_m + \varepsilon_1})$, then

$$0 = \rho(D)w$$

$$= \rho_L(D_{m1}(x^{2\varepsilon_m + \varepsilon_1}))w$$

$$= \rho_L(x^{2\varepsilon_m} D_m - x^{\varepsilon_m + \varepsilon_1} D_1)w$$

$$= \overline{1 \otimes av_1}$$

which is a contradiction.

So $\beta_0 = 0$. i.e. $w = \overline{1 \otimes u_0}$. So $M = \bar{I}_k$ by the simplicity of $\wedge^k V$ as L_0-module. This is to say that $\bar{I}_k$ is simple for $0 < k < m - 1$. For $\bar{I}_{m-1}$ and $\bar{I}_0$, by the exactness of the sequence (6.1), we know $\bar{I}_{m-1} \cong I_m$, while it is obvious that $I_m \cong I_0$, so $\bar{I}_{m-1} \cong I_0$. On the other hand $\bar{I}_0 = I_0 = \{\sum a_\beta E^\beta \otimes 1 \mid \beta \in A(m;\mathbf{n})\}$. One can easily see that all the maximal vector w are of the form $w = 1 \otimes 1$ or $w = D_m \otimes 1$ up to scalars with weights 0 and ω_{m-1} respectively. While one can easily see that each form of w generates I_0 as L-module. This is to say that $\bar{I}_0$ is also simple . Therefore $\mathsf{Ker}\, d_k$ is the unique proper L-submodule of I_k for $0 \leq k < m$. Furthermore it is obvious that $\overline{1 \otimes v_1 \wedge v_2 \wedge \cdots \wedge v_k} \in \bar{I}_k$ is a weight vector of weight ω_k. So $\bar{I}_k \cong L^\chi(\omega_k)$. We finish the proof of (1) and (2). Statement (3) is a direct consequence of Lemma 6.3. $\square$

Applying Theorem 5.3, Theorem 5.4 and Theorem 6.4, we obtain the following results.

THEOREM 6.5. *Assume* ht$\chi = 0$, *then the following statements hold.*

(1) *There are* $p^{m-1} - 1$ *distinct (up to isomorphism) simple* $U_{p^s}(L, \chi)$ *modules. They are represented by* $\{L^\chi(\lambda) \mid \lambda \in \mathbb{F}_p^{m-1}\}$.

(2) $L^\chi(\lambda) \cong Z^\chi(\lambda) := U_{p^s}(L, \chi) \bigotimes_{u(L_0)} L_0(\lambda)$ *if and only if* $\lambda \notin \{\omega_1, \omega_2, \cdots, \omega_{m-1}\}$ *and* $L^\chi(\omega_m) \cong L^\chi(\omega_{m-1})$.

(3) *If* λ *is not exceptional, then* $\dim_F L^\chi(\lambda) = p^{\sum n_i} \dim_F L_0(\lambda)$ *and*

$$
\dim_F L^\chi(\omega_k) = \begin{cases} p^{\sum n_i} \binom{m-1}{k}, & 0 \leq k < m, \\ p^{\sum n_i}, & k = m. \end{cases}
$$

PROOF. We will prove that $L^\chi(\lambda) \ncong L^\chi(\mu)$ unless $\lambda = \mu$ or $\lambda, \mu \in \{\omega_{m-1}, \omega_m\}$, case by case.

Case 1: Both λ and μ are nonexceptional.

According to Theorem 5.3 and Theorem 5.4, in this case $L^\chi(\lambda) \cong \mathbf{Ind}(L_0(\lambda))$, $L^\chi(\mu) \cong \mathbf{Ind}(L_0(\mu))$. Both $L^\chi(\lambda)$ and $L^\chi(\mu)$ are modules in the $\mathfrak{C}$-category. As in the proof of Theorem 5.4, any L-module homomorphism between $L^\chi(\lambda)$ and $L^\chi(\mu)$ is a homomorphism in the $\mathfrak{C}$-category, therefore it is an R-module homomorphism. Assume $\varphi : L^\chi(\lambda) \to L^\chi(\mu)$ is an L-module isomorphism, we will prove $\lambda = \mu$.

Assume v_λ is a maximal vector of $L_0(\lambda)$ with maximal weight λ and suppose $\varphi(1 \otimes v_\lambda) = \sum E^\theta \otimes u_\theta$ where $0 \neq u_\theta \in L_0(\mu)$. Recall that $A(m; \mathbf{n})$ has a natural lexicographical order. If $u_\theta \neq 0$ for some $\theta \neq 0$, set $\theta_0 = \max\{\theta \mid u_\theta \neq 0\} \neq 0$. As φ is also an R-module homomorphism, then

$$
0 = \varphi(\rho_R(x^{\theta_0})(1 \otimes v_\lambda)) = \rho_R(x^{\theta_0})\varphi(1 \otimes v_\lambda) = \rho_R(x^{\theta_0})(\sum E^\theta \otimes u_\theta) = 1 \otimes u_{\theta_0}
$$

a contradiction. Hence $\varphi(1 \otimes v_\lambda) = 1 \otimes u_0$. Then $1 \otimes u_0$ is a maximal vector with maximal weight λ in $L^\chi(\mu)$. Therefore u_0 is a maximal vector with maximal weight λ in $L_0(\mu)$, then $\lambda = \mu$.

Case 2: Both λ and μ are exceptional.

We assume $\lambda, \mu \in \{\omega_1, \cdots, \omega_{m-1}\}$ and $\lambda \neq \mu$. i.e. $\lambda = \omega_i, \mu = \omega_j, i \neq j$. From the proof of Theorem 6.4, we know that the maximal vector of $\bar{I}_k$ is of weight ω_k for $k = 1, \cdots m-2$, while for $k = m-1$, it is ω_{m-1} or ω_0. So $\bar{I}_i \ncong \bar{I}_j$. i.e. $L^\chi(\omega_i) \ncong L^\chi(\omega_j)$.

Case 3: λ is nonexceptional and μ is exceptional.

We can assume $\mu = \omega_i, i \in \{1, \cdots, m-1\}$. We will prove that $L^\chi(\lambda) \ncong L^\chi(\omega_i)$.

Suppose there is an L-module isomorphism $\varphi : L^\chi(\lambda) \to L^\chi(\omega_i)$. We can fix a maximal weight vector of $L_0(\lambda)$ with maximal weight λ, say v_λ. Then $1 \otimes v_\lambda$ is a maximal weight vector of $L^\chi(\lambda)$ with maximal weight λ. So $\varphi(1 \otimes v_\lambda)$ is a maximal weight vector of $L^\chi(\omega_i)$ with maximal weight λ. By the proof of Theorem 6.4, the weight of any maximal vector in $L^\chi(\omega_i)$ is exceptional, Concretely speaking, it is ω_i when $1 \leq i < m-1$, while when $i = m-1$, it is ω_{m-1} or ω_0. So λ is exceptional, a contradiction.

Combining up the arguments for the full cases above, we finish the proof of (1) and (2).

The last statement is a direct sequence of Lemma 6.3. We finish the proof of this Theorem. $\qquad \square$

REMARK 6.6. When $p = 2$, $m = 3$ and $\mathbf{n} = \mathbf{1}$, the results in this theorem coincide with those ones in [**6**].

References

[1] Holmes R. R., *Simple modules with character height at most one for the restricted Witt algebras*, J.Algebra 237 No.2(2001), 446-469.

[2] Holmes R. R. and Zhang Chao-Wen, *Some simple modules for the restricted Cartan type Lie algebras*, Journal of Pure and Applied Algebra 173(2002), 135-165.

[3] Humphreys J. E., *Modular representations of classical Lie algebras and semi-simple algebraic groups*, 19(1971), 51-79.

[4] Jantzen J. C., *Representations of Algebraic Groups* (second edition), American Mathematical Society, 2003.

[5] Nakano D., Projective modules over Lie algebras of Cartan type, Memoirs of AMS No.470 (1992).

[6] Shan Chui-Ping and Jiang Zhi-Hong, *Irreducible representations of S(3,1)*(in chinese), Preprint(2007).

[7] Shen Guang-Yu, *Graded modules of graded Lie algebras of Cartan type, III* , Chinese Ann. Math. Ser.B 9(1988), 404-417.

[8] Shu Bin, *The generalized restricted representations of graded Lie algebras of cartan type*, J.Algebra 194(1997), 157-177.

[9] Shu Bin, *Quasi p-mappings and representations of modular Lie algebras*, in "Proceedings of the international conference on representation theorey", CHEP & Springer-Verlag, Beijing 2000, 375-401.

[10] Shu Bin and Yao Yu-Feng, *Irreducible modules of the generalized Jacobson-Witt algebras*, preprint

[11] Skryabin S., *Independent systems of derivations and Lie algebra representations*, in "Algebra and Analysis, Eds: Archipov / Parshin / Shafarvich." Walter de Gruyter& Co, Berlin-New York, 1994, 115-150.

[12] Strade H. and Farnsteiner R., *Modular Lie algebras and their representations* , Marcel Dekker, New York, 1988.

[13] Wilson R. L., *Classification of generalized witt algebras over algebraically closed filds*, Trans. Amer. Math. Soc. Vol.153(1971), 191-210.

[14] Zhang Chao-Wen, *Representations of the restricted Lie algebras of Cartan-type*, J.Algebra 290(2005), 408-432.

DEPARTMENT OF MATHEMATICS, EAST CHINA NORMAL UNIVERSITY, SHANGHAI 200241, CHINA.
E-mail address: yaoyufeng139@sina.com

DEPARTMENT OF MATHEMATICS, EAST CHINA NORMAL UNIVERSITY, SHANGHAI 200241, CHINA.
E-mail address: bshu@math.ecnu.edu.cn